Vitamin D Handbook

BICENTENNIAL
1807
⊕WILEY
2007
BICENTENNIAL

THE WILEY BICENTENNIAL—KNOWLEDGE FOR GENERATIONS

Each generation has its unique needs and aspirations. When Charles Wiley first opened his small printing shop in lower Manhattan in 1807, it was a generation of boundless potential searching for an identity. And we were there, helping to define a new American literary tradition. Over half a century later, in the midst of the Second Industrial Revolution, it was a generation focused on building the future. Once again, we were there, supplying the critical scientific, technical, and engineering knowledge that helped frame the world. Throughout the 20th Century, and into the new millennium, nations began to reach out beyond their own borders and a new international community was born. Wiley was there, expanding its operations around the world to enable a global exchange of ideas, opinions, and know-how.

For 200 years, Wiley has been an integral part of each generation's journey, enabling the flow of information and understanding necessary to meet their needs and fulfill their aspirations. Today, bold new technologies are changing the way we live and learn. Wiley will be there, providing you the must-have knowledge you need to imagine new worlds, new possibilities, and new opportunities.

Generations come and go, but you can always count on Wiley to provide you the knowledge you need, when and where you need it!

WILLIAM J. PESCE
PRESIDENT AND CHIEF EXECUTIVE OFFICER

PETER BOOTH WILEY
CHAIRMAN OF THE BOARD

Vitamin D Handbook

Structures, Synonyms, and Properties

Edited by

G. W. A. Milne
M. Delander

WILEY-INTERSCIENCE
A John Wiley & Sons, Inc., Publication

Library of Congress Cataloging-in-Publication Data:

Vitamin D handbook : structures, synonyms, and properties / edited
by G. W. A. Milne and M. Delander.
  p. ; cm.
 Includes index.
 ISBN 978-0-470-13983-7 (cloth)
1. Vitamin D—Handbooks, manuals, etc. I. Milne, George W. A.,
1937– II. Delander, M., 1969–
  [DNLM: 1. Vitamin D—analogs & derivatives—Resource Guides. 2.
  Vitamin D—analogs & derivatives—Terminology—English. 3.
  Vitamin D—chemistry—Resource Guides. 4. Vitamin D—chemistry—
Terminology—English. QU 15 V837 2007]
  QP772.V53V578 2007
  612.3'99—dc22   2007029059

Printed in the United States of America.

10 9 8 7 6 5 4 3 2 1

CONTENTS

PREFACE

In the 1920s, it was recognized that rickets, a childhood disease affecting bones, could be ameliorated by exposure to sunlight or administration of cod liver oil. Within a short time, the antirachitic principle in cod liver oil had been isolated and named vitamin D. The process of irradiation, of either the patient or the food consumed by the patient led to antirachitic protection – a discovery that was patented in 1925 by the University of Wisconsin. It is now clear that some materials, such as cod liver oil, contain vitamin D and many other foods contain a provitamin, which is converted upon irradiation to vitamin D. A great deal of chemical work ensued, mainly in Germany, the UK and the US and led in 1936 to the assignment of structure to vitamin D_2 as a C28 steroid, an ergosterol derivative with ring B opened (Record Number 531). Within a year, a second antirachitic compound, formed by irradiation of non-ergosterol steroids was isolated and characterized as a C27 steroid and named vitamin D_3 (Record 351). Reflecting their chemical progenitors, the vitamin D_2 compounds are known collectively as ***ergocalciferols***, and the vitamin D_3 derivatives as ***cholecalciferols***. It has since become clear that vitamin D_3, being derived from the cholesterol family, is more ubiquitous than the ergosterol derived product, vitamin D_2 and much more work has been carried out on the former. It is also well understood that the vitamins themselves are not biologically active but that they are first converted to the $1\alpha,25$-dihydroxyvitamins (386 and 154, respectively) which exhibit the biological behavior associated with the vitamins.

Basic knowledge of vitamin D_3, indeed its birthright, was that it plays a significant role in the management in mammals of both calcium and phosphorus. Studies towards the end of the 20th. Century however revealed intriguing clues that it also was involved in other cellular phenomena. Vitamin D_3, and particularly its metabolites can affect cellular differentiation and this finding has led to an enormous effort to identify and synthesize derivatives in which the cellular differentiating and apoptopic activities have been retained or enhanced while the calcium mobilization properties have been minimized.

The details of the behavior of vitamin D_3 at the molecular level are emerging, but are still far from complete and until a detailed picture is available, such empirical methods are important, both for the light they shed on the compound's biological activity and also as a source of new derivatives. It is often said that over 3,000 derivatives of vitamin D_3 exist. This number appears to include many compounds which are not reported in the literature and numerous others cited only in patents. The number of vitamin D derivatives with published chemical and biochemical data is closer to 1,000, the compounds which are described in this Handbook. These vitamin D derivatives however have not been completely catalogued and this, coupled with the difficulties of chemical nomenclature in this field, has provided the impetus for this compilation of derivatives of vitamins D_2 and D_3. PubChem, a database of 22 million compounds assembled and maintained by the National Library of Medicine (http://pubchem.ncbi.nlm.nih.gov), contains fewer than 1,000 vitamin D derivatives - the exact number depends upon definitions - and the sterol lipids in LipidMaps, a database maintained by a consortium led by the University of California at San Diego (http://www.lipidmaps.org), has about 654 derivatives of vitamin D. The database in this Vitamin D Handbook contains 947 compounds, many, but not all of which are in one or both of these public databases. Entries in the Handbook carry the PubChem and/or LipidMaps identifiers, and thus cross-linking to these databases and to the links they carry is facilitated.

Proprietary Considerations

Every attempt has been made to ensure the accuracy of the information provided in this book. However, the publishers cannot be held responsible for the accuracy of the information, and users are expected to bear in mind the following information:

The reporting of a name in this book cannot imply definitive legality is establishing proprietary usage. Questions concerning legal ownership of a particular name can be resolved by due legal process.

A manufacturer in some countries may manufacture its product under names different from those cited in this book. Similarly manufacture or marketing of a product may be licensed to a separate company in another country under the same or a different name.

We trust that readers will find this compilation contains a wealth of information that is difficult to obtain from any other single source. Individuals wishing to submit new or updated material for inclusion in future editions of this Handbook should contact George W. A. Milne or Michael Delander (addresses below).

Acknowledgements

The editors would like to acknowledge the skilled programming performed by Dr. Ju-yun Li which allowed for accurate formatting and typesetting of this Handbook. We thank Profs. DeLuca and Sicinski (University of Wisconsin) for their helpful comments on the size of the vitamin D database and LipidMaps team members Drs. Eoin Fahy and Manish Sud (University of California, San Diego) for their skilled assistance in checking the chemical structures in this book. GWAM would like to acknowledge the expert assistance and continual support provided by his wife, Kay, without whom none of the deadlines would have been met.

How to Use This Book

Entries in the Vitamin D Handbook are arranged by increasing molecular formula, beginning with $C_{15}H_{15}N_5O_5$ an ending at $C_{60}H_{93}O_6$. To assist in locating a record, indexes of name and synonym, CAS Registry Number (**e.g.** 103909-75-7) and NLM PubChem Compound Identification Number (CID, 5478815) are provided following page 235. Each of the Indexes gives the record number of the entry. A typical record is shown below.

The Record Number (93) is the main identifier for the entry. This is followed by the name of the compound and the CAS Registry Number. If no CAS Registry Number is available,

Record Number

Name

CAS Registry Number or PubChem CID Number

Molecular Formula

Structure

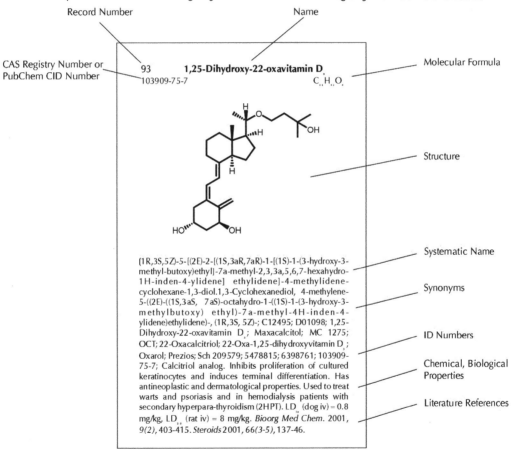

93   **1,25-Dihydroxy-22-oxavitamin D,**
103909-75-7   C.,H.,O.

Systematic Name

Synonyms

ID Numbers

Chemical, Biological Properties

Literature References

(1R,3S,5Z)-5-[(2E)-2-[(1S,3aR,7aR)-1-[(1S)-1-(3-hydroxy-3-methyl-butoxy)ethyl]-7a-methyl-2,3,3a,5,6,7-hexahydro-1H-inden-4-ylidene] ethylidene]-4-methylidene-cyclohexane-1,3-diol.1,3-Cyclohexanediol, 4-methylene-5-((2E)-((1S,3aS, 7aS)-octahydro-1-((1S)-1-(3-hydroxy-3-methylbutoxy) ethyl)-7a-methyl-4H-inden-4-ylidene)ethylidene)-, (1R,3S, 5Z)-; C12495; D01098; 1,25-Dihydroxy-22-oxavitamin D,; Maxacalcitol; MC 1275; OCT; 22-Oxacalcitriol; 22-Oxa-1,25-dihydroxyvitamin D,; Oxarol; Prezios; Sch 209579; 5478815; 6398761; 103909-75-7; Calcitriol analog. Inhibits proliferation of cultured keratinocytes and induces terminal differentiation. Has antineoplastic and dermatological properties. Used to treat warts and psoriasis and in hemodialysis patients with secondary hyperpara-thyroidism (2HPT). LD (dog iv) = 0.8 mg/kg, LD (rat iv) = 8 mg/kg. *Bioorg Med Chem.* 2001, *9(2)*, 403-415. *Steroids* 2001, *66(3-5)*, 137-46.

the Compound Identifier (CID) used in NLM's PubChem is given. This is followed by the structure and a systematic chemical name for the compound. Then other names and synonyms, including other known CAS RNs and CIDs are listed. A summary of the chemical and biological properties of the compound is provided and the record ends with selected leading literature references.

SECTION I

Structures, Chemical Names,
Synonyms and Properties

1   ***4-(2-(6,7-Dimethoxy-4-methyl-3-oxo-3,4-dihydroquinoxalinyl)ethyl)-1,2,4-triazoline-3,5-dione***

132788-52-4 $C_{15}H_{15}N_5O_5$

4-[2-(6,7-dimethoxy-4-methyl-3-oxo-quinoxalin-2-yl)ethyl]-1,2,4-triazole-3,5-dione.
Dmeq-tad; 3H-1,2,4-Triazole-3,5(4H)-dione, 4-(2-(3,4-dihydro-6,7-dimethoxy-4-methyl-3-oxo-2-quinoxalinyl)ethyl)-; 4-(2-(3,4-Dihydro-6,7-dimethoxy-4-methyl-3-oxo-2-quinoxalinyl)ethyl)-3H-1,2,4-triazole-3,5(4H)-dione; 4-(2-(6,7-Dimethoxy-4-methyl-3-oxo-3,4-dihydroquinoxalinyl)ethyl)-1,2,4-triazoline-3,5-dione; 131554; 132788-52-4; Used as a fluorescence label in assays of vitamin D derivatives. ***Anal Biochem.*** 1991, ***194(1)***, 77-81.

2   ***1α-Hydroxy-20-oxo-19,22,23,24,25,26,27-heptanorvitamin D₃***

9547697 $C_{20}H_{30}O_3$

1-[(1S,3aR,4E,7aR)-4-[2-[(3R,5R)-3,5-dihydroxycyclohexylidene]ethylidene]-7a-methyl-2,3,3a,5,6,7-hexahydro-1H-inden-1-yl]ethanone.
LMST03020604;   1R,3R-dihydroxy-19-nor-9,10-seco-5,7E-pregnadien-20-one;   1α-hydroxy-20-oxo-19,22,23,24,25,26,27-heptanorcholecalciferol;   9547697; Synthesized from the corresponding C-22 aldehyde ***via*** oxidative decarboxylation. Has no affinity for progesterone receptor in MCF-7. λ_m = 243, 251.5, 261 nm (EtOH). ***Tet Lett.*** 1994, ***35(15)***, 2295-8.

3   ***20-Oxo-22,23,24,25,26,27-hexanorvitamin D₃;***

86120-56-1 $C_{21}H_{30}O_2$

1-[(1S,3aR,4E,7aR)-4-[(2Z)-2-[(5S)-5-hydroxy-2-methylidene-cyclohexylidene]ethylidene]-7a-methyl-2,3,3a,5,6,7-hexahydro-1H-inden-1-yl]ethanone.
LMST03020002;   20-Oxopregnacalciferol;   3S-dihydroxy-9,10-seco-5Z,7E,10(19)-pregnatrien-20-one; 20-oxo-22,23,24,25,26,27-hexanorcholecalciferol; 9547262; 86120-56-1; Synthetic analog of Vitamin D₃, prepared from the corresponding C-22 aldehyde ***via*** oxidative decarboxylation. Has significant affinity for progesterone receptor in MCF-7 and small though significant bone and soft tissue mobilization activity; shows no significant increase in intestinal calcium transport and possesses no significant antagonistic activity against vitamin D₃. λ_m = 264 nm (EtOH). ***J Steroid Biochem.*** 1982, ***17(5)***, 495-502. ***Tet Lett.*** 1994, ***35(15)***, 2295-8.

4   ***1α-Hydroxy-20-oxo-22,23,24,25,26,27-hexanorvitamin D₃***

9547263 $C_{21}H_{30}O_3$

1-[(1S,3aR,4E,7aR)-4-[(2Z)-2-[(3S,5R)-3,5-dihydroxy-2-methylidene-cyclohexylidene]ethylidene]-7a-methyl-2,3,3a,5,6,7-hexahydro-1H-inden-1-yl]ethanone.
LMST03020003;   1S,3R-dihydroxy-9,10-seco-5Z,7E,10(19)-pregnatrien-20-one; 1α-hydroxy-20-oxo-22,23,24,25,26,27-hexanorcholecalciferol; 9547263; Has no affinity for progesterone receptor in MCF-7. Fails to bind to the vitamin D receptor and has no calcemic activity. $[\alpha]_D^{23}$ = +9.2° (c = 1.00 EtOH); λ_m = 264 nm. ***Bioorg Med Chem Lett.*** 1992, ***2(10)***, 1289-92. ***Tet Lett.*** 1994, ***35(15)***, 2295-8.

5   ***1α,21-Dihydroxy-20-oxo-22,23,24,25,26,27-hexanorvitamin D₃***

9547264 $C_{21}H_{30}O_4$

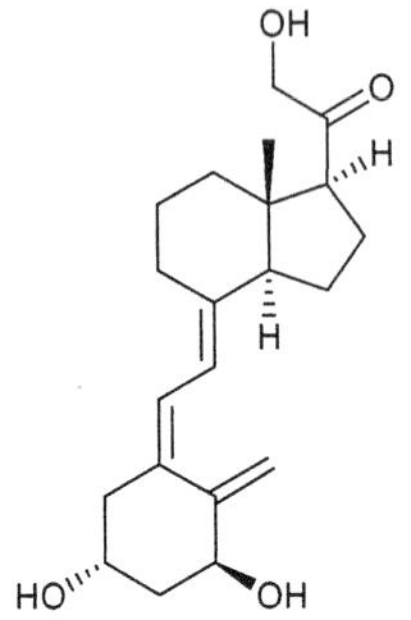

1-[(1S,3aR.4E,7aR)-4-[(2Z)-2-[(3S,5R)-3,5-dihydroxy-2-methylidene-cyclohexylidene]ethylidene]-7a-methyl-2,3,3a,5,6,7-hexahydro-1H-inden-1-yl]-2-hydroxy-ethanone.
LMST03020004; 1S,3R,21-trihydroxy-9,10-seco-5Z,7E,10(19)-pregnatrien-20-one; 1α,21-dihydroxy-20-oxo-22,23,24,25,26,27-hexanorcholecalciferol; 9547264; Binding affinity for chick intestinal cytosolic receptor is less than 1/100,000 compared to 1α,25-dihydroxyvitamin D_3. $[\alpha]_D^{23}$ = +2.81° (c = 0.07 EtOH); λ_m = 262 nm. *Bioorg Med Chem Lett*. 1992, *2(10)*, 1289-92.

6   1α,17α,21-Trihydroxy-20-oxo-22,23,24,25,26,27-hexanorvitamin D_3
9547265 $C_{21}H_{30}O_5$

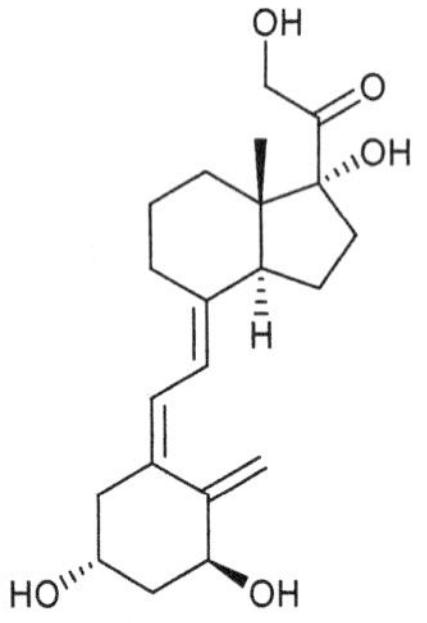

1-[(1R,3aR,4E,7aS)-4-[(2Z)-2-[(3S,5R)-3,5-dihydroxy-2-methylidene-cyclohexylidene]ethylidene]-1-hydroxy-7a-methyl-2,3,3a,5,6,7-hexahydroinden-1-yl]-2-hydroxy-ethanone.
LMST03020005;   1S,3R,17R,21-tetrahydroxy-9,10-seco-5Z,7E,10(19)-pregnatrien-20-one; 1α,17α,21-tri-hydroxy-20-oxo-22,23,24,25,26,27-hexanorcholecalciferol; 9547265; Binding affinity for chick intestinal cytosolic receptor is less than 1/100,000 compared to 1α,25-dihydroxyvitamin D_3. λ_m = 264 nm. *Bioorg Med Chem Lett*. 1992, *2(10)*, 1289-92.

7   1α-Hydroxypregnacalciferol
58702-12-8 $C_{21}H_{32}O_2$

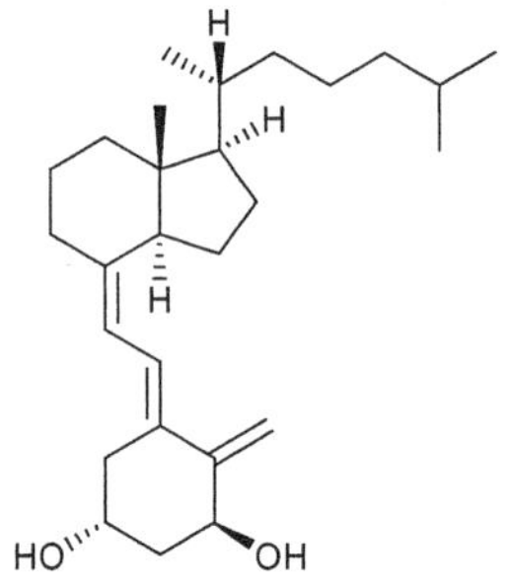

(5Z)-5-[(2Z)-2-(1-ethyl-7a-methyl-2,3,3a,5,6,7-hexahydro-1H-inden-4-ylidene)ethylidene]-4-methylidene-cyclohexane-1,3-diol.
1-Hydroxypregnacalciferol; 1α-(OH)Pregnacalciferol; 1α-Hydroxypregnacalciferol; 1,3-Cyclohexanediol, 5-((1-ethyloctahydro-7a-methyl-4H-inden-4-ylidene)-ethylidene)-4-methylene-, (1S-(1α3aβ,4E(Z(1S*,3R*))-7aα))-; 6438511; 58702-12-8; Compared with 1α,25-dihydroxy-vitamin D_3, shows similar potency with respect to inhibition of cell growth and proliferation in the NCI-H82 and the NCI-H209 SCLC lines. *Res Commun Mol Pathol Pharmacol*. 1997, *98(1)*, 3-18.

8   1α,20S-Dihydroxy-22,23,24,25,26,27-hexanorvitamin D_3
9547266 $C_{21}H_{32}O_3$

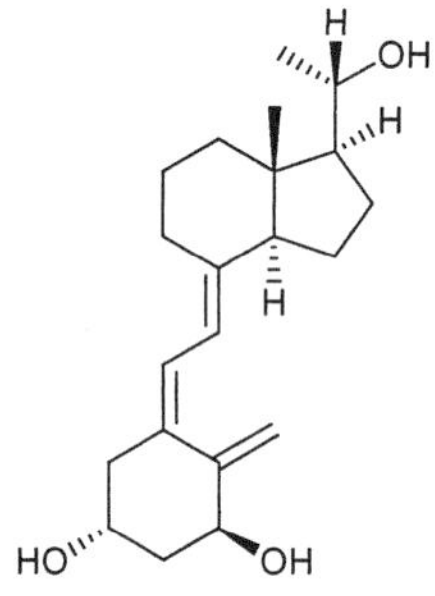

(1R,3S,5Z)-5-[(2E)-2-[(1S,3aR,7aR)-1-(1-hydroxyethyl)-7a-methyl-2,3,3a,5,6,7-hexahydro-1H-inden-4-ylidene]ethylidene]-4-methylidene-cyclohexane-1,3-diol.
LMST03020006; 9,10-seco-5Z,7E,10(19)-pregnatriene-1S,3R,20S-triol; 1α,20S-dihydroxy-22,23,24,25,26,27-hexanorcholecalciferol; 9547266; Compared to 22-oxacalcitriol antiproliferation activity towards HL-60 is <1/217 and binding affinity for vitamin D receptor is 1/929. $[\alpha]_D^{20}$ = +55.1° (c = 0.01 EtOH); λ_m = 263 nm. *Bioorg Med Chem Lett*. 1994, *4(5)*, 753-6; *Chem Pharm Bull*. 1996, *44(12)*, 2280-6.

9   1α-Hydroxy-22-oxo-23,24,25,26,27-pentanorvitamin D_3
9547267 $C_{22}H_{32}O_3$

(2S)-2-[(1R,3aR,4E,7aS)-4-[(2Z)-2-[(3S,5R)-3,5-dihydroxy-2-methylidene-cyclohexylidene]ethylidene]-7a-methyl-2,3,3a,5,6,7-hexahydro-1H-inden-1-yl]propanal.
LMST03020007; 1S,3R-dihydroxy-9,10-seco-23,24-dinor-5Z,7E,10(19)-cholatrien-22-one; 1α-hydroxy-22-oxo-23,24,25,26,27-pentanorvitamin D_3; 1α-hydroxy-22-oxo-23,24,25,26,27-pentanorcholecalciferol; 9547267; 10% as active as 1α,25-dihydroxyvitamin D_3 in HL-60 human promyelocytes differentiation. *J Biol Chem*. 1987, *262(29)*, 14164-71.

10   **22-Oxo-23,24,25,26,27-pentanorvitamin D_3 6RS,19-sulfur dioxide adduct**
9547268 $C_{22}H_{32}O_4S$

(7E)-(3S,6RS)-3-hydroxy-6,19-epithio-23,24-dinor-9,10-seco-5(10),7-choladien-22-al S,S-dioxide.
LMST03020008; 22-oxo-23,24,25,26,27-pentanorcholecalciferol 6RS,19-sulfur dioxide adduct; 3S-hydroxy-6RS,19-epithio-23,24-dinor-9,10-seco-5(10),7E-choladien-22-al S,S-dioxide; 9547268; Prepared from vitamin D_2 sulfur dioxide adduct by ozonolysis. *J Chromatogr B Biomed Appl*. 1995, *672(1)*, 63-71.

11   **1α-Hydroxy-23,24,25,26,27-pentanorvitamin D_3**
9547269   . $C_{22}H_{34}O_2$

(1R,3S,5Z)-5-[(2E)-2-[(1R,3aR,7aS)-7a-methyl-1-propan-2-yl-2,3,3a,5,6,7-hexahydro-1H-inden-4-ylidene]ethylidene]-4-methylidene-cyclohexane-1,3-diol.
LMST03020009; 23,24-dinor-9,10-seco-5Z,7E,10(19)-cholatriene-1S,3R-diol; 1α-hydroxy-23,24,25,26,27-pentanorcholecalciferol; 9547269; One order of magnitude less effective than 1α,25-dihydroxyvitamin D_3 in HL-60 human promyelocytes differentiation. *J Biol Chem*. 1987, *262(29)*, 14164-71.

12   **22-Hydroxy-23,24,25,26,27-pentanorvitamin D_3**
9547270 $C_{22}H_{34}O_2$

(1S,3Z)-3-[(2E)-2-[(1R,3aR,7aS)-1-[(2S)-1-hydroxy-propan-2-yl]-7a-methyl-2,3,3a,5,6,7-hexahydro-1H-inden-4-ylidene]ethylidene]-4-methylidene-cyclohexan-1-ol.
LMST03020010; 23,24-dinor-9,10-seco-5Z,7E,10(19)-cholatrien-3S,22-diol; 22-hydroxy-23,24,25,26,27-pentanorcholecalciferol; 9547270; Prepared from vitamin D_2 by ozonolysis of its sulfur dioxide adduct. *J Chromatogr B Biomed Appl*. 1995, *672(1)*, 63-71.

13   **1α,22-Dihydroxy-23,24,25,26,27-pentanorvitamin D_3**
9547271 $C_{22}H_{34}O_3$

(1R,3S,5Z)-5-[(2E)-2-[(1R,3aR,7aS)-1-[(2S)-1-hydroxy-propan-2-yl]-7a-methyl-2,3,3a,5,6,7-hexahydro-1H-inden-4-ylidene]ethylidene]-4-methylidene-cyclohexane-1,3-diol.
LMST03020011; 23,24-dinor-9,10-seco-5Z,7E,10(19)-cholatriene-1S,3R,22-triol; 1α,22-dihydroxy-23,24,25,26,27-pentanorvitamin D$_3$; 1α,22-dihydroxy-23,24,25,26,27-pentanorcholecalciferol; 9547271; Two orders of magnitude less effective than 1α,25-dihydroxy vitamin D$_3$ in HL-60 human promyelocytic differentiation. *J Biol Chem.* 1987, *262(29)*, 14164-14171; The Receptors (Conn, P. M., ed.), pp1-13, Academic Press (1985).

14   *22-Hydroxy-23,24,25,26,27-pentanor-vitamin D$_3$ 6RS,19-sulfur dioxide adduct*
9547272   C$_{22}$H$_{34}$O$_4$S

(7E)-(3S,6RS)-6,19-epithio-23,24-dinor-9,10-seco-5(10),7-choladiene-3,22-diol S,S-dioxide.
LMST03020012;   22-hydroxy-23,24,25,26,27-penta-norvitamin D$_3$ 6RS,19-sulfur dioxide adduct; 22-hydroxy-23,24,25,26,27-pentanorcholecalciferol 6RS,19-sulfur dioxide adduct; 6RS,19-epithio-23,24-dinor-9,10-seco-5(10),7E-choladiene-3S,22-diol S,S-dioxide; 9547272; Prepared from vitamin D$_2$ sulfur dioxide adduct by ozonolysis followed by reduction. *J Chromatogr B Biomed Appl.* 1995, *672(1)*, 63-71.

15   *1α-Hydroxy-24,25,26,27-tetranorvit-amin D$_3$ 23-carboxylic acid*
71204-89-2   C$_{23}$H$_{34}$O$_4$

(3R)-3-[(1R,3aR,4E,7aS)-4-[(2Z)-2-[(3S,5R)-3,5-dihydroxy-2-methylidene-cyclohexylidene]ethylidene]-7a-methyl-2,3,3a,5,6,7-hexahydro-1H-inden-1-yl]-butanoic acid.
LMST03020013; 1S,3R-dihydroxy-24-nor- 9,10-seco-5Z,7E,10(19)-cholatrien-23-oic acid; calcitroic acid; 9547273; 71204-89-2; Metabolite of 1,25-dihydroxy vitamin D$_3$. Product of renal metabolism of 1,25-dihydroxyvitamin D$_3$ through C-24 oxidation pathway. Calcitroic acid is formed in a second step by cytochrome P450 oxidation. Causes bone mineral mobilization and intestinal calcium transport. Crystals; mp = 122-126°; λ$_m$ = 262 (ε = 18000). *Biochim Biophys Acta.* 1974, *338(2)*, 489-95; *J Biol Chem.* 1973, *248(19)*, 6691-6. *Biochemistry.* 1979, *18(18)*, 3977-83; 1989, *28(4)*, 1763-9.

16   *1α-Hydroxy-2-methylene-19-nor-(20S)-bishomopregnacalciferol*
5289547   C$_{23}$H$_{36}$O$_2$

(1R,3R)-5-[(2E)-2-[(1R,3aR,7aS)-1-[(2S)-butan-2-yl]-7a-methyl-2,3,3a,5,6,7-hexahydro-1H-inden-4-ylidene]-ethylidene]-2-methylidene-cyclohexane-1,3-diol.
2MbisP;   1α-Hydroxy-2-methylene-19-nor-(20S)-bis-homopregnacalciferol; 5-{2-[1-(1-methylpropyl)-7A-methyloctahydroinden-4-ylidene]-ethylidene}-2-methylenecyclohexane-1,3-diol; VD1; 5289547; Binds to the vitamin D receptor with slightly less affinity than 1α,25-dihydroxyvitamin D$_3$. *Biochemistry.* 2004, *43(14)*, 4101-10.

17   *23-Hydroxy-24,25,26,27-tetranorvitamin D$_3$*
88200-28-6   C$_{23}$H$_{36}$O$_2$

(1S,3Z)-3-[(2E)-2-[(1R,3aR,7aS)-1-[(2R)-4-hydroxy-butan-2-yl]-7a-methyl-2,3,3a,5,6,7-hexahydro-1H-inden-4-ylidene]ethylidene]-4-methylidene-cyclohexan-1-ol.
LMST03020019; 24-nor-9,10-seco-Z5,7E,10(19)-cholatriene-3S,23-diol; 23-hydroxy-24,25,26,27-tetranorcholecalciferol; 24,25,26,27-Tetranor-23-hydroxy-vitamin D_3; Tnh-D_3; 9547279; 88200-28-6: Inactive metabolite of vitamin D_3 in rat kidney, formed in the C24 oxidation pathway by cytochrome P450 oxidation of 24-oxo-1α,23,25-trihydroxyvitamin D_3. Prepared by perfusing kidneys from vitamin D_3 replete rats with 24,25-dihydroxyvitamin D_3. *Biochemistry*. 1984, *23(16)*, 3749-54; 1987, *26(1)*, 324-31.

18 **1α-Hydroxy-24,25,26,27-tetranor-vitamin D_3**
9547274 $C_{23}H_{36}O_2$

(1R,3S,5Z)-5-[(2E)-2-[(1R,3aR,7aS)-1-[(2R)-butan-2-yl]-7a-methyl-2,3,3a,5,6,7-hexahydro-1H-inden-4-ylidene]ethylidene]-4-methylidene-cyclohexane-1,3-diol.
LMST03020014; 24-nor-9,10-seco-5Z,7E,10(19)-cholatriene-1S,3R-diol; 1α-hydroxy-24,25,26,27-tetranorcholecalciferol; 9547274; One order of magnitude less effective than 1α,25-dihydroxyvitamin D_3 in HL-60 human promyelocytes differentiation. *J Biol Chem*. 1987, *262(29)*, 14164-71.

19 **22R-Hydroxy-24,25,26,27-tetranor-vitamin D_3**
9547275 $C_{23}H_{36}O_2$

(1S,3Z)-3-[(2E)-2-[(1R,3aR,7aS)-1-[(2S,3R)-3-hydroxy-butan-2-yl]-7a-methyl-2,3,3a,5,6,7-hexahydro-1H-inden-4-ylidene]ethylidene]-4-methylidene-cyclo-hexan-1-ol.
LMST03020015; 24-nor-9,10-seco-5Z,7E,10(19)-cholatriene-3S,22R-diol; 22R-hydroxy-24,25,26,27-tetranorcholecalciferol; 9547275; Prepared from vitamin D_2 sulfur dioxide adduct by ozonolysis followed by Grignard reaction. *J Chromatogr B Biomed Appl*. 1995, *672(1)*, 63-71.

20 **22S-Hydroxy-24,25,26,27-tetranor-vitamin D_3**
9547276 $C_{23}H_{36}O_2$

(1S,3Z)-3-[(2E)-2-[(1R,3aR,7aS)-1-[(2S,3S)-3-hydroxy-butan-2-yl]-7a-methyl-2,3,3a,5,6,7-hexahydro-1H-inden-4-ylidene]ethylidene]-4-methylidene-cyclo-hexan-1-ol.
LMST03020016; 24-nor-9,10-seco-5Z7E,10(19)-cholatriene-3S,22S-diol; 22S-hydroxy-23,24,25,26,27-pentanorcholecalciferol; 9547276; Prepared from vitamin D_2 sulfur dioxide adduct by ozonolysis followed by Grignard reaction. λ_m = 265 nm (EtOH). *J Chromatogr B Biomed Appl*. 1995, *672(1)*, 63-71.

21 **(5E)-22R-Hydroxy-24,25,26,27-tetranor-vitamin D_3**
9547277 $C_{23}H_{36}O_2$

(1S,3E)-3-[(2E)-2-[(1R,3aR,7aS)-1-[(2S,3R)-3-
hydroxybutan-2-yl]-7a-methyl-2,3,3a,5,6,7-hexahydro-
1H-inden-4-ylidene]ethylidene]-4-methylidene-
cyclohexan-1-ol.
LMST03020017; 24-nor-9,10-seco-5E,7E,10(19)-chola-
triene-3S,22R-diol;   (5E)-22R-hydroxy-24,25,26,27-
tetranorcholecalciferol;   9547277;   Prepared   from
vitamin D$_2$ sulfur dioxide adduct by ozonolysis
followed by Grignard reaction. *J Chromatogr B Biomed
Appl.* 1995, *672(1)*, 63-71.

22   *(5E)-22S-Hydroxy-24,25,26,27-tetranor-
vitamin D$_3$*
9547278   C$_{23}$H$_{36}$O$_2$

(1S,3E)-3-[(2E)-2-[(1R,3aR,7aS)-1-[(2S,3S)-3-hydroxy-
butan-2-yl]-7a-methyl-2,3,3a,5,6,7-hexahydro-1H-
inden-4-ylidene]ethylidene]-4-methylidene-
cyclohexan-1-ol.
LMST03020018; 24-nor-9,10-seco-5E,7E,10(19)-chola-
triene-3S,22S-diol;   (5E)-22S-hydroxy-24,25,26,27-
tetranorcholecalciferol;   9547278;   Prepared   from
vitamin D$_2$ sulfur dioxide adduct by ozonolysis
followed by Grignard reaction. *J Chromatogr B Biomed
Appl.* 1995, *672(1)*, 63-71.

23   *1α,23-Dihydroxy-24,25,26,27-tetranor-
vitamin D$_3$*
97903-37-2   C$_{23}$H$_{36}$O$_3$

(1R,3S,5E)-5-[(2E)-2-[(1R,3aR,7aS)-1-[(2R)-4-hydroxy-
butan-2-yl]-7a-methyl-2,3,3a,5,6,7-hexahydro-1H-
inden-4-ylidene]ethylidene]-4-methylidene-cyclohex-
ane-1,3-diol.
1,23-Dihydroxy-24,25,26,27-tetranorvitamin D$_3$; 1,23-
Dtnv-D$_3$; 24-Nor-9,10-secochola-5Z,7E,10(19)-triene-
1α,3β,23-triol; 6438785; 97903-37-2; A metabolite,
via the 24-oxidation pathway, of vitamin D$_3$ and
1α,25-dihydroxyvitamin D$_3$ in rat kidney. Formed
together with three other metabolites, 1,24,25-
trihydroxyvitamin D$_3$, 1,25-dihydroxy-24-oxovitamin
D$_3$, and 1,23,25-trihydroxy-24-oxovitamin D$_3$. Is
metabolized further to calcitrioic acid. *Biochemistry.*
1987, *26(1)*, 324-31; *Endocrinology* 1995, *136(10)*,
4195-203.

24   *1α,23-Dihydroxy-24,25,26,27-tetranor-
vitamin D$_3$*
9547280   C$_{23}$H$_{36}$O$_3$

(1R,3S,5Z)-5-[(2E)-2-[(1R,3aR,7aS)-1-[(2R)-4-hydroxy-
butan-2-yl]-7a-methyl-2,3,3a,5,6,7-hexahydro-1H-
inden-4-ylidene]ethylidene]-4-methylidene-cyclohex-
ane-1,3-diol.
LMST03020020; 24-nor-9,10-seco-5Z,7E,10(19)-chol-
atriene-1S,3R,23-triol; 1α,23-dihydroxy-24,25,26,27-
tetranorcholecalciferol; 9547280; Two orders of
magnitude less effective than 1α,25-dihydroxyvitamin
D$_3$ in HL-60 human promyelocytes differentiation. ; λ$_m$
= 265 nm. *Biochemistry.* 1979, *18(18)*, 3977-83;
1987, *26(1)*, 324-31; *J Biol Chem.* 1987, *262(29)*,
14164-71.

25   *1-Hydroxy-25,26,27-trinorvitamin   D$_3$
24-oic acid*
                                                 C$_{24}$H$_{36}$O$_4$

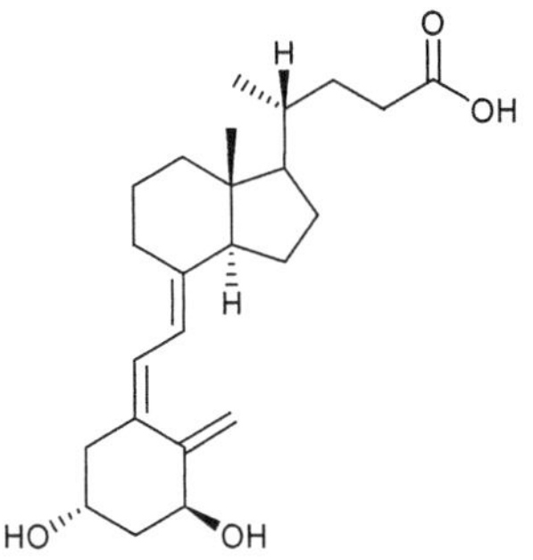

(5Z,7E)-(3S)-1,3-dihydroxy-9,10-seco-5,7,10(19)-chol-atrien-24-oic acid.
Hapten 24-OA; 1-hydroxy-19,25,26,27-tetranor-vitamin D_3 carboxylic acid; Used as a hapten-carrier conjugate of 1α,25-dihydroxyvitamin D_3. *Biol Pharm Bull*. 1997, *20(9)*, 948-53.

26   *25,26,27-Trinorvitamin D_3 24-carboxylic acid*

9547281 $C_{24}H_{36}O_3$

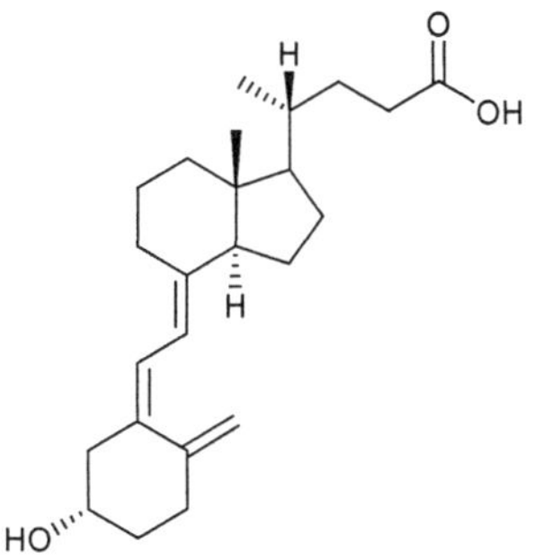

(4R)-4-[(1R,3aR,4E,7aS)-4-[(2Z)-2-[(5S)-5-hydroxy-2-methylidene-cyclohexylidene]ethylidene]-7a-methyl-2,3,3a,5,6,7-hexahydro-1H-inden-1-yl]pentanoic acid.
LMST03020021; 3S-hydroxy-9,10-seco-5Z,7E,10(19)-cholatrien-24-oic acid; cholacalcioic acid; 25,26,27-trinorcholecalciferol 24-carboxylic acid; 9547281; Formed from vitamin D supplemented rat kidney homogenates incubated with 24,25-dihydroxyvitamin D_3. Metabolite *via* the C-24 oxidation pathway, of 25-hydroxyvitamin D_3 which proceeds through 24,25-dihydroxy-, then 25-hydroxy-24-oxo-, 23,25-dihydroxy-24-oxo- 23-hydroxy-24,25,26,27-tetranor-vitamin D_3 and finally, cholacalcioic acid. *Endocrinology* 1995, *136(10)*, 4195-4203. Vitamin D, Basic Research and its Clinical Application. Recent Developments in The Metabolism of Vitamin D., pp 445-458, Walter de Gruyter & Co., Berlin New York.

27   *1α-Hydroxy-25,26,27-trinorvitamin D_3 24-carboxylic acid*

9547282 $C_{24}H_{36}O_4$

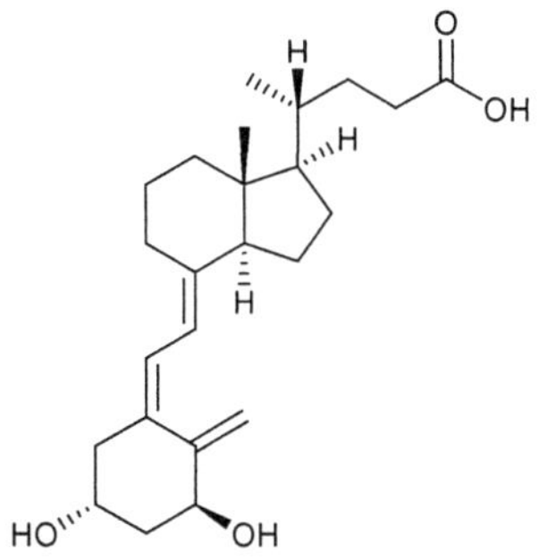

(4R)-4-[(1R,3aR,4E.7aS)-4-[(2Z)-2-[(3S,5R)-3,5-di-hydroxy-2-methylidene-cyclohexylidene]ethylidene]-7a-methyl-2,3,3a,5,6,7-hexahydro-1H-inden-1-yl]-pentanoic acid.
LMST03020022; 1S,3R-dihydroxy-9,10-seco-5Z,7E,10 (19)-cholatrien-24-oic acid; 9547282; Synthesized from methyl-3β-hydroxy-chola-1,5,7-triene-24-oate.
Solid; mp = 117-120°; λ_m = 263 nm (ε = 18200). *J Chem Soc Perkin Trans I*, 1985, 1331.

28   *1α,24-Dihydroxy-25,26,27-trinorvitamin D_3*

9547283 $C_{24}H_{38}O_3$

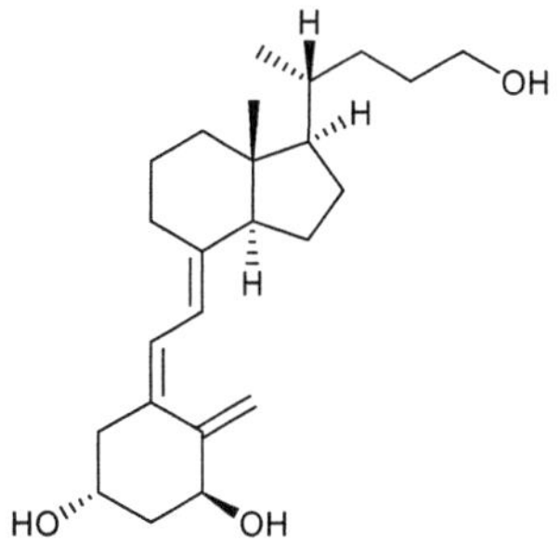

(1R.3S,5Z)-5-[(2E)-2-[(1R,3aR,7aS)-1-[(2R)-5-hydroxy-pentan-2-yl]-7a-methyl-2,3,3a,5,6,7-hexahydro-1H-inden-4-ylidene]ethylidene]-4-methylidene-cyclo-hexane-1,3-diol.
LMST03020023; 9,10-seco-5Z,7E,10(19)-cholatriene-1S,3R,24-triol; 1α,24-dihydroxy-25,26,27-trinor-vitamin D_3; 1α,24-dihydroxy-25,26,27-trinorchole-calciferol; 9547283; One order of magnitude less effective than 1α,25-dihydroxyvitamin D_3 in HL-60 human promyelocytes differentiation. *J Biol Chem*. 1987, *262(29)*, 14164-71.

29   *5α-Cholan-24-oic Acid*

25312-65-6 $C_{24}H_{40}O_2$

(4R)-4-[(5R,8S,9S,10S,13R,14S,17R)-10,13-dimethyl-2,3,4,5,6,7,8,9,11,12,14,15,16,17-tetradecahydro-1H-cyclopenta[a]phenanthren-17-yl]pentanoic acid.
LMST04010002; 5α-cholanic acid; 5α-cholan-24-oic Acid; CHEBI:36237; 65316; 9548797; 25312-65-6; Product of metabolism of cholesterol. Solid; mp = 170°; $[\alpha]_D^{14}$ = 21.74° (CHCl3); *J Lipid Res*, 1980, *21(1)*, 110-17; Dictionary of Steroids, Chapman & Hall, London (1991); Rodd's Chemistry of Carbon Compounds (Vol. II, Part D) (Coffey, S. ed) (1970), pp 168-96.

30   *3α-Hydroxy-5α-cholan-24-oic Acid*
5283803 $C_{24}H_{40}O_3$

(4R)-4-[(3R,5S,8R,9S,10S,13R,14S,17R)-3-hydroxy-10,13-dimethyl-2,3,4,5,6,7,8,9,11,12,14,15,16,17-tetradecahydro-1H-cyclopenta[a]phenanthren-17-yl]pentanoic acid.
LMST04010005; 3α-Hydroxy-5α-cholan-24-oic Acid; 5283803; Bile acid formed by metabolism of cholesterol. Isolated from human urine, feces and bile. Solid; mp = 204-208°; $[\alpha]_D^{26}$ = 26.0° ; *J Lipid Res.*, 1977, *18(3)*, 339-62; *Eur J Clin Invest*. 1979, *9(5)*, 341-8; *FEBS Lett*. 1971, *15(2)*, 161-4.

31   *C-9,11,21-Trisnor-17-methyl-1α,25-dihydroxyvitamin D₃*
                                               $C_{24}H_{42}O_3$

9,11,21,trisnor-17-methylcholecalciferol; Synthesis and biological activities reported. *Bioorg Med Chem Lett*.

2002, *12(12)*, 1629-32.

32   *1α-Hydroxy-26,27-dinorvitamin D₃ 25-carboxylic acid*
9547285 $C_{25}H_{38}O_4$

(5R)-5-[(1R,3aR,4E,7aS)-4-[(2Z)-2-[(3S,5R)-3,5-di-hydroxy-2-methylidene-cyclohexylidene]ethylidene]-7a-methyl-2,3,3a,5,6,7-hexahydro-1H-inden-1-yl]-hexanoic acid.
LMST03020025; 1S,3R-dihydroxy-25-homo-9,10-seco-5Z,7E,10(19)-cholatrien-25-oic acid; 1α-hydroxy-26,27-dinorcholecalciferol 25-carboxylic acid; 9547285; Synthesized from methyl-3β-hydroxy-25-homo-chola-1,5,7-trien-25-oate. Crystals; mp = 97-101°; λ_m = 265 nm (ε = 18300). *J Chem Soc Perkin Trans I*, 1985, 1331.

33   *1α-Hydroxy-21-nor-20-oxavitamin D₃*
9547286 $C_{25}H_{40}O_3$

(1R,3S,5Z)-5-[(2E)-2-[(1S,3aR,7aS)-7a-methyl-1-(4-methylpentoxy)-2,3,3a,5,6,7-hexahydro-1H-inden-4-ylidene]ethylidene]-4-methylidene-cyclohexane-1,3-diol.
LMST03020026; 21-nor-20-oxa-9,10-seco-5Z,7E,10(19)-cholestatriene-1S,3R-diol; 1α-hydroxy-21-nor-20-oxacholecalciferol; 9547286; Synthesized from de-hydroepiandrosterone. $[\alpha]_D^{24}$ = -30.2° (c = 0.86 EtOH); λ_m = 262 nm (EtOH). *Chem Pharm Bull (Tokyo)*. 1986, *34*, 2286-9.

34   *1α,25-Dihydroxy-23,24-dinorvitamin D₃*
9547287 $C_{25}H_{40}O_3$

(1R,3S,5Z)-5-[(2E)-2-[(1R,3aR,7aS)-1-[(2R)-4-hydroxy-4-methyl-pentan-2-yl]-7a-methyl-2,3,3a,5,6,7-hexa hydro-1H-inden-4-ylidene]ethylidene]-4-methylidene-cyclohexane-1,3-diol.
LMST03020027; 23,24-dinor-9,10-seco-5Z,7E,10(19)-cholestatriene-1S,3R,25-triol; 1α,25-dihydroxy-23,24-dinorcholecalciferol; 9547287; Two orders of magnitude less effective than 1α,25-dihydroxyvitamin D₃ in HL-60 human promyelocytes differentiation. *J Biol Chem*. 1987, *262(29)*, 14164-71.

35   *1α,25-Dihydroxy-26,27-dinorvitamin D₃*
9547288   C₂₅H₄₀O₃

(1R,3S,5Z)-5-[(2E)-2-[(1R,3aR,7aS)-1-[(2R)-6-hydroxy-hexan-2-yl]-7a-methyl-2,3,3a,5,6,7-hexahydro-1H-inden-4-ylidene]ethylidene]-4-methylidene-cyclohex-ane-1,3-diol.
LMST03020028; 26,27-dinor-9,10-seco-5Z,7E,10(19)-cholestatriene-1S,3R,25-triol; 1α,25-dihydroxy-26,27-dinorcholecalciferol; 9547288; Two orders of magnitude less effective than 1α,25-dihydroxyvitamin D₃ in HL-60 human promyelocytes differentiation. *Biochemistry*. 1980, *19(11)*, 2515-21; *J Biol Chem*. 1987, *262(29)*, 14164-71.

36   *1α,25-Dihydroxy-21-nor-20-oxavitamin D₃*
106315-28-0   C₂₅H₄₀O₄

(1R,3S,5E)-5-[(2E)-2-[(3aR,7aS)-1-(4-hydroxy-4-methyl-pentoxy)-7a-methyl-2,3,3a,5,6,7-hexahydro-1H-inden-4-ylidene]ethylidene]-4-methylidene-cyclohexane-1,3-diol.
1,25-Dihydroxy-21-nor-20-oxavitamin D₃; 1,25-(OH)₂-21-Nor-20-oxaD₃; 1,3-Cyclohexanediol, 4-methyl-ene-5-((octahydro-1-((4-hydroxy-4-methylpentyl)oxy)-7a-methyl-4H-inden-4-ylidene)ethylidene)-, (1S-(1α, 3aβ,4E(1S*,3R*,5Z),7aα))-; 6439066; 106315-28-0; Has differentiation-inducing activity towards human myeloid leukemia cells approximately one-fifth of that of 1α,25-dihydroxyvitamin D₃, while its binding affinity with chick intestinal cytosolic receptor, is about one-tenth of that of 1α,25-dihydroxyvitamin D₃. Its rather weak effect on serum calcium levels in normal mice at a dosage of 500 μg/kg (intravenous administration) indicates that the essential importance of the 21-methyl moiety may lie in its effect on the regulation of calcium metabolism. *Chem Pharm Bull (Tokyo)*. 1992, *40(3)*, 648-51.

37   *1α,25-Dihydroxy-21-nor-20-oxavitamin D₃*
9547289   C₂₅H₄₀O₄

(1R,3S,5Z)-5-[(2E)-2-[(1S,3aR,7aS)-1-(4-hydroxy-4-methyl-pentoxy)-7a-methyl-2,3,3a,5,6,7-hexahydro-1H-inden-4-ylidene]ethylidene]-4-methylidene-cyclohexane-1,3-diol.
LMST03020029; 21-nor-20-oxa-9,10-seco-5Z,7E,10(19)-cholestatriene-1S,3R,25-triol; 1α,25-dihydroxy-21-nor-20-oxacholecalciferol; 9547289; Synthesized photochemically from dehydroepiandrostertone. Slightly less effective than 1α,25-dihydroxyvitamin D₃ in differentiation inducing activity of HL-60. Binding affinity for chick intestinal receptor is 1/1000 of 1α,25-dihydroxyvitamin D₃. Calcemic activity in normal mice is similar to that of 1α,25-dihydroxyvitamin D₃; Colorless crystals; [α]$_D^{24}$= -44.6° (c = 0.56 EtOH); λ$_m$ = 262 nm (EtOH). *Chem Pharm Bull (Tokyo)*. 1986, *34*,

2286-9; 1992, *40(3)*, 648-51.

38   *1α,25-Dihydroxy-24-nor-22-oxavitamin D₃*

9547290 $C_{25}H_{40}O_4$

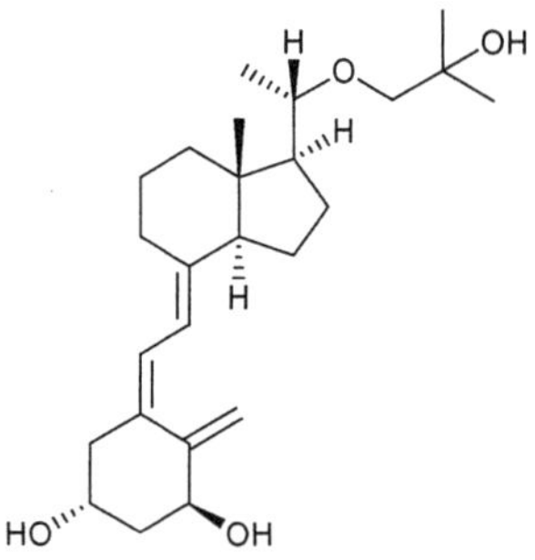

(1R,3S,5Z)-5-[(2E)-2-[(1S,3aR,7aR)-1-[(1S)-1-(2-hydroxy-2-methyl-propoxy)ethyl]-7a-methyl-2,3,3a,5,6,7-hexahydro-1H-inden-4-ylidene]ethylidene]-4-methylidene-cyclohexane-1,3-diol. LMST03020030; 24-nor-22-oxa-9,10-seco-5Z,7E,10 (19)-cholestatriene-1S,3R,25-triol; 1α,25-dihydroxy-24-nor-22-oxacholecalciferol; 9547290; Activity inducing differentiation of human myeloid leukemia cells (HL-60) into macrophases *in vitro* is compared to that of 1α,25-dihydroxyvitamin D₃. Colorless foam; λ_m = 263 nm (EtOH). *Chem Pharm Bull (Tokyo)*. 1992, *40(6)*, 1494-9.

39   *1α-Hydroxy-24-methylsulfonyl-25,26, 27-trinorvitamin D₃*

9547291 $C_{25}H_{40}O_4S$

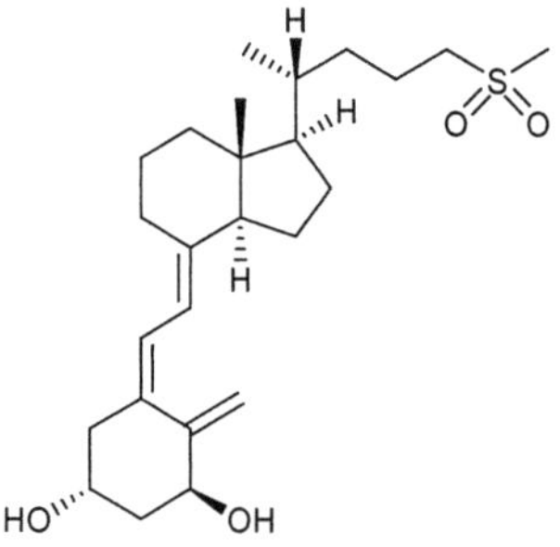

(1R,3S,5Z)-5-[(2E)-2-[(1R,3aR,7aS)-7a-methyl-1-[(2R)-5-methylsulfonylpentan-2-yl]-2,3,3a,5,6,7-hexahydro-1H-inden-4-ylidene]ethylidene]-4-methylidene-cyclohexane-1,3-diol. LMST03020031; 24-methylsulfonyl-9,10-seco-5Z,7E, 10(19)-cholatriene-1S,3R-diol; 1α-hydroxy-24-methyl-sulfonyl-25,26,27-trinorcholecalciferol; 9547291. *J Org Chem*. 1993, *58(1)*, 118-9.

40   *2α-Fluoro-19-nor-22-oxa-1α,25-dihydroxyvitamin D₃*

                         $C_{25}H_{41}FO_4$

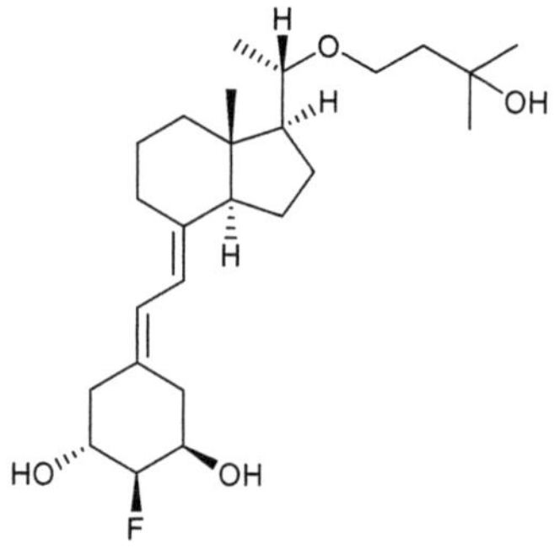

(1R,3S,5Z)-5-[(2E)-2-[(1R,3aR,7aS)-1-[(2R)-6-hydroxy-6-methyl-3-oxa-heptan-2-yl]-7a-methyl-2,3,3a,5,6,7-hexahydro-1H-inden-4-ylidene]ethylidene]-2-fluoro-cyclohexane-1,3-diol. 2α-Fluoro-19-nor-22-oxa-1α,25-dihydroxycholecalciferol; 2-fluoro-19-nor-22-oxa-1,25(OH)2D₃; 2α-fluoro-19-nor-22-oxa-1α,25-dihydroxycholecalciferol; Binds to the vitamin D receptor and shows significant activity in transactivation. Synthesized *via* a asymmetric catalytic carbonyl-ene cyclization; conformation studied by ^{19}FNMR spectroscopy. *Chirality*. 2001, *13(7)*, 366-71.

41   *1α,25-Dihydroxy-18,19-dinorvitamin D₃*

                         $C_{25}H_{42}O_3$

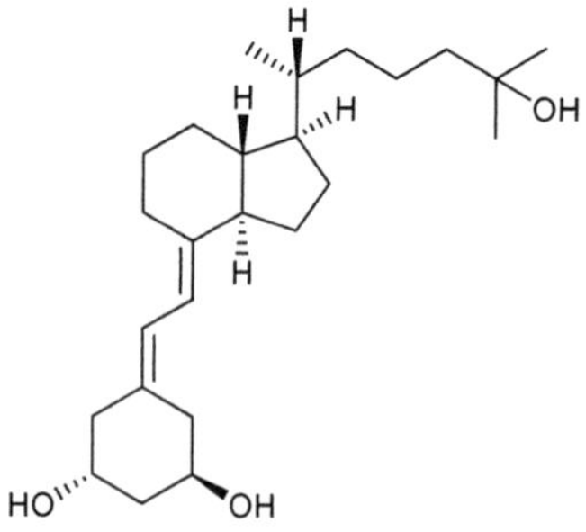

18,19-dinor-1α,25-(OH)₂D₃; Activity in HL60 cell differentiation is similar to that of 1α,25-dihydroxyvitamin D₃, but in calcium mob ilization, it is ten times less active. *J Med Chem*. 1996, *39(22)*, 4497-506.

42   *1α,25-Hydroxy-3-deoxy-19-nor-22-oxa-vitamin D₃*

9547661 $C_{25}H_{42}O_3$

(1S,3Z)-3-[(2E)-2-[(1S,3aR,7aR)-1-[(1S)-1-(3-hydroxy-3-methyl-butoxy)ethyl]-7a-methyl-2,3,3a,5,6,7-hexahydro-1H-inden-4-ylidene]ethylidene]cyclohexan-1-ol. LMST03020566; 19-nor-22-oxa-9,10-seco-5,7E-cholestadiene-1S,25-diol; 1α,25-hydroxy-3-deoxy-19-nor-22-oxacholecalciferol; 9547661; Compared to 1α,25-dihydroxyvitamin D_2, affinity for calf thymus vitamin D receptor is 0.4%, affinity for vitamin D binding protein gave no response, HL-60 cell anfiproliferation is 80%, HL-60 cell differentiation is 50%, transcriptional activity (rat 25-hydroxyvitamin D_3 24-hydroxylase promoter) is 132%. *Biol Pharm Bull.* 1998, *21(12)*, 1300-5.

43   *1β,25-Dihydroxy-3-deoxy-19-nor-22-oxavitamin D_3*

9547662 $C_{25}H_{42}O_3$

(1R,3Z)-3-[(2E)-2-[(1S,3aR,7aR)-1-[(1S)-1-(3-hydroxy-3-methyl-butoxy)ethyl]-7a-methyl-2,3,3a,5,6,7-hexahydro-1H-inden-4-ylidene]ethylidene]cyclohexan-1-ol. LMST03020567; 19-nor-22-oxa-9,10-seco-5,7E-cholestadiene-1R,25-diol; 1β,25-dihydroxy-3-deoxy-19-nor-22-oxacholecalciferol; 9547662; Compared to 1α,25-dihydroxyvitamin D_3, affinity for calf thymus vitamin D receptor gave no response, affinity for vitamin D binding protein is 50%, HL-60 cell anfiproliferation is 20%, HL-60 cell differentiation is 20% and transcriptional activity (rat 25-hydroxyvitamin D_3 24-hydroxylase promoter) is 12%. *Biol Pharm Bull.* 1998, *21*, 1300-5.

44   *25-Hydroxy-19-nor-22-oxa-3-epivitamin D_3*

9547663 $C_{25}H_{42}O_3$

(1R,3E)-3-[(2E)-2-[(1S,3aR,7aR)-1-[(1S)-1-(3-hydroxy-3-methyl-butoxy)ethyl]-7a-methyl-2,3,3a,5,6,7-hexahydro-1H-inden-4-ylidene]ethylidene]cyclohexan-1-ol. LMST03020569; 19-nor-22-oxa-9,10-seco-5,7E-cholestadiene-3R,25-diol; 25-hydroxy-19-nor-22-oxa-3-epicholecalciferol;9547663; Compared to 1α,25-dihydroxyvitamin D_3, affinity for calf thymus vitamin D receptor is 0.1%, HL-60 cell anfiproliferation is 20%, HL-60 cell differentiation is 20%, transcriptional activity (rat 25-hydroxyvitamin D_3 24-hydroxylase promoter) is 74%. *Biol Pharm Bull.* 1998, *21(12)*, 1300-5.

45   *25-Hydroxy-19-nor-22-oxavitamin D_3*

9547664 $C_{25}H_{42}O_3$

(1S,3E)-3-[(2E)-2-[(1S,3aR,7aR)-1-[(1S)-1-(3-hydroxy-3-methyl-butoxy)ethyl]-7a-methyl-2,3,3a,5,6,7-hexahydro-1H-inden-4-ylidene]ethylidene]cyclohexan-1-ol. LMST03020569; 19-nor-22-oxa-9,10-seco-5,7E-cholestadiene-3S,25-diol; 25-hydroxy-19-nor-22-oxacholecalciferol; 9547664; Compared to 1α,25-dihydroxyvitamin D_3, affinity for calf thymus vitamin D receptor gave no response, affinity for vitamin D binding protein gave no response, HL-60 cell anfiproliferation is 50%, HL-60 cell differentiation is 0%, transcriptional activity (rat 25-hydroxyvitamin D_3 24-hydroxylase promoter) is 12%. *Biol Pharm Bull.* 1998, *21(12)*, 1300-5.

46   *1,25-Dihydroxy-2,4-dinor-1,3-secovitamin D_3*

9547292 $C_{25}H_{42}O_3$

(2E)-2-[(2E)-2-[(1R,3aR,7aS)-1-[(2R)-6-hydroxy-6-methyl-heptan-2-yl]-7a-methyl-2,3,3a,5,6,7-hexahydro-1H-inden-4-ylidene]ethylidene]-3-methylidene-butane-1,4-diol.
LMST03020032; A-dinor-(1,2)-(9,10)-diseco-5Z,7E,10(19)-cholestatriene-1,2,25-triol; 1,25-dihydroxy-2,4-dinor-1,3-secovitamin D_3; 1,25-dihydroxy-2,4-dinor-1,3-secocholecalciferol; 9547292; Affinity for chick intestinal receptor is 2% of that of 1α,25-dihydroxyvitamin D_3. *Endocr Rev.* 1995, *16(2)*, 200-57.

47   1α,25-Dihydroxy-19-nor-22-oxavitamin D_3
9547649 $C_{25}H_{42}O_4$

(1R,3R)-5-[(2E)-2-[(1S,3aR,7aR)-1-[(1S)-1-(3-hydroxy-3-methyl-butoxy)ethyl]-7a-methyl-2,3,3a,5,6,7-hexahydro-1H-inden-4-ylidene]ethylidene]cyclohexane-1,3-diol.
LMST03020546; (7E)-(1R,3R)-19-nor-22-oxa-9,10-seco-5,7E-cholestadiene-1R,3R,25-triol; 1α,25-dihydroxy-19-nor-22-oxavitamin D_3; 1α,25-dihydroxy-19-nor-22-oxacholecalciferol; 9547649; Prepared from the 22-oxa homologue of Grundmann ketone. Compared to 1α,25-dihydroxyvitamin D_3, affinity for VDR is 3%, in transactivation of a rat 25-hydroxyvitamin D_3 24-hydroxylase (CYP24) gene is 200%. *Tet Lett.* 1998, *39(21)*, 3359-62.

48   1α,25-Dihydroxy-26,26,26,27,27,27-hexafluoro-16,17,23,23,24,24-hexadehydro-19-norvitamin D_3
9547648 $C_{26}H_{32}F_6O_3$

(1R,3R)-5-[(2E)-2-[(3aR,7aS)-7a-methyl-1-[(2R)-7,7,7-trifluoro-6-hydroxy-6-(trifluoromethyl)hept-4-yn-2-yl]-3a,5,6,7-tetrahydro-3H-inden-4-ylidene]ethylidene]-cyclohexane-1,3-diol.
LMST03020545; 1,25-dihydroxy-16-ene-23-yne-26,27-hexafluoro-19-norvitamin-D_3; 1,25-dihydroxy-16-ene-23-yne-26,27-hexafluoro-19-norcholecalciferol; 1α,25-dihydroxy-26,26,26,27,27,27-hexafluoro-16,17,23,23,24,24-hexadehydro-19-norcholecalciferol; 26,26,26,27,27,27-hexafluoro-19-nor-9,10-seco-5,7E,16-cholestatrien-23-yne-1R,3R,25-triol; Ro 25-6760; 9547648; Potent inhibitor of tumor-derived endothelial cells. Causes a dose dependent growth inhibition of MCF-7 and MDAMB-468 human breast cancer cells at concentrations ranging between 1-100 nM. *Clin Cancer Res.* 1998, *4(11)*, 2869-76. *Endocrinology.* 2002, *143(7)*, 2508-14. *Hoffmann-LaRoche Inc.*

49   25,26-Epoxy-1α-hydroxy-23,23,24,24-tetradehydro-19-norvitamin D_3
9547293 $C_{26}H_{38}O_3$

(1R,3R)-5-[(2E)-2-[(1R,3aR,7aS)-7a-methyl-1-[(2R)-5-(2-methyloxiran-2-yl)pent-4-yn-2-yl]-2,3,3a,5,6,7-hexahydro-1H-inden-4-ylidene]ethylidene]cyclohexane-1,3-diol.
LMST03020033; 25,26-epoxy-19-nor-9,10-seco-5,7E-cholestadien-23-yne-1R,3R-diol; 25,26-epoxy-1α-hydroxy-23,23,24,24-tetradehydro-19-norcholecalciferol; 9547293; Compared to 1α,25-dihydroxyvitamin D_3, binds to human vitamin D binding protein 1% as effectively, calcium mobilization less than 1% and inhibition of proliferation or differentiation induction of human promyeloid leukemia (HL-60), osteosarcoma (MG-63), and breast carcinoma (MCF-7) cells is 12%, 4% and 20%, respectively. *Steroids.* 1995, *60(4)*, 324-32.

50   **1α,26-Dihydroxy-22E,23,24E,25-tetradehydro-27-norvitamin D₃**

9547294 $C_{26}H_{38}O_3$

(1R,3S,5Z)-5-[(2E)-2-[(1R,3aR,7aS)-1-[(2S,3E,5E)-7-hydroxyhepta-3,5-dien-2-yl]-7a-methyl-2,3,3a.5,6,7-hexahydro-1H-inden-4-ylidene]ethylidene]-4-methylidene-cyclohexane-1,3-diol.
LMST03020034;   1α,26-dihydroxy-22E,23,24E,25-tetradehydro-27-norcholecalciferol; 27-nor-9,10-seco-5Z,7E,10(19),22E,24E-cholestapentaene-1S,3R,26-triol;
9547294; Compared to 1α,25-dihydroxyvitamin D₃, inhibition of proliferation of U937 cells, < 80% and induction of differentiation of U937 cells, < 5%. Gene Regulation, Structure-Function Analysis and Clinical Application. Proceedings of the Eighth Workshop on Vitamin D Paris, France July 5-10. Synthesis and Biological Activity of 1α-hydroxylated Vitamin D Analogues with Poly-unsaturated Side Chains. (Norman, A. W. Bouillon, R. Thomasset, M. eds), pp192-193, Walter de Gruyter Berlin New York, (1991).

51   **26,26,26-Trifluoro-25-hydroxy-27-nor-vitamin D₃**

9547284 $C_{26}H_{39}F_3O_2$

(6R)-6-[(1R,3aR,4E,7aS)-4-[(2Z)-2-[(5S)-5-hydroxy-2-methylidene-cyclohexylidene]ethylidene]-7a-methyl-2,3,3a,5,6,7-hexahydro-1H-inden-1-yl]-1,1,1-trifluoro-heptan-2-ol.
LMST03020024; 26,26,26-trifluoro-27-nor-9,10-seco-5Z,7E,10(19)-cholestatriene-3S,25-diol;   26,26,26-trifluoro-25-hydroxy-27-norcholecalciferol; 9547284;
λ_m = 264 nm. *Tet Lett.* 1981, **22(43)**, 4309-12.

52   **19-Nor-14-epi-23-yne-1,25 dihydroxy-**

vitamin D₃

5289504 $C_{26}H_{40}O_3$

(1R,3R)-5-[(2E)-2-[(1R,7aS)-1-[(2R)-6-hydroxy-6-methyl-hept-4-yn-2-yl]-7aS-methyl-2,3,3a,5,6,7-hexahydro-1H-inden-4-ylidene]ethylidene]cyclo-hexane-1,3-diol.
19-Nor-14-epi-23-yne-1,25 dihydroxyvitamin D₃; 19-nor-14-epi-23-yne-1,25-(OH)₂D₃; 5-((Z)-2-((1R,7AS)-hexahydro-1-((S)-6-hydroxy-6-methylhept-4-yn-2-yl)-7A-methyl-1H-inden-4(7AH)-ylidene)ethylidene)cyclo-hexane-1R,3R-diol; TX522; 5289504; Shows 10-fold enhanced antiproliferative activity and is less calcemic than calcitriol. *Mol Pharmacol.* 2005, **67(5)**, 1566-73.

53   **19-Nor-14,20-bisepi-23-yne-1,25 dihydroxyvitamin D₃**

$C_{26}H_{40}O_3$

(1R,3R)-5-[(2E)-2-[(1R,7aS)-1-[(2R)-6-hydroxy-6-methyl-hept-4-yn-2-yl]-7aR-methyl-2,3,3a,5,6,7-hexahydro-1H-inden-4-ylidene]ethylidene]cyclo-hexane-1,3-diol.
19-Nor-14,20-bisepi-23-yne-1α,25-dihydroxyvitamin D₃;   19-Nor-14,20-bisepi-23-yne-1α,25-dihydroxy-cholecalciferol; TX527; Shows 10-fold enhanced antiproliferative activity and is less calcemic than calcitriol. *Mol Pharmacol.* 2005, **67(5)**, 1566-73; *J Steroid Biochem Mol Biol.* 2007, **103(1)**, 51-60.

54   **1α,25-Dihydroxy-17E,20-didehydro-21-norvitamin D₃**

9547295 $C_{26}H_{40}O_3$

(1R,3S,5Z)-5-[(2E)-2-[(1E,3aR,7aS)-1-(5-hydroxy-5-
methyl-hexylidene)-7a-methyl-2,3,3a,5,6,7-
hexahydroinden-4-ylidene]ethylidene]-4-methylidene-
cyclohexane-1,3-diol.
LMST03020035; 21-nor-9,10-seco-5Z,7E,10(19),17E-
cholestatetraene-1S,3R,25-triol; 1α,25-dihydroxy-17E,
20-didehydro-21-norvitamin D_3; 1α,25-dihydroxy-17E,
20-didehydro-21-norcholecalciferol; 9547295; Synthesized from 4-hydroxyhydrindanone *via* introduction of
the side chain and A-ring fragments. *Tet Lett.* 1998,
39(26), 4725-8.

55   *1α,25-Dihydroxy-17Z,20-didehydro-21-
       norvitamin D₃*
       9547296 $C_{26}H_{40}O_3$

(1R,3S,5Z)-5-[(2E)-2-[(1Z,3aR,7aS)-1-(5-hydroxy-5-
methyl-hexylidene)-7a-methyl-2,3,3a,5,6,7-
hexahydroinden-4-ylidene]ethylidene]-4-methylidene-
cyclohexane-1,3-diol.
LMST03020036; 21-nor-9,10-seco-5Z,7E,10(19),17Z-
cholestatetraene-1S,3R,25-triol; 1α,25-dihydroxy-17Z,
20-didehydro-21-norcholecalciferol; 9547296; Synthesized from 4-hydroxyhydrindanone *via* introduction of
the side chain and A-ring fragments. *Tet Lett.* 1998, *39*,
4725.

56   *1α,25-Dihydroxy-16-ene-19-nor-24-oxo-
       vitamin D₃*
                                            $C_{26}H_{40}O_4$

9,10-Secocholesta-5Z,7E,16-triene-24-one, 1α,3β,15-
trihydroxy.
24-Keto-1,15-dihydroxy-16-ene-19-norvitamın D_3; 1α,
25-(OH)2-16-ene-19-nor-24-oxo-D_3; Can decrease
intestinal tumor load in Apc[min] mice without severe
toxic side effects. May have utility as a
chemopreventive agent in groups at high-risk for colon
cancer. *Cancer Res.* 2002, *62(3)*, 741-6.

57   *24-Nor-9,11-seco-11-oxo-3,6-dihydroxy-
       cholesta-7,22-dien-9-one*
                                            $C_{26}H_{40}O_4$

24-Nor-9,11-seco-11-oxo-3β,6α-dihydroxycholesta-
7,22(E)-dien-9-one.
24-NSOD cpd. Analog of vitamin D_3, isolated from the
soft coral *Gersemia fruticosa* and synthesized from
ergosterol. Oil; $[\alpha]_D^{19}$= +26° (MeOH, c = 2.7). *Nat Prod
Lett.* 2001, *15(4)*, 221-8.

58   *25S,26-Epoxy-1α,24R-dihydroxy-27-
       norvitamin D₃*
       9547297 $C_{26}H_{40}O_4$

(1R,3S,5Z)-5-[(2E)-2-[(1R,3aR,7aS)-1-[(2R,5R)-5-hydroxy-5-(oxiran-2-yl)pentan-2-yl]-7a-methyl-2,3,3a,5,6,7-hexahydro-1H-inden-4-ylidene]ethylidene]-4-methylidene-cyclohexane-1,3-diol.

LMST03020037; 25S,26-epoxy-1α,24R-dihydroxy-27-norcholecalciferol; 25S,26-epoxy-27-nor-9,10-seco-5Z,7E,10(19)-cholestatriene-1S,3R,24R-triol; 9547297;

Compared to 1α,25-dihydroxyvitamin D_3, affinity for pig intestinal receptor and human vitamin D binding protein: 3% and 21%, respectively, inhibition of proliferation or differentiation induction of human promyeloid leukemia (HL-60) and osteosarcoma (MG-63) cells: both < 1% and elevation of serum calcium, serum osteocalcin, bone calcium and duodenal calbindin in rachitic chicks: all < 1%. **Bioorg Med Chem Lett**. 1993, *3(9)*, 1863-7; **Steroids**. 1995, *60(4)*, 324-32.

59   *25R,26-Epoxy-1α,24S-dihydroxy-27-norvitamin D_3*

9547298 $C_{26}H_{40}O_4$

(1R,3S,5Z)-5-[(2E)-2-[(1R,3aR,7aS)-1-[(2R,5S)-5-hydroxy-5-(oxiran-2-yl)pentan-2-yl]-7a-methyl-2,3,3a,5,6,7-hexahydro-1H-inden-4-ylidene]ethylidene]-4-methylidene-cyclohexane-1,3-diol.

LMST03020038; 25R,26-epoxy-1α,24S-dihydroxy-27-norcholecalciferol; 25R,26-epoxy-27-nor-9,10-seco-5Z,7E,10(19)-cholestatriene-1S,3R,24S-triol; 9547298;

Compared to 1α,25-dihydroxyvitamin D_3, affinity for pig intestinal receptor and human vitamin D binding protein: 2% and 5%, respectively, inhibition of proliferation or differentiation induction of human promyeloid leukemia (HL-60) and osteosarcoma (MG-63) cells: both < 1% and elevation of serum calcium, serum osteocalcin, bone calcium, and duodenal calbindin in rachitic chicks: all < 1%. **Bioorg Med Chem Lett**. 1993, *3*, 1863; **Steroids**. 1995, *60(4)*, 324-32.

60   *1α,25-Dihydroxy-24-oxo-22-oxavitamin D_3*

9547299 $C_{26}H_{40}O_5$

1-[(1S)-1-[(1S,3aR,4E,7aR)-4-[(2Z)-2-[(3S,5R)-3,5-dihydroxy-2-methylidene-cyclohexylidene]ethylidene]-7a-methyl-2,3,3a,5,6,7-hexahydro-1H-inden-1-yl]ethoxy]-3-hydroxy-3-methyl-butan-2-one.

LMST03020039; 1S,3R,25-trihydroxy-22-oxa-9,10-seco-5Z,7E,10(19)-cholestatrien-24-one; 1α,25-dihydroxy-24-oxo-22-oxacholecalciferol; 9547299; Antiproliferative activity (HL-60) is the same as that of 22-oxacalcitriol. The binding affinity for bovine thymus vitamin D receptor is 1/2 of 22-oxacalcitriol. Colorless oil λ_m = 263 nm (EtOH). **Chem Pharm Bull (Tokyo)**. 1996, *44(12)*, 2280-6.

61   *24,24-Difluoro-1,23,25-trihydroxyvitamin D_3*

                                          $C_{27}H_{42}F_2O_4$

24,24-Difluoro-1,23,25-trihydroxycholecalciferol; Metabolite presumed to be produced by CYP24 in rat kidney from 24,24-difluoro-1,25-dihydroxyvitamin D_3. **J Biol Chem**. 1997, *272(22)*, 14115-19; **Biochem**. 1996, *35(25)*, 8465-72.

62   *24,24-Difluoro-1,25,26-trihydroxyvitamin D_3*

                                          $C_{26}H_{42}F_2O_4$

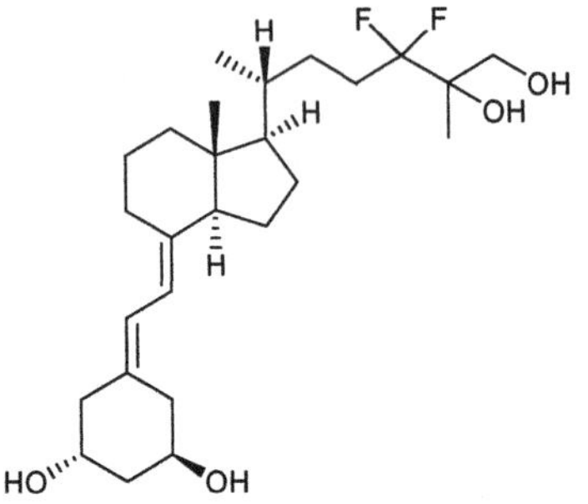

24,24-Difluoro-1,25,26-trihydroxycholecalciferol;
Metabolite produced by CYP24 in rat kidney of 24,24-difluoro-1,25-dihydroxyvitamin D_3. Identified spectroscopically and synthesized. *J Biol Chem.* 1997, *272(22)*, 14115-9.

63 *19-Nor-10-ketovitamin D_3*
62743-72-0 $C_{26}H_{42}O_2$

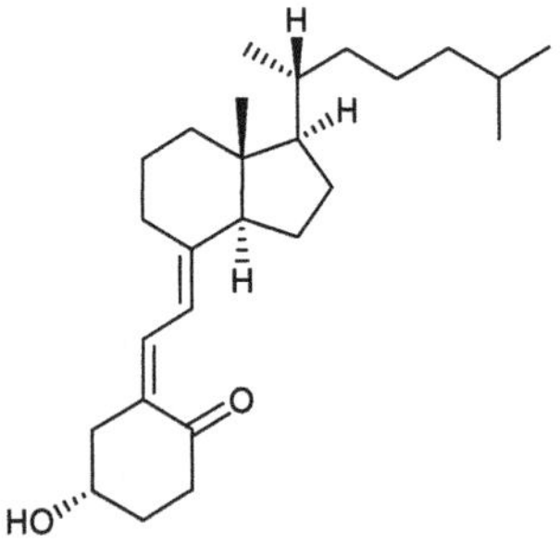

2-[2-[7a-Methyl-1-(6-methylheptan-2-yl)-2,3,3a,5,6,7-hexahydro-1H-inden-4-ylidene]ethylidene]-4-hydroxy-cyclohexan-1-one.
19-Nor-10-ketovitamin D_3; 19-NK-VD$_3$; 5(E)-19-Nor-10-ketocholecalciferol; 19-Nor-9,10-secocholesta-5E, 7E-dien-4-one, 1α-hydroxy-; 124941; 62743-72-0; Metabolite of vitamin D_3 in bovine rumen microbes. The conversion of vitamin D and its metabolites to their 19-nor-10-keto forms is thought to be a detoxification mechanism. *Biochemistry.* 1983, *22(15)*, 3636-40.

64 *10-Ketovitamin D_3*
95480-84-5 $C_{26}H_{42}O_2$

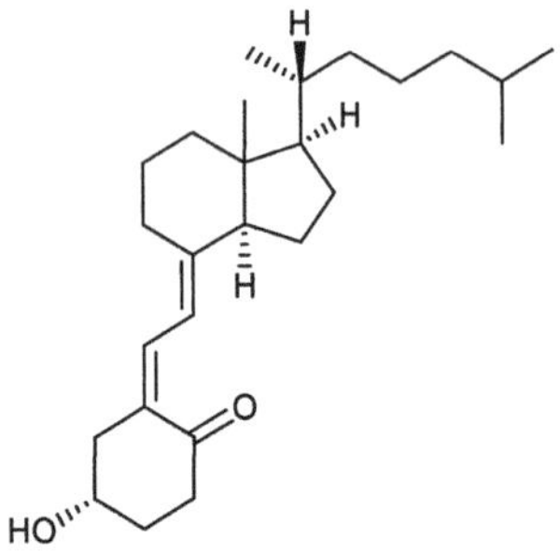

(2Z,4S)-2-[(2E)-2-[(3R,3aR,7aS)-3a-methyl-3-[(2R)-6-methylheptan-2-yl]-2,3,4,6,7,7a-hexahydro-1H-inden-5-ylidene]ethylidene]-4-hydroxy-cyclohexan-1-one.
10-Ketovitamin D_3; 10-Keto-vitamin D_3; 10-Keto-cholecalciferol; Cholecalciferol-10-one; 19-Nor-9,10-secocholesta-5Z,7E-dien-4-one, 1α-hydroxy-; 6438-606; 95480-84-5; Has bone resorbing activity equivalent to that of vitamin D_3. *Arch Biochem Biophys.* 1985, *236(2)*, 555-8.

65 *24-Nor-25-hydroxyvitamin D_3*
36415-31-3 $C_{26}H_{42}O_2$

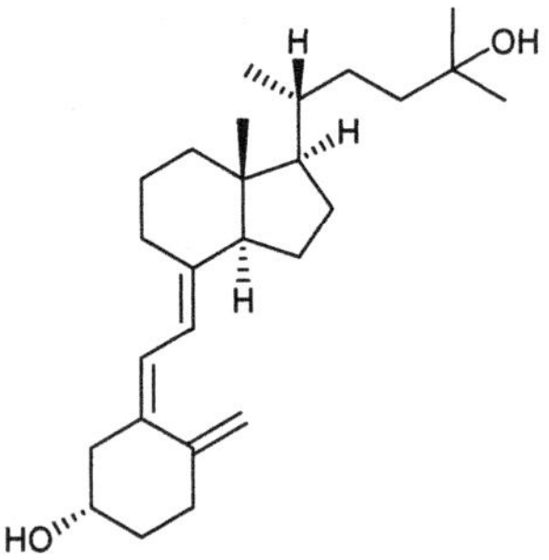

24-Nor-9,10-secoergosta-5,7,10(19)-triene-3β,25-diol.
24-Nor-25-hydroxyvitamin D_3; 24-Nor-25-hydroxy-cholecalciferol; 26,27-Dinor-9,10-secoergosta-5,7,10(19)-triene-3β,25-diol; 36415-31-3; Synthesized and biochemically evaluated. Inhibits the intestinal calcium absorption and bone calcium mobilization induced by vitamin D_3. *J Biol Chem.* 1979, *254(22)*, 11450-6; *J Med. Chem.* 1978, *21(10)*, 1025-9.

66 *24-Oxavitamin D_3*
                                                   $C_{26}H_{42}O_2$

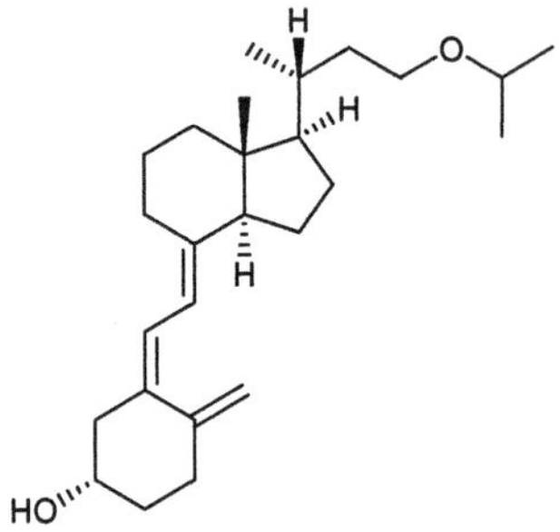

24-oxa-D3; A metabolic weak antagonist of Vitamin D_3. Its affinity for the vitamin D receptor and ability to induce cell differentiation are both reduced compared to calcitriol. *Steroids.* 1995, *60(6)*, 484-90.

67 *24-Nor-25-hydroxy-5,6-trans-vitamin D_3*
                                                   $C_{26}H_{42}O_2$

Synthesized from 24-nor-25-hydroxyvitamin D$_3$. *J Med Chem.* 1978, **21(10)**, 1025-9.

68   *2-Oxa-3-deoxy-25-hydroxyvitamin D$_3$*
113490-39-4   C$_{26}$H$_{42}$O$_2$

(6R)-6-[(1R,3aR,4E,7aS)-7a-methyl-4-[(2Z)-2-(3-methylideneoxan-4-ylidene)ethylidene]-2,3,3a,5,6,7-hexahydro-1H-inden-1-yl]-2-methyl-heptan-2-ol. LMST03020040; 2-Oxa-3-deoxy-25-OH-D3; 2-oxa-9,10-seco-5Z,7E,10(19)-cholestatrien-25-ol; 25-hydroxy-3-deoxy-2-oxavitamin D$_3$; 25-hydroxy-3-deoxy-2-oxacholecalciferol; 6444145; 113490-39-4; Synthesis and properties reported. *J Steroid Biochem Mol Biol.* 1991, **38(6)**, 775-9; *J Org Chem.* 1988, **53(8)**, 1790-6.

69   *1α,25-Dihydroxy-3-deoxy-3-thiavitamin D$_3$*
9547300   C$_{26}$H$_{42}$O$_2$S

(3R,5E)-5-[(2E)-2-[(1R,3aR,7aS)-1-[(2R)-6-hydroxy-6-methyl-heptan-2-yl]-7a-methyl-2,3,3a,5,6,7-hexahydro-1H-inden-4-ylidene]ethylidene]-4-methylidene-thian-3-ol.

LMST03020041; 9,10-seco-3-thia-5Z,7E,10(19)-cholestatriene-1R,25-diol; 1α,25-dihydroxy-3-deoxy-3-thia-cholecalciferol; 9547300; Compared to 1α,25-dihydroxyvitamin D$_3$, intestinal calcium absorption and bone calcium mobilization in vitamin D deficient, rachitic chicks: 20% and <10%, respectively and ability to bind to chick intestinal receptor is 14.5%. *J Org Chem.* 1992, **57(14)**, 3846-54.

70   *1β,25-Dihydroxy-3-deoxy-3-thiavitamin D$_3$*
9547301   C$_{26}$H$_{42}$O$_2$S

(3S,5E)-5-[(2E)-2-[(1R,3aR,7aS)-1-[(2R)-6-hydroxy-6-methyl-heptan-2-yl]-7a-methyl-2,3,3a,5,6,7-hexahydro-1H-inden-4-ylidene]ethylidene]-4-methylidene-thian-3-ol. LMST03020042; 9,10-seco-3-thia-5Z,7E,10(19)-cholestatriene-1S,25-diol; 1β,25-dihydroxy-3-deoxy-3-thia-cholecalciferol; 9547301; Compared to 1α,25-dihydroxyvitamin D$_3$, intestinal calcium absorption and bone calcium mobilization in vitamin D deficient, rachitic chicks: <20% and <10%, respectively and ability to bind to chick intestinal receptor is 1.23%. *J Org Chem.* 1992, **57(14)**, 3846-54.

71   *(5Z)-1,25-Dihydroxy-3-thiavitamin D$_3$*
9547302   C$_{26}$H$_{42}$O$_2$S

(3R,5Z)-5-[(2E)-2-[(1R,3aR,7aS)-1-[(2R)-6-hydroxy-6-methyl-heptan-2-yl]-7a-methyl-2,3,3a,5,6,7-hexahydro-1H-inden-4-ylidene]ethylidene]-4-methylidene-thian-3-ol. LMST03020043; 9,10-seco-3-thia-5Z,7E,10(19)-cholestatriene-1R,25-diol;   (5Z)-1,25-dihydroxy-3-thia-vitamin D$_3$; (5Z)-1,25-dihydroxy-3-thiacholecalciferol; 9547302; Affinity for chicken intestinal receptor is 0.1% of that of 1α,25-dihydroxyvitamin D$_3$. λ_m = 268 nm (ε = 18300 EtOH). *J Steroid Biochem Mol Biol.* 1997, *62(1)*, 73-8.

72   *24-Oxa-1-hydroxyvitamin D$_3$*

$C_{26}H_{42}O_3$

1(OH)24-oxaD3; 24-Oxa-1-hydroxycholecalciferol; A weak vitamin D$_3$ antagonist. Less affinity than calcitriol for vitamin D receptor. Induction of cell differentiation is only 3-19% that of calcitriol. *Steroids.* 1995, *60(6)*, 484-90.

73   *10-Keto-25-hydroxy-19-norvitamin D$_3$*

85925-90-2 $C_{26}H_{42}O_3$

(2E,4R)-2-[(2E)-2-[(1S,3aR,7aS)-1-[(2R)-6-hydroxy-6-methyl-heptan-2-yl]-7a-methyl-2,3,3a,5,6,7-hexahydro-1H-inden-4-ylidene]ethylidene]-4-hydroxy-cyclohexan-1-one.
19-NK-25 Hvd3; 10-Oxo-19-nor-25-hydroxyvitamin D$_3$; 19-Nor-10-oxo-25-hydroxyvitamin D$_3$; 19-Nor-10-keto-25-hydroxyvitamin D$_3$; 5(E)-19-Nor-10-keto-25-hydroxycholecalciferol; 19-Nor-9,10-secocholesta-5E,7E-dien-4-one, 1α,25-dihydroxy-; 6439023; 85925-90-2; Metabolite in bovine rumen microbes of 25-hydroxyvitamin D$_3$. The conversion of vitamin D and its metabolites to their 19-nor-10-keto forms is thought to be a detoxification mechanism. *Biochemistry.* 1983, *22(15)*, 3636-40.

74   *10-Keto-25-hydroxy-19-nor-17-epi-vitamin D$_3$*

86852-07-5 $C_{26}H_{42}O_3$

(2E,4R)-2-[(2E)-2-[(1R,3aR,7aS)-1-[(2R)-6-hydroxy-6-methyl-heptan-2-yl]-7a-methyl-2,3,3a,5,6,7-hexahydro-1H-inden-4-ylidene]ethylidene]-4-hydroxy-cyclohexan-1-one.
25-Hydroxycholecalciferol-10-one; 25-Hydroxy-10-oxocholecalciferol; 10-Keto-25-hydroxyvitamin D$_3$; 10-Khvd; 19-Nor-10-oxo-25-hydroxyvitamin D$_3$; 19-NK-25 HvD$_3$; 19-Nor-9,10-secocholesta-5Z,7E-dien-4-one, 1α,-25-dihydroxy-;19-Nor-10-keto-25-hydroxyvitamin D$_3$; 5(E)-19-Nor-10-keto-25-hydroxycholecalciferol; 10-Oxo-19-nor-25-hydroxyvitamin D$_3$; 6439797; 85925-90-2; 86852-07-5; A metabolite in chick kidney mitochondria of 25-hydroxyvitamin D$_3$, it is not formed from 1,25-dihydroxyvitamin D$_3$. Equivalent to vitamin D$_3$ in *in vitro* bone-resorbing activity. Synthesized from 25-hydroxyvitamin D$_3$. *J Biol Chem.* 1987, *262(29)*, 14164-71; *J Steroid Biochem.* 1989, *33(3)*, 395-403; *Org Chem.* 1989, *48(21)*, 3819-20.

75   *1-Hydroxy-22-oxavitamin D$_3$*

111687-67-3 $C_{26}H_{42}O_3$

(1R,3S,5E)-5-[(2E)-2-[(3aR,7aR)-7a-methyl-1-[(1S)-1-(3-methylbutoxy)ethyl]-2,3,3a,5,6,7-hexahydro-1H-inden-4-ylidene]ethylidene]-4-methylidene-cyclohexane-1,3-diol.
1-Hydroxy-22-oxavitamin D$_3$; 1,3-Cyclohexanediol, 4-methylene-5-(2-(octahydro-7a-methyl-1-(1-(3-methyl-butoxy)ethyl)-4H-inden-4-ylidene)ethylidene)-, (1S-(1α(R*),3aβ,4E(1S*,3R*,5Z),7aα))-; 6439242; 111687-67-3; Synthesis and properties reported. *Chem Pharm Bull (Tokyo).* 1986, *34(10)*, 4410-3.

76   **1-Hydroxy-24-oxovitamin D$_3$**

$C_{27}H_{42}O_3$

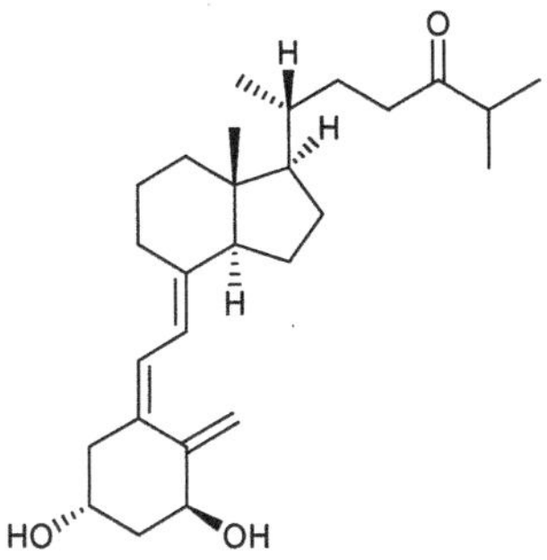

1-Hydroxy-24-oxocholecalciferol; 1α-hydroxy-24-oxo-vitamin D$_3$; Metabolite in rat kidney of 1α,24R-dihydroxyvitamin D$_3$. *Biochem Pharmacol.* 1999, *58(12)*, 1965-73.

77   **1,26-Dihydroxy-27-norvitamin D$_3$**

$C_{26}H_{42}O_3$

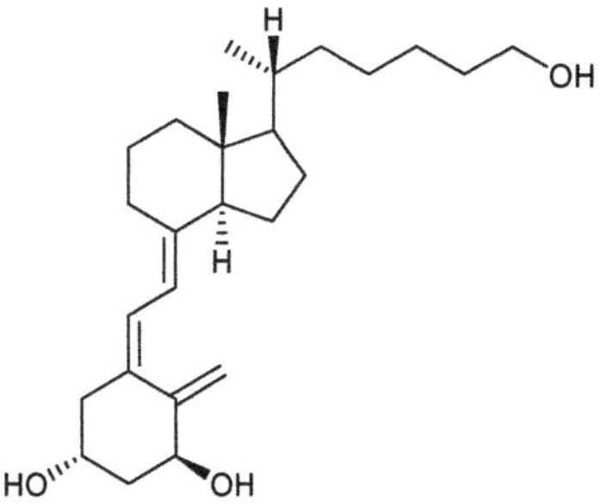

1,26(OH)2-27nVD$_3$; 1α, 26-dihydroxy-27-nor-vitamin D$_3$; Synthetic analog of vitamin D$_3$. Has lower affinity than 1,25-dihydroxyvitamin D$_3$ for the vitamin D receptor. *Steroids.* 1996, *61(10)*, 598-608; *Endrocrinology.* 1991, *128(4)*, 1687-92.

78   **1α,25-Dihydroxy-18-norvitamin D$_3$**

$C_{26}H_{42}O_3$

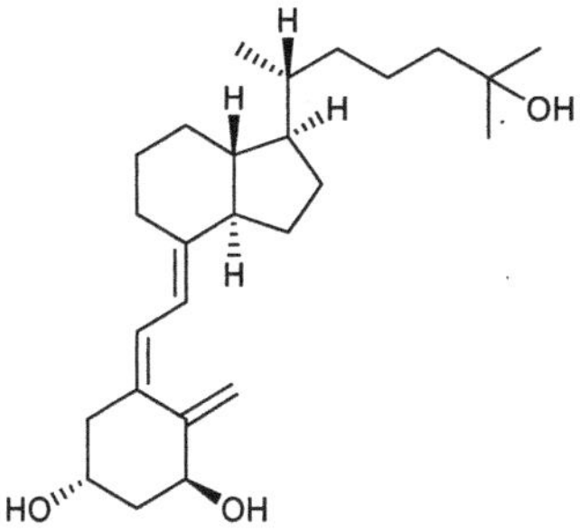

1,25-dihydroxy-18-norvitamin D$_3$; 1,25-dihydroxy-18-norcholecalciferol; 1,25(OH)2-18-nVD$_3$; Synthetic analog of vitamin D$_3$, is 5-10 times as potent as 1,25-dihydroxyvitamin D$_3$ in binding to the vitamin D receptor. *J Med Chem.* 1996, *39(22)*, 4497-506.

79   **1α,25-Dihydroxy-19-nor-3-epipre-vitamin D$_3$**

$C_{26}H_{42}O_3$

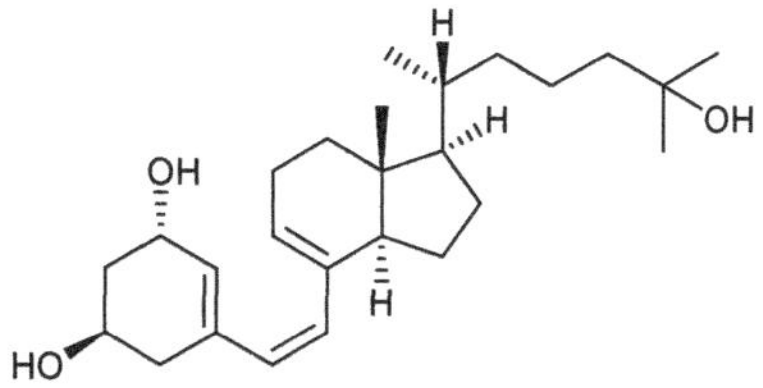

A synthetic 6-s-cis locked analog of previtamin D. Unable to rearrange to the vitamin D form. *J Org Chem.* 2000, *65(18)*, 5647-52.

80   **1α,25-Dihydroxy-19-norprevitamin D$_3$**

144699-06-9 $C_{26}H_{42}O_3$

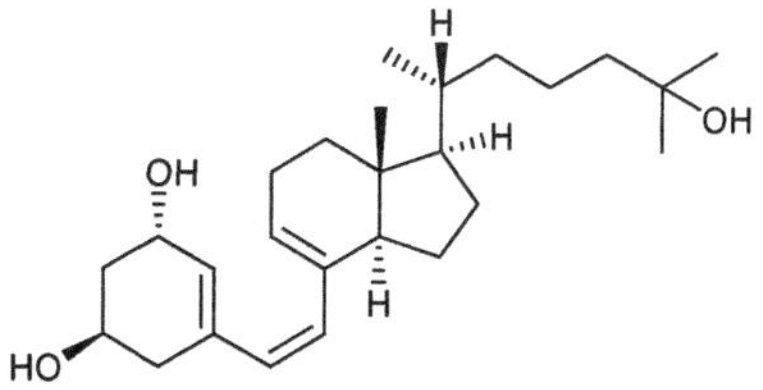

(1S,5R)-3-[(Z)-2-[(1R,3aR,7aS)-1-[(2R)-6-hydroxy-6-methyl-heptan-2-yl]-7a-methyl-1,2,3,3a,6,7-hexa-hydroinden-4-yl]ethenyl]cyclohex-2-ene-1,5-diol. LMST03020047; 19-nor-9,10-seco-5(10),6Z,8-chol-estatriene-1S,3R,25-triol; 1α,25-dihydroxy-19-norpre-vitamin D$_3$; 1α-dihydroxy-19-norprecholecalciferol; 9547306; 144699-06-9; Affinity for the intestinal vitamin D receptor and plasma vitamin D binding protein is 1 and 6% of that of 1α,25-dihydroxyvitamin D$_3$. Little effect on human promyeloid leukemia (HL-60 cell) differentiation and osteocalcin secretion by human osteosarcoma (MG-63) cells and no *in vivo* calcemic effects. *J Bone Miner.Res.* 1993, *8(8)*, 1009-15.

81   **1β,25-Dihydroxy-3-epi-19-nor-pre-vitamin D$_3$**

$C_{26}H_{42}O_3$

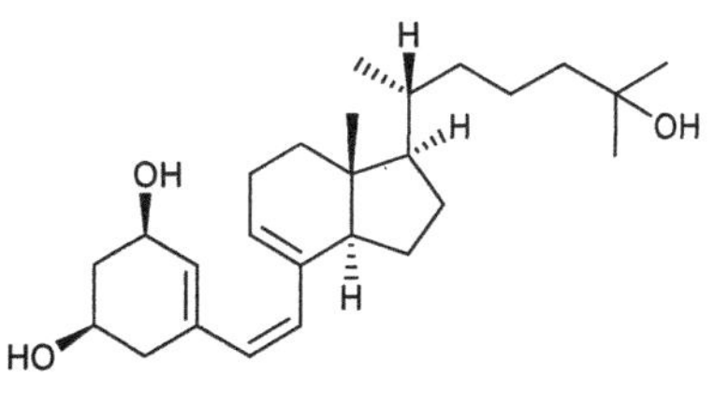

A synthetic 6-s-cis locked analog of previtamin D. Unable to rearrange to the vitamin D form. *J Org Chem*. 2000, *65(18)*, 5647-52.

82   *22-Ethyl-1α,22-dihydroxy-25,26,27-trinorvitamin D₃*

9547436 $C_{26}H_{42}O_3$

(1R,3S,5Z)-5-[(2E)-2-[(1R,3aR,7aS)-1-[(2S)-3-ethyl-3-hydroxy-pentan-2-yl]-7a-methyl-2,3,3a,5,6,7-hexa-hydro-1H-inden-4-ylidene]ethylidene]-4-methylidene-cyclohexane-1,3-diol.
LMST03020255;   22-ethyl-9,10-seco-5Z,7E,10(19)-cholatriene-1S,3R,22-triol;   22-ethyl-1α,22-dihydroxy-25,26,27-trinorcholecalciferol;   9547436;   In   HL-60 human promyelocytes differentiation is more than 2 orders of magnitude less effective than 1α,25-dihydroxyvitamin D₃. *J Biol Chem*. 1987, *262(29)*, 14164-14171; The Receptors (Conn, P. M., ed.), pp1-13, Academic Press (1985).

83   *1α-Hydroxy-22-oxavitamin D₃*

9547303 $C_{26}H_{42}O_3$

(1R,3S,5Z)-5-[(2E)-2-[(1S,3aR,7aR)-7a-methyl-1-[(1S)-1-(3-methylbutoxy)ethyl]-2,3,3a,5,6,7-hexahydro-1H-inden-4-ylidene]ethylidene]-4-methylidene-cyclo-hexane-1,3-diol.
LMST03020044; 1α-hydroxy-22-oxavitamin D₃; 1α-hydroxy-22-oxacholecalciferol; 22-oxa-9,10-seco-5Z,7E,10(19)-cholestatriene-1S,3R-diol;   9547303; Colorless glass; $[\alpha]_D^{24}$ = -30.2° (c = 0.86 EtOH); λ_m = 262 nm (EtOH). *Chem Pharm Bull (Tokyo)*. 1986, *34(10)*, 4410-13.

84   *25-Hydroxy-23-oxavitamin D₃*

9547304 $C_{26}H_{42}O_3$

(1S,3Z)-3-[(2E)-2-[(1R,3aR,7aS)-1-[(2S)-1-(2-hydroxy-2-methyl-propoxy)propan-2-yl]-7a-methyl-2,3,3a,5,6,7-hexahydro-1H-inden-4-ylidene]ethylidene]-4-methylidene-cyclohexan-1-ol.
LMST03020045;   25-hydroxy-23-oxacholecalciferol; 23-oxa-9,10-seco-5Z,7E,10(19)-cholestatriene-3S,25-diol; 9547304; Affinity for chick intestinal receptor is 6% that of 1α,25-dihydroxyvitamin D₃. *Endocr Rev*. 1995, *16(2)*, 200-57.

85   *1α,25-Dihydroxy-24-norvitamin D₃*

9547305 $C_{26}H_{42}O_3$

(1R,3S,5Z)-5-[(2E)-2-[(1R,3aR,7aS)-1-[(2R)-5-hydroxy-5-methyl-hexan-2-yl]-7a-methyl-2,3,3a,5,6,7-hexahydro-1H-inden-4-ylidene]ethylidene]-4-methylidene-cyclohexane-1,3-diol.
LMST03020046; 24-nor-9,10-seco-5Z,7E,10(19)-cholestatriene-1S,3R,25-triol;   1α,25-dihydroxy-24-nor-cholecalciferol; LMST03020046; 9547305; Compared to 1α,25-dihydroxyvitamin D₃, in HL-60 human promyelocytes differentiation is about one order of magnitude less effective. Intestinal calcium absorption in chick is 2.0% and bone calcium mobilization in chick is 2.0%. Competitive binding to the vitamin D receptor in chick intestine is 1.0%. *J Biol Chem*. 1987, *262(29)*, 14164-14171; *Blood* 1991, *75(1)*, 75-92.

86   *1α,25-Dihydroxy-21-norvitamin D₃*

141300-55-2 $C_{26}H_{42}O_3$

(1R,3S,5Z)-5-[(2E)-2-[(1S,3aR,7aR)-1-(5-hydroxy-5-methyl-hexyl)-7a-methyl-2,3,3a,5,6,7-hexahydro-1H-inden-4-ylidene]ethylidene]-4-methylidene-cyclo-hexane-1,3-diol.
LMST03020048; 21-nor-9,10-seco-5Z7E10(19)-chol-estatriene-1S,3R,25-triol; 1α,25-dihydroxy-21-nor-cholecalciferol; 6438742; 9547307; 141300-55-2; Has differentiation-inducing activity towards human myeloid leukemia cells approximately one-fifth of that of 1α,25-dihydroxyvitamin D_3, while in the binding affinity with chick intestinal cytosolic receptor, it is about one-tenth of that of 1α,25-dihydroxyvitamin D_3. The rather weak effect on serum calcium levels in normal mice at a dosage of 500 μg/kg (intravenous administration) indicates that the essential importance of the 21-methyl moiety may lie in its effect on the regulation of calcium metabolism. λ_m = 262 nm (EtOH). *Chem Pharm Bull (Tokyo)*. 1992, *40(3)*, 648-51.

87   *1α,25-dihydroxy-24-norvitamin D_3*

$C_{26}H_{42}O_3$

24-Nor-9,10-seco-5Z,7E,10(19)-cholestatriene-1S,3R,25-triol.
24-nor-1α,25-dihydroxyvitamin D_3; 24-nor-1α,25-di-hydroxycholecalciferol; 1α,25-dihydroxy-24-nor-vitamin D_3; 1α,25-dihydroxy-24-norcholecalciferol; Metabolite of 1α,25-dihydroxyvitamin D_3. Synthesized. White solid; λ_m = 266 nm (ε = 18300, Et_2O). *J Med Chem*. 1978, *21(10)*, 1025-9.

88   *1,25-Dihydroxy-23-thiavitamin D_3*

87480-00-0 $C_{26}H_{42}O_3S$

(1R,3S,5E)-5-[(2E)-2-[(3aR,7aS)-1-[1-(2-hydroxy-2-methyl-propyl)sulfanylpropan-2-yl]-7a-methyl-2,3,3a,5,6,7-hexahydro-1H-inden-4-ylidene]ethyl-idene]-4-methylidene-cyclohexane-1,3-diol.
1,25-DTVD₃; 1α,25-Dihydroxy-23-thiavitamin D_3; 4-methylene-5-(2-(octahydro-1-(2-((2-hydroxy-2-methyl-propyl)thio)-1-methylethyl)-7a-methyl-4H-inden-4-ylidene)ethylidene)-1,3-cyclohexanediol (1R-(1α(S*), 3aβ,4E(Z(1R*,3S*)),7aα))-; 1,3-Cyclohexanediol, 4-methylene-5 (2 (octahydro 1 (2 ((2 hydroxy 2 methyl-propyl)thio)-1-methylethyl)-7a-methyl-4H-inden-4-ylidene)ethylidene)-, (1R-(1α(S*), 3aβ,4E (Z(1R*,3S*)), 7aα))-; 6439813; 87480-00-0; Less active than 1α,25-dihydroxyvitamin D_3 in the differentiation-inducing activity of human myeloid leukemia cells into macrophages *in vitro*. *Chem Pharm Bull (Tokyo)*. 1991, *39(12)*, 3221-4.

89   *1α,25-Dihydroxy-23-thiavitamin D_3*

9547310 $C_{26}H_{42}O_3S$

(1R,3S,5Z)-5-[(2E)-2-[(1R,3aR,7aS)-1-[(2S)-1-(2-hydroxy-2-methyl-propyl)sulfanylpropan-2-yl]-7a-methyl-2,3,3a,5,6,7-hexahydro-1H-inden-4-ylidene]ethylidene]-4-methylidene-cyclohexane-1,3-diol.
LMST03020055; 9,10-seco-23-thia-5Z,7E,10(19)-chol-estatriene-1S,3R,25-triol; 1α,25-dihydroxy-23-thia-cholecalciferol; 9547310; Differentiation inducing activity of HL-60 reported. λ_m = 263 nm (EtOH). *Chem Pharm Bull (Tokyo)*. 1991, *39(12)*, 3221-4.

90   *1α,25-Dihydroxy-22-thiavitamin D_3*

9547308 $C_{26}H_{42}O_3S$

(1R,3S,5Z)-5-[(2E)-2-[(1S,3aR,7aR)-1-[(1S)-1-(3-hydroxy-3-methyl-butyl)sulfanylethyl]-7a-methyl-2,3,3a,5,6,7-hexahydro-1H-inden-4-ylidene]ethyl-idene]-4-methylidene-cyclohexane-1,3-diol. LMST03020053; 9,10-seco-22-thia-5Z,7E,10(19)-chol-estatriene-1S,3R,25-triol; 1α,25-dihydroxy-22-thia-cholecalciferol; 9547308; Synthesized photo-chemically from dehydroepiandrosterone *via* the 5,7-diene. λ_m = 264 nm (EtOH). *Bioorg Med Chem Lett.* 1995, *5(3)*, 279-82.

91   *1α,25-Dihydroxy-22-thia-20-epivitamin*
D₃
9547309 $C_{26}H_{42}O_3S$

(1R,3S,5Z)-5-[(2E)-2-[(1S,3aR,7aR)-1-[(1R)-1-(3-hydroxy-3-methyl-butyl)sulfanylethyl]-7a-methyl-2,3,3a,5,6,7-hexahydro-1H-inden-4-ylidene]-ethylidene]-4-methylidene-cyclohexane-1,3-diol. LMST03020054;   (20R)-9,10-seco-22-thia-5Z,7E,10 (19)-cholestatriene-1S,3R,25-trio; 1α,25-dihydroxy-22-thia-20-epicholecalciferol; 9547309; Synthesized photochemically from dehydroepiandrosertone *via* the 5,7-diene. λ_m = 264 nm (EtOH). *Bioorg Med Chem Lett.* 1995, *5*, 279.

92   *1,23-Dihydroxy-24-oxovitamin D₃*
                                        $C_{27}H_{42}O_4$

1,23-Dihydroxy-24-oxocholecalciferol,   1α,23-di-hydroxy-24-ketovitamin D₃; Metabolite in rat kidney of 1α,24R-dihydroxyvitamin D₃. *Biochemistry.* 1984, *23(7)*, 1473-8; *Biochem Pharmacol.* 1999, *58(12)*, 1965-73.

93   *1,25-Dihydroxy-22-oxavitamin D₃*
103909-75-7 $C_{26}H_{42}O_4$

(1R,3S,5Z)-5-[(2E)-2-[(1S,3aR,7aR)-1-[(1S)-1-(3-hydroxy-3-methyl-butoxy)ethyl]-7a-methyl-2,3,3a,5,6,7-hexahydro-1H-inden-4-ylidene] ethylidene]-4-methylidene-cyclohexane-1,3-diol. 1,3-Cyclohexanediol,   4-methylene-5-((2E)-((1S,3aS, 7aS)-octahydro-1-((1S)-1-(3-hydroxy-3-methylbutoxy) ethyl)-7a-methyl-4H-inden-4-ylidene)ethylidene)-, (1R,3S,   5Z)-;   LMST03020060;   C12495;   D01098; 1α,25-Dihydroxy-22-oxavitamin D₃; Maxacalcitol; MC 1275;   OCT;   22-Oxacalcitriol;   22-Oxa-1,25-dihydroxyvitamin D₃; Oxarol; Prezios; Sch 209579; 5478815; 6398761; 103909-75-7; Calcitriol analog. Inhibits proliferation of cultured keratinocytes and induces terminal differentiation. Has antineoplastic and dermatological properties. Used to treat warts and psoriasis and in hemodialysis patients with secondary hyperpara-thyroidism (2HPT). LD$_{lo}$ (dog iv) = 0.8 mg/kg, LD$_{50}$ (rat iv) = 8 mg/kg. *Bioorg Med Chem.* 2001, *9(2)*, 403-15. *Steroids.* 2001, *66(3-5)*, 137-46.

94   *1,25-Dihydroxy-23-oxavitamin D₃*
140387-52-6 $C_{26}H_{42}O_4$

(1R,3S,5Z)-5-[(2E)-2-[(1R,3aR,7aS)-1-[(2S)-1-(2-hydroxy-2-methyl-propoxy)propan-2-yl]-7a-methyl-2,3,3a,5,6,7-hexahydro-1H-inden-4-ylidene]ethylidene]-4-methylidene-cyclohexane-1,3-diol.

1,25-Dihydroxy-23-oxavitamin D_3; 1,25-DovD$_3$; 1α, 25-Dihydroxy-23-oxavitamin D_3; 23-Oxa-1α,25-dihydroxyvitamin D_3; ZK 150582; 1,3-Cyclohexanediol, 4-methylene-5-(2-(octahydro-1-(2-(2-hydroxy-2-methylpropoxy)-1-methylethyl)-7a-methyl-4H-inden-4-ylidene)ethylidene)-, (1R-(1α(S*),3aβ,4E(1R*,3S*,5Z), 7aα))-; LMST03020063; 6438738; 140387-52-6; Less active than 1α,25-dihydroxyvitamin D_3 in the differentiation-inducing activity of human myeloid leukemia cells into macrophages *in vitro*. *Chem Pharm Bull (Tokyo)*. 1991, **39(12)**, 3221-4.

95 **1α,24R-Dihydroxy-22-oxavitamin D_3**
9547311   C$_{26}$H$_{42}$O$_4$

(1R,3S,5Z)-5-[(2E)-2-[(1S,3aR,7aR)-1-[(1S)-1-[(2R)-2-hydroxy-3-methyl-butoxy]ethyl]-7a-methyl-2,3,3a,5,6,7-hexahydro-1H-inden-4-ylidene]ethylidene]-4-methylidene-cyclohexane-1,3-diol.
LMST03020056; 22-oxa-9,10-seco-5Z,7E,10(19)-cholestatriene-1S,3R,24R-triol; 1α,24R-dihydroxy-22-oxacholecalciferol; 9547311; Induces differentiation of human myeloid leukemia cells (HL-60) into macrophases *in vitro*. Activity, estimated by superoxide anion generation, is approximately the same as that of 1α,25-dihydroxyvitamin D_3. The binding affinity for the chick embryonic intestinal vitamin D_3 receptor is about 1% of that of 1α,25-dihydroxyvitamin D_3. $[\alpha]_D^{24}$ = +46.0° (c = 0.10 EtOH); λ_m = 263 nm (EtOH). *Chem Pharm Bull (Tokyo)*. 1993, **41(9)**, 1659-63.

96 **1α,24R-Dihydroxy-22-oxa-20-epivitamin D_3**

9547312   C$_{26}$H$_{42}$O$_4$

(1R,3S,5Z)-5-[(2E)-2-[(1S,3aR,7aR)-1-[(1R)-1-[(2R)-2-hydroxy-3-methyl-propoxy]ethyl]-7a-methyl-2,3,3a,5,6,7-hexahydro-1H-inden-4-ylidene]ethylidene]-4-methylidene-cyclohexane-1,3-diol.
LMST03020057; (20R)-22-oxa-9,10-seco-5Z,7E,10(19)-cholestatriene-1S,3R,24R-triol; 1α,24R-dihydroxy-22-oxa-20-epivitamin D_3; 1α,24R-dihydroxy-22-oxa-20-epicholecalciferol; 9547312; Ability to induce HL-60 cell differentiation or to bind to the vitamin D receptor is about 10% of that of 1α,25-dihydroxyvitamin D_3. $[\alpha]_D^{23}$ = -30.7° (c = 0.365 EtOH); λ_m = 263 nm (EtOH). *Chem Pharm Bull (Tokyo)*. 1993, **41(9)**, 1659-63.

97 **1α,24S-Dihydroxy-22-oxavitamin D_3**
9547313   C$_{26}$H$_{42}$O$_4$

(1R,3S,5Z)-5-[(2E)-2-[(1S,3aR,7aR)-1-[(1S)-1-[(2S)-2-hydroxy-3-methyl-butoxy]ethyl]-7a-methyl-2,3,3a,5,6,7-hexahydro-1H-inden-4-ylidene]ethylidene]-4-methylidene-cyclohexane-1,3-diol.
LMST03020058; 22-oxa-9,10-seco-5Z,7E,10(19)-cholestatriene-1S,3R,24S-triol; 1α,24S-dihydroxy-22-oxavitamin D_3; 1α,24S-dihydroxy-22-oxacholecalciferol; 9547313; Induces differentiation of human myeloid leukemia cells (HL-60) into macrophases *in vitro* estimated by superoxide anion generation approximately the same as 1α,25-dihydroxyvitamin D_3. Binds to the chick embryonic intestinal vitamin D receptor about 30% as effectively. $[\alpha]_D^{24}$ = +56.0° (c = 0.10 EtOH); λ_m = 263 nm (EtOH). *Chem Pharm Bull (Tokyo)*. 1993, **41(9)**,.1659-63.

98 **1α,24S-Dihydroxy-22-oxa-20-epivitamin D_3**

9547314 $C_{26}H_{42}O_4$

(1R,3S,5Z)-5-[(2E)-2-[(1S,3aR,7aR)-1-[(1R)-1-[(2S)-2-hydroxy-3-methyl-butoxy]ethyl]-7a-methyl-2,3,3a,5,6,7-hexahydro-1H-inden-4-ylidene]ethylidene]-4-methylidene-cyclohexane-1,3-diol.
LMST03020059; (20R)-22-oxa-9,10-seco-5Z,7E,10(19)-cholestatriene-1S,3R,24S-triol; 1α,24S-dihydroxy-22-oxa-20-epicholecalciferol; 9547314; Induces differentiation of human myeloid leukemia cells (HL-60) into macrophases *in vitro* estimated by superoxide anion generation one order of magnitude less than 1α,25-dihydroxyvitamin D_3. Binds to the chick embryonic intestinal vitamin D receptor about 0.3% as effectively. $[\alpha]_D^{23}$ = +24.2° (c = 0.165 EtOH); λ_m = 263 nm (EtOH). *Chem Pharm Bull (Tokyo)*. 1993, *41(9)*, 1659-63.

99   *1β,25-Dihydroxy-22-oxavitamin D_3*
9547315 $C_{26}H_{42}O_4$

(1R,3R,5Z)-5-[(2E)-2-[(1S,3aR,7aR)-1-[(1S)-1-(3-hydroxy-3-methyl-butoxy)ethyl]-7a-methyl-2,3,3a,5,6,7-hexahydro-1H-inden-4-ylidene]ethylidene]-4-methylidene-cyclohexane-1,3-diol.
LMST03020062; 22-oxa-9,10-seco-5Z,7E,10(19)-cholestatriene-1R,3R,25-triol; 1β,25-dihydroxy-22-oxacholecalciferol; 9547315; Antiproliferation activity with HL-60 is about 0.3% that of 1α,25-dihydroxyvitamin D_3 and the binding affinity for both vitamin D receptor and vitamin D binding protein was 0.7%. λ_m = 263 nm (EtOH). *Bioorg Med Chem Lett*. 1994, *4(12)*, 1523-6.

100   *1α,20S,25-Trihydroxy-24-norvitamin D_3*
9547316 $C_{26}H_{42}O_4$

(2S)-2-[(1S,3aR,4E,7aR)-4-[(2Z)-2-[(3S,5R)-3,5-dihydroxy-2-methylidene-cyclohexylidene]ethylidene]-7a-methyl-2,3,3a,5,6,7-hexahydro-1H-inden-1-yl]-5-methyl-hexane-2,5-diol.
LMST03020064; 24-nor-9,10-seco-5Z,7E,10(19)-cholestatriene-1S,3R,20S,25-tetrol; 1α,20S,25-trihydroxy-24-norcholecalciferol; 9547316; Inhibition of U937 cell (human histiocytic lymphoma cell line) proliferation less than 4% that of 1α,25-dihydroxy-vitamin D_3. Binding to the vitamin D receptor from rachitic chicken intestine is less than 0.1% and calcemic activity was not reported. Vitamin D. A Pluripotent Steroid Hormone: Structural Studies, Molecular Endocrinology and Clinical Applications. Proceedings of the Ninth Workshop on Vitamin D, Orlando, Florida (USA) May 28-June 2. Synthesis and Biological Activity of 20-Hydroxylated Vitamin D Analogues. (Norman, A. W. Bouillon, R. Thomasset, M. eds), pp 95-96, Walter de Gruyter Berlin New York (1994).

101   *1,23S,25-Trihydroxy-24-oxo-19-nor-vitamin D_3*

$C_{26}H_{42}O_5$

(6R)-6-[(1R,3aR,4E,7aS)-4-[(2Z)-2-[(3R,5S)-3,5-dihydroxy-cyclohexylidene]ethylidene]-7a-methyl-2,3,3a,5,6,7-hexahydro-1H-inden-1-yl]-2,4S-dihydroxy-2-methyl-heptan-3-one.
1,23S,25-Trihydroxy-24-oxo-19-norvitamin D_3; 1,23S, 25-Trihydroxy-24-oxo-19-norcholecalciferol; Suppresses PTH secretion less than 1α,25-dihydroxyvitamin D_3 and in cell differentiation, is equipotent to 1α,25-dihydroxyvitamin D_3. *Steroids*. 2000, *65(5)* 252-65.

102 **1,23R,25-Trihydroxy-24-oxo-19-norvitamin D₃**

$C_{26}H_{42}O_5$

(6R)-6-[(1R,3aR,4E,7aS)-4-[(2Z)-2-[(3R,5S)-3,5-dihydroxy-cyclohexyliden]   1,23R,25-Trihydroxy-24-oxo-19-norvitamine]ethylidene]-7a-methyl-2,3,3a,5,6,7-hexahydro-1H-inden-1-yl]-2,4R-dihydroxy-2-methyl-heptan-3-one.

1,23R,25-Trihydroxy-24-oxo-19-norvitamin   D₃;
1,23R,25-Trihydroxy-24-oxo-19-norcholecalciferol; Suppresses PTH secretion more effectively than 1α,25-dihydroxyvitamin D₃ and in cell differentiation, is equipotent to 1α,25-dihydroxyvitamin D₃. *Steroids*. 2000, 65(5), 252-65.

103 **1α,24R,25-Trihydroxy-22-oxavitamin D₃**
9547317 $C_{26}H_{42}O_5$

(1R,3S,5Z)-5-[(2E)-2-[(1S,3aR,7aR)-1-[(1S)-1-[(2R)-2,3-dihydroxy-3-methyl-butoxy]ethyl]-7a-methyl-2,3,3a,5,6,7-hexahydro-1H-inden-4-ylidene]ethylidene]-4-methylidene-cyclohexane-1,3-diol.
LMST03020065; 22-oxa-9,10-seco-5Z,7E,10(19)-chol-estatriene-1S,3R,24R,25-tetrol; 1α,24R,25-trihydroxy-22-oxacholecalciferol; 9547317; Antiproliferation activity towards HL-60 is 1/2 and binding affinity for bovine thymus vitamin D receptor is 1/3 that of 22-oxacalcitriol; $[\alpha]_D^{20}$ = +44.0° (c = 0.29 EtOH); λ_m = 262 nm (EtOH). *Bioorg Med Chem Lett*. 1994, **4(5)**, 753-6; *Chem Pharm Bull (Tokyo)*. 1996, **44(12)**, 2280-6.

104 **1α,24S,25-Trihydroxy-22-oxavitamin D₃**
9547318 $C_{26}H_{42}O_5$

(1R,3S,5Z)-5-[(2E)-2-[(1S,3aR,7aR)-1-[(1S)-1-[(2S)-2,3-dihydroxy-3-methyl-butoxy]ethyl]-7a-methyl-2,3,3a,5,6,7-hexahydro-1H-inden-4-ylidene]ethylidene]-4-methylidene-cyclohexane-1,3-diol.
LMST03020066; 22-oxa-9,10-seco-5Z,7E,10(19)-chol-estatriene-1S,3R,24S,25-tetrol;1α,24S,25-trihydroxy-22-oxacholecalciferol;   9547318;   Antiproliferation activity towards HL-60 is 1/3 and binding affinity for bovine thymus vitamin D receptor is 1/22 of 22-oxacalcitriol; $[\alpha]_D^{20}$ = +29.0° (c = 0.19 EtOH); λ_m = 263 nm (EtOH). *Bioorg Med Chem Lett*. 1994, **4(5)**, 753-6; *Chem Pharm Bull (Tokyo)*. 1996, **44(12)**, 2280-6.

105 **1α,25R,26-Trihydroxy-22-oxavitamin D₃**
9547319 $C_{26}H_{42}O_5$

(1R,3S,5Z)-5-[(2E)-2-[(1S,3aR,7aR)-1-[(1S)-1-[(3R)-3,4-dihydroxy-3-methyl-butoxy]ethyl]-7a-methyl-2,3,3a,5,6,7-hexahydro-1H-inden-4-ylidene]ethylidene]-4-methylidene-cyclohexane-1,3-diol.
LMST03020067; 22-oxa-9,10-seco-5Z,7E,10(19)-chol-estatriene-1S,3R,25R,26-tetrol; 1α,25R,26-trihydroxy-22-oxacholecalciferol;   9547319;   Antiproliferation activity towards HL-60 is 1/9 and binding affinity for bovine thymus vitamin D receptor is 1/25 of 22-oxacalcitriol; $[\alpha]_D^{20}$ = +49.4° (c = 0.08 EtOH); λ_m = 263 nm (EtOH). *Bioorg Med Chem Lett*. 1994, **4(5)**, 753-6; *Chem Pharm Bull (Tokyo)*. 1996, **44(12)**, 2280-6.

106 **1α,25S,26-Trihydroxy-22-oxavitamin D₃**
9547320 $C_{26}H_{42}O_5$

(1R,3S,5Z)-5-[(2E)-2-[(1S,3aR,7aR)-1-[(1S)-1-[(3S)-3,4-dihydroxy-3-methyl-butoxy]ethyl]-7a-methyl-2,3,3a,5,6,7-hexahydro-1H-inden-4-ylidene]ethylidene]-4-methylidene-cyclohexane-1,3-diol.
LMST03020068; (5Z,7E)-(1S,3R,25S)-22-oxa-9,10-seco-5Z,7E,10(19)-cholestatriene-1S,3R,25S,26-tetrol; 1α,25S,26-trihydroxy-22-oxacholecalciferol; 9547320; Antiproliferation activity towards HL-60 is 1/7 and binding affinity for bovine thymus vitamin D receptor is 1/30 of 22-oxacalcitriol; $[\alpha]_D^{20}$ = +65.9° (c = 0.09 EtOH); λ_m = 264 nm (EtOH). *Bioorg Med Chem Lett.* 1994, *4(5)*, 753-6; *Chem Pharm Bull (Tokyo).* 1996, *44(12)*, 2280-6.

107   *25-Azavitamin D₃*

63819-61-4 $C_{26}H_{43}NO$

(1S,3Z)-3-[(2E)-2-[(1R,3aR,7aS)-1-[(2R)-5-dimethyl-aminopentan-2-yl]-7a-methyl-2,3,3a,5,6,7-hexahydro-1H-inden-4-ylidene]ethylidene]-4-methylidene-cyclo-hexan-1-ol.
LMST03020069; 25-azacholecalciferol; 25-aza-9,10-seco-5Z,7E,10(19)-cholestatrien-3S-ol; 6443591; 9547321; 63819-61-4; A vitamin D antagonist, inhibits the 25-hydroxylation of vitamin D₃ in the liver. Inhibits bone calcium mobilization and intestinal calcium transport in rats; λ_m = 265 nm (EtOH). *J Biol Chem.* 1979, *254(9)*, 3493-6.

108   *1α,25-Dihydroxy-23-azavitamin D₃*
87407-70-3 $C_{26}H_{43}NO_3$

(1R,3S,5E)-5-[(2E)-2-[1-[1-[(2-hydroxy-2-methyl-propyl)amino]propan-2-yl]-7a-methyl-2,3,3a,5,6,7-hexahydro-1H-inden-4-ylidene]ethylidene]-4-methylidene-cyclohexane-1,3-diol.
1,25-Dihydroxy-23-azavitamin D₃; 1,25-DAVD₃; 1α, 25-Dihydroxy-23-azavitamin D₃; 4-Methylene-5-(2-(octahydro-1-(2-((2-hydroxy-2-methylpropyl)amino)-1-methylethyl)-7a-methyl-4H-inden-4-ylidene)ethyl-idene)-1,3-cyclohexanediol   (2R-1α(S*),3aβ,4E(Z(1R* ,3S*)),7aα))-; 1,3-Cyclohexanediol, 4-methylene-5-(2-(octahydro-1-(2-((2-hydroxy-2-methylpropyl)amino)-1-methylethyl)-7a-methyl-4H-inden-4-ylidene)ethyl-idene)-,   (2R-(1α(S*),3aβ,4E(Z(1R*,3S*)),7aα))-; 6439812; 87407-70-3; Shows little activity in the differentiation-inducing activity of human myeloid leukemia cells in macrophages *in vitro*. *Chem Pharm Bull (Tokyo).* 1991, *39(12)*, 3221-4.

109   *1α,25-Dihydroxy-23-azavitamin D₃*
9547322 $C_{26}H_{43}NO_3$

(1R,3S,5Z)-5-[(2E)-2-[(1R,3aR,7aS)-1-[(2S)-1-[(2-hydroxy-2-methyl-propyl)amino]propan-2-yl]-7a-methyl-2,3,3a,5,6,7-hexahydro-1H-inden-4-ylidene]ethylidene]-4-methylidene-cyclohexane-1,3-diol.
LMST03020070; 23-aza-9,10-seco-5Z,7E,10(19)-cholestatriene-1S,3R,25-triol; 1α,25-dihydroxy-23-azacholecalciferol; 9547322; ; Synthesized from dehydroepiandrosterone *via* 5,7-diene compound by photochemical methodDifferentiation inducing activity of HL-60 reported. λ_m = 263 nm (EtOH). *Chem Pharm Bull (Tokyo).* 1991, *39(12)*, 3221-4.

110   *1α-Hydroxy-24-(dimethylphosphoryl)-25,26,27-trinorvitamin D₃*
9547323 $C_{26}H_{43}O_3P$

(1R,3S,5Z)-5-[(2E)-2-[(1R,3aR,7aS)-1-[(2R)-5-dimethylphosphorylpentan-2-yl]-7a-methyl-2,3,3a,5,6,7-hexahydro-1H-inden-4-ylidene]ethylidene]-4-methylidene-cyclohexane-1,3-diol.
LMST03020071; 24-(dimethylphosphoryl)-9,10-seco-5Z,7E,10(19)-cholatriene-1S,3R-diol; 1α-hydroxy-24-(dimethylphosphoryl)-25,26,27-trinorcholecalciferol;
9547323; Compared to 1α,25-dihydroxyvitamin D_3, intestinal calcium absorption in chick, 0.3%, bone calcium mobilization in chick, 0.15%, competitive binding to the vitamin D receptor in chick intestine, 0.35%. Inhibition of clonal growth of human leukemia cells (HL-60) undetected; $[α]_D$ = +33.1° (c = 0.95 in CH_2Cl_2); $λ_m$ = 265 nm (ε = 14700 EtOH). *Tet Lett.* 1991, *32(36)*, 4643-6.

111   ***24-(Dimethoxyphosphoryl)-25,26,27-tri-norvitamin D_3***
9547324 $C_{26}H_{43}O_4P$

(1S,3Z)-3-[(2E)-2-[(1R,3aR,7aS)-1-[(2R)-5-dimethoxy-phosphorylpentan-2-yl]-7a-methyl-2,3,3a,5,6,7-hexa-hydro-1H-inden-4-ylidene]ethylidene]-4-methylidene-cyclohexan-1-ol.
LMST03020072; 24-(dimethoxyphosphoryl)-9,10-seco-5Z,7E,10(19)-cholatrien-3S-ol; 24-(dimethoxyphosph-oryl)-25,26,27-trinorcholecalciferol; 9547324; $[α]_D$ = +27.0° (c = 0.73 in CH_2Cl_2); $λ_m$ = 264 nm (ε = 14700 EtOH). *Tet Lett.* 1991, *32(36)*, 4643-6.

112   ***1α-Hydroxy-24-(dimethoxyphosphoryl)-25,26,27-trinorvitamin D_3***
9547325 $C_{26}H_{43}O_5P$

(1R,3S,5Z)-5-[(2E)-2-[(1R,3aR,7aS)-1-[(2R)-5-di-methoxyphosphorylpentan-2-yl]-7a-methyl-2,3,3a,5,6,7-hexahydro-1H-inden-4-ylidene]ethylidene]-4-methylidene-cyclohexane-1,3-diol.
LMST03020073; 24-(dimethoxyphosphoryl)-9,10-seco-5Z,7E,10(19)-cholatriene-1S,3R-diol; 1α-hydroxy-24-(dimethoxyphosphoryl)-25,26,27-trinorcholecalciferol;
9547325; Compared to 1α,25-dihydroxyvitamin D_3, intestinal calcium absorption in chick is 0.2%, bone calcium mobilization in chick is 0.13%, competitive binding to the vitamin D receptor in chick intestine is 6.98% and inhibition of clonal growth of human leukemia cells (HL-60) was not detected; $[α]_D$ = +38.3° (c = 0.98 in CH_2Cl_2); $λ_m$ = 211, 264 nm (ε = 17000, 18100 EtOH). *Tet Lett.* 1991, *32(36)*, 4643-6; *Blood*, 1991, *78(1)*, 75-82.

113   ***1α-Hydroxy-19-norvitamin D_3***
                                               $C_{26}H_{44}O_2$

19-Nor-9,10-secocholesta-5Z,7E-diene-3β-ol.
1α-Hydroxy-19-norcholecalciferol; *J Steroid Biochem Mol Biol.* 2004, *89-90(1-5)*, 13-17.

114   ***1α,25-Dihydroxy-3-deoxy-19-norvitamin D_3***
9547657 $C_{26}H_{44}O_2$

(6R)-6-[(1R,3aR,4E,7aS)-4-[(2Z)-2-[(3S)-3-hydroxy-cyclohexylidene]ethylidene]-7a-methyl-2,3,3a,5,6,7-hexahydro-1H-inden-1-yl]-2-methyl-heptan-2-ol. LMST03020562; 19-nor-9,10-seco-5,7E-cholestadiene-1S,25-diol;   1α,25-dihydroxy-3-deoxy-19-norchole-calciferol; 9547657; Synthesized by combination of Grundmann type 8-ketone with an A-ring synthon obtained from (-)-quinic acid. Compared to 1α,25-dihydroxyvitamin D$_3$, affinity for the porcine intestinal vitamin D receptor is 8%, HL-60 cell differentiation is 100%, intestinal calcium transport is 0%, bone calcium mobilization is 0%. Affinity for calf thymus vitamin D receptor: 0.07%, affinity for vitamin D binding protein is 32%, HL-60 cell anfiproliferation is 30%, HL-60 cell differentiation is 30%, and transcriptional activity (rat 25-hydroxyvitamin D$_3$ 24-hydroxylase promoter) is 32%. *Biol Pharm Bull.* 1998, *21(12)*, 1300-5; *J Med Chem.* 1998, *41(23)*, 4662-74.

115   *1β,25-Dihydroxy-3-deoxy-19-norvitamin D$_3$*
9547658   C$_{26}$H$_{44}$O$_2$

(6R)-6-[(1R,3aR,4E,7aS)-4-[(2Z)-2-[(3R)-3-hydroxy-cyclohexylidene]ethylidene]-7a-methyl-2,3,3a,5,6,7-hexahydro-1H-inden-1-yl]-2-methyl-heptan-2-ol. LMST03020563; 19-nor-9,10-seco-5,7E-cholestadiene-1R,25-diol;   1β,25-dihydroxy-3-deoxy-19-norchole-calciferol; 9547658; Compared to 1α,25-dihydroxy-vitamin D$_3$, affinity for calf thymus vitamin D receptor is <0.06%, affinity for vitamin D binding protein is 2350%, HL-60 cell anfiproliferation is 50%, HL-60 cell differentiation is 10%, and transcriptional activity (rat 25-hydroxyvitamin D$_3$ 24-hydroxylase promoter) is 10%. *Biol Pharm Bull.* 1998, *21(12)*, 1300-5.

116   *25-Hydroxy-19-nor-3-epivitamin D$_3$*
9547659   C$_{26}$H$_{44}$O$_2$

(6R)-6-[(1R,3aR,4E,7aS)-4-[(2E)-2-[(3R)-3-hydroxy-cyclohexylidene]ethylidene]-7a-methyl-2,3,3a,5,6,7-hexahydro-1H-inden-1-yl]-2-methyl-heptan-2-ol. LMST03020564; 19-nor-9,10-seco-5,7E-cholestadiene-3R,25-diol;   25-hydroxy-19-nor-3-epicholecalciferol; 9547659; Compared to 1α,25-dihydroxyvitamin D$_3$, affinity for calf thymus vitamin D receptor is 0.06%, affinity for vitamin D binding protein is 570%, HL-60 cell anfiproliferation is 40%, HL-60 cell differentiation is 30%, and transcriptional activity (rat 25-hydroxyvitamin D$_3$ 24-hydroxylase promoter) is 18%. *Biol Pharm Bull.* 1998, *21(12)*, 1300-5.

117   *25-Hydroxy-19-norvitamin D$_3$*
9547660   C$_{26}$H$_{44}$O$_2$

(6R)-6-[(1R,3aR,4E,7aS)-4-[(2E)-2-[(3S)-3-hydroxy-cyclohexylidene]ethylidene]-7a-methyl-2,3,3a,5,6,7-hexahydro-1H-inden-1-yl]-2-methyl-heptan-2-ol. LMST03020565; 19-nor-9,10-seco-5,7E-cholestadiene-3S,25-diol;   25-Hydroxy-19-norvitamin   D$_3$;   25-hydroxy-19-norcholecalciferol; 9547660; Compared to 1α,25-dihydroxyvitamin D$_3$, affinity for calf thymus vitamin D receptor is 0.3%, affinity for vitamin D binding protein is 10%, HL-60 cell anfiproliferation is 80%, HL-60 cell differentiation is 60%, and transcriptional activity (rat 25-hydroxyvitamin D$_3$ 24-hydroxylase promoter) is 78%. *Biol Pharm Bull.* 1998, *21(12)*, 1300-5.

118   *2-Nor-1,3-seco-1α,25-dihydroxyvitamin D$_3$*
113490-37-2   C$_{26}$H$_{44}$O$_3$

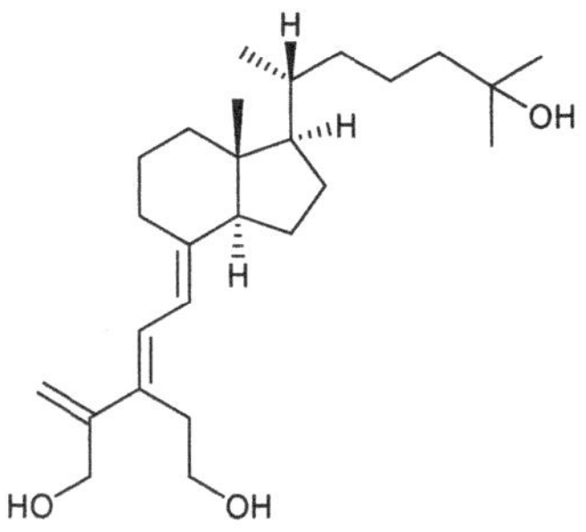

(3E)-3-[(2E)-2-[(1R,3aS,7aR)-1-(6-hydroxy-6-methyl-heptan-2-yl)-7a-methyl-2,3,3a,5,6,7-hexahydro-1H-inden-4-ylidene]ethylidene]-2-methylidene-pentane-1,5-diol.

2-nor-1,3-seco-1,25(OH)2D$_3$; 2-Nor-1,3-seco-1,25-dihydroxyvitamin D$_3$; 2-Nor-1,3-seco-1,25(OH)2 D$_3$; 6444144; 113490-37-2; Inhibits 25-hydroxyvitamin D$_3$ 1α-hydroxylase in isolated mitochondria and induces 25-hydroxyvitamin D$_3$ metabolism in cultured chick kidney cells. *J Steroid Biochem Mol Biol.* 1991, *38(6)*, 775-9.

119 **1,25-Dihydroxy-19-norvitamin D$_3$**
130447-37-9 $C_{26}H_{44}O_3$

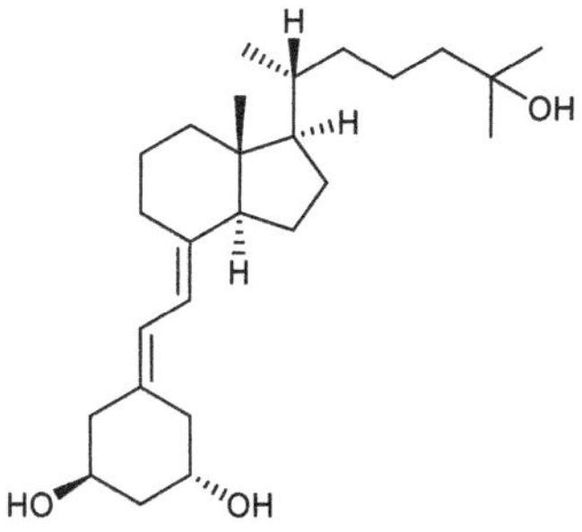

(1S,3S)-5-[(2E)-2-[(1R,3aR,7aS)-1-[(2R)-6-hydroxy-6-methyl-heptan-2-yl]-7a-methyl-2,3,3a,5,6,7-hexahydro-1H-inden-4-ylidene]ethylidene] cyclohexane-1,3-diol.

1,25-Dihydroxy-19-norvitamin D$_3$; 19-Nor-22(E); 1α,-25-(OH)2-19-Nor-vitamin D$_3$; 20-Epi-19-nor-1,25-(OH)2 D$_3$; 19-Nor-9,10-secocholesta-5,7E-diene-1α,3β,25-triol; 6439295; 130447-37-9; Suppresses parathyroid hormone in patients with renal failure and in uremic rats but has less calcemic activity than 1α,25-dihydroxyvitamin D$_3$. It has much less bone resorbing activity *in vivo*, is less effective than 1α,25-dihydroxyvitamin D$_3$ in stimulating mouse marrow osteoclasts to resorb bone. *J Am Soc Nephrol.* 2000, *11(10)*, 1857-64; *Org Lett.* 2001, *3(24)*, 3975-7; *Bioorg Med Chem Lett.* 2002, *12(24)*, 3533-6.

120 **1α,25-Dihydroxy-19-norvitamin D$_3$**
9547326 $C_{26}H_{44}O_3$

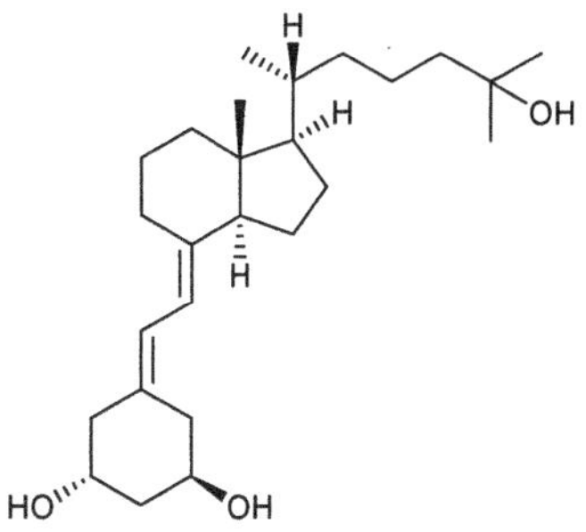

(1R,3R)-5-[(2E)-2-[(1R,3aR,7aS)-1-[(2R)-6-hydroxy-6-methyl-heptan-2-yl]-7a-methyl-2,3,3a,5,6,7-hexahydro-1H-inden-4-ylidene]ethylidene]cyclohexane-1,3-diol.

LMST03020074; 19-nor-9,10-seco-5,7E-cholestadiene-1R,3R,25-triol; 1α,25-dihydroxy-19-norcholecalciferol; 9547326; Compared to 1α,25-dihydroxyvitamin D$_3$, affinity the intestinal vitamin D receptor and plasma vitamin D binding protein is 30% and 20%, respectively. The *in vitro* effects on human promyeloid leukemia (HL-60 cell) differentiation and osteocalcin secretion by human osteosarcoma (MG-63) cells are nearly identical to those of 1α,25-dihydroxyvitamin D$_3$ and the *in vivo* calcemic effect was less than 10% of the reference compound. *J Bone Miner Res.* 1993, *8(8)*, 1009-15.

121 **1,25-Dihydroxy-2-nor-1,2-secovitamin D$_3$**
9547327 $C_{26}H_{44}O_3$

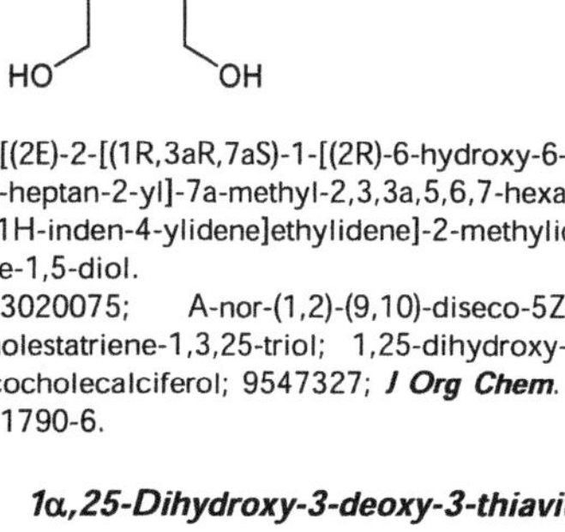

(3Z)-3-[(2E)-2-[(1R,3aR,7aS)-1-[(2R)-6-hydroxy-6-methyl-heptan-2-yl]-7a-methyl-2,3,3a,5,6,7-hexahydro-1H-inden-4-ylidene]ethylidene]-2-methylidene-pentane-1,5-diol.

LMST03020075; A-nor-(1,2)-(9,10)-diseco-5Z,7E,10(19)-cholestatriene-1,3,25-triol; 1,25-dihydroxy-2-nor-1,2-secocholecalciferol; 9547327; *J Org Chem.* 1988, *53(8)*, 1790-6.

122 **1α,25-Dihydroxy-3-deoxy-3-thiavitamin D$_3$ 3-oxide**

$C_{26}H_{44}O_3S$

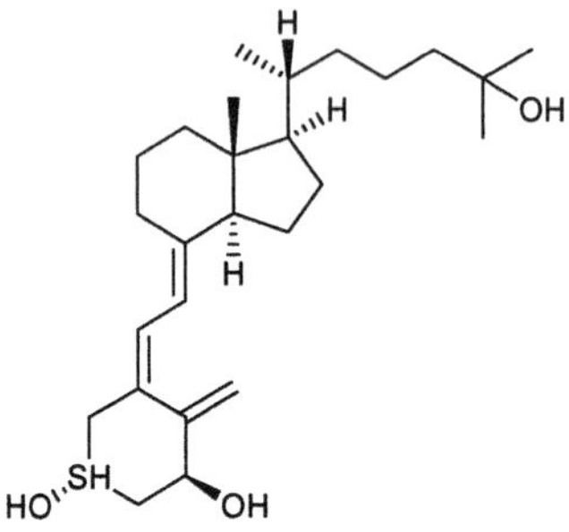

(5E,7E)-(1R,3R)-1,25-dihydroxy-9,10-seco-3-thia-5,7,10(19)-cholestatriene 3-oxide.
LMST03020049;1α,25-dihydroxy-3-deoxy-3-thiacholecalciferol 3-oxide; Affinity for chicken intestinal receptor is 27.0% that of 1α,25-dihydroxyvitamin D_3.
λ_m = 268 nm (ε = 17500 EtOH). *J Steroid Biochem Mol Biol.* 1997, *62(1)*, 73-8.

123   *(5E)-1α,25-Dihydroxy-3-deoxy-3S-thiavitamin D_3 3-oxide*

$C_{26}H_{44}O_3S$

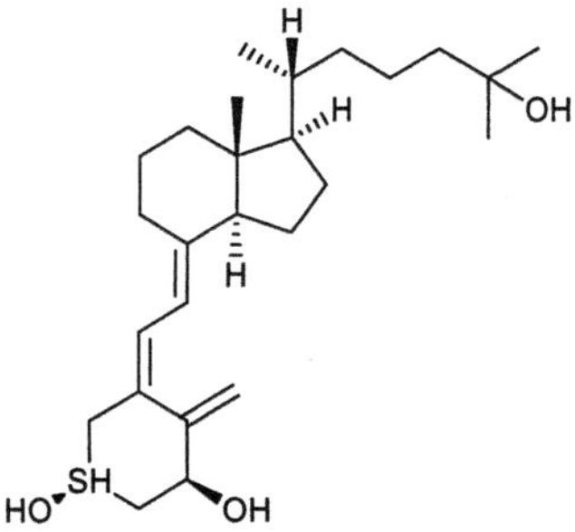

(5E,7E)-(1R,3S)-1,25-dihydroxy-9,10-seco-3-thia-5,7,10(19)-cholestatrien 3-oxide.
LMST03020050;   (5E)-1α,25-dihydroxy-3-deoxy-3S-thiacholecalciferol 3-oxide; Affinity for chicken intestinal   receptor   is   0.8%   that   of   1α,25-dihydroxyvitamin D_3; λ_m = 272 nm (ε = 18000 EtOH). *J Steroid Biochem Mol Biol.* 1997, *62(1)*, 73-8.

124   *(5Z)-1α,25-Dihydroxy-3-deoxy-3-thiavitamin D_3 3-oxide*

$C_{26}H_{44}O_3S$

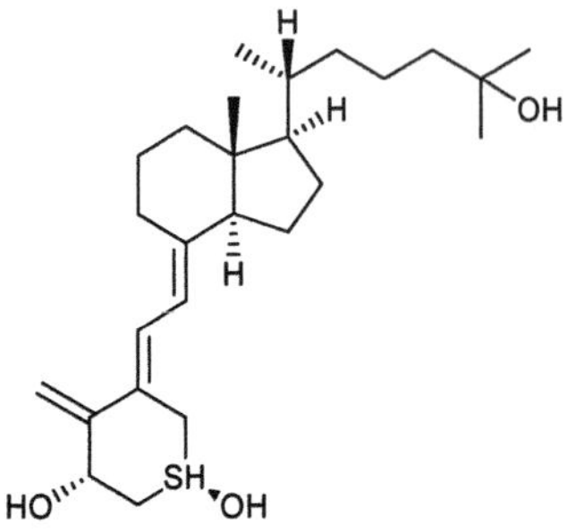

(5Z,7E)-(1R,3R)-1,25-dihydroxy-9,10-seco-3-thia-5,7,10(19)-cholestatriene 3-oxide.
LMST03020051; (5Z)-1α,25-dihydroxy-3-deoxy-3-thiacholecalciferol 3-oxide. *Bioorg Med Chem Lett.* 1995, *5(3)*, 279-82.

125   *(5Z)-1α,25-Dihydroxy-3S-deoxy-3-thiavitamin D_3 3-oxide*

$C_{26}H_{44}O_3S$

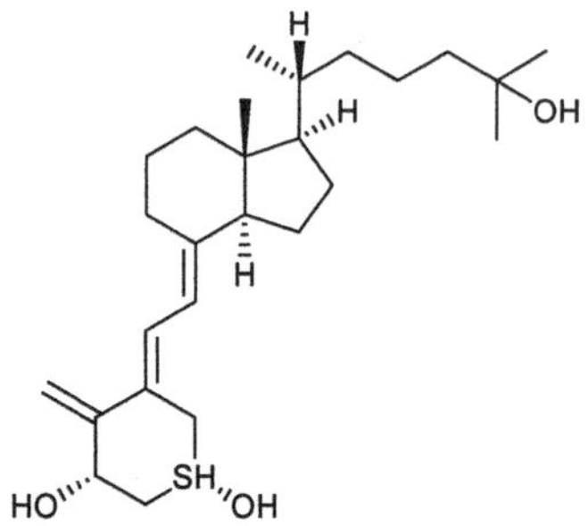

(5Z,7E)-(1R,3S)-1,25-dihydroxy-9,10-seco-3-thia-5,7,10(19)-cholestatrien 3-oxide.
LMST03020052;   (5Z)-1α,25-dihydroxy-3S-deoxy-3-thiacholecalciferol 3-oxide; Affinity for chicken intestinal receptor is 0.9% of that of 1α,25-dihydroxyvitamin D_3 and inhibitory effect on 1α,25-dihydroxyvitamin D_3 1α-hydroxylase activity was evaluated; λ_m = 270 nm (ε = 17800 EtOH). *J Steroid Biochem Mol Biol.* 1997, *62(1)*, 73-8.

126   *Aplisiadiasterol A*

$C_{26}H_{44}O_4$

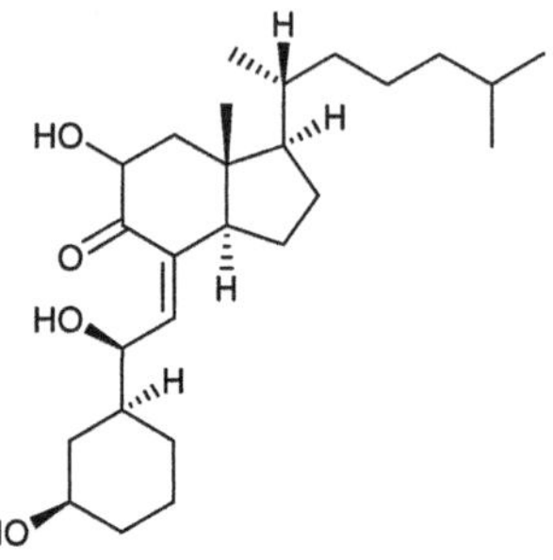

3,6,11-trihydroxy-9,11-secocholest-7-en-9-one; 3β,6β,11-trihydroxy-9,11-seco-5α-cholest-7-en-9-one; Isolated from the Mediterranean *ascidian Aplidium conicum*, structure determined spectroscopically. Cytotoxic against rat glioma (C6) and murine monocyte/macrophage (J774) tumor cells *in vitro*. *J Steroid Biochem Mol Biol.* 1997, *62(1)*, 73-8.

127   *1α,2α,25-Trihydroxy-19-norvitamin D_3*
$C_{26}H_{44}O_4$

19-nor-1α,2α,25-(OH)$_3$D$_3$; LMST03020547; Synthetic vitamin D$_3$ analogue; compared to calcitriol, promotes intestinal calcium transport but not bone calcium mobilization. *J Med Chem*. 1994, *37(22)*, 3730-8.

128 *1α,2α,25-Trihydroxy-19-norvitamin D$_3$*
9547650   C$_{26}$H$_{44}$O$_4$

(1R,2S,3R)-5-[(2E)-2-[(1R,3aR,7aS)-1-[(2R)-6-hydroxy-6-methyl-heptan-2-yl]-7a-methyl-2,3,3a,5,6,7-hexa-hydro-1H-inden-4-ylidene]ethylidene]cyclohexane-1,2,3-triol.
LMST03020547; 19-nor-9,10-seco-5,7E-cholestadiene-1R,2S,3R,25-tetrol; 1α,2α,25-trihydroxy-19-norchole-calciferol; 9547650; Synthesized by combining A-ring synthon obtained from (-)-quinic acid with 25-hydroxylated Grundmann type ketone. Has weak calcemic (intestinal calcium transport and bone calcium-mobilization) activity but potent cell differentiating activity (HL-60 cell) similar to that of 1α,25-dihydroxyvitamin D$_3$; λ$_m$ = 243, 251.5, 261 nm (EtOH). *J Med Chem*. 1994, *37(22)*, 3730-8.

129 *1α,2β,25-Trihydroxy-19-norvitamin D$_3$*
9547650   C$_{26}$H$_{44}$O$_4$

(1R,2R,3R)-5-[(2E)-2-[(1R,3aR,7aS)-1-[(2R)-6-hydroxy-6-methyl-heptan-2-yl]-7a-methyl-2,3,3a,5,6,7-hexa-hydro-1H-inden-4-ylidene]ethylidene]cyclohexane-1,2,3-triol.
LMST03020548; 19-nor-9,10-seco-5,7E-cholestadiene-1R,2R,3R,25-tetrol; 1α,2β,25-trihydroxy-19-norchole-calciferol 19-nor-1α,2β,25-(OH)$_3$D$_3$; 9547650; Vitamin D$_3$ analogue synthesized by combining A-ring synthon obtained from (-)-quinic acid with 25-hydroxylated Grundmann type ketone. Compared to calcitriol, promotes intestinal calcium transport but not bone calcium mobilization. Has potent intestinal calcium transport activity equivalent to that of 1α,25-dihydroxyvitamin D$_3$ but no ability to mobilize calcium from bone. HL-60 cell differentiating activity is about 1/10 of that of 1α,25-dihydroxy-vitamin D$_3$; λ$_m$ = 243, 251.5, 261 nm (EtOH). *J Med Chem*. 1994, *37(22)*, 3730-8.

130 *Aplisiadiasterol B*
                                                  C$_{26}$H$_{44}$O$_5$

3,5,6,11-tetrahydroxy-9,11-secocholest-7-en-9-one; 3β,5α,6β,11-tetrahydroxy-9,11-secocholest-7-en-9-one; Isolated from the Mediterranean ascidian **Aplidium conicum**, structure determined spectro-scopically. Cytotoxic against rat glioma (C6) and murine monocyte/macrophage (J774) tumor cells *in vitro*. *Steroids*. 2003, *68(9)*, 719-23.

131 *1,4-Dihydroxy-3-deoxy-A-homo-19-nor-9,10-secocholesta-4,7-diene*
75946-87-1   C$_{26}$H$_{46}$O$_2$

6-[(2Z)-2-[7a-methyl-1-(6-methylheptan-2-yl)-2,3,3a,
5,6,7-hexahydro-1H-inden-4-ylidene]ethylidene]
cycloheptane-1,4-diol.
1,4-Ddhns;   1,4-Dihydroxy-3-deoxy-A-homo-19-nor-
9,10-secocholesta-4,7-diene;   A-Homo-19-nor-9,10-
secocholesta-5Z,7E-diene-1α,4α-diol;   6444030;
75946-87-1; Biochemical properties reported. *Arch
Biochem Biophys.* 1981, *210(1)*, 238-45.

132   *23,25-Dihydroxy-23-methyl-21-norcholesterol*

71848-98-1 $C_{26}H_{46}O_3$

6-[(3S,8S,9S,10R,13S,14S,17R)-3-hydroxy-10,13-di-
methyl-2,3,4,7,8,9,11,12,14,15,16,17-dodecahydro-
1H-cyclopenta[a]phenanthren-17-yl]-2,4-dimethyl-
hexane-2,4-diol.
SC 31082; SC-31082; 23,25-Dihydroxy-23-methyl-21-
norcholesterol;   21-Norcholest-5-ene-3β,23,25-triol,
23-methyl-;   191986;   71848-98-1;   Inhibitor   of
cholesterol ester formation. *Proc Natl Acad Sci U.S.A.*
1987, *4*, 1877-81; *Physiol Rev.* 2000, *80(1)*, 361-554.

133   *1-Fluoro-25-hydroxy-16-ene-23-yne-26,27-hexadeuterovitamin-D₃*

$C_{27}H_{31}D_6FO_2$

1-F-25(OH)-16-ene-23-yne-D3; AW; Potent inhibitor
of proliferation and inducer of differentiation. Some

300 times less active than 1α,25-dihydroxyvitamin D_3
in mediating calcium absorption and mobilization.
Increases survival of mice with myeloid leukemia.
Antagonized by TEI-9647. *Mol Endocrinol.* 2000,
14(11), 1788-96.

134   *1α,26,26,26,27,27,27-Heptafluoro-25-hydroxy-16,17,23,23,24,24-hexadehydrovitamin D₃*

9547328 $C_{27}H_{31}F_7O_2$

(6R)-6-[(3aR,4E,7aS)-4-[(2Z)-2-[(3S,5R)-3-Fluoro-5-
hydroxy-2-methylidene-cyclohexylidene]ethylidene]-
7a-methyl-3a,5,6,7-tetrahydro-3H-inden-1-yl]-1,1,1-
trifluoro-2-(trifluoromethyl)hept-3-yn-2-ol.
LMST03020076;   1S,26,26,26,27,27,27-heptafluoro-
9,10-seco-5Z,7E,10(19),16-cholestatetraen-23-yne-3R,
25-diol; 1α,26,26,26,27,27,27-heptafluoro-25-hydroxy
-16,17,23,23,24,24-hexadehydrocholecalciferol;
9547328; Compared to 1α,25-dihydroxyvitamin D_3,
intestinal calcium absorption in chick is 0.5%, bone
calcium mobilization in chick is 0.2%, competitive
binding to the vitamin D receptor in chick intetine,
5.3%, and HL-60 cells, 9.2%, inhibition of clonal
growth of human leukemia cells (HL-60), 571%.
Blood. 1991, *78(1)*, 75-82.

135   *26,26,26,27,27,27-Hexafluoro-25-hydroxy-16,17,23,23,24,24-hexadehydrovitamin D₃*

9547329 $C_{27}H_{32}F_6O_2$

(6R)-6-[(3aR,4E,7aS)-4-[(2Z)-2-[(5S)-5-hydroxy-2-
methylidene-cyclohexylidene]ethylidene]-7a-methyl-
3a,5,6,7-tetrahydro-3H-inden-1-yl]-1,1,1-trifluoro-2-
(trifluoromethyl)hept-3-yn-2-ol.
LMST03020077;   26,26,26,27,27,27-hexafluoro-9,10-
seco-5Z,7E,10(19),16-cholestatetraen-23-yne-3S,25-
diol; 26,26,26,27,27,27-hexafluoro-25-hydroxy-16,17,

23,23,24,24-hexadehydrocholecalciferol; 9547329; Compared to 1α,25-dihydroxyvitamin D_3, intestinal calcium absorption in chick is 0.13%, bone calcium mobilization in chick is 2.0%, competitive binding to the vitamin D receptor in chick intestine is 0.58% and HL-60 cells, 2.03%, and inhibition of clonal growth of human leukemia cells (HL-60), 72%. *Blood*. 1991, 78(1), 75-82.

136   *26,26,26,27,27,27-Hexafluoro-1α,25-dihydroxy-16,17,23,23,24,24-hexadehydro-vitamin D_3*

137102-93-3 $C_{27}H_{32}F_6O_3$

(1R,3S,5Z)-5-[(2E)-2-[(3aR,7aS)-7a-methyl-1-[(2R)-7,7,7-trifluoro-6-hydroxy-6-(trifluoromethyl)hept-4-yn-2-yl]-3a,5,6,7-tetrahydro-3H-inden-4-ylidene]ethyl-idene]-4-methylidene-cyclohexane-1,3-diol. LMST03020078; 26,26,26,27,27,27-hexafluoro-9,10-seco-5Z,7E,10(19),16-cholestatetraen-23-yne-1S,3R,25-triol; 26,26,26,27,27,27-hexafluoro-1α,25-dihydroxy-16,17,23,23,24,24-hexadehydrochole-calciferol; 1,25-dihydroxy-16-ene-23-yne-26,27-hexa-fluorocholecalciferol; Ro 24-5531; 1α,25(OH)[2]-16-ene-23-yne-26,27-hexafluorocholecalciferol; 9547330; 137102-93-39; Induces differentiation in HL-60 cells and does not induce hypercalcemia in animal models.

Compared to 1α,25-dihydroxyvitamin D_3, intestinal calcium absorption in chicks is 6.7%, bone calcium mobilization in chicks is 10.4%, competitive binding to the vitamin D receptor in chick intestine is 45%, and to HL-60 cells is 31%, and inhibition of clonal growth of human leukemia cells (HL-60) is 8000%. *J Clin Endocrinol Metab*. 1996, *81(1)*, 93-9; *Blood*. 1991, *75(1)*, 75-92.

137   *26,26,26,27,27,27-Hexafluoro-25-hydroxy-23,23,24,24-tetradehydrovitamin D_3*

9547331 $C_{27}H_{34}F_6O_2$

(6R)-6-[(1R,3aR,4E,7aS)-4-[(2Z)-2-[(5S)-5-hydroxy-2-methylidene-cyclohexylidene]ethylidene]-7a-methyl-2,3,3a,5,6,7-hexahydro-1H-inden-1-yl]-1,1,1-trifluoro-2-(trifluoromethyl)hept-3-yn-2-ol. LMST03020079; 26,26,26,27,27,27-hexafluoro-9,10-seco-5Z,7E,10(19)-cholestatrien-23-yne-3S,25-diol; 26,26,26,27,27,27-hexafluoro-25-hydroxy-23,23,24,24-tetradehydrocholecalciferol; 9547331; Compared to 1α,25-dihydroxyvitamin D_3, affinity for chick intestinal receptor is 0.18% and affinity for human vitamin D binding protein is 470%. *Endocr Rev*. 1995, *16*, 200-57.

138   *26,26,26,27,27,27-Hexafluoro-25-hydroxy-22E,23-didehydrovitamin D_3*

9547332 $C_{27}H_{36}F_6O_2$

(E,6S)-6-[(1R,3aR,4E,7aS)-4-[(2Z)-2-[(5S)-5-hydroxy-2-methylidene-cyclohexylidene]ethylidene]-7a-methyl-2,3,3a,5,6,7-hexahydro-1H-inden-1-yl]-1,1,1-trifluoro-2-(trifluoromethyl)hept-4-en-2-ol. LMST03020081; 26,26,26,27,27,27-hexafluoro-25-hydroxy-22E,23-didehydrocholecalciferol; 26,26,26,27,27,27-hexafluoro-9,10-seco-5Z,7E,10(19),22E-chol-estatetraene-3S,25-diol; 9547332; λ_m = 264 nm (EtOH). *Tet Lett*. 1988, *29(2)*, 227-30.

139   *26,26,26,27,27,27-Hexafluoro-1α,25-dihydroxy-22E,23-didehydrovitamin D_3*

9547333 $C_{27}H_{36}F_6O_3$

(1R,3S,5Z)-5-[(2E)-2-[(1R,3aR,7aS)-7a-methyl-1-[(E,2S)-7,7,7-trifluoro-6-hydroxy-6-(trifluoromethyl)hept-3-en-2-yl]-2,3,3a,5,6,7-hexahydro-1H-inden-4-ylidene]ethylidene]-4-methylidene-cyclohexane-1,3-diol. LMST03020082; 26,26,26,27,27,27-hexafluoro-1α,25-dihydroxy-22E,23-didehydrocholecalciferol; 26,26,26,27,27,27-hexafluoro-9,10-seco-5Z,7E,10(19),22E-cholestatetraene-1S,3R,25-triol; 9547333; Compared to 1α,25-dihydroxyvitamin D₃, inhibition of proliferation of human keratinocytes in culture is 100%, intestinal calcium absorption in rat is 221%, bone calcium mobilization in rat is 110%, competitive binding to rat intestinal vitamin D receptor is 62% and differentiation of human leukemia cells (HL-60) is 3000%. *J Nutr Biochem*. 1993, *4(1)*, 49-57.

140　26,26,26,27,27,27-Hexafluoro-1α,25-dihydroxy-23E,24-didehydrovitamin D₃

9547334　　　　　　　　　　　　$C_{27}H_{36}F_6O_3$

(1R,3S,5Z)-5-[(2E)-2-[(1R,3aR,7aS)-7a-methyl-1-[(E,2R)-7,7,7-trifluoro-6-hydroxy-6-(trifluoromethyl)hept-4-en-2-yl]-2,3,3a,5,6,7-hexahydro-1H-inden-4-ylidene]-ethylidene]-4-methylidene-cyclohexane-1,3-diol. LMST03020083; 26,26,26,27,27,27-hexafluoro-1α,25-dihydroxy-23E,24-didehydrocholecalciferol; 26,26,26,27,27,27-hexafluoro-9,10-seco-5Z,7E,10(19),23E-cholestatetraene-1S,3R,25-triol; 9547334; Ten times more potent that 1α,25-dihydroxyvitamin D₃ in differentiation of HL-60 cells. *Curr Pharm Design*. 1997, *3*, 99-123.

141　1α-Fluoro-25-hydroxy-16,17,23,23,24,24-hexadehydrovitamin D₃

9547335　　　　　　　　　　　　$C_{27}H_{37}FO_2$

(6R)-6-[(3aR,4E,7aS)-4-[(2Z)-2-[(3S,5R)-3-fluoro-5-hydroxy-2-methylidene-cyclohexylidene]ethylidene]-7a-methyl-3a,5,6,7-tetrahydro-3H-inden-1-yl]-2-methyl-hept-3-yn-2-ol. LMST03020084; 1S-fluoro-9,10-seco-5Z,7E,10(19), 16-cholestatetraen-23-yne-3R,25-diol; 1α-fluoro-25-hydroxy-16,17,23,23,24,24-hexadehydrocholecalciferol; 9547335; Compared to 1α,25-dihydroxyvitamin D₃, intestinal calcium absorption in chicks is 0.03%, bone calcium mobilization in chicks is 0.08%,

competitive binding to the vitamin D receptor in chick intestine is 24%, and in human leukemia (HL-60) cells, 35%, inhibition of clonal growth of HL-60 cells is 61% and differentiation of HL-60 cells is 417%. *Cancer Res*. 1990, *50(21)*, 6857-64.

142　26,26,26,27,27,27-Hexafluoro-1α-hydroxyvitamin D₃

9547336　　　　　　　　　　　　$C_{27}H_{38}F_6O_2$

(1R,3S,5Z)-5-[(2E)-2-[(1R,3aR,7aS)-7a-methyl-1-[(2R)-7,7,7-trifluoro-6-(trifluoromethyl)heptan-2-yl]-2,3,3a,5,6,7-hexahydro-1H-inden-4-ylidene]ethylidene]-4-methylidene-cyclohexane-1,3-diol. LMST03020085; 26,26,26,27,27,27-hexafluoro-9,10-seco-5Z,7E,10(19)-cholestatriene-1S,3R-diol; 26,26,26,27,27,27-hexafluoro-1α-hydroxycholecalciferol; 9547336; Binding affinity for chick intestinal and HL-60 cell receptor is 1.4% that of 1α,25-dihydroxyvitamin D₃. In inhibition of proliferation and induction of differentiation of HL-60 cells its activity is equal to that of 1α,25-dihydroxyvitamin D₃. *Arch Biochem Biophys*. 1987, *258(2)*, 421-5.

143　26,26,26,27,27,27-Hexafluoro-25-hydroxyvitamin D₃

9547337　　　　　　　　　　　　$C_{27}H_{38}F_6O_2$

(6R)-6-[(1R,3aR,4E,7aS)-4-[(2Z)-2-[(5S)-5-hydroxy-2-methylidene-cyclohexylidene]ethylidene]-7a-methyl-2,3,3a,5,6,7-hexahydro-1H-inden-1-yl]-1,1,1-trifluoro-2-(trifluoromethyl)heptan-2-ol. LMST03020086; 26,26,26,27,27,27-hexafluoro-9,10-seco-5Z,7E,10(19)-cholestatriene-3S,25-diol; 26,26,26,27,27,27-hexafluoro-25-hydroxycholecalciferol; 9547337; Equipotent with 25-hydroxyvitamin D₃ in stimulating intestinal calcium transport, bone calcium mobilization, bone mineralization, epiphyseal plate calcification, and elevation of serum inorganic phosphorus. Bilateral nephrectomy eliminates the response of intestine and bone suggesting that this compound must be 1α-hydroxylated to be functional; λ_m = 264 nm (ε = 18000 EtOH). *J Chem Soc Chem Comm*. 1980, 459.

144   25-Hydroxy-26,26,26,27,27,27-hexa-fluorovitamin D₃

75303-43-4 $C_{27}H_{38}F_6O_2$

9,10-Secocholesta-5Z,7E,10(19)-triene-3β,25-diol, 26,26,26,27,27,27-hexafluoro-.

25-Hydroxy-26,26,26,27,27,27-hexafluorovitamin D₃; 25-Hydroxy-26,26,26,27,27,27-hexafluorochole-calciferol; 75303-43-4; Has biological activity similar to that of calcifediol. *Arch Biochem Biophys*. 1982, *218(1)*, 134-41.

145   26,26,26,27,27,27-Hexafluoro-1α,24-dihydroxyvitamin D₃

9547339 $C_{27}H_{38}F_6O_3$

(1R,3S,5Z)-5-[(2E)-2-[(1R,3aR,7aS)-7a-methyl-1-[(2R)-7,7,7-trifluoro-5-hydroxy-6-(trifluoromethyl)heptan-2-yl]-2,3,3a,5,6,7-hexahydro-1H-inden-4-ylidene]ethyl-idene]-4-methylidene-cyclohexane-1,3-diol.
LMST03020088; 26,26,26,27,27,27-hexafluoro-9,10-seco-5Z,7E,10(19)-cholestatriene-1S,3R,24-triol; 26,26,26,27,27,27-hexafluoro-1α,24-dihydroxychole-calciferol; 9547339; Binding affinity for chick intestinal and HL-60 cell receptor is 66% and 100% that of 1α,25-dihydroxyvitamin D₃, respectively. Inhibition of proliferation and induction of differentiation of HL-60 cells is ten times higher. *Arch Biochem Biophys*. 1987, *258(2)*, 421-5.

146   26,27-Hexafluoro-1α,25-dihydroxy-vitamin D₃

3323 $C_{27}H_{38}F_6O_3$

5-[2-[7a-methyl-1-[7,7,7-trifluoro-6-hydroxy-6-(trifluoromethyl)heptan-2-yl]-2,3,3a,5,6,7-hexahydro-1H-inden-4-ylidene]ethylidene]-4-methylidene-cyclohexane-1,3-diol.
26,27-hexafluoro-1α,25-dihydroxy-vitamin D₃; 3966620 (ChemDB); in experimental osteoporosis in the rat, prevents the osteoporotic decrease of bone mass by suppressing the elevated bone turnover. *Bone Miner*. 1990, *9(2)*, 101-9

147   Falecalcitriol

83805-11-2 $C_{27}H_{38}F_6O_3$

(1R,3S,5Z)-5-[(2E)-2-[(1R,3aR,7aS)-7a-methyl-1-[(2R)-7,7,7-trifluoro-6-hydroxy-6-(trifluoromethyl)heptan-2-yl]-2,3,3a,5,6,7-hexahydro-1H-inden-4-ylidene] ethylidene]-4-methylidene-cyclohexane-1,3-diol.
Falecalcitriol; Flocalcitriol; Hornel, Fulstan; ST 630; Ro 23-4194; 26,26,26,27,27,27-Hexafluoro-1,25-di-hydroxyvitamin D₃; (+)-(5Z,7E)-26,26,26,27,27,27-Hexafluoro-9,10-secocholesta-5,7,10(19)-triene-1α,3β, 25-triol; 9,10-Secocholesta-5Z,7E,10(19)-triene-1α,3β, 25-triol, 26,26,26,27,27,27-hexafluoro-; 5282190; LMST03020089; 83805-11-2; Calcitriol analog. Dermatologic agent. Used to treat osteoporosis and secondary hyperparathyroidism. Daily oral administration of falecalcitriol at doses lower than those required for calcitriol has been shown to have clinical effects for the treatment of diseases such as hyperparathyroidism due to chronic renal failure, rickets, osteomalacia and hypoparathyroidism. *Bioorg Med Chem*. 2000, *8(8)*, 2157-66; *Chem Pharm Bull (Tokyo)*. 1995, *43(11)*, 1897-901.

148   6,19-Epidioxy-26,26,26,27,27,27-hexafluoro-25-hydroxy-6,19-dihydrovitamin D₃

9547338 $C_{27}H_{38}F_6O_4$

(3S)-10-[(E)-[(1R,3aR,7aS)-7a-methyl-1-[(2R)-7,7,7-trifluoro-6-hydroxy-6-(trifluoromethyl)heptan-2-yl]-2,3,3a,5,6,7-hexahydro-1H-inden-4-ylidene]methyl]-8,9-dioxabicyclo[4.4.0]dec-11-en-3-ol.
LMST03020087;   6,19-epidioxy-26,26,26,27,27,27-hexafluoro-9,10-seco-5(10),7E-cholestadiene-3S,25-diol;   6,19-epidioxy-26,26,26,27,27,27-hexafluoro-25-hydroxy-6,19-dihydrocholecalciferol;   9547338;
Prepared by the reaction of 26,26,26,27,27,27-hexafluoro-25-hydroxyvitamin D_3 with singlet oxygen generated by dye-sensitized photochemical method. Affinity for the vitamin D receptor is 1/1500 that of 1α,25-dihydroxyvitamin D_3, in differentiation of HL-60 cells it is 1/30 as active. *J Med Chem*. 1985, *28(9)*, 1148-53.

149   **26,26,26,27,27,27-Hexafluoro-1α,23S,25-trihydroxyvitamin D$_3$**
9547340 $C_{27}H_{38}F_6O_4$

(4S,6R)-6-[(1R,3aR,4E,7aS)-4-[(2Z)-2-[(3S,5R)-3,5-dihydroxy-2-methylidene-cyclohexylidene]ethylidene]-7a-methyl-2,3,3a,5,6,7-hexahydro-1H-inden-1-yl]-1,1,1-trifluoro-2-(trifluoromethyl)heptane-2,4-diol.
LMST03020090;   26,26,26,27,27,27-hexafluoro-1α,23S,25-trihydroxycholecalciferol;   26,26,26,27,27,27-hexafluoro-9,10-seco-5Z,7E,10(19)-cholestatriene-1S,3R,23S,25-tetrol; 9547340; Binding affinity for chick intestinal receptor is 2-3 times less than that of 1α,25-dihydroxyvitamin D_3. Intestinal calcium transport is comparable to that of 1α,25-dihydroxyvitamin D_3. Vitamin D Molecular, Cellular and Clinical Endocrinology. Proceeding of the Vitamin D Rancho Mirage, California, USA April. Mechanism of Action of 26,27-Hexafluoro-1,25-(OH)2D3. (Norman, A. W., Schaefer, K., Grigoleit, H. G., v. Herrath, D., eds), pp141-142, Walter de Gruyter Berlin New York

(1988).

150   **26,26,26,27,27,27-Hexafluoro-1,23,25-trihydroxyvitamin D$_3$**
114489-80-4 $C_{27}H_{38}F_6O_4$

(6R)-6-[(1R,4E,7aS)-4-[(2Z)-2-[(5R)-3,5-dihydroxy-2-methylidene-cyclohexylidene]ethylidene]-7a-methyl-2,3,3a,5,6,7-hexahydro-1H-inden-1-yl]-1,1,1-trifluoro-2-(trifluoromethyl)heptane-2,4-diol.
LMST03020090;   ST 232;   26,26,26,27,27,27-Hexafluoro-1,23,25-tri-hydroxyvitamin D_3;   9,10-Secocholesta-5Z,7E,10(19)-triene-1α,3β,23S,25-tetrol, 26,26,26,27,27,27-hexa-fluoro-;   9,10-Secocholesta-5Z,7E,10(19)-triene-1α,3β,   23S,25-tetrol-26,26,26,27,27,27-hexafluoro-;   SM-10193;   6439245;   114489-80-4; Inhibits growth inhibition and differentiation of cultured normal human keratinocytes more effectively than 1,25-dihydroxyvitamin D_3 and could be useful in treatment of psioriasis. *Clin Calcium*. 2005, *15(1)*, 29-33; Vitamin D Molecular, Cellular and Clinical Endocrinology. Proceeding of the Vitamin D Rancho Mirage, California, USA April. Mechanism of Action of 26,27-Hexafluoro-1,25-(OH)2D3. (Norman, A. W.; Schaefer, K.; Grigoleit, H. G.; v. Herrath, D.; eds), pp141-142, Walter de Gruyter Berlin New York (1988).

151   **25-Hydroxy-16,17,23,23,24,24-hexadehydrovitamin D$_3$**
9547341 $C_{27}H_{38}O_2$

(6R)-6-[(3aR,4E,7aS)-4-[(2Z)-2-[(5S)-5-hydroxy-2-methylidene-cyclohexylidene]ethylidene]-7a-methyl-3a,5,6,7-tetrahydro-3H-inden-1-yl]-2-methyl-hept-3-yn-2-ol.
LMST03020091;   9,10-seco-5Z,7E,10(19),16-cholestatetraen-23-yne-3S,25-diol;   25-hydroxy-16,17,23,23,24,24-hexadehydrocholecalciferol;   Ro   24-2090;

9547341; Compared to 1α,25-dihydroxyvitamin D_3, inhibition of proliferation of human keratinocytes in culture is 100%, intestinal calcium absorption in rat is 72%, bone calcium mobilization in rat is 0%, competitive binding to rat intestinal vitamin D receptor is 0.07% and differentiation of human leukemia cells (HL-60) is 1%. *J Nutr Biochem*. 1993, *4(1)*, 49-57; *Anal Biochem*. 1996, *243(1)*, 28-40.

152 1,25-Dihydroxy-16-ene-23-yne-vitamin D_3

118694-43-2 $C_{27}H_{38}O_3$

(1R,3S,5Z)-5-[(2E)-2-[(3aR,7aS)-1-[(2R)-6-hydroxy-6-methyl-hept-4-yn-2-yl]-7a-methyl-3a,5,6,7-tetrahydro-3H-inden-4-ylidene]ethylidene]-4-methylidene-cyclohexane-1,3-diol.
Ro 23-7553; LMST03020096; 1,25(OH)2-16-Ene-23-yne-D_3; 1,25-Di-hydroxy-16-ene-23-yne-vitamin D_3; 1,25-Dihydroxy-Δ^{16}-23-yne-vitamin D_3; 1α,25-Dihydroxy-16ene, 23-yne-vitamin D_3; BXL 353; ILEX 23-7553; ILX 23-7553; 9,10-Secocholesta-5Z,7E,10(19),16-tetra-en-23-yne-1α,3β,25-triol; AIDS073069; AIDS080067; AIDS-073069; AIDS-080067; 1,25(OH)2-16-Ene-23-yne-D_3; 1,25-Dihydroxy-16-ene-23-yne-vitamin D_3; 1α,25-Di-hydroxy-16ene, 23yne-vitamin D_3; 6438384; 118694-43-2; Calcitriol analog. Less active than calcitriol in mediation of calcium absorption and mobilization. Ineffective against HIV. Prolongs survival of leukemic mice. Drug candidate for acute myelogenous leukemia. *Proc Natl Acad Sci U.S.A*. 1990, *87(10)*, 3929-32. *Hoffmann-LaRoche Inc.*

153 1α-Hydroxy-24-oxo-26,27-cyclo-22E,23-didehydrovitamin D_3

9547342 $C_{27}H_{38}O_3$

(E,4S)-4-[(1R,3aR,4E,7aS)-4-[(2Z)-2-[(3S,5R)-3,5-dihydroxy-2-methylidene-cyclohexylidene]ethylidene]-7a-methyl-2,3,3a,5,6,7-hexahydro-1H-inden-1-yl]-1-cyclopropyl-pent-2-en-1-one.
LMST03020092; 1α-hydroxy-24-oxo-26,27-cyclo-22,23-didehydrocholecalciferol; 1S,3R-dihydroxy-26,27-cyclo-9,10-seco-5Z,7E,10(19),22E-cholestatetraen-24-one; 9547342; Isolated as a metabolite produced from 1α,24S-dihydroxy-26,27-cyclo-22,23-didehydro-vitamin D_3 (calcipotriol, MC903). *Tetrahedron*. 1987, *43(20)*, 4609-19; *J Biol Chem*. 1994, *269(7)*, 4794-803.

154 24,25-Epoxy-1α-hydroxy-22,22,23,23-tetradehydrovitamin D_3

9547343 $C_{27}H_{38}O_3$

(1R,3S,5Z)-5-[(2E)-2-[(1R,3aR,7aS)-1-[(2S)-4-(3,3-dimethyloxiran-2-yl)but-3-yn-2-yl]-7a-methyl-2,3,3a,5,6,7-hexahydro-1H-inden-4-ylidene]-ethylidene]-4-methylidene-cyclohexane-1,3-diol.
LMST03020093; 24,25-epoxy-9,10-seco-5Z,7E,10(19)-cholestatrien-22-yne-1S,3R-diol; 24,25-epoxy-1α-hydroxy-22,22,23,23-tetradehydrocholecalciferol;
9547343; Compared to 1α,25-dihydroxyvitamin D_3, affinity for pig intestinal receptor and human vitamin D binding protein is 2% and 5% respectively, inhibition of proliferation or differentiation induction of human promyeloid leukemia (HL-60), osteosarcoma (MG-63) and breast carcinoma (MCF-7) cells is 3%, 100% and 20%, respectively and elevation of serum calcium, serum osteocalcin, bone calcium and duodenal calbindin in rachitic chicks are all <1%. *Steroids*. 1995, *60(4)*, 324-32.

155 25,26-Epoxy-1α-hydroxy-23,23,24,24-tetradehydrovitamin D_3

9547344 $C_{27}H_{38}O_3$

(1R,3S,5Z)-5-[(2E)-2-[(1R,3aR,7aS)-7a-methyl-1-[(2R)-5-(2-methyloxiran-2-yl)pent-4-yn-2-yl]-2,3,3a,5,6,7-hexahydro-1H-inden-4-ylidene]ethylidene]-4-methylidene-cyclohexane-1,3-diol.

LMST03020094; 25,26-epoxy-9,10-seco-5Z,7E,10(19)-cholestatrien-23-yne-1S,3R-diol; 25,26-epoxy-1α-hydroxy-23,23,24,24-tetradehydrocholecalciferol;

9547344; Compared to 1α,25-dihydroxyvitamin D_3, affinity for pig intestinal receptor and human vitamin D binding protein is 30% and 6% respectively, inhibition of proliferation or differentiation induction of human promyeloid leukemia (HL-60), osteosarcoma (MG-63) and breast carcinoma (MCF-7) cells is 240%, 150% and 127%, respectively and elevation of serum calcium, serum osteocalcin, bone calcium and duodenal calbindin in rachitic chicks is 1%, 2%, 0.4% and 2%, respectively. **Steroids.** 1995, **60(4)**, 324-32.

156 **25,26-Epoxy-1α-hydroxy-23,23,24,24-tetradehydro-20-epivitamin D_3**
9547345 $C_{27}H_{38}O_3$

(1R,3S,5Z)-5-[(2E)-2-[(1R,3aR,7aS)-7a-methyl-1-[(2S)-5-(2-methyloxiran-2-yl)pent-4-yn-2-yl]-2,3,3a,5,6,7-hexahydro-1H-inden-4-ylidene]ethylidene]-4-methylidene-cyclohexane-1,3-diol.

LMST03020095; (20S)-25,26-epoxy-9,10-seco-5Z,7E,10(19)-cholestatrien-23-yne-1S,3R-diol; 25,26-epoxy-1α-hydroxy-23,23,24,24-tetradehydro-20-epicholecalciferol; 9547345; Compared to 1α,25-dihydroxyvitamin D_3, affinity for pig intestinal receptor and human vitamin D binding protein is 30% and 0%, respectively, inhibition of proliferation or differentiation induction of human promyeloid leukemia (HL-60), osteosarcoma (MG-63), and breast carcinoma (MCF-7) cells is 2000%, 3000% and 4000%, respectively and elevation of serum calcium, serum osteocalcin, bone calcium, and duodenal calbindin in rachitic chicks is 3%, 1%, 1% and 1%, respectively. **Steroids.** 1995, **60(4)**, 324-32.

157 **1α-Hydroxy-25,27-didehydrovitamin D_3 26,23S-lactone**
9547692 $C_{27}H_{38}O_4$

(5S)-5-[(2R)-2-[(1R,3aR,4E,7aS)-4-[(2Z)-2-[(3S,5R)-3,5-dihydroxy-2-methylidene-cyclohexylidene]ethylidene]-7a-methyl-2,3,3a,5,6,7-hexahydro-1H-inden-1-yl]propyl]-3-methylidene-oxolan-2-one.

LMST03020599; 1α-hydroxy-25,27-didehydrocholecalciferol 26,23S-lactone; 1S,3R-dihydroxy-9,10-seco-5Z,7E,10(19),25(27)-cholestatraeno-26,23S-lactone; 23(25)-dehydro-1,25-dihydroxyvitamin D_3-26,23-lactone; 23(25)-dehydro-1,25-Dihydroxycholecalciferol-26-23-lactone; MK; TEI-9647; 9547692; Binding to the vitamin D receptor and affinity for bovine serum vitamin D binding protein are both approximately 10% that of 1α,25-dihydroxyvitamin D_3. Does not induce differentiation of HL-60 cells. Has antagonistic activity. Inhibits osteoclast formation and bone resorption in bone marrow cultures from patients with Paget's disease. **Mol Endocrinol.** 2000, **14(11)**, 1788-96; **Endocrinology.** 2005, **146(4)**, 2023-30.

158 **1α-Hydroxy-25,27-didehydrovitamin D_3 26,23R-lactone**
9547693 $C_{27}H_{38}O_4$

(5R)-5-[(2R)-2-[(1R,3aR,4E,7aS)-4-[(2Z)-2-[(3S,5R)-3,5-dihydroxy-2-methylidene-cyclohexylidene]ethylidene]-7a-methyl-2,3,3a,5,6,7-hexahydro-1H-inden-1-yl]propyl]-3-methylidene-oxolan-2-one.

LMST03020600; 1α-Hydroxy-25,27-didehydrocholecalciferol 26,23R-lactone 1S,3R-dihydroxy-9,10-seco-5Z,7E,10(19),25(27)-cholestatraeno-26,23R-lactone; 9547693; Synthesized from a CD-ring plus side chain fragment and an enyne A-ring precursor. Compared to 1α,25-dihydroxyvitamin D_3, affinity for chick intestinal receptor is 26%, affinity for bovine serum vitamin D binding protein is 2.5%, HL-60 cell differentiation is 0%, shows antagonistic activity. Vitamin D Chemistry, Biology and Clinical Applications of the Steroid Hormone. (Norman, A. W., Bouillon, R., Thomasset, M., eds), pp79-80, Printing and Reprographics University of California, Riverside

(1997).

159   *9,14,19,19,19-Pentadeutero-1α,25-dihydroxyprevitamin D_3*

9547346 $C_{27}H_{39}D_5O_3$

(1S,5R)-3-[(Z)-2-[(1R,3aR,7aR)-3a,5-dideuterio-1-[(2R)-6-hydroxy-6-methyl-heptan-2-yl]-7a-methyl-2,3,6,7-tetrahydro-1H-inden-4-yl]ethenyl]-2-methyl-cyclohex-2-ene-1,5-diol.
LMST03020097;   9,14,19,19,19-pentadeutero-9,10-seco-5(10),6Z,8-cholestatriene-1S,3R,25-triol; 9,14,19,19,19-pentadeutero-1α,25-dihydroxyprecholecalciferol; 9547346; As active as 1α,25-dihydroxyvitamin D_3 in two nongenomic systems, transcaltachia as measured in the perfused chick kidney intestine and $^{45}Ca^{2+}$ uptake through voltage-gated Ca^{2+} channels in ROS 17/2.8 cells but significantly less active both in binding *in vitro* to the plasma vitamin D-binding protein (7%) and to the chick (10%) and pig (4%) intestinal nuclear vitamin D receptors generating genomic biological responses *in vivo* (induction of plasma levels of osteocalcin, <5%) or in culture cells (inhibition of HL-60 cell differentiation, <5%; inhibition of MG-63 proliferation, <2% ; and induction of osteocalcin, <2%). *J Amer. Chem. Soc.* 1991, *113(18)*, 6958-66; *J Biol Chem.* 1993, *268(19)*, 13811-9.

160   *26,26,26-Trifluoro-1α,25R-dihydroxy-22E,23-didehydrovitamin D_3*

9547347 $C_{27}H_{39}F_3O_3$

(1R,3S,5Z)-5-[(2E)-2-[(1R,3aR,7aS)-7a-methyl-1-[(E,2S,6R)-7,7,7-trifluoro-6-hydroxy-6-methyl-hept-3-en-2-yl]-2,3,3a,5,6,7-hexahydro-1H-inden-4-ylidene]ethylidene]-4-methylidene-cyclohexane-1,3-diol.
LMST03020098; 26,26,26-trifluoro-1α,25R-dihydroxy-22E,23-didehydrocholecalciferol;   26,26,26-trifluoro-9,10-seco-5Z,7E,10(19),22E-cholestatetraene-1S,3R,25R-triol; 9547347; In affinity for chick intestinal

receptor, 57% as active as 1α,25-dihydroxyvitamin D_3, in induction of differentiation of HL-60 cells is 300% as active. *Curr Pharm Design.* 1997, *3*, 99-123.

161   *26,26,26-Trifluoro-1α,25S-dihydroxy-22E,23-didehydrovitamin D_3*

9547348 $C_{27}H_{39}F_3O_3$

(1R,3S,5Z)-5-[(2E)-2-[(1R,3aR,7aS)-7a-methyl-1-[(E,2S,6S)-7,7,7-trifluoro-6-hydroxy-6-methyl-hept-3-en-2-yl]-2,3,3a,5,6,7-hexahydro-1H-inden-4-ylidene]ethylidene]-4-methylidene-cyclohexane-1,3-diol.
LMST03020099; 26,26,26-trifluoro-1α,25S-dihydroxy-22E,23-didehydrocholecalciferol;   26,26,26-trifluoro-9,10-seco-5Z,7E,10(19),22E-cholestatetraene-1S,3R,25S-triol; 9547348; In affinity for chick intestinal receptor, 40% as active as 1α,25-dihydroxyvitamin D_3, in induction of differentiation of HL-60 cells is 4000% as active. *Curr Pharm Design.* 1997, *3*, 99-123; *Endocr Rev.* 1995, *16(2)*, 200-57.

162   *Calicoferol A*

9547698 $C_{27}H_{40}O_2$

(1R,3aS,4S,7aR)-4-[2-(5-hydroxy-2-methyl-phenyl)ethyl]-7a-methyl-1-[(E,2S)-6-methylhept-3-en-2-yl]-2,3,3a,4,6,7-hexahydro-1H-inden-5-one.
Calicoferol A; (8S)-3-hydroxy-9,10-seco-1,3,5(10),22E-cholestatraen-9-one; (22E)-(8S)-3-hydroxy-9,10-seco-1,3,5(10),22-cholestatetraen-9-one;   LMST03020605; 9547698; Isolated from a gorgonian of the genus *Calicogorgia.* Toxic to brine shrimp larvae. *Nat Prod Rep.* 1997, *14(2)*, 259-302; *Chem Lett.* 1991, 427.

163   *1α-Hydroxy-22Z,23-didehydrovitamin D_3*

5283680 $C_{27}H_{40}O_2$

(1R,3S,5Z)-5-[(2Z)-2-[(1R,3aR,7aS)-7a-methyl-1-
[(Z,2S)-6-methylhept-3-en-2-yl]-2,3,3a,7-tetrahydro-
1H-inden-4-ylidene]ethylidene]-4-methylidene-
cyclohexane-1,3-diol.
LMST03020152; 9,10-seco-5Z,7E,10(19),22E-cholesta-
tetraene-1S,3R-diol;   1α-hydroxy-9,11,22Z,23-tetra-
dehydro-vitamin D$_3$; 1α-hydroxy-9,11,22Z,23-tetra-
dehydrochole-calciferol; 5283680; Over two orders of
magnitude less effective than 1α,25-dihydroxyvitamin
D$_3$ in induction of HL-60 human promyelocytes
differentiation.

**164   25-Hydroxy-16,17,23,24-tetradehydro-
vitamin D$_3$**

9547349   C$_{27}$H$_{40}$O$_2$

(E,6R)-6-[(3aR,4E,7aS)-4-[(2Z)-2-[(5S)-5-hydroxy-2-
methylidene-cyclohexylidene]ethylidene]-7a-methyl-
3a,5,6,7-tetrahydro-3H-inden-1-yl]-2-methyl-hept-3-
en-2-ol.
LMST03020100; 9,10-seco-5Z,7E,10(19),16,23E-chol-
estapentaene-3S,25-diol;   25-hydroxy-16,17,23,24-
tetradehydrocholecalciferol; 25-(OH)-16,23E-diene-D$_3$;
9547349; Synthetic vitamin D$_3$ derivative. In inhibition
of proliferation of human keratinocytes in culture,
100% as effective as 1α,25-dihydroxyvitamin D$_3$, in
competitive binding to rat intestinal vitamin D
receptor, 0.12% and in differentiation of human
leukemia cells (HL-60), 0.5%. *J Nutr Biochem*. 1993,
4(1), 49-57; *Bioorg Med Chem*. 1998, *6*, 2051.

**165   25-Hydroxy-16,17,23Z,24-
tetradehydrovitamin D$_3$**

9547350   C$_{27}$H$_{40}$O$_2$

(Z,6R)-6-[(3aR,4E,7aS)-4-[(2Z)-2-[(5S)-5-hydroxy-2-
methylidene-cyclohexylidene]ethylidene]-7a-methyl-
3a,5,6,7-tetrahydro-3H-inden-1-yl]-2-methyl-hept-3-
en-2-ol.
LMST03020101; 9,10-seco-5Z,7E,10(19),16,23Z-chol-
estapentaene-3S,25-diol;   25-hydroxy-16,17,23Z,24-
tetradehydrocholecalciferol; 9547350; Affinity for rat
intestinal receptor is 0.8% that of 1α,25-di-
hydroxyvitamin D$_3$. Vitamin D Gene Regulation
Structure-Function Analysis and Clinical Application
(Norman, A.W., Bouillon R. and Thomasset, M., eds),
pp 210-213, Walter de Gruyter, Berlin/New York
(1991).

**166   25-Hydroxy-23,23,24,24-tetradehydro-
vitamin D$_3$**

9547351   C$_{27}$H$_{40}$O$_2$

(6R)-6-[(1R,3aR,4E,7aS)-4-[(2Z)-2-[(5S)-5-hydroxy-2-
methylidene-cyclohexylidene]ethylidene]-7a-methyl-
2,3,3a,5,6,7-hexahydro-1H-inden-1-yl]-2-methyl-hept-
3-yn-2-ol.
LMST03020102; 9,10-seco-5Z,7E,10(19)-cholestatrien-
23-yne-3S,25-diol;   25-hydroxy-23,23,24,24-tetra-
dehydrocholecalciferol; 9547351; Compared to 1α,25-
dihydroxyvitamin D$_3$, inhibition of proliferation of
human keratinocytes in culture is 100%, intestinal
calcium absorption in rat is 71%, bone calcium
mobilization in rat is 0%, competitive binding to rat
intestinal vitamin D receptor is 0.06% and
differentiation of human leukemia cells (HL-60) is 1%.
J Nutr Biochem. 1993, *4(1)*, 49-57.

167   Calcipotriol

5288783   C$_{27}$H$_{40}$O$_3$

(1R,3S,5Z)-5-[(2E)-2-[(1R,3aR,7aS)-1-[(E,2S,5S)-5-cyclopropyl-5-hydroxy-pent-3-en-2-yl]-7a-methyl-2,3,3a,5,6,7-hexahydro-1H-inden-4-ylidene]-ethylidene]-4-methylidene-cyclohexane-1,3-diol.
Calcipotriol;1α,24S-(OH)2-22-ene-26,27-dehydro-vitamin D_3; 17α-24-cyclopropyl-9,10-seco-chola-5Z,7E,10,22E-tetraene-1S,3,24S-triol; Dovobet; Dovonex; Daivobet; MC9; MC903; LMST03020106; 5288783; Has decreased affinity for vitamin D binding protein and is metabolized more rapidly than vitamin D_3, thus has reduced calcemic effects. Used in treatment of plaque-type psoriasis, sometimes in combination with betamethasone dipropionate (Dovobet, Daivobet). *Ann Dermatol Venereol*. 2001, *128(3 Pt 1)*, 229-37; *Am J Clin Dermatol*. 2001, *2(2)*, 95-120. *Leo AB*.

168   24-Epicalcipotriene
112828-00-9 $C_{27}H_{40}O_3$

(1R,3S,5Z)-5-[(2E)-2-[1-[(E,2S,5R)-5-cyclopropyl-5-hydroxy-pent-3-en-2-yl]-7a-methyl-2,3,3a,5,6,7-hexahydro-1H-inden-4-ylidene]ethylidene]-4-methylidene-cyclohexane-1,3-diol.
Calcipotriene; Epicalcipotriene; 17α-24-cyclopropyl-9,10-seco-chola-5Z,7E,10,22E-tetraene-1S,3,24R-triol; LMST03020105; 2522; 6435783; 112828-00-9; Primes leukemia cells for monocytic differentiation. Used alone or in combination with betamethasone to treat psoriasis. *Endocrinology*. 1999, *140(10)*, 4779-88.

169   *Calcipotriene*
112965-21-6 $C_{27}H_{40}O_3$

(1R,3S,5Z)-5-[(2E)-2-[(1R,7aS)-1-[(E,2S,5S)-5-cyclopropyl-5-hydroxy-pent-3-en-2-yl]-7a-methyl-2,3,3a,5,6,7-hexahydro-1H-inden-4-ylidene]-ethylidene]-4-methylidene-cyclohexane-1,3-diol.
24-Cyclopropyl-9,10-secochola-5Z,7E,10(19),22E-tetraene-1α,3β,24S-triol; CCRIS 7700; Calcipotriol; Dovonex; MC 903; Psorcutan; 9,10-Secochola-5Z,7E,10(19),22E-tetraene-1α,3β,24-triol, 24S-cyclopropyl-; Calcipotriol; 6436131; 5282134; 112965-21-6; Has antineoplastic and antipsoriatic properties. Used to treat psoriasis. Has significant antileukemic activity. Compared to 1,25-dihydroxy vitamin D_3, it is 4-12 times more potent in inducing cell differentiation, but 30 times less active in intestinal calcium absorption and 50 times less potent as a bone calcium mobilising agent. *Anticancer Drugs*. 2007, *18(4)*, 447-57; *Drugs Exp Clin Res*. 2005, *31(5-6)*, 175-9; *J Bone Miner Res*. 1991, *6(12)*, 1307-15; *Br J Dermatol*. 2002, *147(6)*, 1270. USP 4,866,048. *Leo AB, Warner-Chilcott Inc*.

170   1α,24R-Dihydroxy-26,27-cyclo-22E,23-didehydrovitamin D_3
9547353 $C_{27}H_{40}O_3$

(1R,3S,5Z)-5-[(2E)-2-[(1R,3aR,7aS)-1-[(E,2S,5R)-5-cyclopropyl-5-hydroxy-pent-3-en-2-yl]-7a-methyl-2,3,3a,5,6,7-hexahydro-1H-inden-4-ylidene]ethylidene]-4-methylidene-cyclohexane-1,3-diol.
LMST03020105; 1α,24R-dihydroxy-26,27-cyclo-22E,23-didehydrovitamin D_3; 1α,24R-dihydroxy-26,27-cyclo-22E,23-didehydrocholecalciferol; 26,27-cyclo-9,10-seco-5Z,7E,10(19),22E-cholestatetraene-1S,3R,24R-triol; 9547353; Solid; mp = 143-145°. *Tetrahedron*. 1987, *43(20)*, 4609-19.

171 **1α,25-Dihydroxy-22,23-diene vitamin D₃**

$C_{27}H_{40}O_3$

1,25(OH)2-22,23-diene-D3; HQ; LMST03020112; Has a low affinity for the vitamin D receptor and fails to induce cell differentiation. *Mol Endocrinol*. 2000, *14(11)*, 1788-96.

172 **1α-Hydroxy-24-oxo-26,27-cyclovitamin D₃**

9547352

$C_{27}H_{40}O_3$

(4R)-4-[(1R,3aR,4E,7aS)-4-[(2Z)-2-[(3S,5R)-3,5-di-hydroxy-2-methylidene-cyclohexylidene]ethylidene]-7a-methyl-2,3,3a,5,6,7-hexahydro-1H-inden-1-yl]-1-cyclopropyl-pentan-1-one.
LMST03020103;   1S,3R-dihydroxy-26,27-cyclo-9,10-seco-5Z,7E,10(19)-cholestatrien-24-one;   1α-hydroxy-24-oxo-26,27-cyclocholecalciferol;   9547352; Produced as a metabolite of 1α,24S-dihydroxy-26,27-cyclo-22,23-didehydrovitamin D₃ (Calcipotriol) and further metabolized to the 23-hydroxylated compound. Affinity for calf thymus receptor is 0.5% that of 1α,25-dihydroxyvitamin D₃, transcriptional activity is 12.5%. *J Biol Chem*. 1994, *269(7)*, 4794-803.

173 **25-Hydroxy-24-oxo-22E,23-didehydro-vitamin D₃**

5283674

$C_{27}H_{40}O_3$

(E,6S)-6-[(1R,3aR,4E,7aS)-4-[(2Z)-2-[(5S)-5-hydroxy-2-methylidene-cyclohexylidene]ethylidene]-7a-methyl-2,3,3a,5,6,7-hexahydro-1H-inden-1-yl]-2-hydroxy-2-methyl-hept-4-en-3-one.
LMST03020104;   3S,25-dihydroxy-9,10-seco-5Z,7E, 10(19),22E-cholestatraen-24-one;   25-hydroxy-24-oxo-22E,23-didehydrocholecalciferol;   5283674; Synthesized from ergosterol *via* introduction of a side chain to the 5,7-diene protected C(22)-aldehyde. ; λ_m = 265 nm (EtOH). *Chem Pharm Bull (Tokyo)*. 1987, *35(3)*, 970-9.

174 **1α,24R-Dihydroxy-26,27-cyclo-22E,23-didehydro-20-epivitamin D₃**

9547354

$C_{27}H_{40}O_3$

(1R,3S,5Z)-5-[(2E)-2-[(1R,3aR,7aS)-1-[(E,2R,5R)-5-cyclopropyl-5-hydroxy-pent-3-en-2-yl]-7a-methyl-2,3,3a,5,6,7-hexahydro-1H-inden-4-ylidene] ethylidene]-4-methylidene-cyclohexane-1,3-diol.
LMST03020107;   1α,24R-dihydroxy-26,27-cyclo-22E, 23-didehydro-20-epicholecalciferol;   (20S)-26,27-cyclo-9,10-seco-5Z,7E,10(19),22-cholestatraene-1S, 3R,24R-triol; 9547354; Inhibition of proliferation of U937 cells, 130% that of 1α,25-dihydroxyvitamin D₃, induction of differentiation of U937 cells is 100% and calciuric effects on normal rats is <1%. Gene Regulation, Structure-Function Analysis and Clinical Application. Proceedings of the Eighth Workshop on Vitamin D Paris, France July 5-10. The 20-Epi Modification in The Vitamin D Series: Selective Enhancement of 'Non-Classical' Receptor-Mediated Effects. (Norman, A. W., Bouillon, R., Thomasset, M., eds), pp163-164, Walter de Gruyter Berlin New York (1991).

175 **1α,24S-Dihydroxy-26,27-cyclo-22E,23-didehydro-20-epivitamin D₃**

9547355 $C_{27}H_{40}O_3$

(1R,3S,5Z)-5-[(2E)-2-[(1R,3aR,7aS)-1-[(E,2R,5S)-5-
cyclopropyl-5-hydroxy-pent-3-en-2-yl]-7a-methyl-
2,3,3a,5,6,7-hexahydro-1H-inden-4-ylidene]
ethylidene]-4-methylidene-cyclohexane-1,3-diol.
LMST03020108;   1α,24S-dihydroxy-26,27-cyclo-22E,
23-didehydro-20-epicholecalciferol;   (20S)-26,27-
cyclo-9,10-seco-5Z,7E,10(19),22E-cholestatetraene-
1S,3R,24S-triol; 9547355; In inhibition of proliferation
of U937 cells is 60% as effective as 1α,25-
dihydroxyvitamin D$_3$, in induction of differentiation of
U937 cells, 10% and calciuric effects on normal rats,
<1%. Gene Regulation, Structure-Function Analysis
and Clinical Application. Proceedings of the Eighth
Workshop on Vitamin D Paris, France July 5-10. The
20-Epi Modification in The Vitamin D Series : Selective
Enhancement of 'Non-Classical' Receptor-Mediated
Effects. (Norman, A. W., Bouillon, R., Thomasset, M.,
eds), pp163-164, Walter de Gruyter Berlin New York
(1991).

176   *1α,25-Dihydroxy-16,17,23E,24-tetra-
dehydrovitamin D$_3$*

9547356 $C_{27}H_{40}O_3$

(1R,3S,5Z)-5-[(2E)-2-[(3aR,7aS)-1-[(E,2R)-6-hydroxy-6-
methyl-hept-4-en-2-yl]-7a-methyl-3a,5,6,7-tetrahydro-
3H-inden-4-ylidene]ethylidene]-4-methylidene-
cyclohexane-1,3-diol.
LMST03020109; 9,10-seco-5Z,7E,10(19),16,23E-chol-
estapentaene-1S,3R,25-triol;   1α,25-dihydroxy-16,17,
23E,24-tetradehydrocholecalciferol;   1,25-(OH)2-16,
23E-diene-D$_3$; 9547356; Synthetic vitamin D$_3$
derivative. In inhibition of proliferation of human
keratinocytes in culture is 100% that of 1α,25-
dihydroxyvitamin D$_3$, competitive binding to rat
intestinal vitamin D receptor is 80% and differentiation

of human leukemia cells (HL-60) is 300%. *J Nutr
Biochem.* 1993, *4(1)*, 49-57; *Bioorg Med Chem.* 1998,
6(11), 2051-9.

177   *1α,25-Dihydroxy-16,17,23Z,24-tetra-
dehydrovitamin D$_3$*

9547357 $C_{27}H_{40}O_3$

(1R,3S,5Z)-5-[(2E)-2-[(3aR,7aS)-1-[(Z,2R)-6-hydroxy-6-
methyl-hept-4-en-2-yl]-7a-methyl-3a,5,6,7-tetrahydro-
3H-inden-4-ylidene]ethylidene]-4-methylidene-
cyclohexane-1,3-diol.
LMST03020110; 9,10-seco-5Z,7E,10(19),16,23Z-chol-
estapentaene-1S,3R,25-triol;   1α,25-dihydroxy-16,17,
23Z,24-tetradehydrocholecalciferol; 9547357; Affinity
for rat intestinal receptor is 145% that of 1α,25-
dihydroxyvitamin D$_3$, affinity for human vitamin D
binding protein is 17% and HL-60 cell differentiation is
100%. *Endocr Rev.* 1995, *16(2)*, 200-57.

178   *1α,25-Dihydroxy-22,23RS,23RS,24-
tetradehydrovitamin D$_3$*

9547358 $C_{27}H_{40}O_3$

(1R,3S,5Z)-5-[(2E)-2-[(1R,3aR,7aS)-1-[(2S)-6-hydroxy-6-
methyl-hepta-3,4-dien-2-yl]-7a-methyl-2,3,3a,5,6,7-
hexahydro-1H-inden-4-ylidene]ethylidene]-4-methyl-
idene-cyclohexane-1,3-diol.
LMST03020111; 9,10-seco-5Z,7E, 10(19),22,23R-chol-
estapentaene-1S,3R,25-triol;   9,10-seco-5Z,7E,10(19),
22,23S-cholestapentaene-1S,3R,25-triol;   1α,25-di-
hydroxy-22,23,23R,24-tetradehydro-cholecalciferol;
1α,25-dihydroxy-22,23,23S,24-tetra-dehydrovitamin
D$_3$; 1α,25-dihydroxy-22,23,23S,24-tetradehydrochole-
calciferol; 3947358; Affinity for chick intestinal
receptor is 52% that of 1α,25-dihydroxyvitamin D$_3$,
affinity for human vitamin D binding protein is 48%.

Endocr Rev. 1995, *16(2)*, 200-57.

179   *1α,25-Dihydroxy-23,23,24,24-tetra-dehydrovitamin D_3*

9547359 $C_{27}H_{40}O_3$

(1R,3S,5Z)-5-[(2E)-2-[(1R,3aR,7aS)-1-[(2R)-6-hydroxy-6-methyl-hept-4-yn-2-yl]-7a-methyl-2,3,3a,5,6,7-hexahydro-1H-inden-4-ylidene]ethylidene]-4-methylidene-cyclohexane-1,3-diol.
LMST03020113; 9,10-seco-5Z,7E,10(19)-cholestatrien-23-yne-1S,3R,25-triol; 1α,25-dihydroxy-23,23,24,24-tetradehydrocholecalciferol; 9547359; Compared to 1α,25-dihydroxyvitamin D_3, inhibition of proliferation of human keratinocytes in culture is 100%, intestinal calcium absorption in rat is 95%, bone calcium mobilization in rat is 15%, competitive binding to rat intestinal vitamin D receptor is 39% and differentiation of human leukemia cells (HL-60) is 300%. *J Nutr Biochem.* 1993, *4(1)*, 49-57.

180   *1α,26a-Dihydroxy-22E,23,24E,25-tetradehydro-26a-homo-27-norvitamin D_3*

9547360 $C_{27}H_{40}O_3$

(1R,3S,5Z)-5-[(2E)-2-[(1R,3aR,7aS)-1-[(2S,3E,5E)-8-hydroxyocta-3,5-dien-2-yl]-7a-methyl-2,3,3a,5,6,7-hexahydro-1H-inden-4-ylidene]ethylidene]-4-methylidene-cyclohexane-1,3-diol.
LMST03020114;   1α,26a-dihydroxy-22E,23,24E,25-tetradehydro-26a-homo-27-norcholecalciferol; 26a-homo-27-nor-9,10-seco-5Z,7E,10(19),22E,24E-cholestapentaene-1S,3R,26a-triol; 9547360; Compared to 1α,25-dihydroxyvitamin D_3, inhibition of proliferation of U937 cells is 300%, induction of differentiation of U937 cells is 200% and calciuric activity is 1%. Gene Regulation, Structure-Function Analysis and Clinical Application. Proceedings of the Eighth Workshop on Vitamin D Paris, France July 5-10. Synthesis and

Biological Activity of 1α-hydroxylated Vitamin D Analogues with Poly-unsaturated Side Chains. (Norman, A. W., Bouillon, R., Thomasset, M., eds), pp192-193, Walter de Gruyter Berlin New York (1991).

181   *Calcifediol lactone*

71302-34-6 $C_{27}H_{40}O_4$

5-[(2R)-2-[(1R,3aR,4E,7aS)-4-[(2Z)-2-[(5S)-5-hydroxy-2-methylidene-cyclohexylidene]ethylidene]-7a-methyl-2,3,3a,5,6,7-hexahydro-1H-inden-1-yl]propyl]-3-hydroxy-3-methyl-oxolan-2-one.
Calcifediol lactone; 25-Hydroxy vitamin D_3-26,23-lactone; 25-Hydroxycholecalciferol-26,23-lactone; 25-Hydroxyvitamin D_3 26,23-lactone; 25-Ohd(3)-26,23-lactone; 9,10-Secocholesta-5Z,7E,10(19)-trien-26-oic acid, 3β,23,25-trihydroxy-, γ-lactone; LMST03020115; 6438386; 71302-34-6; Metabolite of 25-hydroxy-vitamin D_3 and 23S,25-dihydroxyvitamin D_3 *via* the 26,23 lactone pathway. *J Biol Chem.* 1984, *259(2)*, 884-9; *Arch Biochem Biophys.* 1984, *228(1)*, 179-84.

182   *1,25-Dihydroxy-24-oxo-16-ene-vitamin D_3*

                                           $C_{27}H_{40}O_4$

9,10-Secocholesta-5Z,7E,10(19),16-tetraene-24-one, 1α,3β,25-trihydroxy.
1,25-dihydroxy-16-ene-24-oxovitamin D_3; 1,25-dihydroxy-16-enecholecalciferol; A metabolite of a synthetic vitamin D_3 analog, 1α,25-dihydroxy-16-ene vitamin D_3, it is equipotent to its parent in modulating growth and differentiation of human leukemic cells but does not cause hypercalcemia. *J Steroid Biochem Mol Biol.* 1996, *59(5-6)*, 405-12; *Endocrinology.* 1994, *135(6)*, 2818-21.

183   **25R-Hydroxyvitamin D₃ 26,23R-lactone**
9547361 $C_{27}H_{40}O_4$

(3R,5R)-5-[(2R)-2-[(1R,3aR,4E,7aS)-4-[(2Z)-2-[(5S)-5-hydroxy-2-methylidene-cyclohexylidene]ethylidene]-7a-methyl-2,3,3a,5,6,7-hexahydro-1H-inden-1-yl]propyl]-3-hydroxy-3-methyl-oxolan-2-one. LMST03020115; 25R-hydroxycholecalciferol 26,23R-lactone; 3S,25R-dihydroxy-9,10-seco-5Z,7E,10(19)-cholestatrieno-26,23R-lactone; 9547361; Isomer of calcifediol lactone; λ_m = 265 nm. *Chem Pharm Bull (Tokyo)*. 1980, **28**, 2852; 1981, **29**, 2393; *J Org Chem*. 1981, **46(17)**, 3422-8.

184   **25R-Hydroxyvitamin D₃ 26,23S-lactone**
9547362 $C_{27}H_{40}O_4$

(3R,5S)-5-[(2R)-2-[(1R,3aR,4E,7aS)-4-[(2Z)-2-[(5S)-5-hydroxy-2-methylidene-cyclohexylidene]ethylidene]-7a-methyl-2,3,3a,5,6,7-hexahydro-1H-inden-1-yl]-propyl]-3-hydroxy-3-methyl-oxolan-2-one. LMST03020116; 25R-hydroxycholecalciferol 26,23S-lactone; 3S,25R-dihydroxy-9,10-seco-5Z,7E,10(19)-cholestatrieno-26,23S-lactone; 9547362; Isolated and identified from blood plasma of chickens given either maintenance levels or large doses of vitamin D₃. All four possible stereoisomers of 25-hydroxyvitamin D₃ 26,23-lactone were synthesized from norcholadienoic acid. Effect on intestinal calcium transport and bone calcium mobilizing activity of 1α,25-dihydroxyvitamin D₃ in vitamin D-deficient rats reported; λ_m = 264 nm. *Tet Lett*. 1981, **21(49)**, 4667-70; *Biochemistry*. 1979, **22(20)**, 4775-81; *FEBS Lett*. 1982, **139(2)**, 267-70.

185   **25S-Hydroxyvitamin D₃ 26,23R-lactone**
9547363 $C_{27}H_{40}O_4$

(3S,5R)-5-[(2R)-2-[(1R,3aR,4E,7aS)-4-[(2Z)-2-[(5S)-5-hydroxy-2-methylidene-cyclohexylidene]ethylidene]-7a-methyl-2,3,3a,5,6,7-hexahydro-1H-inden-1-yl]propyl]-3-hydroxy-3-methyl-oxolan-2-one. LMST03020117; 25S-hydroxycholecalciferol 26,23R-lactone; 3S,25S-dihydroxy-9,10-seco-5Z,7E,10(19)-cholestatrieno-26,23R-lactone; 9547363; λ_m = 265 nm. *J Org Chem*. 1981, **46(17)**, 3422-8; *Tet Lett*. 1981, **22(27)**, 2591-4.

186   **25S-Hydroxyvitamin D₃ 26,23S-lactone**
9547364 $C_{27}H_{40}O_4$

(3S,5S)-5-[(2R)-2-[(1R,3aR,4E,7aS)-4-[(2Z)-2-[(5S)-5-hydroxy-2-methylidene-cyclohexylidene]ethylidene]-7a-methyl-2,3,3a,5,6,7-hexahydro-1H-inden-1-yl]propyl]-3-hydroxy-3-methyl-oxolan-2-one. LMST03020118; 25S-hydroxycholecalciferol 26,23S-lactone; 3S,25S-dihydroxy-9,10-seco-5Z,7E,10(19)-cholestatrieno-26,23S-lactone; 9547364; Synthesized from C(22) steroid aldehyde *via* iodolactonization of the 22-ene-26-carboxylic acid; λ_m = 265 nm. *J Org Chem*. 1981, **46(17)**, 3422-8.

187   **25R,26-Epoxy-1α,24R-dihydroxy-22E,23-didehydrovitamin D₃**
9547365 $C_{27}H_{40}O_4$

(1R,3S,5Z)-5-[(2E)-2-[(1R,3aR,7aS)-1-[(E,2S,5R)-5-hydroxy-5-[(2R)-2-methyloxiran-2-yl]pent-3-en-2-yl]-7a-methyl-2,3,3a,5,6,7-hexahydro-1H-inden-4-ylidene]ethylidene]-4-methylidene-cyclohexane-1,3-diol. LMST03020119;   25R,26-epoxy-1α,24R-dihydroxy-22E,23-didehydrocholecalciferol;   (5Z,7E,22E)-(1S,3R,24R,25R)-25R,26-epoxy-9,10-seco-5Z,7E,10(19),22E-cholestatetraene-1S,3R,24R-triol; 9547365; Compared to 1α,25-dihydroxyvitamin D₃, affinity for pig intestinal receptor and human vitamin D binding protein is 8% and 2% respectively, inhibition of proliferation or differentiation induction of human promyeloid leukemia (HL-60), osteosarcoma (MG-63) and breast carcinoma (MCF-7) cells is 30%, 140% and 22%, respectively and elevation of serum calcium, serum osteocalcin, bone calcium and duodenal calbindin in rachitic chicks is 1%, 3%, 2% and 1%, respectively. *Bioorg Med Chem Lett*. 1993, *3(9)*, 1863-7; *Steroids*. 1995, *60(4)*, 324-32.

188   **25S,26-Epoxy-1α,24S-dihydroxy-22E,23-didehydrovitamin D₃**
9547366 $C_{27}H_{40}O_4$

(1R,3S,5Z)-5-[(2E)-2-[(1R,3aR,7aS)-1-[(E,2S,5S)-5-hydroxy-5-[(2S)-2-methyloxiran-2-yl]pent-3-en-2-yl]-7a-methyl-2,3,3a,5,6,7-hexahydro-1H-inden-4-ylidene] ethylidene]-4-methylidene-cyclohexane-1,3-diol. LMST03020120; 25S,26-epoxy-1α,24S-dihydroxy-22E,23-didehydrovitamin D₃; 25S,26-epoxy-1α,24S-dihydroxy-22E,23-didehydrocholecalciferol; 25S,26-epoxy-9,10-seco-5Z,7E,10(19),22E-cholestatetraene-1S,3R,24S-triol; 9547366; Relative to 1α,25-dihydroxyvitamin D₃, affinity for pig intestinal receptor and human vitamin D binding protein is 8% and 3% respectively. Inhibition of proliferation or differentiation induction of human promyeloid leukemia (HL-60), osteosarcoma (MG-63) and breast

carcinoma (MCF-7) cells is 15%, 100% and 65%, respectively, and elevation of serum calcium, serum osteocalcin, bone calcium and duodenal calbindin in rachitic chicks are all 1%. *Bioorg Med Chem Lett*. 1993, *3(9)*, 1863-7; *Steroids*. 1995, *60(4)*, 324-32.

189   **25R-Hydroxyvitamin D₃ 26,23S-peroxylactone**
84070-69-9 $C_{27}H_{40}O_5$

(4R,6S)-6-[(2R)-2-[(1R,3aR,4E,7aS)-4-[(2Z)-2-[(5S)-5-hydroxy-2-methylidene-cyclohexylidene]ethylidene]-7a-methyl-2,3,3a,5,6,7-hexahydro-1H-inden-1-yl]propyl]-4-hydroxy-4-methyl-dioxan-3-one. LMST03020121; 25R-hydroxycholecalciferol 26,23S-peroxylactone; 3S,25R-dihydroxy-9,10-seco-5Z,7E,10(19)-cholestatrieno-26,23S-peroxylactone; 25(OH)D₃-Pol; 25-Hydroxyvitamin D₃-26,23-peroxylactone; 25-Hydroxycholecalciferol 23,26-peroxylactone; 1,2-Dioxan-3-one, 4-hydroxy-6-(2-(4-(2-(5-hydroxy-2-methylenecyclohexylidene)ethylidene)octahydro-7a-methyl-1H-inden-1-yl)propyl)-4-methyl-; 9547367; 84070-69-9; An *in vivo* metabolite of vitamin D₃, isolated from rats given large doses of vitamin D₃ and identified as the 23S, 25R isomer by UV, IR, EI and FD mass spectroscopy; λ_m = 264 nm. *J Biol Chem*. 1982, *257(24)*, 14708-13.

190   **1α,25R-Dihydroxyvitamin D₃ 26,23R-lactone**
9547368 $C_{27}H_{40}O_5$

(3R,5R)-5-[(2R)-2-[(1R,3aR,4E,7aS)-4-[(2Z)-2-[(3S,5R)-3,5-dihydroxy-2-methylidene-cyclohexylidene]ethylidene]-7a-methyl-2,3,3a,5,6,7-hexahydro-1H-inden-1-yl]propyl]-3-hydroxy-3-methyl-oxolan-2-one. LMST03020122; (23R,25R)-1α,25R-dihydroxycholecalciferol 26,23R-lactone; 1S,3R,25R-trihydroxy-9,10-seco-5Z,7E,10(19)-cholestatrieno-26,23R-lactone;

9547368; All 4 (23,25) diastereoisomers synthesized. Compared to 1α,25-dihydroxyvitamin D_3, affinity for chick intestinal receptor: (23S,25S)-isomer is 8%; (23R,25R)-isomer, 2% ; (23R,25S)-isomer, 0.2% ; (23S,25R)-isomer, 0.15%. Time course of bone calcium mobilization induced in rats and dose-response of 1,25-(OH)2D3 26,23-lactone diastereomers in intestinal calcium transport and bone calcium mobilization reported. *Arch Biochem Biophys*. 1985, *242(1)*, 82-89; *J Steroid Biochem*. 1986, *25(4)*, 505-10.

191   *1,25-Dihydroxyvitamin D_3-26,23-lactone*

81203-50-1 $C_{27}H_{40}O_5$

(3R,5S)-5-[(2R)-2-[(1R,3aR,4E,7aS)-4-[(2Z)-2-[(3S,5R)-3,5-dihydroxy-2-methylidene-cyclohexylidene]ethylidene]-7a-methyl-2,3,3a,5,6,7-hexahydro-1H-inden-1-yl]propyl]-3-hydroxy-3-methyl-oxolan-2-one. LMST03020123; 1,25-Lactone; Calcitriol-26,23-lactone; 1,25-Dihydroxycholecalciferol-26-23-lactone; 1α,25-Dihydroxyvitamin D3-26,23-lactone; 1,25R (OH)2D3-26,23S-Lactone; 1α,25R-dihydroxyvitamin D3 26,23S-lactone; 1α,25-dihydroxycholecalciferol 26,23-lactone; 1S,3R,25R-trihydroxy-9,10-seco-5Z,7E, 10(19)-cholestatrieno-26,23S-lactone;   6438368; 81203-50-1; Synthesized from 1α-hydroxydehydroepiandrosterone. Metabolite of 1α,25-dihydroxyvitamin D_3. Binds strongly to the vitamin D receptor but does not induce cell differentiation. Increases intestinal calcium transport but decreases serum calcium level in vitamin D-deficient rats fed a low-calcium diet. In terms of affinity for the chick intestinal receptor it is 1/662 as active as 1α,25-dihydroxyvitamin D_3. In vitamin D deficient rats on a low-calcium diet, it has a weak potency to increase intestinal calcium transport but it decreases the serum calcium level: a 500 ng dose decreases the serum calcium level by about 20% at 24 h. The ability of the four 23,25 diastereomers to bind to chick intestinal receptor was compared to the affinity of 1α,25-dihydroxyvitamin D_3. The affinity relative to 1,25-(OH)2D3 is (23S,25S)-isomer, 7.9%; (23R,25R)-isomer, 2.3%; (23R,25S)-isomer, 0.22%; (23S,25R)-isomer, 0.17%; $[\alpha]_D^{25}$ = +21.3° (c = 0.47 Et_2O); λ_m = 264 nm (95% EtOH). *J Org Chem*. 1992, *57(11)*, 3214-7; *Endocrinology*. 1990, *127(2)*, 695-701; *Arch Biochem Biophys*. 1985, *242(1)*, 82-9; *Biochemistry*. 1984, *23(7)*, 1473-8; *Mol Endrocrin*. 2000, *14(11)*, 1788-96.

192   *1α,25S-Dihydroxyvitamin D_3 26,23R-lactone*

9547369 $C_{27}H_{40}O_5$

(3S,5R)-5-[(2R)-2-[(1R,3aR,4E,7aS)-4-[(2Z)-2-[(3S,5R)-3,5-dihydroxy-2-methylidene-cyclohexylidene]ethylidene]-7a-methyl-2,3,3a,5,6,7-hexahydro-1H-inden-1-yl]propyl]-3-hydroxy-3-methyl-oxolan-2-one. LMST03020124; 1α,25-dihydroxycholecalciferol 26, 23R-lactone;   1S,3R,25S-trihydroxy-9,10-seco-5Z,7E, 10(19)-cholestatrieno-26,23R-lactone; 9547369; All 4 (23,25) diastereoisomers synthesized. Compared to 1α,25-dihydroxyvitamin D_3, affinity for chick intestinal receptor: (23S,25S)-isomer is 8%; (23R,25R)-isomer, 2%; (23R,25S)-isomer, 0.2%; (23S,25R)-isomer, 0.15%. Time course of bone calcium mobilization induced in rats and dose-response of 1,25-(OH)2D3 26,23-lactone diastereomers in intestinal calcium transport and bone calcium mobilization reported. *Arch Biochem Biophys*. 1985, *242(1)*, 82-89; *J Steroid Biochem*. 1986, *25(4)*, 505-10.

193   *1α,25S-Dihydroxyvitamin D_3 26,23S-lactone*

9547370 $C_{27}H_{40}O_5$

(3S,5S)-5-[(2R)-2-[(1R,3aR,4E,7aS)-4-[(2Z)-2-[(3S,5R)-3,5-dihydroxy-2-methylidene-cyclohexylidene]ethylidene]-7a-methyl-2,3,3a,5,6,7-hexahydro-1H-inden-1-yl]propyl]-3-hydroxy-3-methyl-oxolan-2-one. LMST03020125; 1α,25-dihydroxycholecalciferol 26, 23S-lactone; 1S,3R,25S-trihydroxy-9,10-seco-5Z,7E,10 (19)-cholestatrieno-26,23S-lactone; 9547370; All 4 (23,25) diastereoisomers synthesized. Compared to 1α,25-dihydroxyvitamin D_3, affinity for chick intestinal receptor: (23S,25S)-isomer is 8%; (23R,25R)-isomer, 2%; (23R,25S)-isomer, 0.2%; (23S,25R)-isomer, 0.15%. Time course of bone calcium mobilization induced in rats and dose-response of 1,25-(OH)2D3 26,23-lactone diastereomers in intestinal calcium transport and bone calcium mobilization reported. *Arch Biochem Biophys*. 1985, *242(1)*, 82-89; *J Steroid Biochem*. 1986, *25(4)*,

505-10.

194   *26,26,26-Trifluoro-25-hydroxyvitamin D₃*

9547376 $C_{27}H_{41}F_3O_2$

(6R)-6-[(1R,3aR,4E,7aS)-4-[(2Z)-2-[(5S)-5-hydroxy-2-methylidene-cyclohexylidene]ethylidene]-7a-methyl-2,3,3a,5,6,7-hexahydro-1H-inden-1-yl]-1,1,1-trifluoro-2-methyl-heptan-2-ol.
LMST03020131; 26,26,26-trifluoro-9,10-seco-5Z,7E,0(19)-cholestatriene-3S,25-diol; 26,26,26-trifluoro-25-hydroxycholecalciferol; 9547376; Prepared from 24-phenylsulfonyl-5-cholen-3-ol *via* reaction with trifluoroacetate; λ_m = 265 nm. *Tet Lett.* 1981, *22(39)*, 3085-8.

195   *26,26,26-Trifluoro-1α,25R-dihydroxyvitamin D₃*

9547377 $C_{27}H_{41}F_3O_3$

(1R,3S,5Z)-5-[(2E)-2-[(1R,3aR,7aS)-7a-methyl-1-[(2R,6R)-7,7,7-trifluoro-6-hydroxy-6-methyl-heptan-2-yl]-2,3,3a,5,6,7-hexahydro-1H-inden-4-ylidene]thylidene]-4-methylidene-cyclohexane-1,3-diol.
LMST03020132; 26,26,26-trifluoro-1α,25R-dihydroxy-holecalciferol; 26,26,26-trifluoro-9,10-seco-5Z,7E,10 19)-cholestatriene-1S,3R,25R-triol; 9547377; Three times as active as 1α,25-dihydroxyvitamin D₃ in differentiation of HL-60 cells and twice the calcemic activity. *Curr Pharm Design.* 1997, *3*, 99-123.

196   *26,26,26-Trifluoro-1α,25S-dihydroxy-vitamin D₃*

9547378 $C_{27}H_{41}F_3O_3$

(1R,3S,5Z)-5-[(2E)-2-[(1R,3aR,7aS)-7a-methyl-1-[(2R,6S)-7,7,7-trifluoro-6-hydroxy-6-methyl-heptan-2-yl]-2,3,3a,5,6,7-hexahydro-1H-inden-4-ylidene]-ethylidene]-4-methylidene-cyclohexane-1,3-diol.
LMST03020133; 26,26,26-trifluoro-1α,25S-dihydroxy-cholecalciferol; 26,26,26-trifluoro-9,10-seco-5Z,7E,10(19)-cholestatriene-1S,3R,25S-triol; 9547378; Compared to 1α,25-dihydroxyvitamin D₃, inhibition of proliferation of human keratinocytes in culture is 100%, intestinal calcium absorption in rat is 197%, bone calcium mobilization in rat is 163%, competitive binding to rat intestinal vitamin D receptor is 199% and differentiation of human leukemia cells (HL-60) is 300%. *J Nutr Biochem.* 1993, *4(1)*, 49-57.

197   *20R-Fluoro-1α,24R-dihydroxy-26,27-cyclovitamin D₃*

9547372 $C_{27}H_{41}FO_3$

(1R,3S,5Z)-5-[(2E)-2-[(1S,3aR,7aR)-1-[(2R,5R)-5-cyclopropyl-2-fluoro-5-hydroxy-pentan-2-yl]-7a-methyl-2,3,3a,5,6,7-hexahydro-1H-inden-4-ylidene]-ethylidene]-4-methylidene-cyclohexane-1,3-diol.
LMST03020127; 20R-fluoro-1α,24R-dihydroxy-26,27-cyclocholecalciferol; 20R-fluoro-26,27-cyclo-9,10-seco-5Z,7E,10(19)-cholestatriene-1S,3R,24R-triol; 9547372; Synthesized with the other three C20, C24 diasteromers from protected (5E)-1α-hydroxy-22-oxo-23,24,25,26,27-pentanorvitamin D₃ *via* fluorination of its silylenol ether and introduction of the side chain fragment by Wittig type reaction as key steps. *Tetrahedron.* 1995, *51(35)*, 9543-50.

198   *20S-Fluoro-1α,24R-dihydroxy-26,27-cyclovitamin D₃*

9547373 $C_{27}H_{41}FO_3$

(1R,3S,5Z)-5-[(2E)-2-[(1S,3aR,7aR)-1-[(2S,5R)-5-cyclopropyl-2-fluoro-5-hydroxy-pentan-2-yl]-7a-methyl-2,3,3a,5,6,7-hexahydro-1H-inden-4-ylidene]-ethylidene]-4-methylidene-cyclohexane-1,3-diol.
LMST03020128; 20S-fluoro-1α,24R-dihydroxy-26,27-cyclocholecalciferol; 20S-fluoro-26,27-cyclo-9,10-seco-5Z,7E,10(19)-cholestatriene-1S,3R,24R-triol; 9547373; Synthesized with the other three C20, C24 diasteromers from protected (5E)-1α-hydroxy-22-oxo-23,24,25,26,27-pentanorvitamin D$_3$ *via* fluorination of its silylenol ether and introduction of the side chain fragment by Wittig type reaction as key steps. *Tetrahedron*. 1995, *51(35)*, 9543-50.

199 **20R-Fluoro-1α,24S-dihydroxy-26,27-cyclovitamin D$_3$**
9547374   C$_{27}$H$_{41}$FO$_3$

(1R,3S,5Z)-5-[(2E)-2-[(1S,3aR,7aR)-1-[(2R,5S)-5-cyclopropyl-2-fluoro-5-hydroxy-pentan-2-yl]-7a-methyl-2,3,3a,5,6,7-hexahydro-1H-inden-4-ylidene]-ethylidene]-4-methylidene-cyclohexane-1,3-diol.
LMST03020129; 20R-fluoro-1α,24S-dihydroxy-26,27-cyclocholecalciferol; 20R-fluoro-26,27-cyclo-9,10-seco-5Z,7E,10(19)-cholestatriene-1S,3R,24S-triol; 9547374; Synthesized with the other three C20, C24 diasteromers from protected (5E)-1α-hydroxy-22-oxo-23,24,25,26,27-pentanorvitamin D$_3$ *via* fluorination of its silylenol ether and introduction of the side chain fragment by Wittig type reaction as key steps. *Tetrahedron*. 1995, *51(35)*, 9543-50.

200 **20S-Fluoro-1α,24S-dihydroxy-26,27-cyclovitamin D$_3$**
9547375   C$_{27}$H$_{41}$FO$_3$

(1R,3S,5Z)-5-[(2E)-2-[(1S,3aR,7aR)-1-[(2S,5S)-5-cyclopropyl-2-fluoro-5-hydroxy-pentan-2-yl]-7a-methyl-2,3,3a,5,6,7-hexahydro-1H-inden-4-ylidene]-ethylidene]-4-methylidene-cyclohexane-1,3-diol.
LMST03020130; 20S-fluoro-1α,24S-dihydroxy-26,27-cyclocholecalciferol; 20S-fluoro-26,27-cyclo-9,10-seco-5Z,7E,10(19)-cholestatriene-1S,3R,24S-triol; 9547375. *Tetrahedron*, 1995, *51(35)*, 9543-50.

201 **1,24,26-Trihydroxy-Δ^{22}-vitamin D$_3$**
101558-90-1   C$_{27}$H$_{42}$O$_4$

(E,2S,6S)-6-[(1R,3aR,4Z,7aS)-4-[(2Z)-2-[(3S,5R)-3,5-dihydroxy-2-methylidene-cyclohexylidene]ethylidene]-7a-methyl-2,3,3a,5,6,7-hexahydro-1H-inden-1-yl]-2-methyl-hept-4-ene-1,2-diol.
Ro 23-4319; 1,24,26-Trihydroxy-Δ^{22}-vitamin D$_3$; Ro-23-4319; 9,10-Secocholesta-5Z,7E,10(19),22E-tetra-ene-1α,3β,25S,26-tetrol; 6438910; 101558-90-1; Shows antiproliferative activity in human colon adenocarcinoma-derived CaCo-2 cells. *Biochem Biophys Res Commun*. 1991, *179(1)*, 57-62.

202 **1α,25-Difluorovitamin D$_3$**
78609-64-0   C$_{27}$H$_{42}$F$_2$O

(1S,3Z,5R)-3-[(2E)-2-[(1R,3aR,7aS)-1-[(2R)-6-fluoro-6-methyl-heptan-2-yl]-7a-methyl-2,3,3a,5,6,7-hexahydro-1H-inden-4-ylidene]ethylidene]-5-fluoro-4-methylidene-cyclohexan-1-ol.

1α,25-Difluorovitamin D_3; 9,10-Secocholesta-5Z,7E, 10(19)-trien-3β-ol, 1α,25-difluoro-; 6444057; 78609-64-0; An inactive vitamin D analog. **Arch Biochem Biophys**. 1981, **209(2)**, 579-83; The 16-Ene Analogs of 1,25-Dihydroxycholecalciferol. Synthesis and Biological Activity. In Vitamin D Gene Regulation Structure-Function Analysis and Clinical Application (Norman, A.W., Bouillon R. and Thomasset, M., eds), pp 210-213, Walter de Gruyter, Berlin/New York (1991).

203   *1α,25-Difluorovitamin D_3*
9547381 $C_{27}H_{42}F_2O$

(1R,3Z,5S)-3-[(2E)-2-[(1R,3aR,7aS)-1-[(2R)-6-fluoro-6-methyl-heptan-2-yl]-7a-methyl-2,3,3a,5,6,7-hexahydro-1H-inden-4-ylidene]ethylidene]-5-fluoro-4-methylidene-cyclohexan-1-ol.

LMST03020136; 1α,25-difluorocholecalciferol; 1S,25-difluoro-9,10-seco-5Z,7E,10(19)-cholestatrien-3R-ol; 9547381; Synthesized from 1α,25-dihydroxyvitamin D_3 3-acetate, by fluorination with diethylaminosulfurtrifluoride followed by deprotection. Fails to cause intestinal calcium transport and elevation of serum calcium level in vitamin D deficient rats. At a 1000-fold excess administration, fails to block the expression of vitamin D activity; (Stereochemistry at C-1 is ambiguous); λ_m = 265 nm. **Arch Biochem Biophys**. 1981, **209(2)**, 579-83.

204   *4,4-Difluorovitamin D_3*
9547379 $C_{27}H_{42}F_2O$

(1S,3E)-3-[(2E)-2-[(1R,3aR,7aS)-7a-methyl-1-[(2R)-6-methylheptan-2-yl]-2,3,3a,5,6,7-hexahydro-1H-inden-4-ylidene]ethylidene]-2,2-difluoro-4-methylidene-cyclohexan-1-ol.

LMST03020134; 4,4-difluorocholecalciferol; 4,4-di-fluoro-9,10-seco-5E,7E,10(19)-cholestatrien-3S-ol;

9547379; λ_m = 270 nm (ε = 19400 95% EtOH). Vitamin D Chemistry, Biology and Clinical Applications of the Steroid Hormone. Proceedings of the Tenth Workshop on Vitamin D Strasbourg, France-May. Synthesis of Fluorovitamin D Analogs for Conformational Analysis of Ligand Bound to Vitamin D Receptor. (Norman, A. W., Bouillon, R., Thomasset, M., eds), pp24-25, Printing and Reprographics University of California, Riverside (1997).

205   *(5Z)-4,4-Difluorovitamin D_3*
9547380 $C_{27}H_{42}F_2O$

(1S,3Z)-3-[(2E)-2-[(1R,3aR,7aS)-7a-methyl-1-[(2R)-6-methylheptan-2-yl]-2,3,3a,5,6,7-hexahydro-1H-inden-4-ylidene]ethylidene]-2,2-difluoro-4-methylidene-cyclohexan-1-ol.

LMST03020135; (5Z)-4,4-difluorocholecalciferol; 4,4-difluoro-9,10-seco-5Z,7E,10(19)-cholestatrien-3S-ol; 9547380; Prepared from 4,7-cholestadien-3-one *via* electrophilic difluorination at C(4) followed by reduction giving 4,4-difluoroprovitamin D_3; λ_m = 277 nm (ε = 23000 95% EtOH). Vitamin D Chemistry, Biology and Clinical Applications of the Steroid Hormone. Proceedings of the Tenth Workshop on Vitamin D Strasbourg, France-May. Synthesis of Fluorovitamin D Analogs for Conformational Analysis of Ligand Bound to Vitamin D Receptor. (Norman, A. W., Bouillon, R., Thomasset, M., eds), pp24-25, Printing and Reprographics University of California, Riverside (1997).

206   *23,23-Difluoro-25-hydroxyvitamin D_3*
95826-03-2 $C_{27}H_{42}F_2O_2$

4,4-difluoro-6-[(4Z)-4-[(2Z)-2-(5-hydroxy-2-methyl-idene-cyclohexylidene)ethylidene]-7a-methyl-2,3,3a, 5,6,7-hexahydro-1H-inden-1-yl]-2-methyl-heptan-2-ol.

23,23-F2-OH-D_3; 23,23-Difluoro-25-hydroxyvitamin D_3; 6438648; 95826-03-2; 5-10 times less active than 25-hydroxyvitamin D_3 in stimulating intestinal calcium transport, bone calcium mobilization, increasing serum phosphorus, mineralization of rachitic bone, and binding to the plasma transport protein in rats. It is converted to 23,23-difluoro-1α, 25-dihydroxyvitamin D_3 by chick renal 25-hydroxyvitamin D-1-hydroxylase. *Arch Biochem Biophys*. 1985, *241(1)*, 173-8.

207   ***24,24-Difluoro-1α-hydroxyvitamin D_3***
9547382 $C_{27}H_{42}F_2O_2$

(1R,3S,5Z)-5-[(2E)-2-[(1R,3aR,7aS)-1-[(2R)-5,5-difluoro-6-methyl-heptan-2-yl]-7a-methyl-2,3,3a,5,6,7-hexa-hydro-1H-inden-4-ylidene]ethylidene]-4-methylidene-cyclohexane-1,3-diol.
LMST03020137; 24,24-difluoro-9,10-seco-5Z,7E,10 (19)-cholestatriene-1S,3R-diol; 24,24-difluoro-1α-hydroxycholecalciferol; 9547382; Shows higher activity than 24,24-difluoro-1α,25-dihydroxyvitamin D_3 in intestinal calcium absorption; [α]$_D$ = +5.6° (c = 0.15 CHCl$_3$); λ$_m$ = 264 nm (ε = 18100 EtOH). *Chem Pharm Bull (Tokyo)*.1996, *44*, 62.

208   ***4,4-Difluoro-1α-hydroxyvitamin D_3***
9547383 $C_{27}H_{42}F_2O_2$

(1S,3R,5E)-5-[(2E)-2-[(1R,3aR,7aS)-7a-methyl-1-[(2R)-6-methylheptan-2-yl]-2,3,3a,5,6,7-hexahydro-1H-inden-4-ylidene]ethylidene]-4,4-difluoro-6-methylidene-cyclohexane-1,3-diol.
LMST03020138; 4,4-difluoro-9,10-seco-5E,7E,10(19)-cholestatriene-1S,3S-diol; 4,4-difluoro-1α-hydroxy-cholecalciferol; 9547383; Prepared from 4,7-

cholestadien-3-one *via* electrophilic difluorination at C(4) followed by reduction giving 4,4-difluoroprovitamin D_3; λ$_m$ = 271 nm (95% EtOH). Vitamin D Chemistry, Biology and Clinical Applications of the Steroid Hormone. Proceedings of the Tenth Workshop on Vitamin D Strasbourg, France-May. Synthesis of Fluorovitamin D Analogs for Conformational Analysis of Ligand Bound to Vitamin D Receptor. (Norman, A. W., Bouillon, R., Thomasset, M., eds), pp24-25, Printing and Reprographics University of California, Riverside (1997).

209   ***(5Z)-4,4-Difluoro-1α-hydroxyvitamin D_3***
9547384 $C_{27}H_{42}F_2O_2$

(1S,3S,5Z)-5-[(2E)-2-[(1R,3aR,7aS)-7a-methyl-1-[(2R)-6-ylidene]ethylidene]-2,3,3a,5,6,7-hexahydro-1H-inden-4-ylidene]ethylidene]-4,4-difluoro-6-methylidene-cyclohexane-1,3-diol.
LMST03020139; 4,4-difluoro-9,10-seco-5Z,7E,10(19)-cholestatriene-1S,3S-diol; (5Z)-4,4-difluoro-1α-hydroxycholecalciferol; 9547384; Prepared from 4,7-cholestadien-3-one *via* electrophilic difluorination at C(4) followed by reduction giving 4,4-difluoro-provitamin D_3. Vitamin D Chemistry, Biology and Clinical Applications of the Steroid Hormone. Proceedings of the Tenth Workshop on Vitamin D Strasbourg, France-May. Synthesis of Fluorovitamin D Analogs for Conformational Analysis of Ligand Bound to Vitamin D Receptor. (Norman, A. W., Bouillon, R., Thomasset, M., eds), pp24-25, Printing and Reprographics University of California, Riverside (1997).

210   ***23,23-Difluoro-25-hydroxyvitamin D_3***
9547385 $C_{27}H_{42}F_2O_2$

(6R)-6-[(1R,3aR,4E,7aS)-4-[(2Z)-2-[(5S)-5-hydroxy-2-methylidene-cyclohexylidene]ethylidene]-7a-methyl-2,3,3a,5,6,7-hexahydro-1H-inden-1-yl]-4,4-difluoro-2-

methyl-heptan-2-ol.
LMST03020140; 23,23-difluoro-9,10-seco-5Z,7E,10
(19)-cholestatriene-3S,25-diol; 23,23-difluoro-25-
hydroxycholecalciferol; 9547385; Prepared from 6β-
methoxy-3α,5-cyclo-23,24-dinor-5α-cholan-22-ol *via*
methyl 23,23-difluorocholan-24-oate. as key intermediate. Is 5-10 times less active than 25-
hydroxyvitamin D_3 in stimulating intestinal calcium
transport, bone calcium mobilization, increasing
serum phosphorus, mineralization of rachitic bone, and
binding to the plasma transport protein in rats; λ_m =
265 nm (EtOH). *Tet Lett*. 1984, *25(21)*, 4933-6; *Arch
Biochem Biophys*. 1985, *241(1)*, 173-8.

211   **24,24-Difluoro-25-hydroxyvitamin D_3**
71603-41-3 $C_{27}H_{42}F_2O_2$

(6R)-6-[(1R,3aR,4E,7aS)-4-[(2Z)-2-[(5S)-5-hydroxy-2-
methylidene-cyclohexylidene]ethylidene]-7a-methyl-
2,3,3a,5,6,7-hexahydro-1H-inden-1-yl]-3,3-difluoro-2-
methyl-heptan-2-ol.
LMST03020141; 24,24-difluoro-9,10-seco-5Z,7E,10
(19)-cholestatriene-3S,25-diol; 24,24-difluoro-25-
hydroxycholecalciferol; 24,24-Dfhv; 24,24-Difluoro-
25-hydroxyvitamin D_3; 9,10-Secocholesta-5Z,7E,10
(19)-triene-3β,25-diol, 24,24-difluoro-; 6441337;
9547386; Synthesized from lithocholic acid *via*
fluorination of 24-oxo-25-carboxylic acid ester with
DAST (diethylaminosulfur trifluoride) as a key step. In
calcium mobilization and transport, behaves
essentially identically to 1α,25-dihydroxyvitain D_3.
Equally as active as 25-hydroxyvitamin D_3 in all
known functions of vitamin D. Equivalent to 25-
hydroxyvitamin D_3 in the stimulation of growth,
intestinal calcium absorption, elevation of serum
calcium and serum phosphorus, healing of rachitic
cartilage and mineralization of rachitic bone; λ_m =
265 nm (log ε 4.24, 95% EtOH). *Tet Lett*. 1979,
20(21), 1859-62; *J Biol Chem*. 1979, *254(15)*, 7163-7.

212   **1α-25-Dihydroxy-23,23-difluorovitamin
D_3**
98040-59-6 $C_{27}H_{42}F_2O_3$

(5Z)-5-[(2Z)-2-[1-(4,4-difluoro-6-hydroxy-6-methyl-
heptan-2-yl)-7a-methyl-2,3,3a,5,6,7-hexahydro-1H-
inden-4-ylidene]ethylidene]-4-methylidene-cyclo-
hexane-1,3-diol.
25-OH-DF-D_3; 1,25-Dihydroxy-23,23-difluorovitamin
D_3; 1α-25-Dihydroxy-23,23-difluorovitamin D_3; 9,10-
Secocholesta-5Z,7E,10(19)-triene-1α,3β,25-triol,
23,23-difluoro-; 6438798; 98040-59-6; Some 5-10
times less active than 25-hydroxyvitamin D_3 in
stimulating intestinal calcium transport, bone calcium
mobilization, increasing serum phosphorus,
mineralization of rachitic bone, and binding to the
plasma transport protein in rats. *Arch Biochem
Biophys*. 1985, *241(1)*, 173-8.

213   **24,24-Difluoro-1,25-dihydroxyvitamin
D_3**
72696-49-2 $C_{27}H_{42}F_2O_3$

(1S,3R,5Z)-5-[(2E)-2-[(1R,3aR,7aS)-1-[(2R)-5,5-difluoro-
6-hydroxy-6-methyl-heptan-2-yl]-7a-methyl-2,3,3a,5,
6,7-hexahydro-1H-inden-4-ylidene]ethylidene]-4-
methylidene-cyclohexane-1,3-diol.
CCRIS 5802; 24,24-Difluoro-1,25-dihydroxyvitamin
D_3; 24,24-Difluoro-1α,25-dihydroxyvitamin D_3;
6438701; 72606-49-2; Found to be 5-10 times more
potent than 1,25-dihydroxyvitamin D_3 in the known *in
vivo* vitamin D responsive systems, including intestinal
calcium transport, bone calcium mobilization,
calcification of epiphyseal plate cartilage, and
elevation of plasma calcium and phosphorus
concentrations. *Ann Nutr Metab*. 1986, *30(1)*, 9-14;
Am J Physiol. 1983, *244(2)*, E159-63; *Biochem
Biophys Res Commun*. 1980, *96(4)*, 1800-3.

214 ***24,24-Difluoro-1α,25-dihydroxyvitamin D₃***

6476068

$C_{27}H_{42}F_2O_3$

(1R,3S,5Z)-5-[(2E)-2-[(1R,3aR,7aS)-1-[(2R)-5,5-difluoro-6-hydroxy-6-methyl-heptan-2-yl]-7a-methyl-2,3,3a,5,6,7-hexahydro-1H-inden-4-ylidene]ethylidene]-4-methylidene-cyclohexane-1,3-diol.
AIDS109013; AIDS-109013; 9,10-Secocholesta-5Z,7E,10-triene-1S,3R,25-triol, 24,24-difluoro-; 24,24-difluoro-1α,25-dihydroxyvitamin D₃; 24,24-difluoro-1α,25-dihydroxycholecalciferol; LMST03020145; 6476068; Regulates HIV-1 replication. *Chem Pharm Bull.* 1979, **27**, 3196-8; 1992, **40**, 1662-4; *Arch Biochem Biophys.* 1983, **220**, 90-4; USP 4201881.

215 ***4,4-Difluoro-1α,25-dihydroxyvitamin D₃***

9547387

$C_{27}H_{42}F_2O_3$

(1S,3S,5E)-5-[(2E)-2-[(1R,3aR,7aS)-1-[(2R)-6-hydroxy-6-methyl-heptan-2-yl]-7a-methyl-2,3,3a,5,6,7-hexahydro-1H-inden-4-ylidene]ethylidene]-4,4-difluoro-6-methylidene-cyclohexane-1,3-diol.
LMST03020142; 4,4-difluoro-9,10-seco-5E,7E,10(19)-cholestatriene-1S,3S,25-triol; 4,4-difluoro-1α,25-dihydroxycholecalciferol; 9547387; Synthesized from 25-hydroxy-4,7-cholestadien-3-one *via* electrophilic difluorination at C(4) followed by reduction giving 4,4-difluoro-25-hydroxyprovitamin D₃. NMR spectra reported. Affinity for the vitamin D receptor is much less than that of calcitriol; λ_m = 271 nm (95% EtOH). *Chem Pharm Bull (Tokyo).* 2002, **50(4)**, 475-83.

216 ***(5Z)-4,4-Difluoro-1α,25-dihydroxy-vitamin D₃***

9547388

$C_{27}H_{42}F_2O_3$

(1S,3S,5Z)-5-[(2E)-2-[(1R,3aR,7aS)-1-[(2R)-6-hydroxy-6-methyl-heptan-2-yl]-7a-methyl-2,3,3a,5,6,7-hexahydro-1H-inden-4-ylidene]ethylidene]-4,4-difluoro-6-methylidene-cyclohexane-1,3-diol.
LMST03020143; 4,4-difluoro-9,10-seco-5Z,7E,10(19)-cholestatriene-1S,3S,25-triol; (5Z)-4,4-difluoro-1α,25-dihydroxycholecalciferol; 9547388; Synthesized from 25-hydroxy-4,7-cholestadien-3-one *via* electrophilic difluorination at C(4) followed by reduction giving 4,4-difluoro-25-hydroxyprovitamin D₃.

217 ***23,23-Difluoro-1α,25-dihydroxyvitamin D₃***

9547389

$C_{27}H_{42}F_2O_3$

(1R,3S,5Z)-5-[(2E)-2-[(1R,3aR,7aS)-1-[(2R)-4,4-difluoro-6-hydroxy-6-methyl-heptan-2-yl]-7a-methyl-2,3,3a,5,6,7-hexahydro-1H-inden-4-ylidene]ethylidene]-4-methylidene-cyclohexane-1,3-diol.
LMST03020144; 23,23-difluoro-9,10-seco-5Z,7E,10(19)-cholestatriene-1S,3R,25-triol; 23,23-difluoro-1α,25-dihydroxycholecalciferol; 9547389; Prepared from 23,23-difluoro-25-hydroxyvitamin D₃ by *in vitro* incubation with vitamin D-deficient chick kidney homogenates. Affinity for chick intestinal receptor is 1/7 that of 1α,25-dihydroxyvitamin D₃, intestinal calcium transport, approximately as potent as 1α,25-dihydroxyvitamin D₃. *Arch Biochem Biophys.* 1985, **241(1)**, 173-8.

218 ***6RS,19-Epidioxy-24,24-difluoro-25-hydroxy-6,19-dihydrovitamin D₃***

9547390

$C_{27}H_{42}F_2O_4$

(3S)-10-[(E)-[(1R,3aR,7aS)-1-[(2R)-5,5-difluoro-6-hydroxy-6-methyl-heptan-2-yl]-7a-methyl-2,3,3a,5,6,7-hexahydro-1H-inden-4-ylidene]methyl]-8,9-dioxa-bicyclo[4.4.0]dec-11-en-3-ol.
LMST03020146; 6RS,19-epidioxy-24,24-difluoro-25-hydroxy-6,19-dihydrocholecalciferol; 6RS,19-epidioxy-24,24-difluoro-9,10-seco-5(10),7E-cholestadiene-3S,25-diol; 9547390; Major product of the reaction of 24,24-difluoro-25-hydroxyvitamin D$_3$ with singlet oxygen generated by dye-sensitized photochemical method. Has 1/1000 of the affinity of 1α,25-dihydroxyvitamin D$_3$ for the vitamin D receptor in HL-60 cells and is 1/45 as active in cell differentiation. *J Med Chem*. 1985, *28(9)*, 1148-53.

219   *24,25-Didehydrovitamin D$_3$*
5283675 $C_{27}H_{42}O$

(1S,3Z)-3-[(2E)-2-[(1R,3aR,7aS)-7a-methyl-1-[(2R)-6-methylhept-5-en-2-yl]-2,3,3a,5,6,7-hexahydro-1H-inden-4-ylidene]ethylidene]-4-methylidene-cyclo-hexan-1-ol.
LMST03020147; 24,25-didehydrovitamin D$_3$; 24,25-didehydrocholecalciferol; 9,10-seco-5Z,7E,10(19),24-cholestatetraen-3S-ol; 5283675. *Mol Cell Endocrinol*. 2002, *197(1-2)*, 1-13; Chemistry. 2002, 8(12), 2747-52; *J Cell Biochem*. 2006, *99(6)*, 1572-81.

220   *25-Dehydrovitamin D$_3$*
63819-60-3 $C_{27}H_{42}O$

(1R,3Z)-3-[(2E)-2-[(1R,3aR,7aS)-7a-methyl-1-[(2R)-6-methylhept-6-en-2-yl]-2,3,3a,5,6,7-hexahydro-1H-inden-4-ylidene]ethylidene]-4-methylidene-cyclo-hexan-1-ol.
25-Dehydrovitamin D$_3$; 25-Dehydrocholecalciferol; 9,10-Secocholesta-5Z,7E,10(19),25-tetraen-3β-ol; 6443590; 63819-60-3; Oil; λ_m = 265 nm (EtOH). *Biochem J.* 1979, *182*, 1-9; *Med Res Rev*. 2006, *7(3)*, 333-66.

221   *24-Dehydrovitamin D$_3$*
63819-59-0 $C_{27}H_{42}O$

(1R,3Z)-3-[(2E)-2-[(1R,3aR,7aS)-7a-methyl-1-[(2R)-6-methylhept-5-en-2-yl]-2,3,3a,5,6,7-hexahydro-1H-inden-4-ylidene]ethylidene]-4-methylidene-cyclo-hexan-1-ol.
24-Dehydrovitamin D$_3$; 24-Dehydrocholecalciferol; 9,10-Secocholesta-5Z,7E,10(19),24-tetraen-3β-ol; 6443822; 63819-59-0; Potent inhibitor of rat liver microsomal vitamin D-25-hydroxylase. Metabolized to 1α,25-dihydroxyvitamin D$_3$; Oil; λ_m = 265 nm (EtOH). *Biochem J.* 1979, *182(1)*, 1-9.

222   *24-Dehydroprevitamin D$_3$*
70574-97-9 $C_{27}H_{42}O$

9,10-Secocholesta-5(10),6Z,8,24-tetraen-3β-ol.
24-Dehydroprevitamin D$_3$; 9,10-Secocholesta-5(10),6,
8,24-tetraen-3-ol; 7-dehydrodesmosterol; 11970180;
70574-97-9; A provitamin D that is converted to
vitamin D$_3$ when the skin is exposed to sunlight. Upon
irradiation, gives previtamin D$_3$ and 24-dehydro-
previtamin D$_3$. *Biochem J.* 1979, *182(1)*, 1-9; *J Biol
Chem.* 1985, *260(22)*, 12181-4.

223   24-Dehydroprovitamin D$_3$
1715-86-2   C$_{27}$H$_{42}$O

(3S,9R,10R,13S,14R,17R)-10,13-dimethyl-17-[(2R)-6-
methylhept-5-en-2-yl]-2,3,4,9,11,12,14,15,16,17-
decahydro-1H-cyclopenta[a]phenanthren-3-ol.
7-Dehydrodesmosterol; 24-Dehydroprovitamin D$_3$;
Cholesta-5,7,24-trien-3β-ol; Cholesta-5,7,24-trien-3β-
ol; C05107; 440558; 143982-69-8; 1715-86-2;
Converted by irradiation to 24-dehydroprevitamin D$_3$. *J
Biol Chem.* 1985, *260(22)*, 12181-84.

224   19-Nor-10-ketovitamin D$_2$
85925-89-9   C$_{27}$H$_{42}$O$_2$

(2E,4R)-2-[(2E)-2-[(1S,3aR,7aS)-1-[(E,2S,5R)-5,6-di-
methylhept-3-en-2-yl]-7a-methyl-2,3,3a,5,6,7-hexa-
hydro-1H-inden-4-ylidene]ethylidene]-4-hydroxy-
cyclohexan-1-one.
19-Nor-10-ketovitamin D$_2$; 19-NK-VD2; 5(E)-19-Nor-
10-ketoergocalciferol; 19-Nor-9,10-secoergosta-5E,7E,
22E-trien-4-one, 1α-hydroxy-; 6439724; 85925-89-9;
Metabolite in bovine rumen microbes of 25-
hydroxyvitamin D$_3$. The conversion of vitamin D and
its metabolites to their 19-nor-10-keto forms is thought
to be a detoxification mechanism; Crystals; λ$_m$ = 312
nm. *Biochemistry* 1983, *22(15)*, 3636-40.

**225   1α,25-Dihydroxy-9,11-didehydro-3-
deoxy-vitamin D$_3$**

5283682   C$_{27}$H$_{42}$O$_2$

(6R)-6-[(1R,3aR,4Z,7aS)-4-[(2Z)-2-[(3S)-3-hydroxy-2-
methylidene-cyclohexylidene]ethylidene]-7a-methyl-2,
3,3a,7-tetrahydro-1H-inden-1-yl]-2-methyl-heptan-2-
ol.
LMST03020154;   9,10-seco-5Z,7E,10(19),9(11)-chol-
estatetraene-1S,25-diol;   1α,25-dihydroxy-9,11-di-
dehydro-3-deoxyvitamin D$_3$; 1α,25-dihydroxy-9,11-
didehydro-3-deoxycholecalciferol;   5283682;
Synthesized from de-A,B-cholesta-9(11),25-dien-8-one
and 1-ethynyl-3-hydroxy-2-methyl-1-cyclohexene *via*
the 9,10-seco-5(10),6,7,9(11),25-cholestapentaen-1-ol
derivative. *J Org Chem.* 1989, *54(17)*, 4072-83.

226   1α-Hydroxy-22E,23-didehydrovitamin D$_3$
5283676   C$_{27}$H$_{42}$O$_2$

(1R,3S,5Z)-5-[(2E)-2-[(1R,3aR,7aS)-7a-methyl-1-[(E,2S)-
6-methylhept-3-en-2-yl]-2,3,3a,5,6,7-hexahydro-1H-
inden-4-ylidene]ethylidene]-4-methylidene-cyclo-
hexane-1,3-diol.
LMST03020148;   9,10-seco-5Z,7E,10(19),22E-chol-
estatetraene-1S,3R-diol;   1α-hydroxy-22E,23-di-
dehydrovitamin D$_3$; 1α-hydroxy-22E,23-didehydro-
cholecalciferol; 5283676; Synthetic derivative. About
one order of magnitude less effective than 1α,25-
dihydroxyvitamin D$_3$ in HL-60 human promyelocytes
differentiation. *Bioorg Chem.* 1987, *15(2)*, 152-66; *J
Biol Chem.* 1987, *262(29)*, 14164-71.

227   1β-Hydroxy-22E,23-didehydrovitamin D$_3$
5283677   C$_{27}$H$_{42}$O$_2$

(1R,3R,5Z)-5-[(2E)-2-[(1R,3aR,7aS)-7a-methyl-1-[(E,2S)-6-methylhept-3-en-2-yl]-2,3,3a,5,6,7-hexahydro-1H-inden-4-ylidene]ethylidene]-4-methylidene-cyclo-hexane-1,3-diol.
LMST03020149; 9,10-seco-5Z,7E,10(19),22E-chol-estatetraene-1R,3R-diol; 1β-hydroxy-22E,23-di-dehydrovitamin D$_3$; 1β-hydroxy-22E,23-didehydro-cholecalciferol; 5283677; Synthetic vitamin D$_3$ derivative. *Bioorg Chem.* 1987, *15(2)*, 152-66.

228   1α-Hydroxy-22Z,23-didehydrovitamin D$_3$
5283678   C$_{27}$H$_{42}$O$_2$

(1R,3S,5Z)-5-[(2E)-2-[(1R,3aR,7aS)-7a-methyl-1-[(Z,2S)-6-methylhept-3-en-2-yl]-2,3,3a,5,6,7-hexahydro-1H-inden-4-ylidene]ethylidene]-4-methylidene-cyclo-hexane-1,3-diol.
LMST03020150; 9,10-seco-5Z,7E,10(19),22Z-chol-estatetraene-1S,3R-diol; 1α-hydroxy-22Z,23-di-dehydrovitamin D$_3$; 1α-hydroxy-22Z,23-didehydro-cholecalciferol; 5283678; Synthetic vitamin D$_3$ derivative. *Bioorg Chem.* 1987, *15(2)*, 152-66.

229   1β-Hydroxy-22Z,23-didehydrovitamin D$_3$
5283679   C$_{27}$H$_{42}$O$_2$

(1R,3R,5Z)-5-[(2E)-2-[(1R,3aR,7aS)-7a-methyl-1-[(Z,2S)-6-methylhept-3-en-2-yl]-2,3,3a,5,6,7-hexahydro-1H-inden-4-ylidene]ethylidene]-4-methylidene-cyclo-hexane-1,3-diol.
LMST03020151; 9,10-seco-5Z,7E,10(19),22Z-chol-estatetraene-1R,3R-diol; 1β-hydroxy-22Z,23-di-dehydrovitamin D$_3$; 1β-hydroxy-22Z,23-didehydro-cholecalciferol; 5283679; Synthetic vitamin D$_3$ derivative. *Bioorg Chem.* 1987, *15(2)*, 152-66.

230   1α-Hydroxy-24,25-didehydrovitamin D$_3$
5283681   C$_{27}$H$_{42}$O$_2$

(1R,3S,5Z)-5-[(2E)-2-[(1R,3aR,7aS)-7a-methyl-1-[(2R)-6-methylhept-5-en-2-yl]-2,3,3a,5,6,7-hexahydro-1H-inden-4-ylidene]ethylidene]-4-methylidene-cyclo-hexane-1,3-diol.
LMST03020153; 9,10-seco-5Z,7E,10(19),24-chol-estatetraene-1S,3R-diol; 1α-hydroxy-24,25-didehydro-vitamin D$_3$; 1α-hydroxy-24,25-didehydrochole-calciferol; 5283681.

231   25-Hydroxy-16,17-didehydrovitamin D$_3$
5283683   C$_{27}$H$_{42}$O$_2$

(6R)-6-[(3aR,4E,7aS)-4-[(2Z)-2-[(5S)-5-hydroxy-2-meth-ylidene-cyclohexylidene]ethylidene]-7a-methyl-3a,5,6,7-tetrahydro-3H-inden-1-yl]-2-methyl-heptan-2-ol.
LMST03020155; 9,10-seco-5Z,7E,10(19),16-chol-estatetraene-3S,25-diol; 25-hydroxy-16,17-didehydro-vitamin D_3; 25-hydroxy-16,17-didehydrocholecal-ciferol; 5283683; Compared to 1α,25-dihydroxy-vitamin D_3, inhibition of proliferation of human keratinocytes in culture is 10%, competitive binding to rat intestinal vitamin D receptor is 0.23% and differentiation of human leukemia cells (HL-60) is 3%. *J Nutr Biochem*. 1993, *4(1)*, 49-57.

232   *23,24-Didehydro-25-hydroxyvitamin D_3*
5283684 $C_{27}H_{42}O_2$

(E,6R)-6-[(1R,3aR,4E,7aS)-4-[(2Z)-2-[(5S)-5-hydroxy-2-methylidene-cyclohexylidene]ethylidene]-7a-methyl-2,3,3a,5,6,7-hexahydro-1H-inden-1-yl]-2-methyl-hept-3-en-2-ol.
LMST03020156; 9,10-seco-5Z,7E,10(19),23-cholesta-tetraene-3S,25-diol; 23E,24-didehydro-25-hydroxy-vitamin D_3; 23,24-didehydro-25-hydroxycholecal-ciferol; 5283684; Isolated and identified from blood plasma of chicks given large doses of vitamin D_3. *Biochemistry*. 1981, *20(26)*, 7385-91.

233   *Calicoferol E*
5283685 $C_{27}H_{42}O_2$

(1R,3aS,4S,7aR)-4-[2-(5-hydroxy-2-methyl-phenyl)ethyl]-7a-methyl-1-[(2R)-6-methylheptan-2-yl]-2,3,3a,4,6,7-hexahydro-1H-inden-5-one.
LMST03020157; Calicoferol E; (8S)-3-hydroxy-9,10-seco-1,3,5(10)-cholestatrien-9-one; 5283685; Isolated with Calicoferol C from an undescribed gorgonian of the genus *Muricella*. Synthesized. Exhibits potent antiviral activity and brine-shrimp lethality; $[\alpha]_D$ = +21.6° (c = 0.6 CHCl₃). *Tet Lett*. 1989, *30*, 7079; 1998,

39, 4741.

234   *1-Oxoprevitamin D_3*
5283686 $C_{27}H_{42}O_2$

(5R)-3-[(Z)-2-[(1R,3aR,7aS)-7a-methyl-1-[(2R)-6-methylheptan-2-yl]-1,2,3,3a,6,7-hexahydroinden-4-yl]ethenyl]-5-hydroxy-2-methyl-cyclohex-2-en-1-one.
LMST03020158; 1-oxoprevitamin D_3; 1-oxoprecholecalciferol; 3R-hydroxy-9,10-seco-5(10),6Z,8-cholestatriene-1-one; 5283686; Synthesized from 1α-hydroxyvitamin D_3 by selective (MnO₂) oxidation of the allylic 1-hydroxyl group; λ_m = 237, 288 nm (ε = 6800, 7100 EtOH). *J Chem Soc Chem Comm*. 1977, 890.

235   *23-Dehydro-25-hydroxyvitamin D_3*
80463-19-0 $C_{27}H_{42}O_2$

(E,6R)-6-[(1R,3aR,4E,7aS)-4-[(2Z)-2-[(5R)-5-hydroxy-2-methylidene-cyclohexylidene]ethylidene]-7a-methyl-2,3,3a,5,6,7-hexahydro-1H-inden-1-yl]-2-methyl-hept-3-en-2-ol.
23-Dehydro-25-hydroxyvitamin D_3; 9,10-Secochol-esta-5Z,7E,10(19),23-tetraene-3β,25-diol; 6439568; 80463-19-0; Kidney metabolite of vitamin D_3 and 25-hydroxyvitamin D_3. *Biochemistry*. 1981, *20(26)*, 7385-91.

236   *1,3-Dihydroxy-20-(3'-cyclopropyl-propyl)-9,10-secopregna-5,7,10(19)-triene*
120336-94-9 $C_{27}H_{42}O_2$

(1S,3R,5Z)-5-[(2E)-2-[(3aR,7aS)-1-[(2R)-5-cyclopropyl-pentan-2-yl]-7a-methyl-2,3,3a,5,6,7-hexahydro-1H-inden-4-ylidene]ethylidene]-4-methylidene-cyclo-hexane-1,3-diol.

MC 969; MC-969; 1,3-Dihydroxy-20-(3'-cyclopropyl-propyl)-9,10-secopregna-5,7,10(19)-triene; 9,10-Seco-chola-5Z,7E,10(19)-triene-1α,3β-diol, 24-cyclopropyl-; 6439055; 120336-94-9; Inhibits 25-hydroxylation of vitamin D$_3$ more effectively than 1α-hydroxyvitamin D$_3$ itself. *Biochem Pharmacol.* 1990, *40(2)*, 333-41.

237   *1β-Hydroxy-22Z,23-didehydrovitamin D$_3$*
5283679   C$_{27}$H$_{42}$O$_2$

((1S,3R,5Z)-5-[(2E)-2-[(1R,3aR,7aR)-7a-methyl-1-[(Z,2S)-6-methylhept-3-en-2-yl]-2,3,3a,5,6,7-hexahydro-1H-inden-4-ylidene]ethylidene]-4-methylidene-cyclohexane-1,3-diol.

LMST03020151; 1β-hydroxy-22Z,23-didehydrochol-ecalciferol; 4525693; 5283769; 1S,3R,5Z)-5-[(2E)-2-[(1R,3aR,7aR)-7a-methyl-1-[(Z,2S)-6-methylhept-3-en-2-yl]-2,3,3a,5,6,7-hexahydro-1H-inden-4-ylidene]ethylidene]-4-methylidene-cyclohexane-1,3-diol.

LMST03020151; 1β-hydroxy-22Z,23-didehydrochol-ecalciferol; 4525693; 5283769; Synthesized by coupling the C(22) steroid aldehyde with the appropriate Wittig reagent. *Bioorg Chem.* 1987, *15*, 152-66.

238   *1α-Hydroxy-22-oxovitamin D$_3$*
5283687   C$_{27}$H$_{42}$O$_3$

(2S)-2-[(1R,3aR,4E,7aS)-4-[(2Z)-2-[(3S,5R)-3,5-di-hydroxy-2-methylidene-cyclohexylidene]ethylidene]-7a-methyl-2,3,3a,5,6,7-hexahydro-1H-inden-1-yl]-6-methyl-heptan-3-one.

LMST03020159; 1α-hydroxy-22-oxovitamin D$_3$; 1α-hydroxy-22-oxocholecalciferol; 1S,3R-dihydroxy-9,10-seco-5Z,7E,10(19)-cholestatrien-22-one; 5283687.

239   *25-Hydroxy-23-oxovitamin D$_3$*
5283688   C$_{27}$H$_{42}$O$_3$

(6R)-6-[(1R,3aR,4E,7aS)-4-[(2Z)-2-[(5S)-5-hydroxy-2-methylidene-cyclohexylidene]ethylidene]-7a-methyl-2,3,3a,5,6,7-hexahydro-1H-inden-1-yl]-2-hydroxy-2-methyl-heptan-4-one.

LMST03020163; 25-hydroxy-23-oxovitamin D$_3$; 25-hydroxy-23-oxocholecalciferol; 3S,25-dihydroxy-9,10-seco-5Z,7E,10(19)-cholestatrien-23-one; 5283688; Prepared by epoxidation of the corresponding olefinic precursor which was constructed from the Inhoffen-Lythgoe diol. Compared to 1α,25-dihydroxyvitamin D$_3$, induction of differentiation of human promyeloid leukemia (HL-60), osteosarcoma (MG-63) and breast carcinoma (MCF-7) cells is 6%, 40% and 18%, respectively, elevation of serum calcium, serum osteocalcin, bone calcium and duodenal calbindin in rachitic chicks is 1%, 11%, 2% and 2%, respectively, and affinity for pig intestinal receptor and human vitamin D binding protein is 16% and 27%, respectively. *Steroids.* 1995, *60(4)*, 324-32.

240   *25-Hydroxy-24-oxovitamin D$_3$*
5283689   C$_{27}$H$_{42}$O$_3$

(6R)-6-[(1R,3aR,4E,7aS)-4-[(2Z)-2-[(5S)-5-hydroxy-2-methylidene-cyclohexylidene]ethylidene]-7a-methyl-2,3,3a,5,6,7-hexahydro-1H-inden-1-yl]-2-hydroxy-2-methyl-heptan-3-one.

LMST03020164; 25-hydroxy-24-oxovitamin D_3; 25-hydroxy-24-oxocholecalciferol; 3S,25-dihydroxy-9,10-seco-5Z,7E,10(19)-cholestatrien-24-one; 5283689; Synthesized in five steps from 22-phenylsulfonyl-23,24-dinor-5,7-choladien-3β-ol tetrahydropyranyl ether and 1,2-epoxy-3-hydroxy-3-methyl butane. Isolated from chicken kidney homogenates incubated with 25-hydroxyvitamin D_3, from rat kidneys perfused with pharmacological concentration of 25-hydroxyvitamin D_3 or from blood plasma of chicks given large doses of vitamin D_3. *In vitro*, 25-hydroxy-24-oxovitamin D_3 is hydroxylated at C(23) to give 23S,25-dihydroxy-24-oxovitamin D_3 and then cleaved at 23,24-bond to afford 23-hydroxy-24,25,26,27-tetranorvitamin D_3. Biochemistry and influences of vitamin D_3 metabolites on medullary bone formation reported. *Biochemistry*. 1984, *23(16)*, 3749-54; *J Biol Chem*. 1982, *257(7)*, 3732-8; 1983, *258(21)*, 12920-8.

241   *25-Hydroxy-1-oxo-3-epiprevitamin D₃*
5283690 $C_{27}H_{42}O_3$

(5S)-3-[(Z)-2-[(1R,3aR,7aS)-1-[(2R)-6-hydroxy-6-methyl-heptan-2-yl]-7a-methyl-1,2,3,3a,6,7-hexa-hydroinden-4-yl]ethenyl]-5-hydroxy-2-methyl-cyclo-hex-2-en-1-one.

LMST03020165; 3S,25-dihydroxy-9,10-seco-5(10),6Z,8-cholestatrien-1-one; 25-hydroxy-1-oxo-3-epiprevitamin D_3; 25-hydroxy-1-oxo-3-epiprecholecalciferol; 5283690; Synthesized from 1β,25-dihydroxy-3-epi-vitamin D_3 by Des-Martin oxidation; λ_m = 242, 298 nm (ε = 10000, 11200, 95% EtOH). *J Org Chem*. 1993, *58(1)*, 118-23.

242   *1α,23S-Dihydroxy-25,26-didehydro-vitamin D₃*

5283692 $C_{27}H_{42}O_3$

(1R,3S,5Z)-5-[(2E)-2-[(1R,3aR,7aS)-1-[(2R,4S)-4-hydroxy-6-methyl-hept-6-en-2-yl]-7a-methyl-2,3,3a,5,6,7-hexahydro-1H-inden-4-ylidene]-ethylidene]-4-methylidene-cyclohexane-1,3-diol.

LMST03020167; 9,10-seco-5Z,7E,10(19),25-cholesta-tetraene-1S,3R,23S-triol; 1α,23S-dihydroxy-25,26-di-dehydrovitamin D_3; 1α,23S-dihydroxy-25,26-di-dehydrocholecalciferol; 5283692; In inhibition of proliferation of human keratinocytes in culture, is 100% as active as 1α,25-dihydroxyvitamin D_3, in competitive binding to rat intestinal vitamin D receptor, it is 0.048% as active. *J Nutr Biochem*. 1993, *4(1)*, 49-57.

243   *1α,24R-Dihydroxy-22E,23-didehydro-vitamin D₃*
5283693 $C_{27}H_{42}O_3$

(1R,3S,5Z)-5-[(2E)-2-[(1R,3aR,7aS)-1-[(E,2S,5R)-5-hydroxy-6-methyl-hept-3-en-2-yl]-7a-methyl-2,3,3a,5,6,7-hexahydro-1H-inden-4-ylidene]ethylidene]-4-methylidene-cyclohexane-1,3-diol.

LMST03020168; 9,10-seco-5Z,7E,10(19),22E-cholesta-tetraene-1S,3R,24R-triol; 1α,24R-dihydroxy-22E,23-di-dehydrovitamin D_3; 1α,24R-dihydroxy-22E,23-di-dehydrocholecalciferol; 5283693; Synthesized from dinorcholenic acid acetate. Affinity for chick the intestinal receptor is 10% of that of 1α,25-dihydroxyvitamin D_3. Activities in increasing intestinal calcium transport, serum calcium and inorganic phosphorus concentration and bone ash were compared with 1α,25-dihydroxyvitamin D_3 and its 24S-epimer. This compound is less potent than either its 24S-epimer or 1α,25-dihydroxyvitamin D_3 in these activities; λ_m = 265 nm (EtOH). *Chem Pharm Bull*

(Tokyo). 1984, *32(10),* 3866-72.

244   *1α,24S-Dihydroxy-22E,23-didehydro-vitamin D₃*
5283694 $C_{27}H_{42}O_3$

(1R,3S,5Z)-5-[(2E)-2-[(1R,3aR,7aS)-1-[(E,2S,5S)-5-hydroxy-6-methyl-hept-3-en-2-yl]-7a-methyl-2,3,3a,5,6,7-hexahydro-1H-inden-4-ylidene]ethylidene]-4-methylidene-cyclohexane-1,3-diol.
LMST03020169; 9,10-seco-5Z,7E,10(19),22E-cholesta-tetraene-1S,3R,24S-triol; 1α,24S-dihydroxy-22E,23-di-dehydrovitamin   D₃;   1α,24S-dihydroxy-22E,23-di-dehydrocholecalciferol; 5283694; Synthesized starting from dinorcholenic acid acetate. Affinity for chick the intestinal receptor is approximately the same of that of 1α,25-dihydroxyvitamin D₃. Less potent than 1α,25-dihydroxyvitamin D₃ but more potent than its 24R-epimer in increasing intestinal calcium transport, serum   calcium   and   inorganic   phosphorus concentration and bone ash; λ_m = 265 nm (EtOH).
Chem Pharm Bull (Tokyo). 1984, *32(10),* 3866-72.

245   *1α,25-Dihydroxy-9,11-didehydrovitamin D₃*
5283695 $C_{27}H_{42}O_3$

(1R,3S,5Z)-5-[(2Z)-2-[(1R,3aR,7aS)-1-[(2R)-6-hydroxy-6-methyl-heptan-2-yl]-7a-methyl-2,3,3a,7-tetrahydro-1H-inden-4-ylidene]ethylidene]-4-methylidene-cyclo-hexane-1,3-diol.
LMST03020170;   9,10-seco-5Z,7E,9(11),10(19)-chol-estatetraene-1S,3R,25-triol;   1α,25-dihydroxy-9,11-di-dehydrovitamin   D₃; 1α,25-dihydroxy-9,11-didehydro-cholecalciferol; 5283695; Synthesized from de-A,B-cholesta-9(11),25-dien-8-one   and   1-ethynyl-3S,5R-dihydroxy-2-methyl-1-cyclohexene derivative. *J Org*

Chem. 1989, *54(17),* 4072-83.

246   *1α,25-Dihydroxy-16,17-didehydro-vitamin D₃*
5283696 $C_{27}H_{42}O_3$

(1R,3S,5Z)-5-[(2E)-2-[(3aR,7aS)-1-[(2R)-6-hydroxy-6-methyl-heptan-2-yl]-7a-methyl-3a,5,6,7-tetrahydro-3H-inden-4-ylidene]ethylidene]-4-methylidene-cyclo-hexane-1,3-diol.
LMST03020171;   9,10-seco-5Z,7E,10(19),16-cholesta-tetraene-1S,3R,25-triol;   1α,25-dihydroxy-16,17-di-dehydrovitamin   D₃;   1α,25-dihydroxy-16,17-di-dehydrocholecalciferol; 5283696; Compared to 1α,25-dihydroxyvitamin D₃, inhibition of proliferation of human keratinocytes in culture is 10%, competitive binding to rat intestinal vitamin D receptor is 240% and differentiation of human leukemia cells (HL-60) is 500%. *J Nutr Biochem.* 1993, *4(1),* 49-57.

247   *1α,25-Dihydroxy-22,23-didehydro-vitamin D₃*
5283697 $C_{27}H_{42}O_3$

(1R,3S,5Z)-5-[(2E)-2-[(1R,3aR,7aS)-1-[(E,2S)-6-hydroxy-6-methyl-hept-3-en-2-yl]-7a-methyl-2,3,3a,5,6,7-hexa-hydro-1H-inden-4-ylidene]ethylidene]-4-methylidene-cyclohexane-1,3-diol.
LMST03020172; 9,10-seco-5Z,7E,10(19),22E-cholesta-tetraene-1S,3R,25-triol;   1α,25-dihydroxy-22,23-di-dehydrovitamin   D₃;   1α,25-dihydroxy-22,23-di-dehydrocholecalciferol; 5283697; Compared to 1α,25-dihydroxyvitamin D₃, inhibition of proliferation of human keratinocytes in culture is 10%, intestinal calcium absorption in rat is 92%, bone calcium mobilization in rat is 133%, competitive binding to rat intestinal vitamin D receptor is 122%, and differentiation of human leukemia cells (HL-60) is

125%; λ_m = 266 nm (ε = 17300 MeOH); $[\alpha]_D^{20}$ = +57.2° (c = 0.14 EtOH). *Tet Lett.* 1992, **33(26)**, 3741-4; *J Nutr Biochem.* 1993, **4(1)**, 49-57.

248   **1α,25-Dihydroxy-23E,24-didehydro-vitamin D$_3$**

5283698 $C_{27}H_{42}O_3$

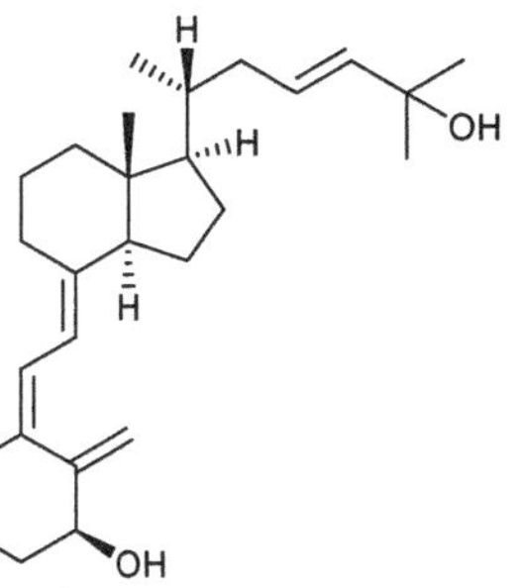

(1R,3S,5Z)-5-[(2E)-2-[(1R,3aR,7aS)-1-[(E,2R)-6-hydroxy-6-methyl-hept-4-en-2-yl]-7a-methyl-2,3,3a,5,6,7-hexa-hydro-1H-inden-4-ylidene]ethylidene]-4-methylidene-cyclohexane-1,3-diol.
LMST03020173; 9,10-seco-5Z,7E,10(19),23E-cholesta-tetraene-1S,3R,25-triol; 1α,25-dihydroxy-23E,24-didehydrovitamin D$_3$; 1α,25-dihydroxy-23E,24-didehydrocholecalciferol; 5283698.

249   **1α,25-Dihydroxy-6,7-didehydro-previtamin D$_3$**

9547394 $C_{27}H_{42}O_3$

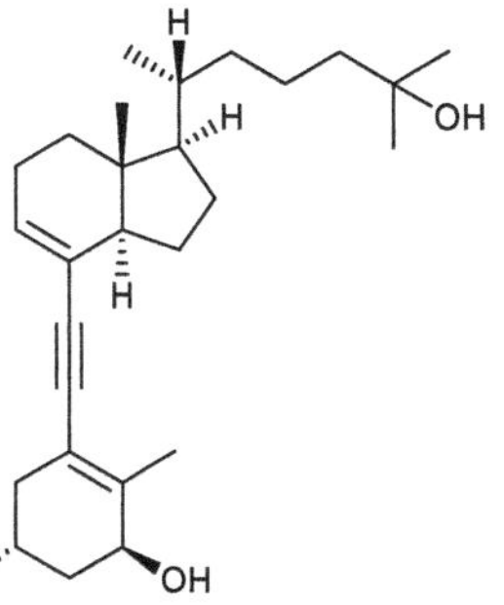

(1S,5R)-3-[2-[(1R,3aR,7aS)-1-[(2R)-6-hydroxy-6-methyl-heptan-2-yl]-7a-methyl-1,2,3,3a,6,7-hexahydroinden-4-yl]ethynyl]-2-methyl-cyclohex-2-ene-1,5-diol.
LMST03020174; 9,10-seco-5(10),8-cholestadien-6-yne-1S,3R,25-triol; 1α,25-dihydroxy-6,7-didehydro-precholecalciferol; 9547394.

250   **1β,25-Dihydroxy-6,7-didehydro-3-epi-previtamin D$_3$**

9547395 $C_{27}H_{42}O_3$

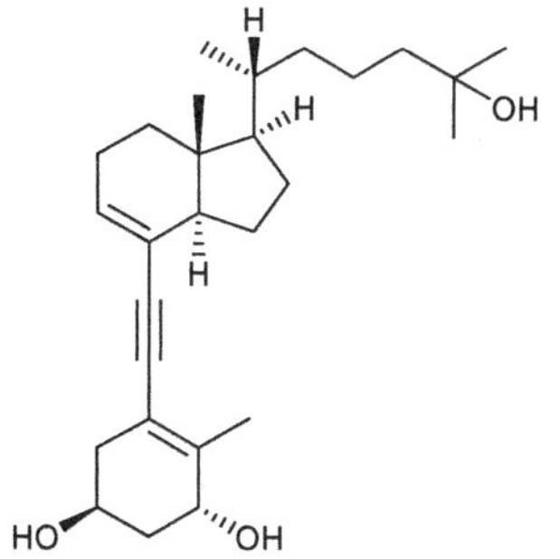

(1R,5S)-3-[2-[(1R,3aR,7aS)-1-[(2R)-6-hydroxy-6-methyl-heptan-2-yl]-7a-methyl-1,2,3,3a,6,7-hexahydroinden-4-yl]ethynyl]-2-methyl-cyclohex-2-ene-1,5-diol.
LMST03020175; 9,10-seco-5(10),8-cholestadien-6-yne-1R,3S,25-triol; 1β,25-dihydroxy-6,7-didehydro-3-epiprecholecalciferol; 9547395.

251   **1α,25-Dihydroxy-17S,20S-methano-21-norvitamin D$_3$**

9547396 $C_{27}H_{42}O_3$

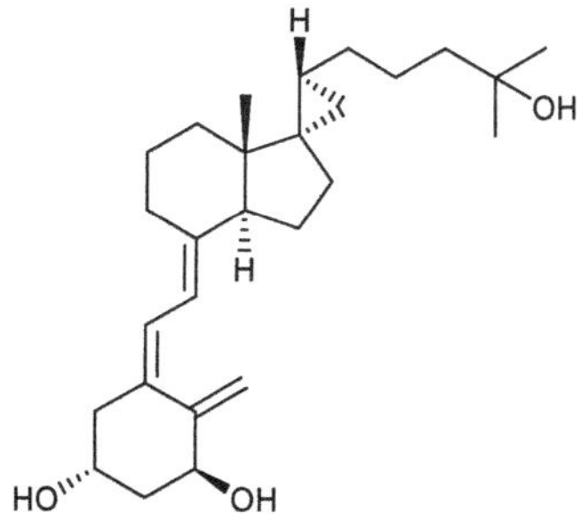

(17S,20S)-1α,25-dihydroxy-17,20-methano-21-norcholecalciferol.
LMST03020176; 1α,25-dihydroxy-17S,20S-methano-21-norcholecalciferol; 17S,20S-methano-21-nor-9,10-seco-5Z,7E,10(19)-cholestatriene-1S,3R,25-triol; 9547396; Synthesized from 4-hydroxyhydrindanone by introduction of the side chain and A-ring fragment. Stereoselective introduction of 17,20-methano-bridge was achieved by the reaction of 17E(20)-ene CD-fragment with dichlorocarbene followed by reduction with sodium in ethanol. *Tet Lett.* 1998, **39(26)**, 4725-8.

252   **1α,25-Dihydroxy-17S,20R-methano-21-norvitamin D$_3$**

9547397 $C_{27}H_{42}O_3$

(17S,20R)-1α,25-dihydroxy-17,20-methano-21-norcholecalciferol.
LMST03020177; 1α,25-dihydroxy-17S,20R-methano-21-norvitamin D_3; 1α,25-dihydroxy-17S,20R-methano-21-norcholecalciferol; 9547397; Synthesized from 4-hydroxyhydrindanone by introduction of the side chain and A-ring fragment. Stereoselective introduction of 17,20-methano-bridge was achieved by the reaction of 17E(20)-ene CD-fragment with dichlorocarbene followed by reduction with sodium in ethanol. *Tet Lett.* 1998, *39(26)*, 4725-8.

253   **24R,25-Dihydroxy-22E,23-didehydro-vitamin D_3**
      5283699 $C_{27}H_{42}O_3$

(E,3R,6S)-6-[(1R,3aR,4E,7aS)-4-[(2Z)-2-[(5S)-5-hydroxy-2-methylidene-cyclohexylidene]ethylidene]-7a-methyl-2,3,3a,5,6,7-hexahydro-1H-inden-1-yl]-2-methyl-hept-4-ene-2,3-diol.
LMST03020178; 9,10-seco-5Z,7E,10(19),22E-cholesta-tetraene-3S,24R,25-triol;24R,25-dihydroxy-22E,23-di-dehydrovitamin D_3; 24R,25-dihydroxy-22E,23-di-de-hydrocholecalciferol; 5283699; $\lambda_m = 265$ nm (EtOH). *Chem Pharm Bull (Tokyo)*. 1987, *35(3)*, 970-9.

254   **24S,25-Dihydroxy-22E,23-didehydro-vitamin D_3**
      5283700 $C_{27}H_{42}O_3$

(E,3S,6S)-6-[(1R,3aR,4E,7aS)-4-[(2Z)-2-[(5S)-5-hydroxy-2-methylidene-cyclohexylidene]ethylidene]-7a-methyl-2,3,3a,5,6,7-hexahydro-1H-inden-1-yl]-2-methyl-hept-4-ene-2,3-diol.
LMST03020179; 9,10-seco-5Z,7E,10(19),22E-cholesta-tetraene-3,24,25-triol; 24S,25-dihydroxy-22E,23-di-dehydrovitamin D_3; 24S,25-dihydroxy-22E,23-di-de-hydrocholecalciferol; 5283700; $\lambda_m = 265$ nm (EtOH). *Chem Pharm Bull (Tokyo)*. 1987, *35(3)*, 970-9.

255   **Calicoferol B**
                                    $C_{27}H_{42}O_3$

(1R,2S,3aS,7aS)-2-hydroxy-4-[2-(5-hydroxy-2-methyl-phenyl)ethyl]-7a-methyl-1-[(2R)-6-methylheptan-2-yl]-2,3,3a,4,6,7-hexahydro-1H-inden-5-one.
LMST03020180; Calicoferol B; (8S)-3,16S-dihydroxy-9,10-seco-1,3,5(10)-cholestatrien-9-one; Isolated from the gorgonian Calicogorgia so. (1991). Is lethal to brine shrimp. Synthesized, 2002; Colorless oil; $[\alpha]_D^{20} = -16.2°$ (c = 0.09 $CHCl_3$). *Chem Lett.* 1991, 20, 427; *J Org Chem.* 2002, *67(14)*, 4821-7.

256   **1,24-Dihydroxy-22-dehydrovitamin D_3**
      95270-41-0 $C_{27}H_{42}O_3$

(1R,3S,5E)-5-[(2E)-2-[(1R,3aS,7aR)-1-[(E,2S,5R)-5-hydroxy-6-methyl-hept-3-en-2-yl]-7a-methyl-2,3,3a,4,6,7-hexahydro-1H-inden-5-ylidene]ethylidene]-4-methylidene-cyclohexane-1,3-diol.
LMST03020169; 1,24-Dihydroxy-22-dehydrovitamin D_3; 1,24-OH-22-DH-D_3; 1,24-Dihydroxy-22-dehydro-cholecalciferol; 6438598; 95270-41-0; Synthesis and biological activity reported. *Chem Pharm Bull (Tokyo).* 1984, *32(10)*, 3866-72.

257 *25-Hydroxy-6,19-epoxyvitamin D_3*
96999-68-7 $C_{27}H_{42}O_3$

3-[(E)-1-(6-hydroxy-6-methyl-heptan-2-yl)-7a-methyl-2,3,3a,5,6,7-hexahydro-1H-inden-4-ylidene]methyl]-4,5,6,7-tetrahydroisobenzofuran-5-ol.
25-Hydroxy-6,19-epoxyvitamin D_3; 25-HE-Vitamin D_3; 9,10-Secocholesta-5,7E,10(19)-triene-3β,25-diol, 6R,19-epoxy-; 6438681; 96999-68-7; More active than 25-hydroxyvitamin D_3 in inducing differentiation of HL-60 cells and in bone resorption. *J Med Chem.* 1985, *28(9)*, 1153-8.

258 *1,25-Dihydroxy-16-ene vitamin D_3*
124409-58-1 $C_{27}H_{42}O_3$

(5R)-3-[(2E)-2-[(3aR,7aS)-1-[(2R)-6-hydroxy-6-methyl-heptan-2-yl]-7a-methyl-3a,5,6,7-tetrahydro-3H-inden-4-ylidene]ethyl]-4-methylidene-cyclohexene-1,5-diol.
1,25-Dihydroxy-16-ene-vitamin D_3; 1,25-(OH)2-16-Ene-D_3; (5Z)-9,10-Secocholesta-3,7E,10(19),16-tetra-ene-1α,3β,25-triol; (7E)-9,10-Secocholesta-3,7E,10(19),16-tetraene-1α,3β,25-triol; 6439178; 124409-58-1; More potent than Ro 23-7553 in clonogenic growth assays, but less potent in cell differentiation. *J Steroid Biochem Mol Biol.* 1996, *59(5-6)*, 405-12; *Leuk Res.* 1994, *18(6)*, 453-63.

259 *23-Keto-25-hydroxyvitamin D_3*
83353-84-8 $C_{27}H_{42}O_3$

(6R)-6-[(1R,3aR,4E,7aS)-4-[(2Z)-2-[(5R)-5-hydroxy-2-methylidene-cyclohexylidene]ethylidene]-7a-methyl-2,3,3a,5,6,7-hexahydro-1H-inden-1-yl]-2-hydroxy-2-methyl-heptan-4-one.
23-Keto-25-hydroxyvitamin D_3; 25-Hydroxy-23-oxo-cholecalciferol; 25-Hydroxy-23-oxovitamin D_3; 9,10-Secocholesta-5Z,7E,10(19)-trien-23-one, 3β,25-dihydroxy-; 6439646; 83353-84-8; A metabolite of 23,25-dihydroxyvitamin D_3, generated with kidney homogenates prepared from vitamin D treated chicks. Approximately 20-fold less effective than vitamin D_3 in producing intestinal-calcium transport. Has greater affinity than 25-hydroxyvitamin D_3 to both the rat plasma vitamin D binding protein and the 1,25-dihydroxyvitamin D specific cytosol receptor, but fails to stimulate bone calcium resorption. *Biochemistry.* 1983, *22(2)*, 245-50.

260 *25-Hydroxy-24-oxovitamin D_3*
74886-61-6 $C_{27}H_{42}O_3$

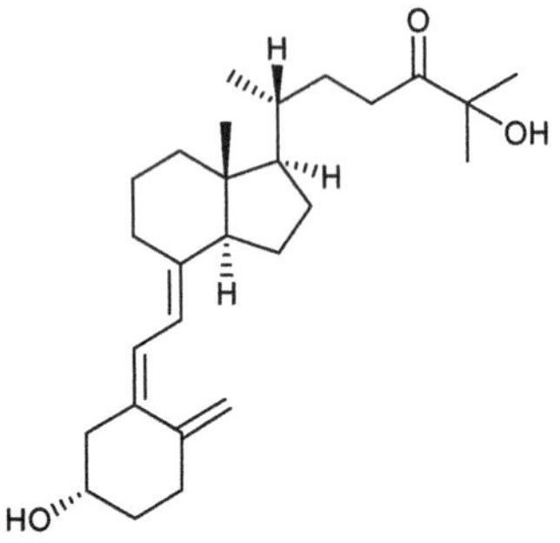

(6R)-6-[(1R,3aR,4E,7aS)-4-[(2Z)-2-[(5R)-5-hydroxy-2-methylidene-cyclohexylidene]ethylidene]-7a-methyl-2,3,3a,5,6,7-hexahydro-1H-inden-1-yl]-2-hydroxy-2-methyl-heptan-3-one.

25-Hydroxy-24-oxocholecalciferol; 24-Keto-25-hydroxyvitamin D_3; 25-Hydroxy-24-oxovitamin D_3; 9,10-Secocholesta-5Z,7E,10(19)-trien-24-one, 3β,25-dihydroxy-; 6443868; 74886-61-6; An *in vivo* metabolite of vitamin D_3 and 25-hydroxyvitamin D_3. Isolated from the blood plasma of chicks given large doses of vitamin D_3. Like 24,25-dihydroxyvitamin D_3, enhances intestinal calcium transport and bone calcium mobilization activities in vitamin D-deficient chicks. Biological activity similar to that of 24,25-dihydroxyvitamin D_3, enhancing intestinal calcium transport and bone calcium mobilization activities in vitamin D-deficient chicks. *Biochemistry* 1981, *20(26)*, 7385-91.

261   ***1,25-Dihydroxy-5,6-trans-16-ene-vitamin D_3***

$C_{27}H_{42}O_3$

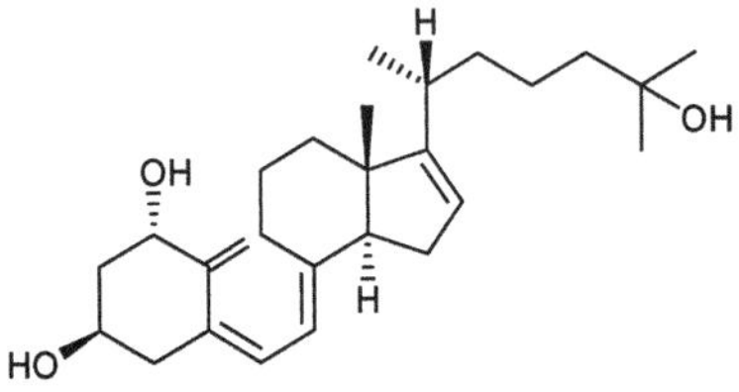

9,10-Secocholesta-5E,7E,10(19),16-tetraene-1α,3β,20-triol.
1,25(OH)2,5,5-16-D_3; A non-hyper-calcemic analog of vitamin D_3. *J Cell Physiol.* 2002, *191(2)*, 198-207.

262   ***1,25-Dihydroxy-22-ene-vitamin D_3***

$C_{27}H_{42}O_3$

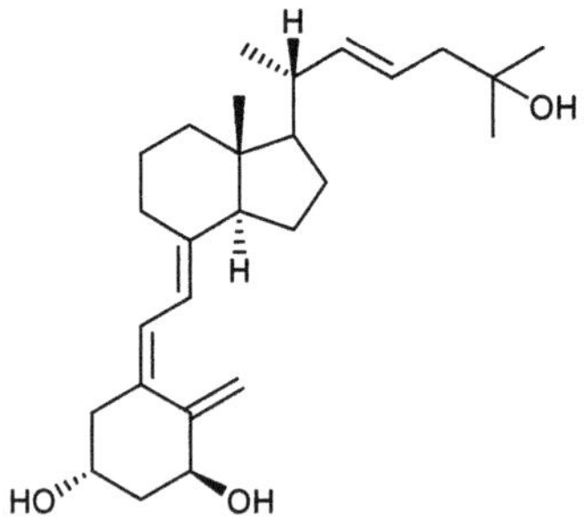

9,10-Secocholesta-5Z,7E,10(19)-22-tetraene-3β,20-diol.

The C_{22}-C_{23} double bond is reduced during the metabolism of the compound. *Steroid Biochem Mol Biol.* 2001, *78(2)*, 167-76.

263   ***18-(3-Hydroxy-3-methylbutyl)-20-methyl-9,10-secopregna-5,7,10(19),20-tetra-ene-1,3-diol***

$C_{27}H_{42}O_3$

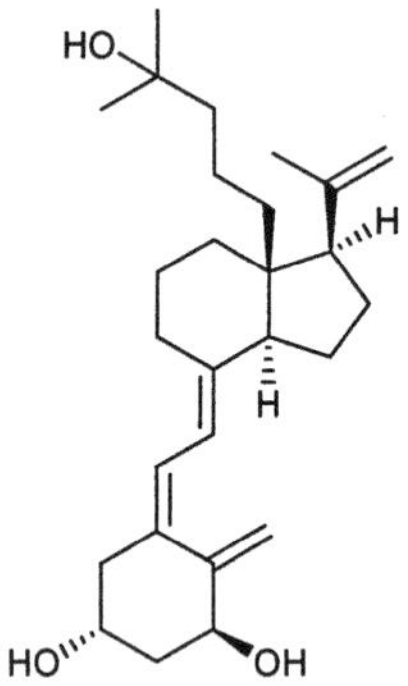

Synthesized photochemically. *J Org Chem.* 2002, *67(14)*, 4707-14.

264   ***8β,25-Dihydroxy-9,10-secocholesta-4,6E,10-trien-3-one***
109947-25-3 $C_{27}H_{42}O_3$

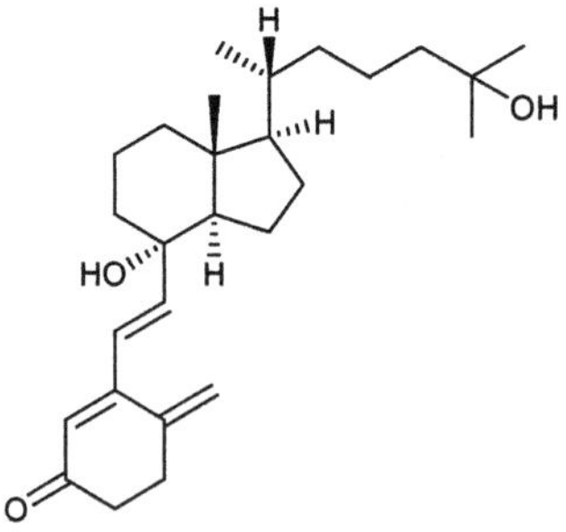

3-[(E)-2-[(1R,3aR,4R,7aS)-4-hydroxy-1-[(2R)-6-hydroxy-6-methyl-heptan-2-yl]-7a-methyl-2,3,3a,5,6,7-hexa-hydro-1H-inden-4-yl]ethenyl]-4-methylidene-cyclo-hex-2-en-1-one.
LMST03020166;   8S,25-dihydroxy-9,10-seco-4,6E,10

(19)-cholestatrien-3-one; 8β,25-dihydroxy-9,10-seco-cholesta-4,6E,10-trien-3-one; Dsco/*bi; 8,25-(OH)2-3-Oxoneo-D$_3$; 8,25-Dihydroxy-9,10-seco-4,6,10(19)-cholestatrien-3-one; 9,10-Secocholesta-4,6,10(19)-trien-3-one, 8α,25-dihydroxy-; 5283691; 6439321; 109947-25-3; Isolated from phagocytic cells. Identified spectroscopically and synthesized. Synthesized from 25-hydroxyvitamin D$_3$ *via* regio- and stereoselective 7,8-epoxidation. The C(18) stereochemistry of the natural metabolite is S. Produced from calcitriol, presumably by action of a dioxygenase and isolated from various phagocytic cells alveolar macrophages, murine myeloid leukemia cells (M1), human promyelocytic leukemia cells (HL-60); human monoblast-like lymphoma cells (U937)] and in rat liver microsomes incubated with 25-hydroxyvitamin D$_3$. Solid; λ$_m$ = 295 nm (log ε = 4.2, EtOH). *J Biol Chem*. 1987, *262(27)*, 12939-44; *FEBS Lett*. 1987, *218(2)*, 200-4.

265   *16-Dehydrocalcitriol*

C$_{27}$H$_{42}$O$_3$

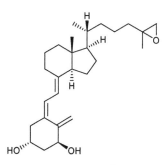

1α,25-dihydroxy-16-dehydrovitamin D$_3$; 16-dehydro-1α,25-dihydroxyvitamin D$_3$; 16-dehydro-1α,25-dihydroxycholecalciferol; LMST03020171; VD 2708; More effective than calcitriol in osteocalcin formation. *Eur J Biochem*. 1999, *261(3)*, 706-13. *Leo AB*.

266   *16-Dehydroepicalcitriol*

C$_{27}$H$_{42}$O$_3$

1α,25-dihydroxy-16-dehydro-20-epivitamin D$_3$; 20-epi-16-dehydro-1α,25-dihydroxyvitamin D$_3$; 20-epi-16-dehydro-1α,25-dihydroxycholecalciferol; VD 2668; More effective than calcitriol in osteocalcin formation. *Eur J Biochem*. 1999, *261(3)*, 706-13. *Leo AB*.

267   *24,25-Epoxy-1α-hydroxyvitamin D$_3$*

9547391   C$_{27}$H$_{42}$O$_3$

(1R,3S,5Z)-5-[(2E)-2-[(1R,3aR,7aS)-1-[(2R)-4-(3,3-dimethyloxiran-2-yl)butan-2-yl]-7a-methyl-2,3,3a,5,6,7-hexahydro-1H-inden-4-ylidene]ethylidene]-4-methylidene-cyclohexane-1,3-diol.
LMST03020160; 24,25-epoxy-9,10-seco-5Z,7E,10(19)-cholestatriene-1S,3R-diol; 24,25-epoxy-1α-hydroxy-cholecalciferol; 9547391; Synthesized by epoxidation of the corresponding olefinic precursor which was constructed from Inhoffen-Lythgoe diol. Compared to 1α,25-dihydroxyvitamin D$_3$, inhibition of proliferation or differentiation induction of human promyeloid leukemia (HL-60), osteosarcoma (MG-63) and breast carcinoma (MCF-7) cells are 6%, 40% and 18%, respectively, elevation of serum calcium, serum osteocalcin, bone calcium and duodenal calbindin in rachitic chicks is 1%, 11%, 2% and 2%, respectively and affinity for pig intestinal receptor and human vitamin D binding protein is 16% and 27%, respectively. *Steroids*. 1995, *60(4)*, 324-32.

268   *25,26-Epoxy-1α-hydroxyvitamin D$_3$*

9547392   C$_{27}$H$_{42}$O$_3$

(1R,3S,5Z)-5-[(2E)-2-[(1R,3aR,7aS)-7a-methyl-1-[(2R)-5-(2-methyloxiran-2-yl)pentan-2-yl]-2,3,3a,5,6,7-hexa-hydro-1H-inden-4-ylidene]ethylidene]-4-methylidene-cyclohexane-1,3-diol.
LMST03020161; 25,26-epoxy-9,10-seco-5Z,7E,10(19)-cholestatriene-1S,3R-diol; 25,26-epoxy-1α-hydroxy-cholecalciferol; 9547392; Synthesized by epoxidation of the corresponding olefinic precursor which was constructed from Inhoffen-Lythgoe diol. Compared to 1α,25-dihydroxyvitamin D$_3$, inhibition of proliferation or differentiation induction of human promyeloid leukemia (HL-60), osteosarcoma (MG-63) and breast carcinoma (MCF-7) cells is 37%, 18% and 22%,

respectively, elevation of serum calcium, serum osteocalcin, bone calcium and duodenal calbindin in rachitic chicks is 14%, 25%, 9% and 17%, respectively, and affinity for pig intestinal receptor and human vitamin D binding protein is 27% and 76%, respectively. *Steroids.* 1995, *60(4)*, 324-32.

269   *24,25-Epoxy-1α-hydroxy-23,23-dimethyl-26,27-dinorvitamin D₃*

9547393 $C_{27}H_{42}O_3$

(1R,3S,5Z)-5-[(2E)-2-[(1R,3aR,7aS)-7a-methyl-1-[(2R)-4-methyl-4-(oxiran-2-yl)pentan-2-yl]-2,3,3a,5,6,7-hexahydro-1H-inden-4-ylidene]ethylidene]-4-methyl-idene-cyclohexane-1,3-diol.
LMST03020162; 24,25-epoxy-23,23-dimethyl-26,27-dinor-9,10-seco-5Z,7E,10(19)-cholestatriene-1S,3R-diol; 24,25-epoxy-1α-hydroxy-23,23-dimethyl-26,27-dinorcholecalciferol; 9547393; Synthesized by epoxidation of the corresponding olefinic precursor which was constructed from Inhoffen-Lythgoe diol.

Compared to 1α,25-dihydroxyvitamin D₃, inhibition of proliferation or differentiation induction of human promyeloid leukemia (HL-60), osteosarcoma (MG-63) and breast carcinoma (MCF-7) cells is 0%, 5% and 0%, respectively, elevation of serum calcium, serum osteocalcin, bone calcium and duodenal calbindin in rachitic chicks are all < 0.1% and affinity for pig intestinal receptor and human vitamin D binding protein is 1% and 0%, respectively. *Steroids.* 1995, *60(4)*, 324-332.

270   *25R,26-Epoxy-1α,24R-dihydroxyvitamin D₃*

9547399 $C_{27}H_{42}O_4$

(1R,3S,5Z)-5-[(2E)-2-[(1R,3aR,7aS)-1-[(2R,5R)-5-hydroxy-5-[(2R)-2-methyloxiran-2-yl]pentan-2-yl]-7a-methyl-2,3,3a,5,6,7-hexahydro-1H-inden-4-ylidene]ethylidene]-4-methylidene-cyclohexane-1,3-diol.
LMST03020181; 25R,26-epoxy-1α,24-dihydroxy-cholecalciferol; 25R,26-epoxy-9,10-seco-5Z,7E,10(19)-cholestatriene-1S,3R,24R-triol; 9547399; Synthesized by Sharpless kinetic epoxidation of the corresponding olefinic precursor. Compared to 1α,25-dihydroxy-vitamin D₃, affinity for pig intestinal receptor and human vitamin D binding protein is 20% and 73%, respectively, inhibition of proliferation or differentiation induction of human promyeloid leukemia (HL-60), osteosarcoma (MG-63) and breast carcinoma (MCF-7) cells are 12%, 160% and 45%, respectively, and elevation of serum calcium, serum osteocalcin, bone calcium and duodenal calbindin in rachitic chicks are 1%, 3%, 1% and < 1%, respectively. *Bioorg Med Chem Lett.* 1993, *3(9)*, 1863-7; *Steroids.* 1995, *60(4)*, 324-32.

271   *25S,26-Epoxy-1α,24R-dihydroxyvitamin D₃*

9547400 $C_{27}H_{42}O_4$

(1R,3S,5Z)-5-[(2E)-2-[(1R,3aR,7aS)-1-[(2R,5R)-5-hydroxy-5-[(2S)-2-methyloxiran-2-yl]pentan-2-yl]-7a-methyl-2,3,3a,5,6,7-hexahydro-1H-inden-4-ylidene]ethylidene]-4-methylidene-cyclohexane-1,3-diol.
LMST03020182; 25S,26-epoxy-1α,24R-dihydroxy-cholecalciferol; 25S,26-epoxy-9,10-seco-5Z,7E,10(19)-cholestatriene-1S,3R,24R-triol; 9547400; Synthesized by Sharpless kinetic epoxidation of the corresponding olefinic precursor. Compared to 1α,25-dihydroxy-vitamin D₃, affinity for pig intestinal receptor and human vitamin D binding protein is 9% and 30%, respectively, inhibition of proliferation or differentiation induction of human promyeloid leukemia (HL-60), osteosarcoma (MG-63) and breast carcinoma (MCF-7) cells is 33%, 4% and 34%, respectively, and elevation of serum calcium, serum osteocalcin, bone calcium and duodenal calbindin in rachitic chicks is 1%, 2%, 4% and 1%, respectively. *Bioorg Med Chem Lett.* 1993, *3(9)*, 1863-7; *Steroids.* 1995, *60(4)*, 324-32.

272   *25R,26-Epoxy-1α,24S-dihydroxyvitamin D₃*

9547401 $C_{27}H_{42}O_4$

(1R,3S,5Z)-5-[(2E)-2-[(1R,3aR,7aS)-1-[(2R,5S)-5-hydroxy-5-[(2R)-2-methyloxiran-2-yl]pentan-2-yl]-7a-methyl-2,3,3a,5,6,7-hexahydro-1H-inden-4-ylidene]ethylidene]-4-methylidene-cyclohexane-1,3-diol.
LMST03020183; 25R,26-epoxy-1α,24S-dihydroxy-cholecalciferol; 25R,26-epoxy-9,10-seco-5Z,7E,10(19)-cholestatriene-1S,3R,24S-triol; 9547401; Synthesized by Sharpless kinetic epoxidation of the corresponding olefinic precursor. Compared to 1α,25-dihydroxy-vitamin D$_3$, affinity for pig intestinal receptor and human vitamin D binding protein is 10% and 7%, respectively, inhibition of proliferation or differentiation induction of human promyeloid leukemia (HL-60), osteosarcoma (MG-63) and breast carcinoma (MCF-7) cells are 44%, 7% and 20%, respectively, and elevation of serum calcium, serum osteocalcin, bone calcium and duodenal calbindin in rachitic chicks are 3%, 2%, 1% and 2%, respectively. *Bioorg Med Chem Lett*. 1993, *3(9)*, 1863-7; *Steroids*. 1995, *60(4)*, 324-32.

273   *25S,26-Epoxy-1α,24S-dihydroxyvitamin D$_3$*
9547402   C$_{27}$H$_{42}$O$_4$

(1R,3S,5Z)-5-[(2E)-2-[(1R,3aR,7aS)-1-[(2R,5S)-5-hydroxy-5-[(2S)-2-methyloxiran-2-yl]pentan-2-yl]-7a-methyl-2,3,3a,5,6,7-hexahydro-1H-inden-4-ylidene]ethylidene]-4-methylidene-cyclohexane-1,3-diol.
LMST03020184; 25S,26-epoxy-1α,24S-dihydroxy-cholecalciferol; 25S,26-epoxy-9,10-seco-5Z,7E,10(19)-cholestatriene-1S,3R,24S-triol; 9547402; Synthesized by Sharpless kinetic epoxidation of the corresponding olefinic precursor. Compared to 1α,25-dihydroxy-vitamin D$_3$, affinity for pig intestinal receptor and human vitamin D binding protein is 5% and 3%, respectively, inhibition of proliferation or differentiation induction of human promyeloid

leukemia (HL-60), osteosarcoma (MG-63) and breast carcinoma (MCF-7) cells is 7%, 15% and 20%, respectively, and elevation of serum calcium, serum osteocalcin, bone calcium and duodenal calbindin in rachitic chicks is 2%, < 1%, < 1% and < 1%, respectively. *Bioorg Med Chem Lett*. 1993, *3*, 1863; *Steroids*. 1995, *60(4)*, 324-32.

274   *1α,25-Dihydroxy-18-oxovitamin D$_3$*
5283702   C$_{27}$H$_{42}$O$_4$

(5Z,7E)-(1S,3R)-1,3,25-trihydroxy-9,10-seco-5,7,10(19)-cholestatrien-18-al.
LMST03020185; 1S,3R,25-trihydroxy-9,10-seco-5Z,7E,10(19)-cholestatrien-18-al; 1α,25-dihydroxy-18-oxocholecalciferol; 5293702.

275   *1α,25-Dihydroxy-24-oxovitamin D$_3$*
5283703   C$_{27}$H$_{42}$O$_4$

(6R)-6-[(1R,3aR,4E,7aS)-4-[(2Z)-2-[(3S,5R)-3,5-di-hydroxy-2-methylidene-cyclohexylidene]ethylidene]-7a-methyl-2,3,3a,5,6,7-hexahydro-1H-inden-1-yl]-2-hydroxy-2-methyl-heptan-3-one.
LMST03020186; 1S,3R,25-trihydroxy-9,10-seco-5Z,7E,10(19)-cholestatrien-24-one; 1α,25-dihydroxy-24-oxo-vitamin D$_3$; 1α,25-dihydroxy-24-oxocholecalciferol; 5283703; Isolated from homogenates of either chick small intestine mucoca or rat kidney incubated with either 1,25-dihydroxyvitamin D$_3$ or 1,24,25-trihydroxy vitamin D$_3$. Metabolite in kidney of 25-hydroxy-24-oxovitamin D$_3$. Competes with 1α,25-dihydroxy-vitamin D$_3$ for the 3.7 S cytosolic receptors present in intestine and thymus; λ_m = 265 nm. *J Biol Chem*. 1983, *258(1)*, 457-63; 1983, *258(22)*, 13458-65.

276   *23S,25-Dihydroxy-24-oxovitamin D₃*
5283704 $C_{27}H_{42}O_4$

(4S,6R)-6-[(1R,3aR,4E,7aS)-4-[(2Z)-2-[(5S)-5-hydroxy-2-methylidene-cyclohexylidene]ethylidene]-7a-methyl-2,3,3a,5,6,7-hexahydro-1H-inden-1-yl]-2,4-dihydroxy-2-methyl-heptan-3-one.
LMST03020187; 23S,25-dihydroxy-24-oxovitamin D₃; 23S,25-dihydroxy-24-oxocholecalciferol; 3S,23S,25-trihydroxy-9,10-seco-5Z,7E,10(19)-cholestatrien-24-one; 5283704; Formed in the kidney as a metabolite of vitamin D₃. Isolated from chicken kidney homogenate incubated with 25-hydroxy-24-oxovitamin D₃ and from rat kidneys perfused with pharmacological concentration of 25-hydroxyvitamin D₃. Synthesized from 3β-hydroxy-5,7-cholestadien-24-one. *In vitro*, 23S,25-dihydroxy-24-oxovitamin D₃ is cleaved at the 23,24-bond to afford 23-hydroxy-24,25,26,27-tetra-norvitamin D₃ and *in vivo*, it is conjugated at the 23-hydroxyl group as the β-glucuronide to be excreted to bile. λ_m = 265 nm. *J Biol Chem*. 1983, *258(1)*, 457-463; *Biochem*. 1983, *22(8)*, 1798-805.

277   *25R-Hydroxyvitamin D₃ 26,23S-lactol*
9547403 $C_{27}H_{42}O_4$

(3R,5S)-5-[(2R)-2-[(1R,3aR,4E,7aS)-4-[(2Z)-2-[(5S)-5-hydroxy-2-methylidene-cyclohexylidene]ethylidene]-7a-methyl-2,3,3a,5,6,7-hexahydro-1H-inden-1-yl]propyl]-3-methyl-oxolane-2,3-diol.
LMST03020188; 25R-hydroxycholecalciferol 26,23S-lactol; 3S,25R-dihydroxy-9,10-seco-Z,7E,10(19)-cholestatrieno-26,23S-lactol; 9547403; Synthesized from 25R-hydroxyvitamin D₃ 26,23S-lactone by reduction with diisobutyl alminium hydride. Produced by incubating vitamin D supplemented chick kidney homogenate with 23S,25R,26-trihydroxyvitamin D₃;

λ_m = 265 nm. *J Org Chem*. 1982, *47(24)*, 4770-2; *J Biol Chem*. 1984, *259(2)*, 884-9.

278   *1α,24R,25-Trihydroxy-22E,23-didehydrovitamin D₃*
5283705 $C_{27}H_{42}O_4$

(E,3R,6S)-6-[(1R,3aR,4E,7aS)-4-[(2Z)-2-[(3S,5R)-3,5-dihydroxy-2-methylidene-cyclohexylidene]ethylidene]-7a-methyl-2,3,3a,5,6,7-hexahydro-1H-inden-1-yl]-2-methyl-hept-4-ene-2,3-diol.
LMST03020189; 1α,24R,25-trihydroxy-22E,23-di-dehydrovitamin D₃; 1α,24R,25-trihydroxy-22E,23-di-dehydrocholecalciferol; 9,10-seco-5Z,7E,10(19),22E-cholestatetraene-1S,3R,24R,25-tetrol; 5283705; Affects intestinal calcium transport, serum calcium concentration, serum inorganic phosphorus concentration and bone ash; λ_m = 265 nm (EtOH). *Chem Pharm Bull (Tokyo)*. 1984, *32(10)*, 3866-72.

279   *1α,24S,25-Trihydroxy-22E,23-didehydro-vitamin D₃*
5283706 $C_{27}H_{42}O_4$

(E,3S,6S)-6-[(1R,3aR,4E,7aS)-4-[(2Z)-2-[(3S,5R)-3,5-dihydroxy-2-methylidene-cyclohexylidene]ethylidene]-7a-methyl-2,3,3a,5,6,7-hexahydro-1H-inden-1-yl]-2-methyl-hept-4-ene-2,3-diol.
LMST03020190; 1α,24S,25-trihydroxy-22E,23-di-dehydrovitamin D₃; 1α,24S,25-trihydroxy-22E,23-di-dehydrocholecalciferol; 9,10-seco-5Z,7E,10(19),22E-cholestatetraene-1S,3R,24S,25-tetrol; 5283706; Affects intestinal calcium transport, serum calcium concentration, serum inorganic phosphorus concentration and bone ash; λ_m = 265 nm (EtOH). *Chem Pharm Bull (Tokyo)*. 1984, *32(10)*, 3866-72.

280   *1α,25R,26-Trihydroxy-22E,23-di-dehydrovitamin D₃*

5283707

$C_{27}H_{42}O_4$

(E,2R,6S)-6-[(1R,3aR,4E,7aS)-4-[(2Z)-2-[(3S,5R)-3,5-dihydroxy-2-methylidene-cyclohexylidene]ethylidene]-7a-methyl-2,3,3a,5,6,7-hexahydro-1H-inden-1-yl]-2-methyl-hept-4-ene-1,2-diol.
LMST03020192;   1α,25R,26-trihydroxy-22E,23-di-dehydrovitamin D₃; 1α,25R,26-trihydroxy-22E,23-di-dehydrocholecalciferol;   9,10-seco-5Z,7E,10(19),22E-cholestatetraene-1S,3R,25R,26-tetrol; 5283707.

281   *1α,25S,26-Trihydroxy-22E,23-di-dehydrovitamin D₃*

5283708

$C_{27}H_{42}O_4$

(E,2S,6S)-6-[(1R,3aR,4E,7aS)-4-[(2Z)-2-[(3S,5R)-3,5-dihydroxy-2-methylidene-cyclohexylidene]ethylidene]-7a-methyl-2,3,3a,5,6,7-hexahydro-1H-inden-1-yl]-2-methyl-hept-4-ene-1,2-diol.
LMST03020192;   1α,25S,26-trihydroxy-22E,23-di-dehydrovitamin D₃; 1α,25S,26-trihydroxy-22E,23-di-dehydrocholecalciferol;   9,10-seco-5Z,7E,10(19),22E-cholestatetraene-1S,3R,25S,26-tetrol; 5293708; Prepared by a convergent method in which the CD ring synthon and the A-ring precursor were combined by Wittig-Horner reaction. Compared to 1α,25-dihydroxyvitamin D₃, Vitamin D receptor binding (chick intestine) is 5.5%, inhibition of cell (HL-60) proliferation is 50%, induction cell (HL-60) differentiation is 100% and ^{45}Ca retention in kidney in rats is <10%. Vitamin D Chemical, Biochemical and Clinical Update. Proceedings of the Sixth Workshop on Vitamin D Merano, Italy, March, 1985.

282   *Δ²²-1,25S,26-Trihydroxyvitamin D₃*

144300-56-1

$C_{27}H_{42}O_4$

(E)-6-[(4Z)-4-[(2Z)-2-(3,5-dihydroxy-2-methylidene-cyclohexylidene)ethylidene]-7a-methyl-2,3,3a,5,6,7-hexahydro-1H-inden-1-yl]-2-methyl-hept-4-ene-1,2-diol.
LMST03020191;   1,25S,26-Trihydroxy-Δ²²-cholalciferol; Δ²²-1,25S,26-Trihydroxyvitamin D₃; Δ²²-1,25S,26-OHD₃; 9,10-Seco-cholesta-5Z,7E,10(19),22E-tetraene-1,3β,25S,26-tetrol;   6438767;   144300-56-1; Nonhypercalcemic analog of vitamin D₃. Induces the human osteocalcin gene promoter stably transfected into rat osteosarcoma cells (ROSCO-2). *J Bone Miner Res.*, 1991, *6(8)*, 893-9.

283   *1β,25-Dihydroxy-24-oxo-vitamin D₃*

76338-50-6

$C_{27}H_{42}O_4$

6R)-6-[(1R,3aR,4E,7aS)-4-[(2Z)-2-[(3R,5S)-3,5-di-hydroxy-2-methylidene-cyclohexylidene]ethylidene]-7a-methyl-2,3,3a,5,6,7-hexahydro-1H-inden-1-yl]-2-hydroxy-2-methyl-heptan-3-one.
1,25-Dihydroxy-24-oxocholecalciferol;   24-KdhvD₃;
24-Keto-1,25-dihydroxyvitamin D₃; 6438808; 76338-50-6; A metabolite in rat intestinal mucosa homogenates of 1,25-dihydroxyvitamin D₃. Less active than vitamin D₃ in bone calcium mobilization. Has affinity equivalent to that of 1,24R,25-trihydroxy-vitamin D₃ for the 3.7 S cytosolic receptor specific for 1,25-dihydroxyvitamin D₃ in the intestine and thymus. *J Biol Chem.* 1983, *258(22)*, 13458-65.

284   *1,25-Dihydroxy-23-oxo-vitamin D₃*

82095-23-6

$C_{27}H_{42}O_4$

(6R)-6-[(1R,3aR,4E,7aS)-4-[(2Z)-2-[(3R,5S)-3,5-dihydroxy-2-methylidene-cyclohexylidene]ethylidene]-7a-methyl-2,3,3a,5,6,7-hexahydro-1H-inden-1-yl]-2-hydroxy-2-methyl-heptan-4-one.

1,25-Dov; 23-Keto-1,25-dihydroxyvitamin D$_3$; 1,25-Dihydroxy-23-oxo-vitamin D$_3$; 1,25-Dihydroxy-23-oxocholecalciferol; 9,10-Secocholesta-5Z,7E,10(19)-trien-23-one, 1α,3β,25-trihydroxy-; 6439608; 82095-23-6; Vitamin D$_3$ metabolite with high affinity for the 1,25-dihydroxyvitamin D$_3$ receptor. Metabolite of 1,25-dihydroxyvitamin D$_3$ in homogenates of small intestinal mucosa of vitamin D$_3$-replete chicks. A co-metabolite is 1α,25,26-trihydroxy-23-oxo-vitamin D$_3$. *J Biol Chem*. 1982, *257(9)*, 5097-102.

285 *Calcifediol lactol*

83136-06-5   C$_{27}$H$_{42}$O$_4$

(3R,5S)-5-[(2R)-2-[(1R,3aR,4E,7aS)-4-[(2Z)-2-[(5R)-5-hydroxy-2-methylidene-cyclohexylidene]ethylidene]-7a-methyl-2,3,3a,5,6,7-hexahydro-1H-inden-1-yl]propyl]-3-methyl-oxolane-2,3-diol.

Calcifediol lactol; (23S,25R)-25-Hydroxyvitamin D$_3$ 26,23-lactol; 25-Hydroxyvitamin D$_3$ 26,23-lactol; 9,10-Secocholesta-5Z,7E,10(19)-triene-3β,25R,26-triol, 23S,26-epoxy-; LMST03020188; 6439640; 83136-06-5; Isolated from the incubation mixture of 23S,25R,26-trihydroxy-vitamin D$_3$ with kidney homogenates prepared from vitamin D-supplemented chicks. Converted to the lactone in kidney homogenates regardless of the vitamin D status of the animals; λ_m = 265 nm (95% EtOH). *J Biol Chem*. 1984, *259(2)*, 884-9.

286 *1α,25,26-Trihydroxy-22-enevitamin D$_3$*

6505210   C$_{27}$H$_{42}$O$_4$

(E,2S,6S)-6-[(1R,3aR,4E,7aS)-4-[(2E)-2-[(3S,5R)-3,5-dihydroxy-2-methylidene-cyclohexylidene]ethylidene]-7a-methyl-2,3,3a,5,6,7-hexahydro-1H-inden-1-yl]-2-methyl-hept-4-ene-1,2-diol.

1α,25,26-trihydroxy-22-enecholecalciferol; 6505210. *J Biol Chem*. 1984, *259(2)*, 884-9.

287 *Δ^{22},1,25S,26-Trihydroxyvitamin D$_3$*

144300-56-1   C$_{27}$H$_{42}$O$_4$

(E)-6-[(4Z)-4-[(2Z)-2-(3,5-dihydroxy-2-methylidene-cyclohexylidene)ethylidene]-7a-methyl-2,3,3a,5,6,7-hexahydro-1H-inden-1-yl]-2-methyl-hept-4-ene-1,2-diol.

Δ^{22}-1,25S,26-OHD$_3$; Δ^{22}-1,25S,26-Trihydroxyvitamin D$_3$; 1,25S,26-Trihydroxy-Δ^{22}-cholecalciferol; 6438767; 144300-56-1; Induces osteocalcin promotion, similarly to 1,25-dihydroxyvitamin D$_3$. Can effect cellular differentiation but lacks hypercalcemic activity *in vitro*. *J Bone Miner Res*. 1991, *6(8)*, 893-9.

288 *3β-Hydroxy-5α,6α-epoxy-9-oxo-9,10-seco-5-cholest-7-en-11-al*

C$_{27}$H$_{42}$O$_4$

3-hydroxy-5,6-epoxy-9-oxo-9,10-seco-5-cholest-7-en-11-al; 3-HEOSC; Isolated from the marine sponge *spongia matamata*; Colorless needles; mp = 171-3°;

$[\alpha]_D$ = -6.7° (c = 0.24, $CHCl_3$). *J Nat Prod.* 1997, *60(2)*, 195-8.

289   *23,25-Dihydroxy-24-oxovitamin D₃*
84164-55-6 $C_{27}H_{42}O_4$

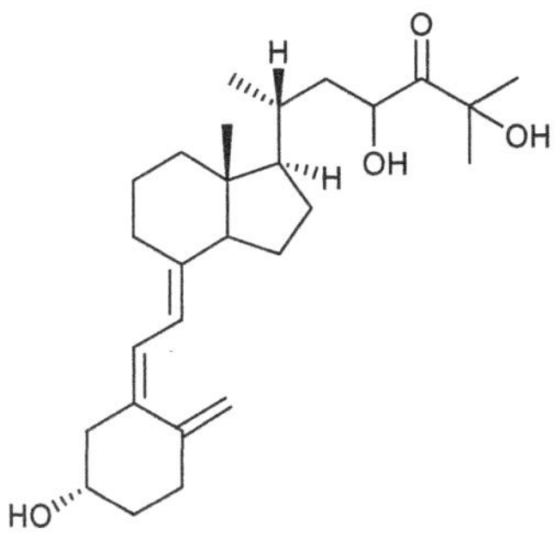

(6R)-6-[(1R,4E,7aS)-4-[(2Z)-2-[(5S)-5-hydroxy-2-methylidene-cyclohexylidene]ethylidene]-7a-methyl-2,3,3a,5,6,7-hexahydro-1H-inden-1-yl]-2,4-dihydroxy-2-methyl-heptan-3-one.
23,25-Dihydroxy-24-oxovitamin D₃; 23,25-DO-Vitamin D₃; 23,25-Dihydroxy-24-keto-cholecalciferol; 23,25-Dihydroxy-24-oxo-vitamin D₃; 23,25-Dihydroxy 24-oxocholecalciferol; 9,10-Secochlesta-5Z,7E,10(19)-trien-24-one, 3β,23,25-trihydroxy-; LMST03020187; 6439689; 84164-55-6; Major metabolite of 24R,25-dihydroxyvitamin D₃ in bile and kidney. *J Biol Chem.* 1983, *258(1)*, 457-63.

290   *1,25-Dihydroxyvitamin D₃-23,26-lactol*
87680-15-7 $C_{27}H_{42}O_5$

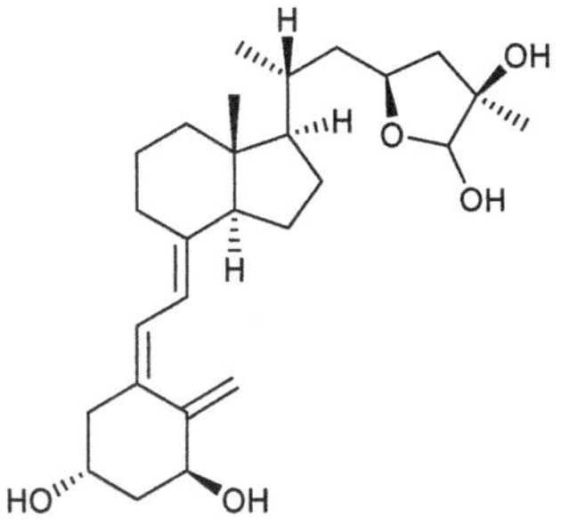

LMST03020193; 1,25-Dihydroxyvitamin D₃-23,26-lactol; Calcitriol-23,26-lactol; 87680-15-7; Affinity for the chick intestinal vitamin D receptor is 2% that of 1α,25-dihydroxyvitamn D₃. *Endocrinology,* 1990, *127*, 695-701; *J Biol Chem.* 1987, *262(15)*, 7165-70.

291   *1α,25R-Dihydroxyvitamin D₃ 26,23S-lactol*
9547404 $C_{27}H_{42}O_5$

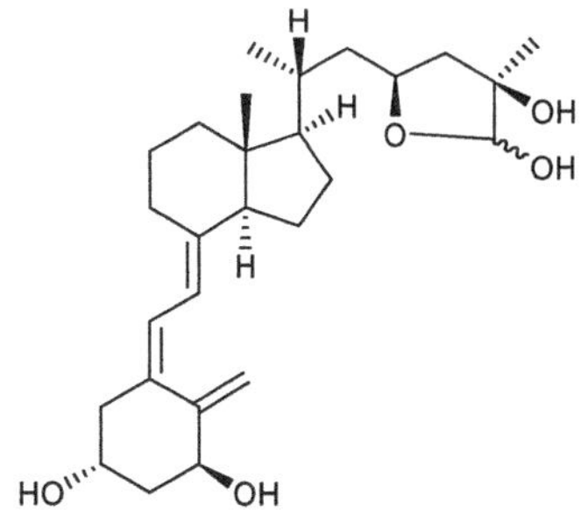

(3R,5S)-5-[(2R)-2-[(1R,3aR,4E,7aS)-4-[(2Z)-2-[(3S,5R)-3,5-dihydroxy-2-methylidene-cyclohexylidene]ethylidene]-7a-methyl-2,3,3a,5,6,7-hexahydro-1H-inden-1-yl]propyl]-3-methyl-oxolane-2,3-diol.
LMST03020193; 1α,25R-dihydroxycholeciferol 26, 23S-lactol; 1S,3R,25R-trihydroxy-9,10-seco- 5Z,7E,10,(19)-cholestatrieno-26,23S-lactol; 9547404; Is 1/59 as active as 1α,25-dihydroxyvitamin D₃ in its affinity for the chick intestinal receptor. *Endocrinology.* 1990, *127(2)*, 695-701.

292   *1α,23R,25-Trihydroxy-24-oxovitamin D₃*
9547405 $C_{27}H_{42}O_5$

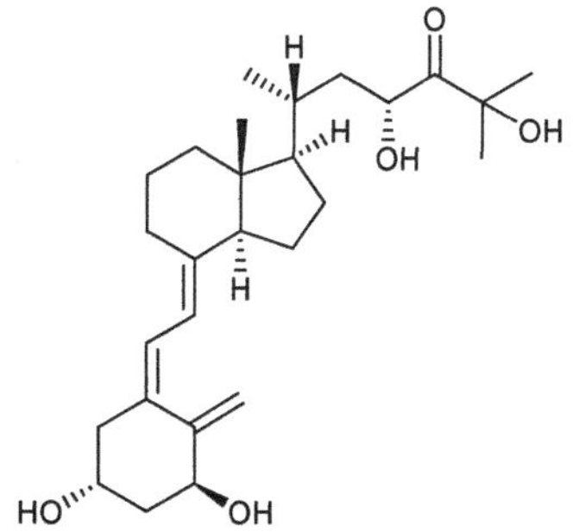

(4R,6R)-6-[(1R,3aR,4E,7aS)-4-[(2Z)-2-[(3S,5R)-3,5-dihydroxy-2-methylidene-cyclohexylidene]ethylidene]-7a-methyl-2,3,3a,5,6,7-hexahydro-1H-inden-1-yl]-2,4-dihydroxy-2-methyl-heptan-3-one.
LMST03020194; 1α,23R,25-trihydroxy-24-oxocholecalciferol; 1S,3R,23R,25-tetrahydroxy-9-10-seco-5Z, 7E,10(19)-cholestatrien-24-one; 9547405; Synthesized by Pd-catalyzed coupling of the appropriate CD ring fragment with A ring enyne, prepared from quinic acid, as the key step. Suppresses PTH secretion significantly at 10^{-12} M and is equipotent with 1,25-dihydroxy-vitamin D₃, though the affinity of the former for the receptor in parathyroid cells was 10 times less effective than that of the latter. *Biochemistry* 1997, *36(31)*, 9429-37.

293   *1α,23S,25-Trihydroxy-24-oxovitamin D₃*
9547406 $C_{27}H_{42}O_5$

Vitamin D Handbook

(4S,6R)-6-[(1R,3aR,4E,7aS)-4-[(2Z)-2-[(3S,5R)-3,5-dihydroxy-2-methylidene-cyclohexylidene]ethylidene]-7a-methyl-2,3,3a,5,6,7-hexahydro-1H-inden-1-yl]-2,4-dihydroxy-2-methyl-heptan-3-one.
LMST03020195; 1α,23S,25-trihydroxy-24-oxocholecalciferol; 1S,3R,23S,25-tetrahydroxy-9-10-seco-5Z,7E,10(19)-cholestatrien-24-one; 9547406; Isolated from homogenates of either chick small intestine mucosa or rat kidney incubated with either 1α,25-dihydroxyvitamin D_3 or 1,24,25-trihydroxyvitamin D_3. Synthesized by Pd-catalyzed coupling of the appropriate CD ring fragment with A ring enyne, prepared from quinic acid, as the key step. Metabolized to 1-OH-tetranorvitamin D_3 23-carboxylic acid (calcitroic acid) *via* 1,23-(OH)2-tetranorvitamin D_3. Suppresses PTH secretion significantly at 10^{-12} M and is equipotent with 1α,25-dihydroxyvitamin D_3, though the affinity for the receptor in parathyroid cells was 10 times less effective than that of the latter; λ_m = 265 nm. *J Biol Chem.* 1983, *258(22)*, 13458-65; *Biochemistry.* 1979, *18(18)*, 3977-83; 1987, *26(1)*, 324-31; 1997, *36(31)*, 9429-37.

294   *1,25,26-Trihydroxy-23-oxo-vitamin D_3*
82095-24-7 $C_{27}H_{42}O_5$

(6R)-6-[(1R,3aR,4E,7aS)-4-[(2Z)-2-[(3R,5S)-3,5-dihydroxy-2-methylidene-cyclohexylidene]ethylidene]-7a-methyl-2,3,3a,5,6,7-hexahydro-1H-inden-1-yl]-1,2-dihydroxy-2-methyl-heptan-4-one.
1,25,26-Trihydroxy-23-oxo-vitamin D_3; 1,25,26-Tov; 1,25,26-Trihydroxy-23-oxocholecalciferol; 9,10-Seco-cholesta-5Z,7E,10(19)-trien-23-one, 1α,3β,25,26-tetra-hydroxy-; 6439609; 82095-24-7; Metabolite of 1,25-dihydroxyvitamin D_3. *J Biol Chem.* 1982, *257(9)*, 5097-102.

295   *1,23,25-Trihydroxy-24-oxo-vitamin D_3*
87147-48-6 $C_{27}H_{42}O_5$

(6R)-6-[(1R,3aR,4E,7aS)-4-[(2Z)-2-[(3R,5S)-3,5-dihydroxy-2-methylidene-cyclohexylidene]ethylidene]-7a-methyl-2,3,3a,5,6,7-hexahydro-1H-inden-1-yl]-2,4-dihydroxy-2-methyl-heptan-3-one.
LMST03020194; 1,23,25-Trihydroxy-24-oxo-vitamin D_3; 1,23,25-Tov; 1,23,25-Trihydroxy-24-oxocholecalciferol; 9,10-Seco-cholesta-5Z,7E,10(19)-trien-24-one, 1α,3β,23,25-tetra-hydroxy-; 6439809; 87147-48-6; A major natural metabolite of 1α,25-dihydroxyvitamin D_3. Suppresses parathyroid hormone (PTH) secretion in primary cultures of bovine parathyroid cells. *J Biol Chem.* 1983, *258(22)*, 13458-65.

296   *1,23,25-Trihydroxy-24-oxo-vitamin D_3*
87678-01-1 $C_{27}H_{42}O_5$

(6R)-6-[(1R,4E,7aS)-4-[(2Z)-2-[(5R)-3,5-dihydroxy-2-methylidene-cyclohexylidene]ethylidene]-7a-methyl-2,3,3a,5,6,7-hexahydro-1H-inden-1-yl]-2,4-dihydroxy-2-methyl-heptan-3-one.
24-Keto-1,23,25-trihydroxyvitamin D_3; 1,23,25-Trihydroxy-24-ketovitamin D_3; 6439818; 87678-01-1; A major 1,25-dihydroxyvitamin D_3 metabolite, produced by intestinal mucosa cells isolated from rats dosed chronically with 1,25-dihydroxyvitamin D_3. Biosynthesis is *via* 1,24,25-trihydroxyvitamin D_3 and 24-keto-1,25-dihydroxyvitamin D_3, which are physiological metabolites of 1α,25-dihydroxyvitamin D_3. *Biochemistry.* 1983, *22(25)*, 5848-53.

297   *2α-Chloro-1β,25-dihydroxyvitamin D₃*

9547407 $C_{27}H_{43}ClO_3$

(1R,2S,3S,5Z)-5-[(2E)-2-[(1R,3aR,7aS)-1-[(2R)-6-
hydroxy-6-methyl-heptan-2-yl]-7a-methyl-2,3,3a,5,6,7-
hexahydro-1H-inden-4-ylidene]ethylidene]-2-chloro-4-
methylidene-cyclohexane-1,3-diol.
LMST03020196;   2S-chloro-9,10-seco-5Z,7E,10(19)-
cholestatriene-1S,3R,25-triol;   2α-chloro-1β,25-di-
hydroxycholecalciferol;   9,10-Seco-2α-chlorocholesta-
5,7,10(19)-trien-1β,3β,25-triol; 9547407; Synthesized
from   1β,2β-epoxy-25-hydroxy-7-dehydrocholesterol
via   reaction with hydrogen chloride followed by
photochemical transformation. Binds to a site in the
vitamin D binding protein and competes with 1-
epicalcitriol for this site. Affinity for rat intestinal VDR
is 10 % that of 25-hydroxyvitamin D₃; in a gene
transcription (VDRE-CAT assay), it is nearly 10,000
times   weaker   than   1α,25-dihydroxyvitamin D₃.
Crystals; mp = 173-179°; λₘ = 264 nm (ε = 15300
MeOH). **Steroids**. 1998, **63(1)**, 28-36.

298   *2β-Chloro-1α,25-dihydroxyvitamin D₃*

9547694 $C_{27}H_{43}ClO_3$

(1R,2R,3R,5Z)-5-[(2E)-2-[(1R,3aR,7aS)-1-[(2R)-6-
hydroxy-6-methyl-heptan-2-yl]-7a-methyl-2,3,3a,5,6,7-
hexahydro-1H-inden-4-ylidene]ethylidene]-2-chloro-4-
methylidene-cyclohexane-1,3-diol.
LMST03020601;   2β-chloro-1α,25-dihydroxychole-
calciferol; 2R-chloro-9,10-seco-5Z,7E,10(19)-cholesta-
triene-1R,3R,25-triol; 9547694; Synthesized photochemically from a C(22) steroid precursor. Compared
to 1α,25-dihydroxyvitamin D₃, affinity for chicken
intestine vitamin D receptor is 100%, affinity for sheep
serum vitamin D binding protein is 140%, inhibition of

proliferation of C3H10T(BMP-4) cells is 0.9% (FCS-
free), 11% (10% FCS); CAT assay (osteocalcin) is
130%, inhibition of adipogenesis is 8%. **Steroids,**
1998, **63(12)**, 633-43.

299   *3-Fluoro-9,10-secocholesta-5,7,10(19)-
triene*

53839-02-4 $C_{27}H_{43}F$

(1R,3aR,4E,7aS)-4-[(2Z)-2-[(5S)-5-fluoro-2-
methylidene-cyclohexylidene]ethylidene]-7a-methyl-1-
[(2R)-6-methylheptan-2-yl]-2,3,3a,5,6,7-hexahydro-
1H-indene.
3-Fluoro-9,10-secocholestatrien-5,7,10(19); 3α-Fluoro-
9,10-secocholestatrien-5,7,10(19); 9,10-Secocholesta-
5,7,10(19)-triene, 3α-fluoro-; 6452902; 53839-02-4.
Ukr Biochim Zh. 1974, **46(5)**, 627-30.

300   *3-Deoxy-3β-fluorovitamin D₃*

$C_{27}H_{43}F$

9,10-Secocholesta-3β-fluoro-5Z,7E,10(19)-triene.
Has essentially no vitamin D activity. Inactive in the
chick duodenum and shows very poor binding affinity
to the chick intestinal cytosol receptor. **J Steroid
Biochem.** 1985, **22(4)**, 469-474.

301   *(10Z)-19-Fluorovitamin D₃*

9547409 $C_{27}H_{43}FO$

(1S,3Z,4Z)-3-[(2E)-2-[(1R,3aR,7aS)-7a-methyl-1-[(2R)-
6-methylheptan-2-yl]-2,3,3a,5,6,7-hexahydro-1H-
inden-4-ylidene]ethylidene]-4-(fluoromethylidene)-
cyclohexan-1-ol.
LMST03020198; (10Z)-19-fluorocholecalciferol; 19-
fluoro-9,10-seco-5Z,7E,10Z(19)-cholestatrien-3S-ol;
9547409; Synthesized from vitamin D₃ *via* elimination
of the exocyclic methylene and introduction of a
fluoromethylene group; λ$_m$ = 263 nm (95% EtOH).
Chem Pharm Bull (Tokyo). 2000, **48(10)**, 1484-93.

302   **(10E)-19-Fluorovitamin D₃**
9547410   C₂₇H₄₃FO

(1S,3E,4Z)-3-[(2E)-2-[(1R,3aR,7aS)-7a-methyl-1-[(2R)-6-
methylheptan-2-yl]-2,3,3a,5,6,7-hexahydro-1H-inden-
4-ylidene]ethylidene]-4-(fluoromethylidene)cyclo-
hexan-1-ol.
LMST03020200; (5E,10Z)-19-fluorocholecalciferol; 19-
fluoro-9,10-seco-5E,7E,10Z(19)-cholestatrien-3S-ol;
9547411; Synthesized from vitamin D₃ by
regioselective electrophilic 19-fluorination of its sulfur
dioxide adduct; λ$_m$ = 270 nm (EtOH). **Tet Lett**. 1996,
37(37), 6753-4.

304   **(5E,10E)-19-Fluorovitamin D₃**
9547412   C₂₇H₄₃FO

(1S,3Z,4E)-3-[(2E)-2-[(1R,3aR,7aS)-7a-methyl-1-[(2R)-6-
methylheptan-2-yl]-2,3,3a,5,6,7-hexahydro-1H-inden-
4-ylidene]ethylidene]-4-(fluoromethylidene)cyclo-
hexan-1-ol.
LMST03020199; (10E)-19-fluorocholecalciferol; 19-
fluoro-9,10-seco-5Z,7E,10E(19)-cholestatrien-3S-ol;
9547410; Synthesized from vitamin D₃ by
regioselective electrophilic 19-fluorination of its sulfur
dioxide adduct; λ$_m$ = 260 nm (ε = 20900 EtOH). **Tet
Lett**. 1996, **37(37)**, 6753-4. **Chem Pharm Bull (Tokyo)**.
2000, **48(10)**, 1484-93.

303   **(5E,10Z)-19-Fluorovitamin D₃**
9547411   C₂₇H₄₃FO

(1S,3E,4E)-3-[(2E)-2-[(1R,3aR,7aS)-7a-methyl-1-[(2R)-6-
methylheptan-2-yl]-2,3,3a,5,6,7-hexahydro-1H-inden-
4-ylidene]ethylidene]-4-(fluoromethylidene)cyclo-
hexan-1-ol.
LMST03020201; (5E,10E)-19-fluorocholecalciferol; (5E,
7E)-(3S)-19-fluoro-9,10E-seco-5,7,10(19)-cholestatrien-
3-ol; 9547412; Synthesized from vitamin D₃ by
regioselective electrophilic 19-fluorination of its sulfur
dioxide adduct; λ$_m$ = 209, 269 nm (EtOH). **Tet Lett**.
1996, **37(37)**, 6753-4.

305   **22-Fluorovitamin D₃**
110536-31-7   C₂₇H₄₃FO

(1S,3Z)-3-[(2E)-2-[(1R,3aR,7aS)-1-[(2S)-3-fluoro-6-methyl-heptan-2-yl]-7a-methyl-2,3,3a,5,6,7-hexahydro-1H-inden-4-ylidene]ethylidene]-4-methylidene-cyclohexan-1-ol.

LMST03020202; 22-fluorocholecalciferol; 22-fluoro-9,10-seco-5Z,7E,10(19)-cholestatrien-3S-ol; 22-Fluorovitamin D_3; 9,10-Secocholesta-5Z,7E,10(19)-trien-3β-ol, 22-fluoro-; 6443891; 9547413; 110536-31-7; Induces calcium binding protein in duodenal organ culture but fails to increase active intestinal calcium transport and does not mobilize bone and soft tissue calcium. Is not an antagonist of 1,25-dihydroxyvitamin D_3. The presence of a fluorine group at the C-22 position inhibits the binding of the vitamin to rat vitamin D binding protein when compared to the binding of its hydrogen analog, vitamin D_3; λ_m = 264 nm (EtOH). *Steroids.* 1986, *48(1-2)*, 93-108.

306   6-Fluorovitamin D_3

91625-75-1 $C_{27}H_{43}FO$

(1S,3E)-3-[(2E)-2-[(1R,3aR,7aS)-7a-methyl-1-[(2R)-6-methylheptan-2-yl]-2,3,3a,5,6,7-hexahydro-1H-inden-4-ylidene]-1-fluoro-ethylidene]-4-methylidene-cyclohexan-1-ol.

LMST03020197; 6-fluorocholecalciferol; 6-fluoro-9,10-seco-5E,7E,10(19)-cholestatrien-3S-ol; 6-Fluorovitamin D_3; 6-Fluoro-vitamin D_3; 9,10-Secocholesta-5E,7E,10(19)-trien-3β-ol, 6-fluoro-; 6438542; 9547408; 91625-75-1; Synthesized from 6-fluorocholesteryl acetate *via* conventional photochemical conversion of the corresponding 7-dehydro derivative. An antagonist of the biological actions of vitamin D_3 and its metabolites which interacts with the intestinal receptor for 1α,25-dihydroxyvitamin D_3. Competes with 1α,25-dihydroxyvitamin D_3 for binding to the chick intestinal receptor, but is only 0.26% as effective. Shows no biological activity with either intestinal calcium

absorption or bone calcium mobilization at doses up to 130 nmol but significantly inhibits vitamin D-mediated intestinal calcium absorption; λ_m = 268 nm (ε = 10300 C_6H_{16}). *Arch Biochem Biophys.* 1984, *233(1)*, 127-32.

307   2-Fluorovitamin D_3

103638-37-5 $C_{27}H_{43}FO$

(1S,2S,5Z)-5-[(2E)-2-[(1R,3aR,7aS)-7a-methyl-1-[(2R)-6-methylheptan-2-yl]-2,3,3a,5,6,7-hexahydro-1H-inden-4-ylidene]ethylidene]-2-fluoro-4-methylidene-cyclohexan-1-ol.

2-Fluorovitamin D_3; 9,10-Secocholesta-5Z,7E,10(19)-trien-3β-ol, 2α-fluoro-; 6438974; 103638-37-5; Synthesized (1986). *Chem Pharm Bull (Tokyo).* 1986, *34(4)*, 1568-72.

308   25-Fluorovitamin D_3

63819-58-9 $C_{27}H_{43}FO$

(1R,3Z)-3-[(2E)-2-[(1R,3aR,7aS)-1-[(2R)-6-fluoro-6-methyl-heptan-2-yl]-7a-methyl-2,3,3a,5,6,7-hexahydro-1H-inden-4-ylidene]ethylidene]-4-methylidene-cyclohexan-1-ol.

25-Fluorocholeciferol; 9,10-Secocholesta-5Z,7E,10(19)-trien-3β-ol, 25-fluoro-; 6443589; 63819-58-9. *Tet Let.* 1977, *18(27)*, 2315-6; USP 5391777.

309   1-Fluorovitamin D_3

69879-46-5 $C_{27}H_{43}FO$

(1R,3Z)-3-[(2E)-2-[(1R,3aR,7aS)-7a-methyl-1-[(2R)-6-methylheptan-2-yl]-2,3,3a,5,6,7-hexahydro-1H-inden-4-ylidene]ethylidene]-5-fluoro-4-methylidene-cyclohexan-1-ol.
1-Fluorovitamin D_3; 9,10-Secocholesta-5Z,7E,10(19)-trien-3α-ol, 1-fluoro-; 6443613; 69879-46-5; Synthetic vitamin D_3 derivative. Synthesized in unlabelled and tritium-labelled forms. Metabolized to 1α,25-dihydroxyvitamin D_3. *Biochem J.* 1979, *182(1)*, 1-9.

310   *(5Z,10Z)-19-Fluoro-1α-hydroxyvitamin-D₃*
9547414 $C_{27}H_{43}FO_2$

(1R,3S,4Z,5Z)-5-[(2E)-2-[(1R,3aR,7aS)-7a-methyl-1-[(2R)-6-methylheptan-2-yl]-2,3,3a,5,6,7-hexahydro-1H-inden-4-ylidene]ethylidene]-4-(fluoromethylidene)cyclohexane-1,3-diol.
LMST03020203;   19-fluoro-9,10-seco-5Z,7E,10Z(19)-cholestatriene-1S,3R-diol;   (5Z,10Z)-19-fluoro-1α-hydroxycholecalciferol; 9547414;   λ$_m$ = 262 nm (EtOH).

311   *(5Z,10E)-19-Fluoro-1α-hydroxyvitamin-D₃*
9547415 $C_{27}H_{43}FO_2$

311   *(5Z,10E)-19-Fluoro-1α-hydroxyvitamin-D₃*
9547415 $C_{27}H_{43}FO_2$

(1R,3S,4E,5Z)-5-[(2E)-2-[(1R,3aR,7aS)-7a-methyl-1-[(2R)-6-methylheptan-2-yl]-2,3,3a,5,6,7-hexahydro-1H-inden-4-ylidene]ethylidene]-4-(fluoromethylidene)-cyclohexane-1,3-diol.
LMST03020204;   19-fluoro-9,10-seco-5Z,7E,10E(19)-cholestatriene-1S,3R-diol;   (5Z,10E)-19-fluoro-1α-hydroxycholecalciferol; 9547415;   λ$_m$ = 264 nm (EtOH).

312   *(5E,10Z)-19-Fluoro-1α-hydroxyvitamin-D₃*
9547416 $C_{27}H_{43}FO_2$

(1S,3R,4Z,5E)-5-[(2E)-2-[(1R,3aR,7aS)-7a-methyl-1-[(2R)-6-methylheptan-2-yl]-2,3,3a,5,6,7-hexahydro-1H-inden-4-ylidene]ethylidene]-4-(fluoromethylidene)-cyclohexane-1,3-diol.
LMST03020205;   19-fluoro-9,10-seco-5E,7E,10Z(19)-cholestatriene-1S,3R-diol;   (5E,10Z)-19-fluoro-1α-hydroxycholecalciferol; 9547416;   λ$_m$ = 269 nm (EtOH).

313   *(5E,10E)-19-Fluoro-1α-hydroxyvitamin-D₃*
9547417 $C_{27}H_{43}FO_2$

(1S,3R,4E,5E)-5-[(2E)-2-[(1R,3aR,7aS)-7a-methyl-1-[(2R)-6-methylheptan-2-yl]-2,3,3a,5,6,7-hexahydro-1H-inden-4-ylidene]ethylidene]-4-(fluoromethylidene)-cyclohexane-1,3-diol.
LMST03020206; 19-fluoro-9,10-seco-5E,7E,10E(19)-cholestatriene-1S,3R-diol; (5E,10E)-19-fluoro-1α-hydroxycholecalciferol; 9547417; λ_m = 272 nm (EtOH).

314   *25-Fluoro-1α-hydroxyvitamin D₃*
64164-40-5 $C_{27}H_{43}FO_2$

(1R,3S,5Z)-5-[(2E)-2-[(1R,3aR,7aS)-1-[(2R)-6-fluoro-6-methyl-heptan-2-yl]-7a-methyl-2,3,3a,5,6,7-hexa-hydro-1H-inden-4-ylidene]ethylidene]-4-methylidene-cyclohexane-1,3-diol.
LMST03020207; 25-fluoro-9,10-seco-5Z,7E,10(19)-cholestatriene-1S,3R-diol; 25-fluoro-1α-hydroxychole-calciferol; 1-Hydroxy-25-fluorovitamin D_3; 1α-Hydroxy-25-fluorovitamin D_3; 9,10-Secocholesta-5Z,7E,10(19)-triene-1α,3β-diol, 25-fluoro-; 6443824; 9547418; 64164-40-5; Synthetic vitamin D_3 derivative. Is 315 times less effective than 1α,25-dihydroxyvitamin D_3 binding to chick intestinal cytosol protein in competitive protein-binding assay, about 50 times less active in intestinal calcium transport and bone mineral mobilization and 40 times less antirachitic. Only 2% as active as 1α,25-dihydroxyvitamin D_3 in stimulation of intestinal calcium transport but otherwise has 1α,25-dihydroxyvitamin D_3 activity in the rat. *Biochemistry*. 1978, 17(12), 2387-92.

315   *25-Fluoro-24R-hydroxyvitamin D₃*
9547419 $C_{27}H_{43}FO_2$

(3R,6R)-6-[(1R,3aR,4E,7aS)-4-[(2Z)-2-[(5S)-5-hydroxy-2-methylidene-cyclohexylidene]ethylidene]-7a-methyl-2,3,3a,5,6,7-hexahydro-1H-inden-1-yl]-2-fluoro-2-methyl-heptan-3-ol.
LMST03020208; 25-fluoro-24-hydroxychole-ciferol; 25-fluoro-9,10-seco-5Z,7E,10(19)-cholesta-triene-3S,24R-diol; 9547419.

316   *1α-Fluoro-25-hydroxyvitamin D₃*
9547420 $C_{27}H_{43}FO_2$

(6R)-6-[(1R,3aR,4E,7aS)-4-[(2Z)-2-[(3S,5R)-5-fluoro-3-hydroxy-2-methylidene-cyclohexylidene]ethylidene]-7a-methyl-2,3,3a,5,6,7-hexahydro-1H-inden-1-yl]-2-methyl-heptan-2-ol.
LMST03020209; 1α-fluoro-25-hydroxycholecalciferol; (5Z,7E)-(1S)-fluoro-9,10-seco-5Z,7E,10(19)-cholesta-triene-3R,25-diol; 9547420; Prepared *via* the 3,5-cyclovitamin D intermediate. In HL-60 human promyelocytes differentiation is one order of magnitude less active than 1α,25-dihydroxyvitamin D_3. *J Nutr Biochem*. 1993, *4(1)*, 49-57.

317   *24-Fluoro-25-hydroxyvitamin D₃*
9547421 $C_{27}H_{43}FO_2$

(6R)-6-[(1R,3aR,4E,7aS)-4-[(2Z)-2-[(5S)-5-hydroxy-2-methylidene-cyclohexylidene]ethylidene]-7a-methyl-2,3,3a,5,6,7-hexahydro-1H-inden-1-yl]-3-fluoro-2-methyl-heptan-2-ol.
LMST03020210; 24-fluoro-25-hydroxycholecalciferol; (5Z,7E)-(3S)-24-fluoro-9,10-seco-5Z,7,10(19)-cholesta-triene-3,25-diol; 9547421; Synthesized from vitamin D_2 *via* ozonolysis of its sulfur dioxide adduct. *J Chromatogr B Biomed Appl*. 1995, *672(1)*, 63-71.

318   ***1-Fluoro-25-hydroxyvitamin D$_3$***
74041-09-1   C$_{27}$H$_{43}$FO$_2$

(6R)-6-[(1R,3aR,4E,7aS)-4-[(2Z)-2-[(5S)-3-fluoro-5-
hydroxy-2-methylene-cyclohexylidene]ethylidene]-
7a-methyl-2,3,3a,5,6,7-hexahydro-1H-inden-1-yl]-2-
methyl-heptan-2-ol.
1α-Fluorocalcifediol; 1-Fluoro-25-hydroxyvitamin D$_3$;
1α-Fluoro-25-hydroxycholecalciferol;   6443860;
74041-09-1; Synthesized (1984). *Chem Pharm Bull
(Tokyo)* 1984, *32(9)*, 3525-31.

319   ***24-Hydroxy-25-fluorovitamin D$_3$***
                                                C$_{27}$H$_{43}$FO$_2$

(6R)-6-[(3aR,4E,7aS)-4-[(2Z)-2-[(5R)-5-hydroxy-2-
methylene-cyclohexylidene]ethylidene]-7a-methyl-
2,3,3a,5,6,7-hexahydro-1H-inden-1-yl]-2-fluoro-2-
methyl-heptan-3-ol.
24-Hydroxy-25-fluorocholecalciferol; 9,10-Secochol-
esta-5Z,7E,10(19)-triene-3β,24-ol, 25-fluoro; More
than 500 times less potent than 1α,25-
dihydroxyvitamin D$_3$ in binding to the chick intestinal
cytosol binding protein. Similar in binding activity to
25-hydroxyvitamin D$_3$. A potent mediator of calcium
metabolism - similar to 25-hydroxyvitamin D$_3$ in
promotion of calcium and phosphorus metabolism. *J
Biol Chem.* 1979, *254(6)*, 2017-22.

320   ***25-Fluoro-1α,24R-dihydroxyvitamin D$_3$***
9547422   C$_{27}$H$_{43}$FO$_3$

(1R,3S,5Z)-5-[(2E)-2-[(1R,3aR,7aS)-1-[(2R,5R)-6-fluoro-
5-hydroxy-6-methyl-heptan-2-yl]-7a-methyl-2,3,3a,5,
6,7-hexahydro-1H-inden-4-ylidene]ethylidene]-4-
methylidene-cyclohexane-1,3-diol.
LMST03020211;   25-fluoro-9,10-seco-5Z,7E,10(19)-
cholestatriene-1S,3R,24R-triol;   25-fluoro-1α,24R-di-
hydroxycholecalciferol; 9547422; Compared to 1α,25-
dihydroxyvitamin D$_3$, inhibition of proliferation of
human keratinocytes in culture is 100%, and
competitive binding to rat intestinal vitamin D receptor
is 0.165%. *J Nutr Biochem.* 1993, *4(1)*, 49-57.

321   ***(10Z)-19-Fluoro-1α,25-dihydroxyvitamin
D$_3$***
9547423   C$_{27}$H$_{43}$FO$_3$

(1R,3S,4Z,5Z)-5-[(2E)-2-[(1R,3aR,7aS)-1-[(2R)-6-
hydroxy-6-methyl-heptan-2-yl]-7a-methyl-2,3,3a,5,6,7-
hexahydro-1H-inden-4-ylidene]ethylidene]-4-(fluoromethylidene)cyclohexane-1,3-diol.
LMST03020212; 19-fluoro-9,10-seco-5Z,7E,10Z(19)-
cholestatriene-1S,3R,25-triol; (10Z)-19-fluoro-1α,25-
dihydroxycholecalciferol; 9547423; Synthesized from
(5E)-1α,25-dihydroxyvitamin D$_3$ *via* elimination of the
exocyclic methylene and introduction of a
fluoromethylene group or from vitamin D$_2$, *via* side
chain cleavage, side chain introduction, 1α-
hydroxylation, and 19-fluorination of 6,19-sulfur-
dioxide adduct of the resulting 1α,25-dihydroxyvitamin
D$_3$ derivative; λ$_m$ = 261 nm (95% EtOH). *Chem Pharm
Bull (Tokyo).* 2000, *48(10)*, 1484-93; 2001, *49(3)*,
312-7; Vitamin D Chemistry, Biology and Clinical
Applications of the Steroid Hormone. Proceedings of
the Tenth Workshop on Vitamin D Strasbourg, France-
May. Synthesis of Fluorovitamin D Analogs for
Conformational Analysis of Ligand Bound to Vitamin D
Receptor. (Norman, A. W., Bouillon, R., Thomasset,
M., eds), pp24-25, Printing and Reprographics

University of California, Riverside (1997).

322 *(10E)-19-Fluoro-1α,25-dihydroxyvitamin D₃*

9547424 $C_{27}H_{43}FO_3$

(1R,3S,4E,5Z)-5-[(2E)-2-[(1R,3aR,7aS)-1-[(2R)-6-hydroxy-6-methyl-heptan-2-yl]-7a-methyl-2,3,3a,5,6,7-hexahydro-1H-inden-4-ylidene]ethylidene]-4-(fluoromethylidene)cyclohexane-1,3-diol. LMST03020213; 19-fluoro-9,10-seco-5Z,7E,10E(19)-cholestatriene-1S,3R,25-triol; (10E)-19-fluoro-1α,25-dihydroxycholecalciferol; 9547424; Prepared from (5E)-1α,25-dihydroxyvitamin D₃ *via* elimination of the exocyclic methylene and introduction of a fluoromethylene group; λ_m = 264 nm (95% EtOH). *Chem Pharm Bull (Tokyo)*. 2000, **48(10)**, 1484-93; 2001, **49(3)**, 312-7; Proceedings of the Tenth Workshop on Vitamin D Strasbourg, France-May. Synthesis of Fluorovitamin D Analogs for Conformational Analysis of Ligand Bound to Vitamin D Receptor. (Norman, A. W., Bouillon, R., Thomasset, M., eds), pp24-25, Printing and Reprographics University of California, Riverside (1997).

323 *(5E,10Z)-19-Fluoro-1α,25-dihydroxy-vitamin D₃*

9547425 $C_{27}H_{43}FO_3$

(1R,3S,4Z,5E)-5-[(2E)-2-[(1R,3aR,7aS)-1-[(2R)-6-hydroxy-6-methyl-heptan-2-yl]-7a-methyl-2,3,3a,5,6,7-hexahydro-1H-inden-4-ylidene]ethylidene]-4-(fluoromethylidene)cyclohexane-1,3-diol. LMST03020214; 19-fluoro-9,10-seco-5E,7E,10Z(19)-cholestatriene-1S,3R,25-triol; (5E,10Z)-19-fluoro-1α,25-dihydroxycholecalciferol; 9547425; Prepared from (5E)-1α,25-dihydroxyvitamin D₃ *via* elimination of the exocyclic methylene and introduction of a

fluoromethylene group; λ_m = 269 nm (95% EtOH). Proceedings of the Tenth Workshop on Vitamin D Strasbourg, France-May. Synthesis of Fluorovitamin D Analogs for Conformational Analysis of Ligand Bound to Vitamin D Receptor. (Norman, A. W., Bouillon, R., Thomasset, M., eds), pp24-25, Printing and Reprographics University of California, Riverside (1997).

324 *(5E,10E)-19-Fluoro-1α,25-dihydroxy-vitamin D₃*

9547426 $C_{27}H_{43}FO_3$

(1R,3S,4E,5E)-5-[(2E)-2-[(1R,3aR,7aS)-1-[(2R)-6-hydroxy-6-methyl-heptan-2-yl]-7a-methyl-2,3,3a,5,6,7-hexahydro-1H-inden-4-ylidene]ethylidene]-4-(fluoromethylidene)cyclohexane-1,3-diol. LMST03020215; 19-fluoro-9,10-seco-5E,7E,10E(19)-cholestatriene-1S,3R,25-triol; (5E,10E)-19-fluoro-1α,25-dihydroxycholecalciferol; 9547426; Synthesized from (5E)-1α,25-dihydroxyvitamin D₃ *via* replacement of the exocyclic methylene by a fluoromethylene group; λ_m = 269 nm (95% EtOH). Proceedings of the Tenth Workshop on Vitamin D Strasbourg, France-May. Synthesis of Fluorovitamin D Analogs for Conformational Analysis of Ligand Bound to Vitamin D Receptor. (Norman, A. W., Bouillon, R., Thomasset, M., eds), pp24-25, Printing and Reprographics University of California, Riverside (1997).

325 *24R-Fluoro-1α,25-dihydroxyvitamin D₃*

86677-62-5 $C_{27}H_{43}FO_3$

(1R,3S,5Z)-5-[(2E)-2-[(1R,3aR,7aS)-1-[(2R,5R)-5-fluoro-6-hydroxy-6-methyl-heptan-2-yl]-7a-methyl-2,3,3a,5,6,7-hexahydro-1H-inden-4-ylidene]ethylidene]-4-methylidene-cyclohexane-1,3-diol. LMST03020216; 24R-fluoro-9,10-seco-5Z,7E,10(19)-

cholestatriene-1S,3R,25-triol;   24R-fluoro-1α,25-di-
hydroxycholecalciferol;   24-Fluoro-1,25-dihydroxy-
cholecalciferol;   24-F-1,25(OH)2D$_3$;   24-Fluoro-
calcitriol;   9,10-Secocholesta-5Z,7E,10(19)-triene-1α,
3β,25-triol, 24R-fluoro-; 6439790; 9547427; 86677-
62-5; Synthetic vitamin D$_3$ derivative. Compared to
1α,25-dihydroxyvitamin D$_3$, inhibition of proliferation
of human keratinocytes in culture is 100%, intestinal
calcium absorption in the rat is 65%, bone calcium
mobilization in the rat is 40%, competitive binding to
rat intestinal vitamin D receptor is 115% and
differentiation of human leukemia cells (HL-60) is
30%. Has less than half the activity of 1α,25-
dihydroxyvitamin D$_3$ on plasma calcium, bone weight
and duodenal calcium binding protein in chicken, on
calcium excretion *via* egg shell in Japanese quails and
on mobilization of calcium from the bone. Has been
used to reduce the incidence of parturient paresis in
dairy cows; $[α]_D^{25}$ = +67.9° (c = 0.524 MeOH); $λ_m$ =
213, 265 nm (ε = 12500, 16385 EtOH). *J Org Chem*.
1988, *53(5)*, 1040-6; *J Nutr Biochem*. 1993, *4(1)*, 49-
57.

326   **2β-Fluoro-1α,25-dihydroxyvitamin D$_3$**
9547695   C$_{27}$H$_{43}$FO$_3$

(1R,2R,3R,5Z)-5-[(2E)-2-[(1R,3aR,7aS)-1-[(2R)-6-
hydroxy-6-methyl-heptan-2-yl]-7a-methyl-2,3,3a,5,6,7-
hexahydro-1H-inden-4-ylidene]ethylidene]-2-fluoro-4-
methylidene-cyclohexane-1,3-diol.
LMST03020602;   2β-fluoro-1α,25-dihydroxycholecal-
ciferol;   2R-fluoro-9,10-seco-5Z,7E,10(19)-cholesta-
triene-1R,3R,25-triol;   9547695;   Synthesized
photochemically from a C(22) steroid precursor.
Compared to 1α,25-dihydroxyvitamin D$_3$, affinity for
chicken intestine vitamin D receptor is 667%, affinity
for sheep serum vitamin D binding protein is 160%,
inhibition of proliferation of C3H10T(BMP-4) cells is
133% (FCS-free), 900% (10% FCS); CAT assay
(osteocalcin) is 130%, inhibition of adipogenesis is
900%. *Steroids*, 1998, *63(12)*, 633-43.

327   **1,24-Dihydroxy-25-fluorovitamin D$_3$**
71699-09-7   C$_{27}$H$_{43}$FO$_3$

(1S,3R,5Z)-5-[(2E)-2-[(1R,3aR,7aS)-1-[(2R,5R)-6-fluoro-
5-hydroxy-6-methyl-heptan-2-yl]-7a-methyl-2,3,3a,5,
6,7-hexahydro-1H-inden-4-ylidene]ethylidene]-4-
methylidene-cyclohexane-1,3-diol.
1,24-Dihydroxy-25-fluorovitamin   D$_3$;   1α,24-Di-
hydroxy-25-fluorovitamin D$_3$; 9,10-Secocholesta-5Z,
7E,10(19)-triene-1α,3β,24R-triol, 25-fluoro-; 6443843;
71699-09-7;   Metabolite of   24R-hydroxy-25-
fluorovitamin D$_3$ and 1α-hydroxy-25-fluorovitamin D$_3$.
Arch Biochem Biophys. 1979, *197(1)*, 193-8.

328   **24-Fluoro-1,25-dihydroxyvitamin D$_2$**
135776-86-2   C$_{28}$H$_{43}$FO$_3$

9,10-Secoergosta-5Z,7E,10(19),22E-tetraene-1α,3lb,25-
triol, 24-fluoro- .
24-Fluoro-1,25-dihydroxyvitamin D$_2$; 24-Fluoro-9,10-
secoergosta-5Z,7E,10(19),22E-tetraene-1α,3β,25-triol;
24-F-1,25-(OH)2D$_2$;   24-Fluoro-1α,25-dihydroxy-
vitamin D$_2$; LMST03010011; BXL 628; Ro 26-9228;
135776-86-2; A fluorine atom at C24 of either 1,25-
dihydroxyvitamin D$_2$ or its 24-epimer have no
potentiating effect. This is in contrast with the cases of
24- and 26,27-multifluorinated analogs of active
vitamin D$_3$. *Biochim Biophys Acta*. 1994, *1213(3)*,
302-308.

329   **19-Fluoro-(10E)-1,25-dihydroxyvitamin
D$_3$**
                                        C$_{27}$H$_{43}$FO$_3$

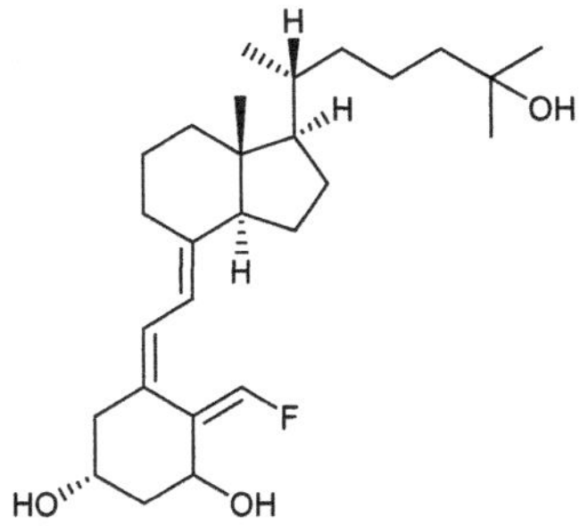

9,10-Secocholesta-5Z,7E,10(19)E-trien-1,25-diol, 19-fluoro.
LMST03020212; 19-Fluoro-1,25-dihydroxycholecalciferol; 19-fluoro-calcitriol; Synthesis and NMR spectra reported. Affinity for the vitamin D receptor is much less than that of calcitriol. *Chem Pharm Bull (Tokyo)*. 2001, **49(3)**, 312-317; 2002, **50(4)**, 475-83.

330   *19-Fluoro-(10Z)-1,25-dihydroxyvitamin*

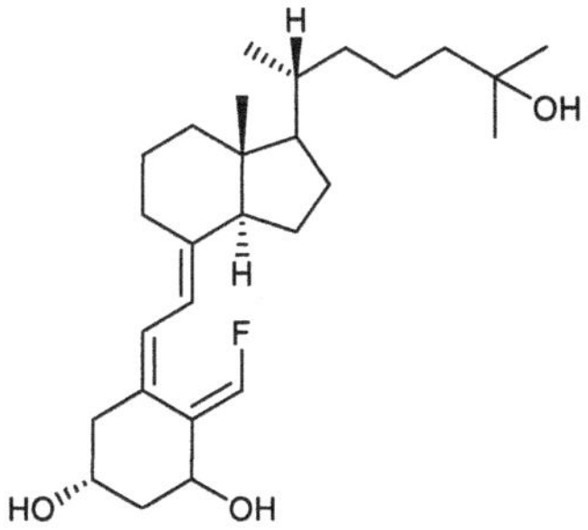

D$_3$

$C_{27}H_{43}FO_3$

9,10-Secocholesta-5Z,7E,10(19)Z-trien-1,25-diol, 19-fluoro.
LMST03020213; 19-Fluoro-1,25-dihydroxycholecalciferol; 19-fluoro-calcitriol; Synthesis and NMR spectra reported. Affinity for the vitamin D receptor is much less than that of calcitriol. *Chem Pharm Bull (Tokyo)*. 2001, **49(3)**, 312-7; 2002, **50(4)**, 475-83.

331   *3-Deoxy-3-azido-25-hydroxyvitamin D$_3$*
108345-00-2 $C_{27}H_{43}N_3O$

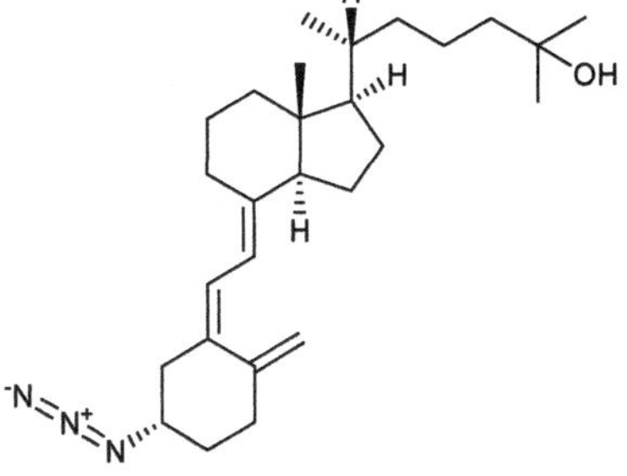

9,10-Secocholesta-5Z,7E,10(19)-trien-25-ol, 3β-azido-.
3-Deoxy-3-azido-25-hydroxyvitamin D$_3$; 3β-Azido-9,10-secocholesta-5Z,7E,10(19)-trien-25-ol; 3-Dahv-D$_3$; 3-Deoxy-3-azido-25-hydroxycholecalciferol; 108345-00-2; 6439165; Used for photoaffinity labeling of serum vitamin D binding protein. *Biochemistry*. 1987, **26(13)**, 3957-64.

332   *1α,25-Dihydroxy-24-oxo-23-azavitamin D$_2$*
9547223 $C_{27}H_{43}NO_4$

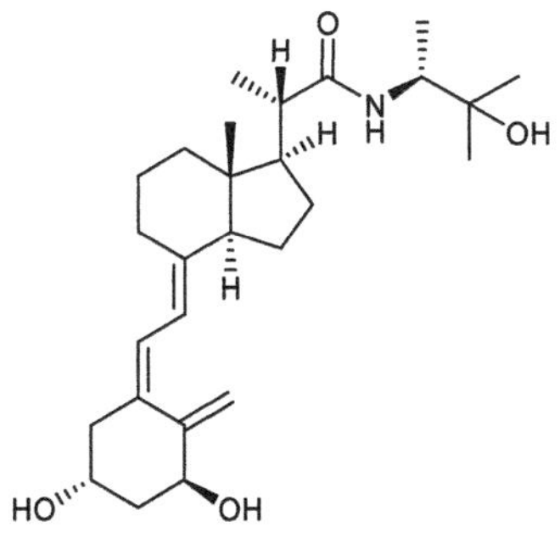

(2S)-2-[(1R,3aR,4E,7aS)-4-[(2Z)-2-[(3S,5R)-3,5-di-hydroxy-2-methylidene-cyclohexylidene]ethylidene]-7a-methyl-2,3,3a,5,6,7-hexahydro-1H-inden-1-yl]-N-[(2R)-3-hydroxy-3-methyl-butan-2-yl]propanamide.
LMST03010007; 1α,25-dihydroxy-24-oxo-23-azaergocalciferol; 9547223; Has no affinity for the vitamin D receptor; Solid; [α]$_D$ = +1.5° (c = 0.0023, EtOH); λ$_m$ = 265 nm (ε = 18500, EtOH). *Chem Pharm Bull (Tokyo)*. 1997, **45(1)**, 185-8.

333   *1,24-Dihydroxy-25-nitrovitamin D$_3$*
214678-06-5 $C_{27}H_{43}NO_5$

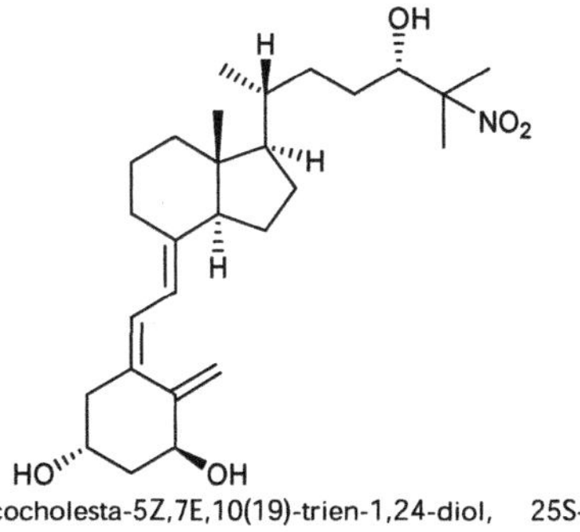

9,10-Secocholesta-5Z,7E,10(19)-trien-1,24-diol, 25S-nitro.
1α,24R-dihydroxy-25-nitrovitamin D$_3$; 1,24-diOH-NO2D3; 214678-06-5; Synthesized and biological properties studied based on VDR binding affinity and HL-60 cell differentiation activity. *Bioorg Med Chem Lett*. 1999, **9(3)**, 381-4.

334   *1,24-Dihydroxy-25-nitrovitamin D$_3$*
214678-00-9 $C_{27}H_{43}NO_5$

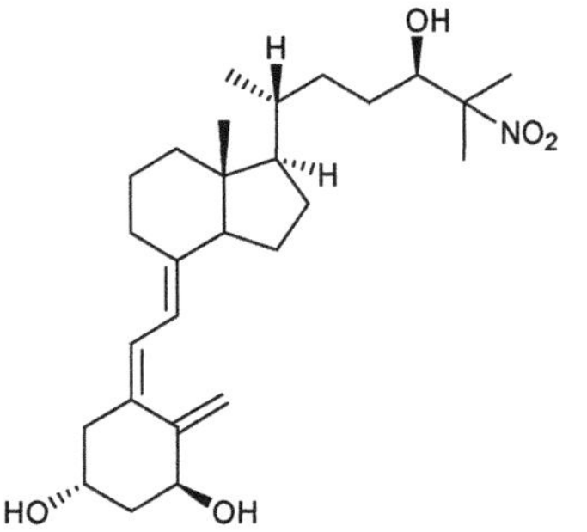

9,10-Secocholesta-5Z,7E,10(19)-trien-1,24-diol, 25R-nitro.

1α, 24S-dihydroxy-25-nitrovitamin D_3; 1,24-diOH-NO2D3; 214678-00-9; Synthesized and biological properties studied based on VDR binding affinity and HL-60 cell differentiation activity. *Bioorg Med Chem Lett.* 1999, *9(3)*, 381-4.

335   **3-Deoxyvitamin D_3**

$C_{27}H_{44}$

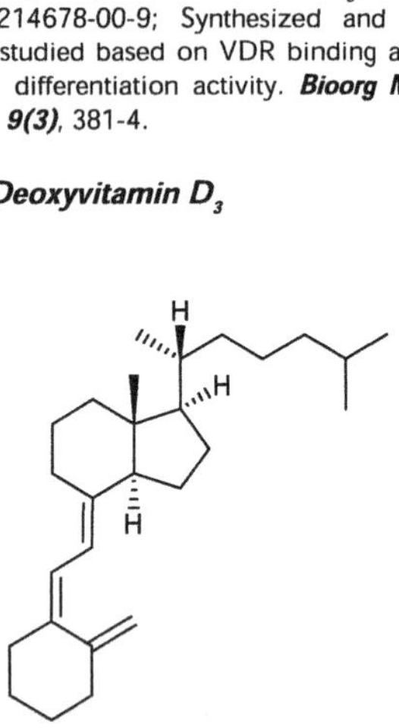

9,10-Secocholesta-5Z,7E,10(19)-triene.
Has essentially no vitamin D activity. Inactive in the chick duodenum and shows very poor binding affinity to the chick intestinal cytosol receptor. *J Steroid Biochem.* 1985, *22(4)*, 469-74.

336   **Fluor-vigantoletten 1000**
58542-37-3

$C_{27}H_{44}FNaO$

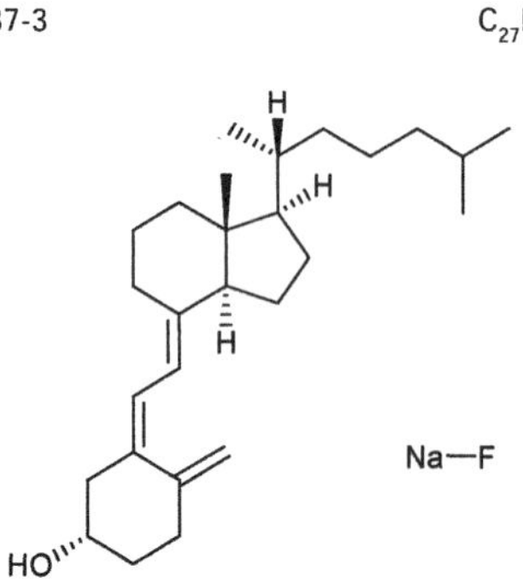

Na—F

Sodium (1R,3Z)-3-[(2E)-2-[(1R,3aR,7aS)-7a-methyl-1-[(2R)-6-methylheptan-2-yl]-2,3,3a,5,6,7-hexahydro-1H-inden-4-ylidene]ethylidene]-4-methylidene-cyclo-hexan-1-ol fluoride.
Fluor-vigantoletten 1000; 9,10-Secocholesta-5Z,7E,10

(19)-trien-3β-ol, mixt. with sodium fluoride; 6453648; 58542-37-3; used in treatment of childhood dental caries. *Monatsschr Kinderheilkd.* 1979, *127(4)*, 192-5.

337   **1α-Hydroxy-3-deoxyvitamin D_3**
5283709

$C_{27}H_{44}O$

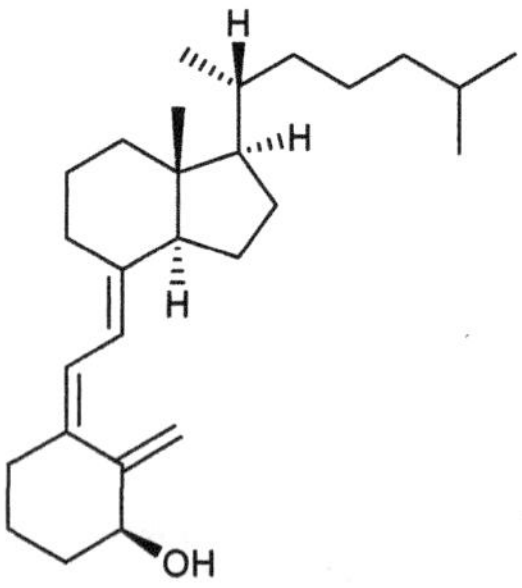

(1S,3Z)-3-[(2E)-2-[(1R,3aR,7aS)-7a-methyl-1-[(2R)-6-methylheptan-2-yl]-2,3,3a,5,6,7-hexahydro-1H-inden-4-ylidene]ethylidene]-2-methylidene-cyclohexan-1-ol. LMSTO3020217; (1S)-9,10-seco-5Z,7E,10(19)-chol-estatrien-1S-ol; 1α-hydroxy-3-deoxyvitamin D_3; 1α-hydroxy-3-deoxycholecalciferol; 5283709; Synthesized from either 1α,2α-epoxy-6-ethylenedioxycholestan-3-one or from 1α-acetoxycholesterol tosylate. Intestinal calcium transport and bone calcium mobilization activity reported; λ_m = 264.5 nm. *Biochem Biophys Res Commun.* 1974, *59(3)*, 845-9.

338   **3-Epivitamin D_3**
5283710

$C_{27}H_{44}O$

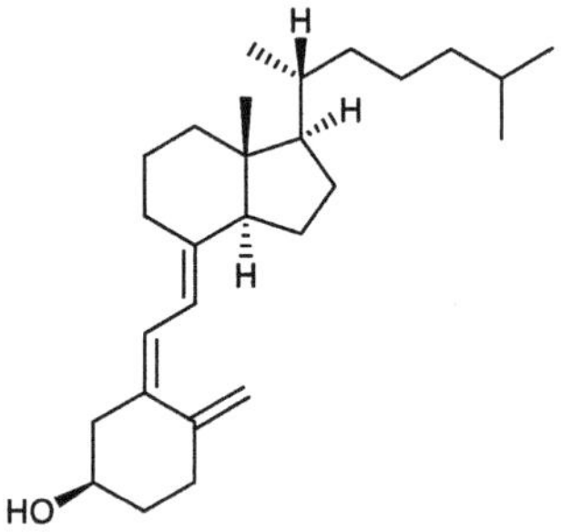

(1R,3Z)-3-[(2E)-2-[(1R,3aR,7aS)-7a-methyl-1-[(2R)-6-methylheptan-2-yl]-2,3,3a,5,6,7-hexahydro-1H-inden-4-ylidene]ethylidene]-4-methylidene-cyclohexan-1-ol. LMST03020219; ZINC03947571; 3-epicholecalciferol; 9,10-seco-5Z,7E,10(19)-cholestatrien-3R-ol; 5283710; Prepared from vitamin D_3 by inversion of the configuration at C(3); $[\alpha]_D$ = -0.5° (c = 0.9 C_6H_6). *J Org Chem.* 1976, *41(6)*, 1067-9; *Biochemistry* 1980, *19(17)*, 3933-7.

339   **(5E)-Vitamin D_3**
5283711

$C_{27}H_{44}O$

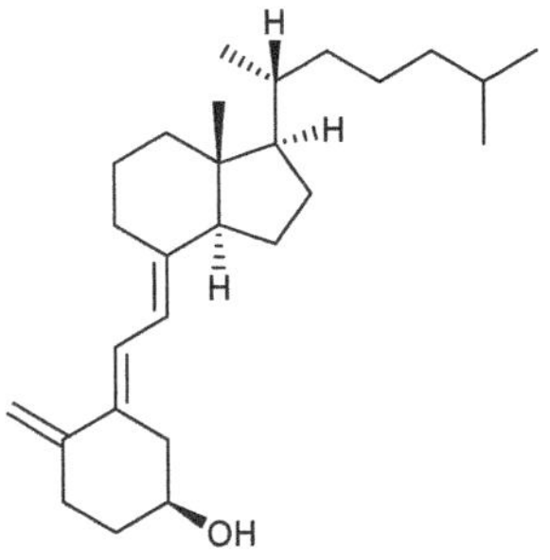

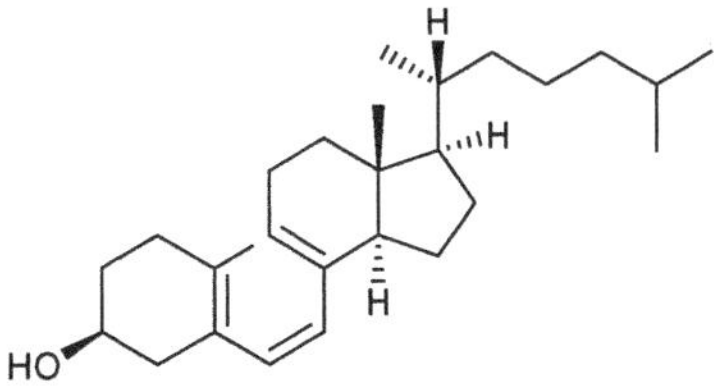

(1S)-3-[(Z)-2-[(1R,3aR,7aS)-7a-methyl-1-[(2R)-6-methyl-heptan-2-yl]-1,2,3,3a,6,7-hexahydroinden-4-yl]-ethenyl]-4-methyl-cyclohex-3-en-1-ol.
LMST03020222; EINECS 214-634-2; Previtamin D(3); C07711; 9,10-Secocholesta-5(10),6Z,8-trien-3β-ol; 5281058; 1173-13-3; Formed by irradiation of 7-dehydrocholesterol. Converted by irradiation to vitamin D$_3$; $[\alpha]_D^{20}$ = +54° (CHCl$_3$); λ$_m$ = 260 nm (ε 9250). *Analyst*. 1957, *82(970)*, 2-7.

342   *Tachysterol$_3$*
5283713   C$_{27}$H$_{44}$O

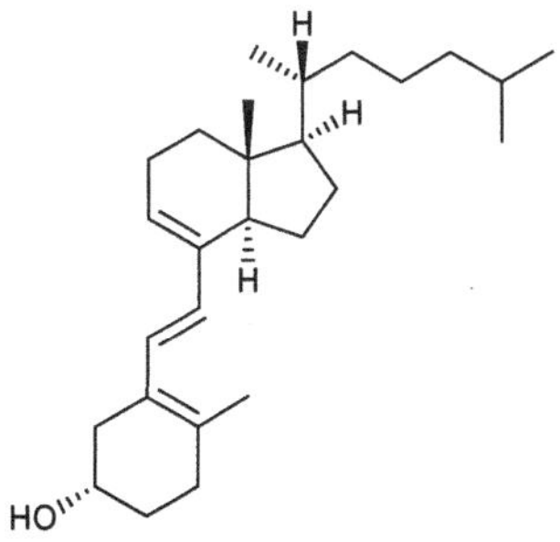

(1S)-3-[(E)-2-[(1R,3aR,7aS)-7a-methyl-1-[(2R)-6-methylheptan-2-yl]-1,2,3,3a,6,7-hexahydroinden-4-yl]-ethenyl]-4-methyl-cyclohex-3-en-1-ol.
LMST03020223; Tachysterol_3; 9,10-seco-5(10),6E,8-cholestatrien-3S-ol; 5283713; *Tetrahedron*. 1960, *11(4)*, 276-84; USP 7211172.

343   *Isotachysterol$_3$*
5283714   C$_{27}$H$_{44}$O

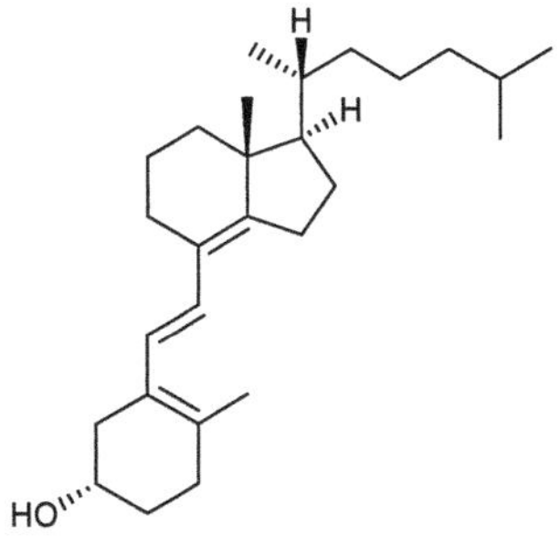

(1S)-3-[(E)-2-[(1R,7aR)-7a-methyl-1-[(2R)-6-methyl-heptan-2-yl]-1,2,3,5,6,7-hexahydroinden-4-yl]ethenyl]-4-methyl-cyclohex-3-en-1-ol.
LMST03020224; Isotachysterol$_3$; 9,10-seco-5(10),6E, 8(14)-cholestatrien-3S-ol; 5283714; A minor product formed by overirradiation of 7-dehydrocholesterol in ethanol. Prepared by treatment of vitamin D$_3$ with

(1S,3E)-3-[(2E)-2-[(1R,3aR,7aS)-7a-methyl-1-[(2R)-6-methylheptan-2-yl]-2,3,3a,5,6,7-hexahydro-1H-inden-4-ylidene]ethylidene]-4-methylidene-cyclohexan-1-ol.
LMST03020220; (5E)-vitamin D$_3$; (5E)-cholecalciferol; (5E)-calciol; 9,10-seco-5E,7E,10(19)-cholestatrien-3S-ol; Spectrum_001163; SpecPlus_000180; Spectrum-2_001369; Spectrum3_000764; Spectrum4_001201; KBioGR_001602; KBioSS_001643; DivK1c_006276; SPBio_001298; KBio1_001220; 5283711; 6708595; Synthesized from vitamin D$_3$ by treatment with liquid SO$_2$ to give the D$_3$-SO2 adduct which upon heating in refluxing EtOH in the presence of NaHCO$_3$ gives (5E)-vitamin-D$_3$. *J Org Chem*. 1977, *42(13)*, 2284-91; *Tet Lett*. 1981, *22(39)*, 3085-8; *Chem Lett*. 1979, 583.

340   *(5E)-3-Epivitamin D$_3$*
5283712   C$_{27}$H$_{44}$O

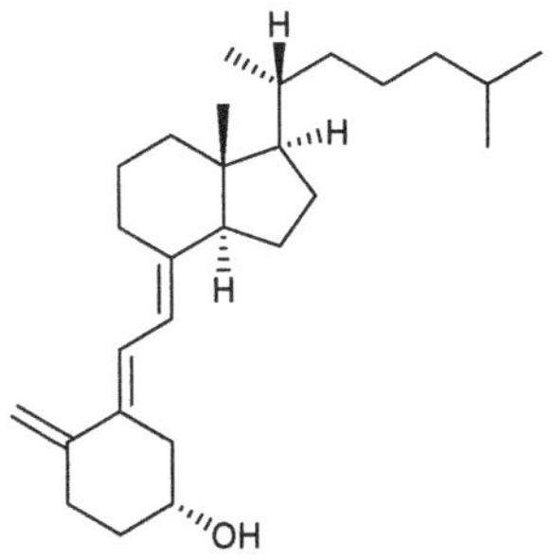

(1R,3E)-3-[(2E)-2-[(1R,3aR,7aS)-7a-methyl-1-[(2R)-6-methylheptan-2-yl]-2,3,3a,5,6,7-hexahydro-1H-inden-4-ylidene]ethylidene]-4-methylidene-cyclohexan-1-ol.
LMST03020221; (5E)-3-epivitamin D$_3$; (5E)-3-epi-cholecalciferol; 9,10-seco-5E,7E,10(19)-cholestatrien-3R-ol; Prestwick0_000429; Prestwick1_000429; SPBio_002357; 5283712; 6713938; Prepared from (5E)-vitamin D$_3$ by inversion of the configuration at C(3); $[\alpha]_D$ = -30.0° (c = 0.85 C$_6$H$_6$); λ$_m$ = 272 nm (ε = 22000). *J Org Chem*. 1977, *42(21)*, 3325-30.

341   *Previtamin D$_3$*
1173-13-3   C$_{27}$H$_{44}$O

Lewis acids; λ_m = 277, 288, 301 nm (EtOH). *Yakugaku Zasshi*. 1969, *89(7)*, 919-24; *J R Neth Chem Soc*. 1977, *96*, 104.

344   *(5E)-Isovitamin D₃*
5314031 $C_{27}H_{44}O$

(1S,5E)-5-[(2E)-2-[(1R,3aR,7aS)-7a-methyl-1-[(2R)-6-methylheptan-2-yl]-2,3,3a,5,6,7-hexahydro-1H-inden-4-ylidene]ethylidene]-4-methyl-cyclohex-3-en-1-ol. LMST03020225; (5E)-isocholecalciferol; 9,10-seco-1(10),5E,7E-cholestatrien-3S-ol; 5314031; A minor product by overirradiation of 7-dehydrocholesterol in ethanol. Prepared by treatment of vitamin D₃ with Lewis acids; λ_m = 276, 287, 298 nm (EtOH). *Yakugaku Zasshi*. 1969, *89(7)*, 919-24.

345   *9,10-Seco-5Z,8,10(19)-cholestatrien-3S-ol*
5283715 $C_{27}H_{44}O$

(1S,3Z)-3-[2-[(1R,3aR,7aS)-7a-methyl-1-[(2R)-6-methylheptan-2-yl]-1,2,3,3a,6,7-hexahydroinden-4-yl]ethylidene]-4-methylidene-cyclohexan-1-ol. LMST03020226; Toxisterol₃ D1; 9,10-seco-5Z,8,10(19)-cholestatrien-3S-ol; 5283715; An overirradiation product of 7-dehydrocholesterol; λ_m = 217 nm (ε 7200 EtOH). *Tet Lett*. 1975, 16(*7*), 427-30; *J R Neth Chem Soc*. 1977, *96*, 104.

346   *Toxisterol₃ E1*
9547428 $C_{27}H_{44}O$

(3S,6R,9R)-9(10->6)abeo-5(10),7-cholestadien-3-ol. LMST03020227; (6R,9R)-9(10->6)abeo-5(10),7-cholestadien-3S-ol; 9547428; An overirradiation product of 7-dehydrocholesterol; $[\alpha]_D^{22}$ = +155° (CHCl₃). No UV absorption above 210 nm. *J R.Neth Chem Soc*. 1977, *96*, 104.

347   *9,10-Seco-5(10),6S,7-cholestatrien-3S-ol*
5283716 $C_{27}H_{44}O$

(1S)-3-[2-[(1R,3aR,7aS)-7a-methyl-1-[(2R)-6-methylheptan-2-yl]-2,3,3a,5,6,7-hexahydro-1H-inden-4-ylidene]ethenyl]-4-methyl-cyclohex-3-en-1-ol. LMST03020228; 9,10-seco-5(10),6S,7-cholestatrien-3S-ol; 5823716; A minor irradiation (>250 nm) product of vitamin D₃ in ethanol; λ_m = 230 nm. *Recueil*. 1982, *91*, 1459.

348   *9,10-Seco-5(10),6R,7-cholestatrien-3S-ol*
5283716 $C_{27}H_{44}O$

(1S)-3-[2-[(1R,3aR,7aR)-7a-methyl-1-[(2R)-6-methylheptan-2-yl]-2,3,3a,5,6,7-hexahydro-1H-inden-4-ylidene]ethenyl]-4-methyl-cyclohex-3-en-1-ol. LMST03020229; 9,10-seco-5(10),6R,7-cholestatrien-3S-ol; 5283716; A minor irradiation (>250 nm) product of vitamin D_3 in ethanol; λ_m = 230 nm. *Recueil*. 1982, **91**, 1459.

349   *6S,19-Cyclo-9,10-seco-5(10),7E-cholestadien-3S-ol*

9547429 $C_{27}H_{44}O$

(3S,8S)-8-[(E)-[(1R,3aR,7aS)-7a-methyl-1-[(2R)-6-methylheptan-2-yl]-2,3,3a,5,6,7-hexahydro-1H-inden-4-ylidene]methyl]bicyclo[4.2.0]oct-9-en-3-ol. LMST03020230;   6S,19-cyclo-9,10-seco-5(10),7E-cholestadien-3S-ol; 9547429; A minor irradiation (>250 nm) product of vitamin D_3 in EtOH. *Recueil*. 1982, **91**, 1459.

350   *7-Dehydrocholesterol*

434-16-2 $C_{27}H_{44}O$

(3S,9R,10R,13S,14R,17R)-10,13-dimethyl-17-[(2R)-6-methylheptan-2-yl]-2,3,4,9,11,12,14,15,16,17-decahydro-1H-cyclopenta[a]phenanthren-3-ol. Cholesta-5,7-dien-3-ol; 5,7-Cholestandien-3β-ol; Cholesta-5,7-dien-3β-ol; 7-Dehydrocholesterin; 7-Dehydrocholesterol; 7,8-Didehydrocholesterol; Caswell No. 902E; Cholesterol, 7-dehydro-; Dehydrocholesterin; Dehydrocholesterol; (3β)-7-Dehydrocholesterol; 7-Dehydrocholesterol; EINECS 207-100-5; EPA Pesticide Chemical Code 208700; LMST01010069; NSC 18159; Provitamin D_3; Vitamin D_3; $\Delta^{5,7}$-Cholesterol; Δ^{7}-Cholesterol; $\Delta^{5,7}$-Cholesterol; Δ^{7}-Cholesterol; 7,8-Didehydrocholesterol; Cholesta-5,7-dien-3β-ol; 439423; 434-16-2; Biochemical precursor to cholesterol. Reduced by Δ^{7}-sterol reductase (EC 1.3.1.21) to cholesterol. Deficiency of this enzyme is linked to Smith-Lemli-Opitz syndrome an autosomal recessive disorder due to an inborn error of cholesterol metabolism; Solid, mp = 150.5°. *Z Physiol Chem*. 1937, **245**, 168 -170 (1937); *Braz J Med Biol Res*. 2003, **36(10)**, 1327-32.

351   *Vitamin D₃*

67-97-0   10157 $C_{27}H_{44}O$

(1S,3Z)-3-[(2E)-2-[(1R,3aR,7aS)-7a-methyl-1-[(2R)-6-methylheptan-2-yl]-2,3,3a,5,6,7-hexahydro-1H-inden-4-ylidene]ethylidene]-4-methylidene-cyclohexan-1-ol. 7-Dehydrocholesterol, irradiated; 7-Dehydro-cholesterol, activated; 9,10-Secocholesta-5,7,10(19)-trien-3b-ol; 9,10-secocholesta-5Z,7E,10(19)-trien-3S-ol; 9, 10-Seco-5Z,7E,10(19)-cholestatrien-3-ol; 9,10-Seco-cholesta-5Z,7E,10(19)-trien-3b-ol; Activated 7-dehydrocholesterol; Activated ergosterol; Arachitol; Calciol; CC; CCRIS 5813, 6286; CHEBI:28940; Cholecalciferol; Cholecalciferol, D_3; Cholecalciferolum; Colecalciferol; Colecalciferolo; Colecalciferolum; Colecalcipherol; D_3-Vigantol; D_3-Vicotrat; Delsterol; Deparal; Duphafral D_3 1000; Ebivit; EINECS 200-673-2, 215-797-2; EPA Pesticide Chemical Code 202901; HSDB 820; Irradiated 7-dehydrocholesterol; LMST03020001; NEO Dohyfral D_3; NSC 375571; Oleovitamin D_3; Prestwick_63; Quintox; Rampage; Ricketon; Trivitan; Vi-de-3-hydrosol; Vigantol; Vigorsan; Vitamin D_3; VITAMIN D; Vitinc Dan-Dee-3; ZINC04474460; 5280795; 67-97-0; 1406-16-2; 3965-99-9; 57651-82-8; 8024-19-9; 8050-67-7; An antirachitic vitamin that undergoes metabolic conversion before exerting biological effects. In an oily solution, used as a veterinary preparation to prevent rickets and osteomalacia. Derivative of 7-dehydroxycholesterol formed by UV radiation breaking of the C9-C10 bond. It differs from vitamin D_2 in having a single bond between C22 and C23 and lacking a methyl group at C24; Solid; mp = 84-85°; $[\alpha]_D^{20}$ = 84.8° (c = 1.6 Me$_2$CO), 51.9° (c = 1.6 CHCl$_3$); λ_m = 264.5 nm (E$_{1\,cm}^{1\%}$ 450-490 EtOH or C$_6$H$_{14}$); soluble in most organic solvents, insoluble in H_2O. *Chem Rev*. 1995, **95**, 1877-952; *Arch Dermatol*. 1989, **125(2)**, 231-4; *Chem Ind*. 1957, 1148-9; *J Nutr*. 1974, **104(8)**, 1056-60; *Endocr Rev*. 1995, **16(2)**, 200-57; *Comp Med Chem*. 1990, **3**, 1129; *Rec Results in Cancer Res*. 2003, **164**, 55-82. *Eli Lilly Asia Pacific Pte. Ltd*.

352   1*3,14,20-Epivitamin D₃*

3965-99-9 $C_{27}H_{44}O$

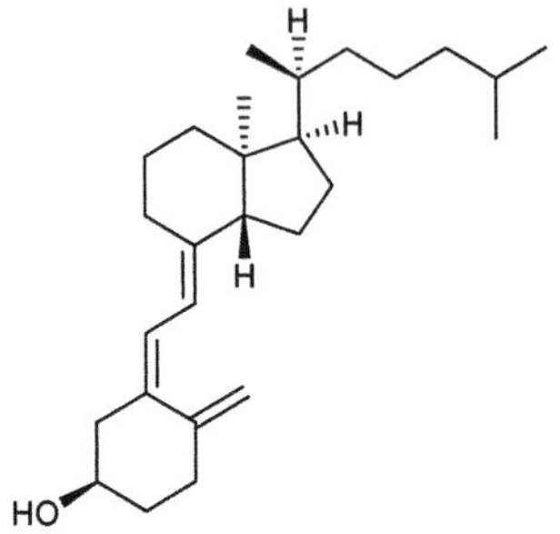

(1R,3Z)-3-[(2E)-2-[(1R,3aS,7aR)-7a-methyl-1-[(2R)-6-methylheptan-2-yl]-2,3,3a,5,6,7-hexahydro-1H-inden-4-ylidene]ethylidene]-4-methylidene-cyclohexan-1-ol. ZINC01530170; 2735; 6560150; 3965999.

353   *13,14,17,20-Epi-9,10-secocholesta-5,7, 10(19)-trien-3α-ol*
6604662 $C_{27}H_{44}O$

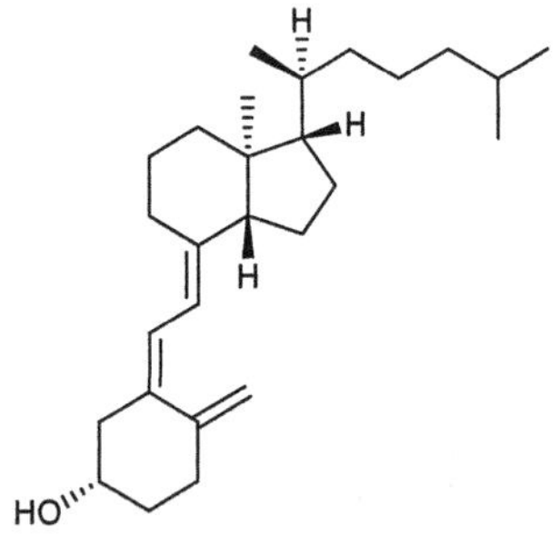

(1R,3Z)-3-[(2E)-2-[(1S,3aS,7aR)-7a-methyl-1-[(2S)-6-methylheptan-2-yl]-2,3,3a,5,6,7-hexahydro-1H-inden-4-ylidene]ethylidene]-4-methylidene-cyclohexan-1-ol. TNP00266; NCGC00017328-01; 6604662.

354   *3-Deoxy-25-hydroxyvitamin D_3*
$C_{27}H_{44}O$

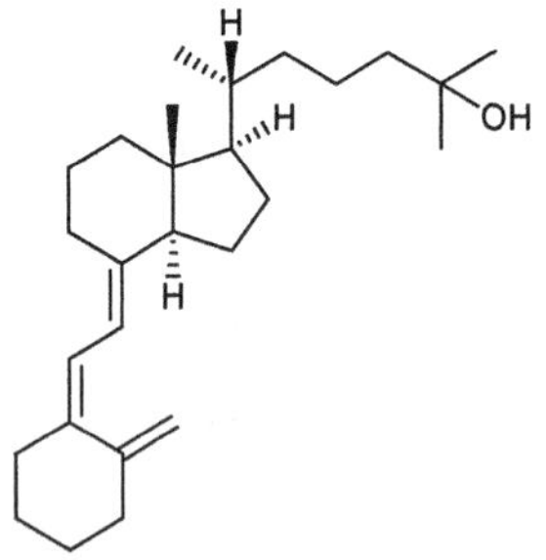

Synthesis and biological activity reported. *FEBS Lett.* 1979, *97(2)*, 241-4.

355   *Alfacalcidol*
41294-56-8 $C_{27}H_{44}O_2$

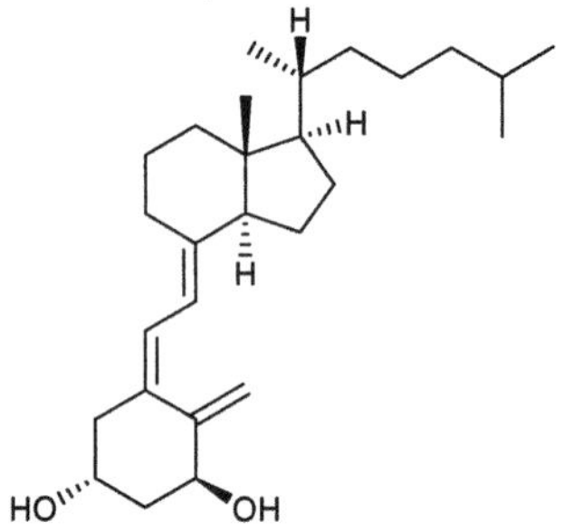

(1S,3R,5Z)-5-[(2E)-2-[(1R,7aS)-7a-methyl-1-[(2R)-6-methylheptan-2-yl]-2,3,3a,5,6,7-hexahydro-1H-inden-4-ylidene]ethylidene]-4-methylidene-cyclohexane-1,3-diol.

LMST03020231; 1α-hydroxyvitamin D_3; 1α-hydroxy cholecalciferol; 9,10-seco-5Z,7E,10(19)-cholestatriene-1S,3R-diol; Alfacalcidol; Alphacalcidol; Oxydevit; Alfarol; Etalpha; Sinovul; α-Calcidol; One-Alpha; Un-Alpha; 1-Hydroxyvitamin D_3; Alpha D_3; 1-Hydroxy-cholecalciferol; 1α-Hydroxy-vitamin D_3; Alfacalcid-olum; 1α-Hydroxycholecalciferol; Alfacalcidol; C12969; D01518; CCRIS 3341; EINECS 255-297-1; EB 644; 5282181; 6441773; 6434322; 69556-15-6; Synthesized from cholesterol, 1,4,6-cholestatrien-3-one, 1,4-cholestadien-3-one, or vitamin D_3. Intestinal calcium transport and bone mineral mobilization response in anephric rats has been reported. Compared to 1α,25-dihydroxyvitamin D_3, differentiation of HL-60 human promyelocytes is about 2 orders of magnitude less effective. Metabolised to 1α,25-dihydroxyvitamin D_3. Used (0.75 mg/day) in treatment of senile psoriasis. Metabolite of vitamin D_3. In human promyelocytic leukemia cells, metabolized to (5Z)- and (5E)-(24R)-19-nor-10-oxo-24-hydroxycalcidiol. Used to treat bone loss. In contrast to vitamin D, alfacalcidol has pleiotropic effects improving bone and muscle metabolism and clinical symptoms in patients with rheumatoid arthritis; Crystals; mp = 134-136°; $[\alpha]_D^{25}$ = +28.0° (c = 0.6 Et_2O); λ_m = 264-265 nm (ε = 18000 Et_2O). *Tet Lett.* 1972, *13(40)*, 4147-50; *J Amer Chem Soc.* 1973, *95(8)*, 2748-9; *J Chem Soc Perkin I.* 1974, 2654-7; 1979, 1695; *Steroids.* 1974, *23(1)*, 75-92; *Proc Natl Acad Sci U.S.A.* 1978, *75(5)*, 2080-1; *Science.* 1973, *180(82)*, 190-1; *J Biol Chem.* 1987, *262(29)*, 14164-71.

356   *1α-Hydroxy-3-epivitamin D_3*
5283717 $C_{27}H_{44}O_2$

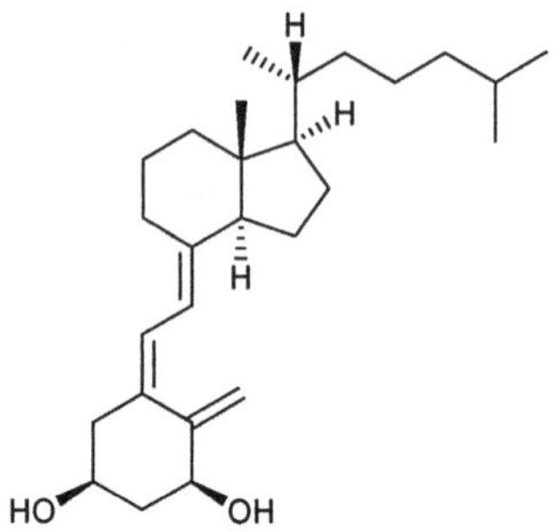

(1S,3S,5Z)-5-[(2E)-2-[(1R,3aR,7aS)-7a-methyl-1-[(2R)-6-methylheptan-2-yl]-2,3,3a,5,6,7-hexahydro-1H-inden-4-ylidene]ethylidene]-4-methylidene-cyclohexane-1,3-diol. LMST03020232; 9,10-seco-5Z,7E,10(19)-cholestatriene-1S,3S-diol; 1α-hydroxy-3-epivitamin D_3; 1α-hydroxy-3-epicholecalciferol; 5283717; Synthesized from 3-epivitamin D_3 or from 1α-hydroxycholesterol. Relative efficacy reported for competing with 1α,25-dihydroxyvitamin D_3 for binding to the vitamin D receptor; λ_m = 263 nm. *Tet Lett.* 1975, **49(16)**, 4317-20; *Bioorg Chem.* 1985, **13(1)**, 62-75.

357   *(5E)-1α-Hydroxyvitamin D_3*
5283718 $C_{27}H_{44}O_2$

(1R,3S,5E)-5-[(2E)-2-[(1R,3aR,7aS)-7a-methyl-1-[(2R)-6-methylheptan-2-yl]-2,3,3a,5,6,7-hexahydro-1H-inden-4-ylidene]ethylidene]-4-methylidene-cyclohexane-1,3-diol. LMST03020233; (5E)-1α-hydroxyvitamin D_3; (5E)-1α-hydroxycholecalciferol; 9,10-seco-5E,7E,10(19)-cholestatriene-1S,3R-diol; 5283719; Synthesized from vitamin D_3 *via* SeO_2 oxidation of its 3,5-cyclovitamin D derivative as the key step. Relative efficacy reported for competing with 1α,25-dihydroxyvitamin D_3 for binding to the vitamin D receptor; λ_m = 273 nm. *Bioorg Chem.* 1985, **13(1)**, 62-75.

358   *(5E)-1α-Hydroxy-3-epivitamin D_3*
5283719 $C_{27}H_{44}O_2$

(1S,3S,5E)-5-[(2E)-2-[(1R,3aR,7aS)-7a-methyl-1-[(2R)-6-methylheptan-2-yl]-2,3,3a,5,6,7-hexahydro-1H-inden-4-ylidene]ethylidene]-4-methylidene-cyclohexane-1,3-diol. LMST03020234;   9,10-seco-5E,7E,10(19)-cholestatri-

ene-1S,3S-diol; (5E)-1α-hydroxy-3-epivitamin D_3; (5E)-1α-hydroxy-3-epicholecalciferol; 5283719; Synthesized from 3-epivitamin D_3. Relative efficacy reported for competing with 1α,25-dihydroxyvitamin D_3 for binding to the vitamin D receptor; λ_m = 271 nm. *Bioorg Chem.* 1985, **13(1)**, 62-75.

359   *1β-Hydroxyvitamin D_3*
5283720 $C_{27}H_{44}O_2$

(1R,3R,5Z)-5-[(2E)-2-[(1R,3aR,7aS)-7a-methyl-1-[(2R)-6-methylheptan-2-yl]-2,3,3a,5,6,7-hexahydro-1H-inden-4-ylidene]ethylidene]-4-methylidene-cyclohexane-1,3-diol. LMST03020235; 1β-hydroxyvitamin D_3; 1β-hydroxycholecalciferol; 9,10-seco-5Z,7E,10(19)-cholestatriene-1R,3R-diol; 5283720; Relative to 1α,25-dihydroxyvitamin D_3, binding affinity for intestinal vitamin D receptor is 1/1.65x10^5; λ_m = 263 nm (ε 17000 EtOH). *J Chem Soc Chem Comm.* 1977, 890; *J Org Chem.* 1977, **42(22)**, 3597-9.

360   *1β-Hydroxy-3-epivitamin D_3*
65445-14-9 $C_{27}H_{44}O_2$

(1S,3R,5Z)-5-[(2E)-2-[(1R,3aR,7aS)-7a-methyl-1-[(2R)-6-methylheptan-2-yl]-2,3,3a,5,6,7-hexahydro-1H-inden-4-ylidene]ethylidene]-4-methylidene-cyclohexane-1,3-diol. LMST03020236; BRN 4715875; 1β-hydroxy-3-epivitamin D_3; 1β-hydroxy-3-epicholecalciferol; 9,10-seco-5Z,7E,10(19)-cholestatriene-1R,3S-diol; 5283721; 65445-14-9; Synthesized from 3-epivitamin D_3 *via* SeO_2 oxidation of its 3,5-cyclovitamin D derivative. Binding affinity for intestinal vitamin D receptor relative to 1α,25-dihydroxyvitamin D_3 is reported; λ_m

= 264 nm. **Bioorg Chem**. 1985, **13(1)**, 62-75.

361   *(5E)-1β-Hydroxyvitamin D₃*
5283722 $C_{27}H_{44}O_2$

(1R,3R,5E)-5-[(2E)-2-[(1R,3aR,7aS)-7a-methyl-1-[(2R)-6-methylheptan-2-yl]-2,3,3a,5,6,7-hexahydro-1H-inden-4-ylidene]ethylidene]-4-methylidene-cyclo-hexane-1,3-diol.

LMST03020237; (5E)-1β-hydroxyvitamin D₃; (5E)-1β-hydroxycholecalciferol; 9,10-seco-5E,7E,10(19)-cholestatriene-1R,3R-diol; 5283722; Synthesized from 3-epivitamin D₃ **via** SeO₂ oxidation of its 3,5-cyclovitamin D derivative. Binding affinity for intestinal vitamin D receptor relative to 1α,25-dihydroxyvitamin D₃ is reported; λ_m = 271 nm. **Bioorg Chem**. 1985, **13(1)**, 62-75.

362   *(5E)-1β-Hydroxy-3-epivitamin D₃*
5283723 $C_{27}H_{44}O_2$

(1S,3R,5E)-5-[(2E)-2-[(1R,3aR,7aS)-7a-methyl-1-[(2R)-6-methylheptan-2-yl]-2,3,3a,5,6,7-hexahydro-1H-inden-4-ylidene]ethylidene]-4-methylidene-cyclohexane-1,3-diol.
LMST03020238; 9,10-seco-5E,7E,10(19)-cholestatriene-1R,3S-diol; (5E)-1β-hydroxy-3-epivitamin D₃; (5E)-1β-hydroxy-3-epicholecalciferol; 5283723; Synthesized from 3-epivitamin D₃ **via** SeO₂ oxidation of its 3,5-cyclovitamin D derivative. Binding affinity for intestinal vitamin D receptor relative to 1α,25-dihydroxyvitamin D₃ is reported; λ_m = 273 nm. **Bioorg Chem**. 1985, **13(1)**, 62-75.

363   *2α-Hydroxyvitamin D₃*
5283724 $C_{27}H_{44}O_2$

(1R,2R,4Z)-4-[(2E)-2-[(1R,3aR,7aS)-7a-methyl-1-[(2R)-6-methylheptan-2-yl]-2,3,3a,5,6,7-hexahydro-1H-inden-4-ylidene]ethylidene]-5-methylidene-cyclohexane-1,2-diol.
LMST03020239; 2α-hydroxyvitamin D₃; 2α-hydroxycholecalciferol; 9,10-seco-5Z,7E,10(19)-cholestatriene-2R,3R-diol; 5283724. EP 1280546.

364   *2β-Hydroxyvitamin D₃*
5283725 $C_{27}H_{44}O_2$

(1S,2R,4Z)-4-[(2E)-2-[(1R,3aR,7aS)-7a-methyl-1-[(2R)-6-methylheptan-2-yl]-2,3,3a,5,6,7-hexahydro-1H-inden-4-ylidene]ethylidene]-5-methylidene-cyclohexane-1,2-diol.
LMST03020240; 2β-hydroxyvitamin D₃; 2β-hydroxycholecalciferol; 9,10-seco-5Z,7E,10(19)-cholestatriene-2S,3R-diol; 5283725. EP 1280546.

365   *1α,25-Dihydroxy-3-deoxyvitamin D₃*
5283726 $C_{27}H_{44}O_2$

(6R)-6-[(1R,3aR,4E,7aS)-4-[(2Z)-2-[(3S)-3-hydroxy-2-methylidene-cyclohexylidene]ethylidene]-7a-methyl-2,3,3a,5,6,7-hexahydro-1H-inden-1-yl]-2-methyl-heptan-2-ol.
LMST03020241; 9,10-seco-5Z,7E,10(19)-cholestatriene-1S,25-diol; 1α,25-dihydroxy-3-deoxyvitamin D₃; 1α,25-dihydroxy-3-deoxycholecalciferol; 5283726; Prepared from 1α,25-dihydroxycholesterol *via* reductive deoxygenation of 3-tosyloxy group followed by conventional photochemical transformation. Affinity for chick intestinal receptor is 1/8 that of 1α,25-dihydroxyvitamin D₃. *Biochem Biophys Res Commun*. 1975, *65(1)*, 24-30.

366   *Astrogorgiadiol B*
5283727 $C_{27}H_{44}O_2$

(1R,3aS,4S,5R,7aR)-4-[2-(5-hydroxy-2-methyl-phenyl)-ethyl]-7a-methyl-1-[(2R)-6-methylheptan-2-yl]-1,2,3,3a,4,5,6,7-octahydroinden-5-ol.
LMST03020242; Astrogorgiadiol B; (8S)-9R,10-seco-1,3,5(10)-cholestatriene-3,9-diol; 5283727; Isolated from a gorgonian of the genus **Astrogorgia**. Synthesized by the coupling of upper-half fragment derived from Grundmann's ketone with lower-half fragment of benzyl iodide derivative. Inhibits cell division in fertilized starfish eggs; [α]$_D$ = -4.6° (c = 0.2 CHCl₃). *Tet Lett*. 1989, *30(50)*, 7079-82; 1998, *39(26)*, 4741-50; *Nat Prod Rep*. 1997, *14(2)*, 259-302.

367   *22S-Hydroxyvitamin D₃*
5283728 $C_{27}H_{44}O_2$

(2S,3S)-2-[(1R,3aR,4E,7aS)-4-[(2Z)-2-[(5S)-5-hydroxy-2-methylidene-cyclohexylidene]ethylidene]-7a-methyl-2,3,3a,5,6,7-hexahydro-1H-inden-1-yl]-6-methyl-heptan-3-ol.
LMST03020243; 22S-hydroxyvitamin D₃; 22S-

hydroxycholecalciferol; 9,10-seco-5Z,7E,10(19)-cholestatriene-3S,22S-diol; 5283728; Synthesized from 23,24-dinor-5-cholenoic acid *via* the Grignard reaction of the corresponding C-22 aldehyde with side chain fragment. Displays no vitamin D agonist activity in the intestine or in bone *in vivo* and did not block the activity of vitamin D₃ or 25-hydroxyvitamin D₃. Affinity for chick intestinal receptor is 6 orders of magnitude less than that of 1α,25-dihydroxyvitamin D₃, 3 orders of magnitude less than that of 25-hydroxyvitamin D₃; λ$_m$ = 265 nm (EtOH). *J Steroid Biochem*. 1987, *28(2)*, 147-53.

368   *24-Hydroxyvitamin D₃*
56720-87-7 $C_{27}H_{44}O_2$

(3R,6R)-6-[(1R,3aR,4E,7aS)-4-[(2Z)-2-[(5S)-5-hydroxy-2-methylidene-cyclohexylidene]ethylidene]-7a-methyl-2,3,3a,5,6,7-hexahydro-1H-inden-1-yl]-2-methyl-heptan-3-ol.
LMST03020244; 24-Hydroxyvitamin D₃; 24-Hydroxycalcidiol; 9,10-Secocholesta-5Z,7E,10(19)-triene-3β,24R-diol; 5283729; 56720-87-7; Isolated from the blood plasma of chickens given large doses of vitamin D₃. Synthesized as mixture of C(24) epimers from 3β-acetoxy-27-nor-5-cholesten-25-one. Metabolite of vitamin D₃. In human promyelocytic leukemia cells, metabolized to (5Z)- and (5E)-(24R)-19-nor-10-oxo-24-hydroxycalcidiol. Biochemistry. 1981, *20(8)*, 2350-3; *Arch Biochem Biophys*. 1983, *225(2)*, 986-92; *Biochem Biophys Res Comm*. 1965, *67(3)*, 965-71; EP 1208843.

369   *24S-Hydroxyvitamin D₃*
5283730 $C_{27}H_{44}O_2$

(3S,6R)-6-[(1R,3aR,4E,7aS)-4-[(2Z)-2-[(5S)-5-hydroxy-2-methylidene-cyclohexylidene]ethylidene]-7a-methyl-2,3,3a,5,6,7-hexahydro-1H-inden-1-yl]-2-methyl-heptan-3-ol.
LMST03020245; 24S-hydroxyvitamin D_3; 24S-hydroxycholecalciferol; 9,10-seco-5Z,7E,10(19)-cholestatriene-3S,24S-diol; 5283730. *Biochem Biophys Res Comm.* 1965, *67(3)*, 965-71.

370   *25-Hydroxy Vitamin D_3*
19356-17-3 $C_{27}H_{44}O_2$

(6R)-6-[(1R,3aR,4E,7aS)-4-[(2Z)-2-[(5S)-5-hydroxy-2-methylidene-cyclohexylidene]ethylidene]-7a-methyl-2,3,3a,5,6,7-hexahydro-1H-inden-1-yl]-2-methyl-heptan-2-ol.
LMST03020246; Calcifediol; Calcifediol anhydrous; 9,10-Seco-5Z,7E,10(19)-cholestatrien-3β,25-diol; 25-Hydroxycholecalciferol; 25-Hydroxyvitamin D; 25-Hydroxyvitamin D_3; 5,6-cis-25-Hydroxyvitamin D_3; 5,6-trans-25-Hydroxycholecalciferol; 5,6-trans-9,10-Seco-5,7,10(19)-cholestatrien-3β,25-diol; Calcidiol; Calcifediol; Calcifediolum; Calcifidiol; Cholecalciferol, 25-hydroxy-; Delakmin; EINECS 242-990-9; Ro 8-8892; U 32070 E; Calcifediol; 9,10-Secocholesta-5Z,7E,10(19)-triene-3β,25-diol; 5280447; 5283731; 5353325; 19356-17-3; The major circulating metabolite of vitamin D_3 (Calciferol). It is produced in the liver and is the best indicator of the body's vitamin D stores. It is effective in the treatment of rickets and osteomalacia, both in azotemic and non-azotemic patients. Calcifediol also has mineralizing properties. The vitamin D prohormone, found in human plasma at concentrations of 10-40 ng/ml, it is the precursor of 1,25-dihydroxy vitamin D_3 and 24,25-dihydroxy vitamin D_3. Synthesized from 25-hydroxycholesterol acetate and 3β-acetoxy-27-nor-cholest-5-en-25-one by a standard photochemical method *via* 5,7-cholestadiene-3β,25-diol. Formed by hydroxylation of vitamin D_3 by vitamin D 25-hydroxylase (CYP27) in the liver, it is the major circulating metabolite of vitamin D_3 produced in the liver and the best indicator of the body's vitamin D stores. It is effective in the treatment of rickets and osteomalacia, both in azotemic and non-azotemic patients. Calcifediol also has mineralizing properties. It is further metabolized by 1α-hydroxylation in the kidney to give 1α,25-dihydroxyvitamin D_3 under vitamin D deficient conditions and by 24-hydroxylation to give 24R,25-dihydroxyvitamin D_3 in tissues possessing the vitamin D receptor under vitamin D supplemented conditions or 26-hydroxylation to give 25,26-dihydroxyvitamin

D_3; Crystals; mp = 96-106°; $[\alpha]_D^{25}$ = +88.1° (c = 0.5 EtOH); λ_m = 265 nm (ε 18000 95% EtOH). *Biochem* 1969, *8(2)*, 671-5; *Ann Rev Biochem.* 1976, *45*, 631-6; *J Biol Chem.* 1988, *263(28)*, 14256-60; *J Org Chem.* 1996, *61(1)*, 118-24.

371   *25-Hydroxy-14-epivitamin D_3*
5283732 $C_{27}H_{44}O_2$

(6R)-6-[(1R,3aS,4E,7aS)-4-[(2Z)-2-[(5S)-5-hydroxy-2-methylidene-cyclohexylidene]ethylidene]-7a-methyl-2,3,3a,5,6,7-hexahydro-1H-inden-1-yl]-2-methyl-heptan-2-ol.
LMST03020247; 9,10-seco-5Z,7E,10(19)-cholestatriene-3S,25-diol; 25-hydroxy-14-epicholecalciferol; 5283732; Synthetic vitamin D_3 analogue. Undergoes sigmatropic hydrogen shift to the previtamin form like the natural 14-isomer. Has essentially no vitamin D activity. Synthesized by Wittig-Horner coupling of des-AB 14-epivitamin D 8-ketone with A-ring phosphine oxide and ozonolysis followed by oxidation at C-25 and epimerization at C-14. Compared to 1α,25-dihydroxyvitamin D_3, affinity for chick intestinal receptor is 0.08%, affinity for human vitamin D binding protein is 3450%, intestinal calcium absorption in vitamin D-deficient chicks is < 0.1%, and bone calcium mobilization in vitamin D-deficient chicks, < 1%; λ_m = 264 nm (ε 18100 95% EtOH). *J Med Chem.* 1994, *37(15)*, 2387-93.

372   *7-Dehydrocholesterol 5,6-oxide*
95841-71-7 $C_{27}H_{44}O_2$

(3β,5β,6β)-5,6-Epoxycholest-7-en-3-ol.
7-Dehydrocholesterol 5,6-oxide; 7-Dehydrocholesterol 5,6-epoxide; 7-Dhco; 125859; 95841-71-7; An inhibitor of microsomal cholesterol oxide hydrolase; *J Biol Chem.* 1986, *261(6)*, 2510-3.

373   *5,6-trans-25-Hydroxyvitamin D_3*
36149-00-5 $C_{27}H_{44}O_2$

6-[(4Z)-4-[(2E)-2-(5-hydroxy-2-methylidene-cyclo-hexylidene)ethylidene]-7a-methyl-2,3,3a,5,6,7-hex-ahydro-1H-inden-1-yl]-2-methyl-heptan-2-ol.
5,6-trans-25-Hydroxyvitamin D_3; 25-Hydroxy-5,6-trans-vitamin D_3; 5,6-trans-25-hydroxycholecalciferol; 25-Hydroxy-5,6-trans-cholecalciferol; 9,10-Seco-cholesta-5E,7E,10(19)-triene-3β,25-diol; EINECS 252-891-2; 6368826; 36149-00-5; Formed by irradiation of vitamin D_3.

374 *Calcifediol*
6433735 $C_{27}H_{44}O_2$

(6R)-6-[(1R,4E,7aS)-4-[(2Z)-2-[(5R)-5-hydroxy-2-methylidene-cyclohexylidene]ethylidene]-7a-methyl-2,3,3a,5,6,7-hexahydro-1H-inden-1-yl]-2-methyl-heptan-2-ol.
Calcifediol; Calcifidiol; Calcidiol; Delakmin; Cal-cifediol anhydrous; 25-Hydroxyvitamin D; 25-hydroxyvitamin D_3; 25-Hydroxycholecalciferol; Calcifediolum; Cholecalciferol, 25-hydroxy-; LMST03020246; 6433735; The major circulating metabolite of vitamin D_3, it is produced in the liver and is the best indicator of the body's vitamin D stores. It is effective in the treatment of rickets and osteomalacia, both in azotemic and non-azotemic patients. Calcifediol also has mineralizing properties. *Chirality.* 1999, *11(9)*, 701-6.

375 *11-Hydroxyvitamin D_3*
89321-96-0 $C_{27}H_{44}O_2$

(3R,3aS,5R,7E,7aR)-7-[(2Z)-2-[(5R)-5-hydroxy-2-methylidene-cyclohexylidene]ethylidene]-3a-methyl-3-[(2R)-6-methylheptan-2-yl]-2,3,4,5,6,7a-hexahydro-1H-inden-5-ol.
11-Hydroxyvitamin D_3; 11α-Hydroxyvitamin D_3; 9,10-Secocholesta-5Z,7E,10(19)-triene-3β,11α-diol; 6439865; 89321-96-0; Increases calcium transport at a dose of 500 pmol/rat but fails to increase bone calcium mobilization. Under the same conditions, corre-sponding doses of vitamin D_3 and 25-hydroxyvitamin D_3 increase bone calcium mobilization and intestinal calcium transport. Shows significantly greater binding efficiency to vitamin D binding protein than 1α-hydroxyvitamin D_3, 1,25-dihydroxyvitamin D_3, or vitamin D_3. *Biochemistry.* 1984, *23(9)*,1983-7.

376 *26-Hydroxy Vitamin D_3*
69556-15-6 $C_{27}H_{44}O_2$

(6R)-6-[(1R,3aR,4E,7aS)-4-[(2Z)-2-[(5R)-5-hydroxy-2-methylidene-cyclohexylidene]ethylidene]-7a-methyl-2,3,3a,5,6,7-hexahydro-1H-inden-1-yl]-2-methyl-heptan-1-ol.
26-Hydroxycholecalciferol; (25R)-26-Hydroxychole-calciferol; (25S)-26-Hydroxycholecalciferol; 26-Hydroxyvitamin D_3; 9,10-Secocholesta-5Z,7E,10(19)-triene-3,26-diol; 6441773; 69556-15-6. *J Steroid Biochem.* 1980, *13(7)*, 803-7.

377 *24-Hydroxy Vitamin D_3*
58239-34-2 $C_{27}H_{44}O_2$

(6R)-6-[(1R,3aR,4E,7aS)-4-[(2Z)-2-[(5R)-5-hydroxy-2-methylidene-cyclohexylidene]ethylidene]-7a-methyl-2,3,3a,5,6,7-hexahydro-1H-inden-1-yl]-2-methyl-heptan-3-ol.

24-Hydroxycholecalciferol; 24-Hydroxyvitamin D$_3$; 9,10-Secocholesta-5Z,7E,10(19)-triene-3β,24-diol; LMST03020245; 6446381; 58239-34-2; A metabolite of vitamin D$_3$. Used to treat psoriasis. USP 4871723.

378   ***22-Hydroxyvitamin D$_3$***
110927-46-3   C$_{27}$H$_{44}$O$_2$

(2S,3S)-2-[(1R,3aR,4E,7aS)-4-[(2Z)-2-[(5R)-5-hydroxy-2-methylidene-cyclohexylidene]ethylidene]-7a-methyl-2,3,3a,5,6,7-hexahydro-1H-inden-1-yl]-6-methyl-heptan-3-ol.

22-Hydroxyvitamin D$_3$; 22-Hydroxycholecalciferol; 9,10-Secocholesta-5Z,7E,10(19)-triene-3β,22S-diol; 6449826; 110927-46-3; Displays no vitamin D agonist activity in the intestine or in bone *in vivo* and does not block the activity of vitamin D$_3$ or 25-hydroxyvitamin D$_3$. It is a weak vitamin D$_3$ agonist in the chick embryonal duodenum *in vitro*. It does not antagonize the activity of 1,25-dihydroxyvitamin D$_3$. *J Steroid Biochem.* 1987, *28(2)*, 147-53.

379   ***1-Hydroxyprevitamin D$_3$***
41461-13-6   C$_{27}$H$_{44}$O$_2$.

(1R,5S)-3-[(Z)-2-[(1R,3aR,7aS)-7a-methyl-1-[(2R)-6-methylheptan-2-yl]-1,2,3,3a,6,7-hexahydroinden-4-yl]ethenyl]-2-methyl-cyclohex-2-ene-1,5-diol.

1-Hydroxyprevitamin D$_3$; 1α-Hydroxyprevitamin D$_3$; Oxydevit; Pre-1α-hydroxyvitamin D$_3$; 9,10-Secocholesta-5(10),6Z,8-triene-1α,3β-diol; 6450023; 41461-13-6; Formed by irradiation of 1α-hydroxyprovitamin D$_3$. Has immunomodulatory action in children with chronic glomerulonephritis. Reported to inhibit progression of systemic osteoporosis. *Proc Natl Acad Sci U S A.* 1982, *79(16)*, 5115-6; *Pediatriia.* 1991, *(7)*, 75-8.

380   ***25-Hydroxyvitamin D$_3$***
6506392   C$_{27}$H$_{44}$O$_2$

(6R)-6-[(1R,3aR,4E,7aS)-4-[(2E)-2-[(5S)-5-hydroxy-2-methylidene-cyclohexylidene]ethylidene]-7a-methyl-2,3,3a,5,6,7-hexahydro-1H-inden-1-yl]-2-methyl-heptan-2-ol.

25-Hydroxyvitamin D$_3$; CBiol_001814; KBioGR_000-131; KBioSS_000131; KBio2_000131; KBio2_002699; KBio2_005267; KBio3_000261; KBio3_000262; Bio1_000100; Bio1_000589; Bio1_001078; Bio2_000131; Bio2_000611; 166583; 6506392; *Proc Natl Acad Sci U.S.A.* 1997, 94(24), 12920-5; USP 7166585.

381   ***25-Hydroxy-3-epivitamin D$_3$***
                          C$_{27}$H$_{44}$O$_2$

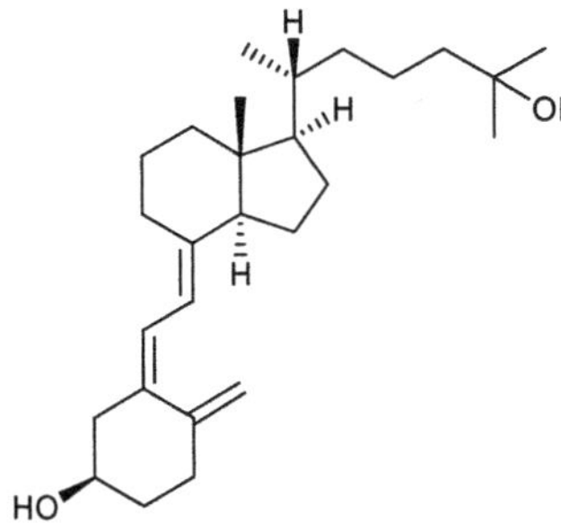

(1S,3R,5Z)-5-[(2E)-2-[(1R,3aR,7aS)-1-[(2R)-6-hydroxy-6-methyl-heptan-2-yl]-7a-methyl-2,3,3a,5,6,7-hexahydro-1H-inden-4-ylidene]ethylidene]-4-methylidene-cyclohexan-1-ol.
(5Z,7E)-(1R,3S)-9,10-seco-5,7,10(19)-cholestatriene-3,25-diol;25-hydroxy-3-epivitamin D$_3$; 25-hydroxy-3-epicholecalciferol; A major metabolite of 25-hydroxyvitamin D$_3$. *J Biol Chem.* 2004, **279(16)**, 15897-907.

382   *20-Hydroxyvitamin D$_3$*

C$_{27}$H$_{44}$O$_2$

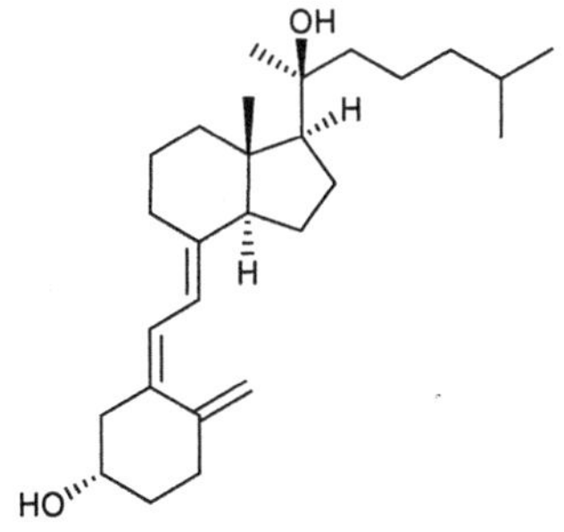

9,10-Secocholesta-5Z,7E,10(19)-triene-3β,20-diol.
20-Hydroxyvitamin D$_3$; Mitochondrial metabolite of vitamin D$_3$. *Proc Natl Acad Sci U.S.A.* 2003, **100(25)**, 14754-9.

383   *Cholesta-5,7-diene-1,3-diol*
43217-89-6   C$_{27}$H$_{44}$O$_2$

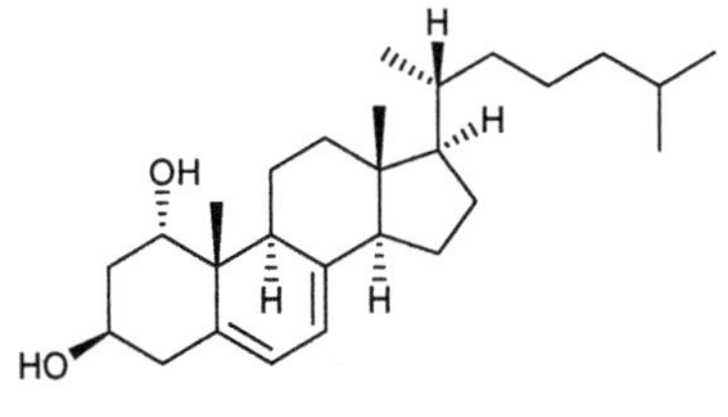

(1S,3R,9R,10R,13S,14R,17R)-10,13-dimethyl-17-[(2R)-6-methylheptan-2-yl]-2,3,4,9,11,12,14,15,16,17-decahydro-1H-cyclopenta[a]phenanthrene-1,3-diol.
CDABD;   Cholesta-5,7-diene-1α,3β-diol;   193829;   43217-89-6;   Irradiated, forms 1α-hydroxyprevitamin D$_3$. *J Nutr Sci Vitaminol (Tokyo).* 1980, **26**(6), 545-56.

384   *1α,24-Dihydroxyvitamin D$_3$*
60965-80-2   C$_{27}$H$_{44}$O$_3$

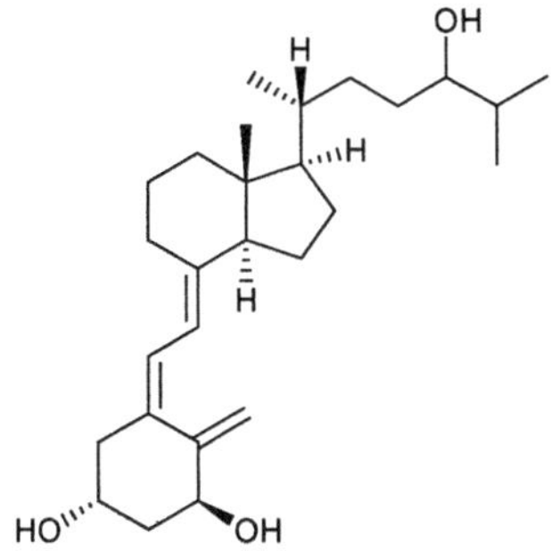

(1R,3S,5Z)-5-[(2E)-2-[(1R,3aR,7aS)-1-[(2R)-5-hydroxy-6-methyl-heptan-2-yl]-7a-methyl-2,3,3a,5,6,7-hexahydro-1H-inden-4-ylidene]ethylidene]-4-methylidene-cyclohexane-1,3-diol.
LMST03020256; 1α,24-Dihydroxyvitamin D$_3$; 1,24-Dihydroxychole-calciferol; 1,24-Dihydroxyvitamin D$_3$; 1α,24-Di-hydroxycholecalciferol; 1α,24-Dihydroxy-vitamin D$_3$; 9,10-Secocholesta-5Z,7E,10(19)-triene-1α,3β,24-triol; Tacalcitol; 6437387; 60965-80-2; Calcitriol analog. Has been used as an antipsoriatic and anti-inflammatory agent. Used to treat hypocalcemia and psoriasis. Used systemically to treat hyperparathyroidism. Has been used to treat the skin disease vitiligo. *Steroids.* 2004, **69(10)**, 629-35; *Acta Biomed.* 2005, **76(1)**, 13-19.

385   *1α,24-Dihydroxyvitamin D$_3$*
93129-94-3   C$_{27}$H$_{44}$O$_3$

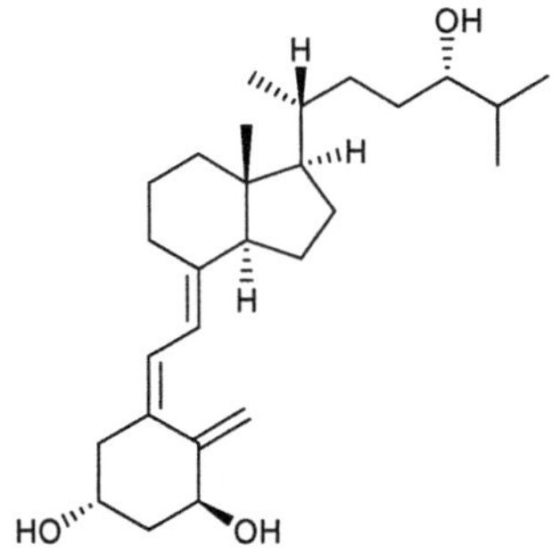

(1R,3S)-5-[2-[(1R,3aR,7aS)-1-[(2R,5S)-5-hydroxy-6-methyl-heptan-2-yl]-7a-methyl-2,3,3a,5,6,7-hexahydro-1H-inden-4-ylidene]ethylidene]-4-methylidene-cyclohexane-1,3-diol.
Tacalcitol; Curaderm; Bonalfa; Bonalfa high; Curaderm; D01472; 1α,24-Dihydroxyvitamin D$_3$; C12919; PRI-2191; 443981; 5282177; 93129-94-3; (See also 400). Used topically in ointment form in a concentration of 4 µg/g, for the treatment of mild and/or moderate psoriasis (involvement of <20% of the surface of the skin).

386   *1α,25-Dihydroxyvitamin D$_3$*
32222-06-3   C$_{27}$H$_{44}$O$_3$

(1R,3S,5Z)-5-[(2E)-2-[(1R,3aR,7aS)-1-[(2R)-6-hydroxy-6-methyl-heptan-2-yl]-7a-methyl-2,3,3a,5,6,7-hexahydro-1H-inden-4-ylidene]ethylidene]-4-methylidene-cyclohexane-1,3-diol.

LMST03020258; Calcitriol; 1,25-(OH)2D$_3$; 1,25-DHCC; 1,25-Dihydroxycholecaliferol; 1,25-Dihydroxyvitamin D; 1α,25-Dihydroxycholecalciferol; 1α,25-Dihydroxyvitamin D; CCRIS 5522; Calcijex; Calcitriol; Calcitriolum; Cholecalciferol, 1α,25-dihydroxy- DN 101; DN-101; Dihydroxyvitamin D$_3$; EINECS 250-963-8; HSDB 3482; MC 1288; Ro 21-5535; Ro 215535; Rocaltrol; 9,10-Secocholesta-5,7,10(19)-triene-1,3,25-triol; 9,10-Seco-5Z,7E,10(19)-cholestatriene-1α,3β,25-triol; Soltriol; Topitriol; U 49562; 5280453; 32222-06-3; The physiologically active hormonal form of vitamin D$_3$, islated from chicken intestine and kidney and from various plants. It is formed primarily in the kidney by enzymatic hydroxylation of calcifediol (25-hydroxycholecalciferol). Its production is stimulated by low blood calcium levels and parathyroid hormone. Calcitriol increases intestinal absorption of calcium and phosphorus, and in concert with parathyroid hormone increases bone resorption. Dissociation constant (K$_d$) for vitamin D receptor is 8.2 x 10^{-11}, for mammalian VDR is 10^{-11} - 10^{-10}M. Concentration in human plasma is 20-65 pg/ml. Possesses antiproliferative activity in myeloid leukemia cells but also has significant hypercalcemia-inducing effects, which works against its use as an anticancer agent ; Solid, mp = 113°; [α]$_D$ = +29° (Et$_2$O); λ$_m$ - 264 nm, (ε = 18000 Et$_2$O); LD50 (mus ip) = 1.9 mg/kg, (mus orl) = 1.35 mg/kg, (mus sc) = 0.145 mg/kg, (rat ip) > 5 mg/kg, (rat iv) = 0.105 mg/kg, (rat orl)= 0.62 mg/kg, (rat sc) = 0.066 mg/kg. *Ann Rev Biochem*. 1976, **45**, 631-6; *Nature*. 1970, **228(5273)**, 764-6; *J Chem Soc Chem Comm*. 1984, 203; *Biochemistry*, 1971, **10(14)**, 2799-804; *Mol Pharm*. 2002, **62(4)**, 788-94.

387   **Rocaltrol**

32511-63-0   C$_{27}$H$_{44}$O$_3$

(1R,5Z)-5-[(2E)-2-[(1R,3aR,7aS)-1-[(2R)-6-hydroxy-6-methyl-heptan-2-yl]-7a-methyl-2,3,3a,5,6,7-hexahydro-1H-inden-4-ylidene]ethylidene]-4-methylidene-cyclohexane-1,3-diol.

Rocaltrol; 9,10-Sechocholesta-5Z,7E,10(19)-triene-1,3β,25-triol; LMST03020258; 6437079; 32511-63-0; Geometrical isomer of calcitriol. A synthetic vitamin D analog which is active in the regulation of absorption and utilization of calcium. *Chirality*. 1999, **11(9)**, 701-6; *J Steroid Biochem Mol Biol*. 1992, **43(7)**, 677-82; *J Biol Chem*. 1982, **257(9)**, 5097-102.

388   *1β,25-Dihydroxyvitamin D$_3$*

5283741   C$_{27}$H$_{44}$O$_3$

(1R,3R,5Z)-5-[(2E)-2-[(1R,3aR,7aS)-1-[(2R)-6-hydroxy-6-methyl-heptan-2-yl]-7a-methyl-2,3,3a,5,6,7-hexahydro-1H-inden-4-ylidene]ethylidene]-4-methylidene-cyclohexane-1,3-diol.

LMST03020266; 1β,25-dihydroxyvitamin D$_3$; 1β,25-dihydroxycholecalciferol; (5Z,7E)-(1R,3R)-9,10-seco-5,7,10(19)-cholestatriene-1,3,25-triol; 5293741; Synthesized from 1α,25-dihydroxyvitamin D$_3$ by selective oxidation of the 1-hydroxyl group followed by LiAlH$_4$ reduction. Binding affinity for intestinal vitamin D receptor is 1/3000 that of 1α,25-dihydroxyvitamin D$_3$. Has been used in dental monomers; λ$_m$ = 263 nm. *J Chem Soc Chem Comm*. 1977, 890; *J Biol Chem*. 1993, **268(27)**, 20022-30.

389   *1β,25-Dihydroxy-3-epivitamin D$_3$*

5283742   C$_{27}$H$_{44}$O$_3$

(1S,3R,5Z)-5-[(2E)-2-[(1R,3aR,7aS)-1-[(2R)-6-hydroxy-6-methyl-heptan-2-yl]-7a-methyl-2,3,3a,5,6,7-hexa-hydro-1H-inden-4-ylidene]ethylidene]-4-methylidene-cyclohexane-1,3-diol.

LMST03020267; (5Z,7E)-(1R,3S)-9,10-seco-5,7,10(19)-cholestatriene-1,3,25-triol;1β,25-dihydroxy-3-epivitamin D$_3$; 1β,25-dihydroxy-3-epicholecalciferol; 5283742; Metabolite of both 25-hydroxyvitamin D$_3$ and 25-hydroxy-3-epivitamin D$_3$. Synthesized from R-carvone. Its affinity for the chick intestinal vitamin D receptor is 1/500 that of 1α,25-dihydroxyvitamin D$_3$; λ$_m$ = 264 nm (ε = 17000 95% EtOH). *J Org Chem*. 1993, **58(7)**, 1895-9; *J Biol Chem*. 1993, **268(27)**, 20022-30; 2004, **279(16)**, 15897-907.

390   *1α,25-Dihydroxy-14-epivitamin D$_3$*
5283737   C$_{27}$H$_{44}$O$_3$

(1R,3S,5Z)-5-[(2E)-2-[(1R,3aS,7aS)-1-[(2R)-6-hydroxy-6-methyl-heptan-2-yl]-7a-methyl-2,3,3a,5,6,7-hexa-hydro-1H-inden-4-ylidene]ethylidene]-4-methylidene-cyclohexane-1,3-diol.

LMST03020260; 1,25-(OH)$_2$-14-epi-D$_3$; (5Z,7E)-(1S,3R,14R)-9,10-seco-5,7,10(19)-cholestatriene-1,3,25-triol; 1α,25-dihydroxy-14-epivitamin D$_3$; 1α,25-dihydroxy-14-epicholecalciferol; 5283737; Vitamin D$_3$ analogue synthesized from des-AB 14-epivitamin D 8-ketone with A-ring phosphine oxide. Undergoes sigmatropic hydrogen shift to the previtamin form like the natural 14-isomer. Has essentially essentially no vitamin D activity. Compared to 1α,25-dihydroxyvitamin D$_3$, its affinity for chick intestinal receptor is 15%, affinity for human vitamin D binding protein is 12%, and intestinal calcium absorption in vitamin D-deficient chicks is 3.9%; λ$_m$ = 266 nm (ε = 17700 95% EtOH). *J Med Chem*. 1994, **37(15)**, 2387-93.

391   *1α,25-Dihydroxy-20-epivitamin D$_3$*
5283739   C$_{27}$H$_{44}$O$_3$

(1R,3S,5Z)-5-[(2E)-2-[(1R,3aR,7aS)-1-[(2S)-6-hydroxy-6-methyl-heptan-2-yl]-7a-methyl-2,3,3a,5,6,7-hexa-hydro-1H-inden-4-ylidene]ethylidene]-4-methylidene-cyclohexane-1,3-diol.

LMST03020262; 1,25-Dihydroxyvitamin D$_3$; (20S)-9,10-seco-5Z,7E,10(19)-cholestatriene-1S,3R,25-triol; 1α,25-dihydroxy-20-epivitamin D$_3$; 1α,25-dihydroxy-20-epicholecalciferol; VDX; 5-{2-[1-(5-Hydroxy-1,5-dimethyl-hexyl)-7a-methyl-octahydro-inden-4-ylidene]-ethylidene}-4-methylene-cyclohexane-1,3-diol; 20-epi-1α,25-dihydroxyvitamin D$_3$; 20-epi-1α,25-dihydroxycholecalciferol; 20-Epicalcitriol; MC 1288; 5283739; Calcitriol analog synthesized from 17-epicholesterol. Used systemically for treatment of immune disorders. Promotes formation of osteocalcin mRNA slightly more than calcitriol, but osteocalcin secretion is unchanged. Powerful inhibitor of clonal growth of HL-60, HTLV-1 and leukemic clonogenic cells. Affinity for the vitamin D receptor is slightly higher than that of 1α,25-dihydroxyvitamin D$_3$.

Compared to 1α,25-dihydroxyvitamin D$_3$, inhibition of U937 cell (human histiocytic lymphoma cell line) proliferation is 5000%, induction of U937 cell differentiation is 2666%, binding to the vitamin D receptor from rachitic chicken intestine is 120%, calcemic activity determined by the increase in urinary calcium excretion in rats is 230%, inhibitory effects on murine thymocyte activation is 730800%. *Biochem Pharmacol*. 1991, **42(8)**, 1569-75; *J Org Chem*. 2005, **70(21)**, 8513-21; *Eur J Biochem*. 1999, **261(3)**, 706-13; *Blood*. 1994, **84(6)**, 1960-7.

392   *1,25-Dihydroxyvitamin D$_3$*
2524   C$_{27}$H$_{44}$O$_3$

5-[2-[1-(6-hydroxy-6-methyl-heptan-2-yl)-7a-methyl-2,3,3a,5,6,7-hexahydro-1H-inden-4-ylidene]ethylidene]-4-methylidene-cyclohexane-1,3-diol.
1,25-Dihydroxyvitamin D_3; LMST03020265; 2524. Relative to $1\alpha,25$-dihydroxyvitamin D_3, is 6% as active in intestinal calcium transport, 1.5% as active in bone calcium mobilization and its affinity for the vitamin D receptor is 13%. ***Steroids***, 1980, **35**, 419-25.

393   *(5E)-1α,25-Dihydroxyvitamin D₃*
5283740 $C_{27}H_{44}O_3$

(1R,3S,5E)-5-[(2E)-2-[(1R,3aR,7aS)-1-[(2R)-6-hydroxy-6-methyl-heptan-2-yl]-7a-methyl-2,3,3a,5,6,7-hexa-hydro-1H-inden-4-ylidene]ethylidene]-4-methylidene-cyclohexane-1,3-diol.
LMST03020265; 9,10-seco-5E,7E,10(19)-cholestatri-ene-1S,3R,25-triol; (5E)-1α,25-dihydroxycalc-iferol; 5283740; Is approximately 1/16 and 1/64 as active as $1\alpha,25$-dihydroxyvitamin D_3 in intestinal calcium transport and bone calcium mobilization, respectively. The affinity for intestinal vitamin D receptor is 13% of that of $1\alpha,25$-dihydroxyvitamin D_3. ***Steroids***, 1980, **35(4)**, 419-25.

394   *1,25-Dihydroxyvitamin D₃*
5371993 $C_{27}H_{44}O_3$

(5E)-5-[(2Z)-2-[1-(6-hydroxy-6-methyl-heptan-2-yl)-7a-methyl-2,3,3a,5,6,7-hexahydro-1H-inden-4-ylidene]ethylidene]-4-methylidene-cyclohexane-1,3-diol.
9,10-Secocholesta-5,7,10(19)-triene-1,3β,25-triol; 5371993.

395   *6R,19-Epidioxy-6,19-dihydrovitamin D₃*
9547430 $C_{27}H_{44}O_3$

(3S,10R)-10-[(E)-[(1R,3aR,7aS)-7a-methyl-1-[(2R)-6-methylheptan-2-yl]-2,3,3a,5,6,7-hexahydro-1H-inden-4-ylidene]methyl]-8,9-dioxabicyclo[4.4.0]dec-11-en-3-ol.
LMST03020248; 6R,19-epidioxy-9,10-seco-5(10),7E-cholestadien-3S-ol; 6R,19-epidioxy-6,19-dihydrochole-calciferol; 9547430; A major product by the reaction of vitamin D_3 with singlet oxygen generated by a dye-sensitized photochemical method. Activities in stimulating intestinal calcium transport and in increasing serum calcium and inorganic phosphorus evaluated compared with vitamin D_3. *J Nutr Sci Vitaminol*. 1979, **25(5)**, 455-8; *J Org Chem*. 1983, **48(20)**, 3477-83.

396   *6S,19-Epidioxy-6,19-dihydrovitamin D₃*
9547431 $C_{27}H_{44}O_3$

(3S,10S)-10-[(E)-[(1R,3aR,7aS)-7a-methyl-1-[(2R)-6-methylheptan-2-yl]-2,3,3a,5,6,7-hexahydro-1H-inden-4-ylidene]methyl]-8,9-dioxabicyclo[4.4.0]dec-11-en-3-ol.
LMST03020249; 6S,19-epidioxy-9,10-seco-5(10),7E-cholestadien-3S-ol; 6S,19-epidioxy-6,19-dihydro-vitamin D_3; 6S,19-epidioxy-6,19-dihydrocholecalc-iferol; 9547431; Produced as a major product in the reaction of vitamin D_3 with singlet oxygen generated by a dye-sensitized photochemical method. Activities in stimulating intestinal calcium transport and in increasing serum calcium and inorganic phosphorus evaluated and compared with those of vitamin D_3. *J Nutr Sci Vitaminol*. 1979, **25(5)**, 455-8; *J Org Chem*. 1983, **48(20)**, 3477-83.

397   *6R-Hydroperoxy-9,10-seco-4,7E,10(19)-cholestatrien-3S-ol*
9547434 $C_{27}H_{44}O_3$

(1S)-3-[(1R,2E)-2-[(1R,3aR,7aS)-7a-methyl-1-[(2R)-6-methylheptan-2-yl]-2,3,3a,5,6,7-hexahydro-1H-inden-4-ylidene]-1-hydroperoxy-ethyl]-4-methylidene-cyclohex-2-en-1-ol.
LMST03020252; 6R-hydroperoxy-9,10-seco-4,7E,10(19)-cholestatrien-3S-ol; 9547434; Formed as a minor product by the reaction of vitamin D_3 with singlet oxygen generated by a dye-sensitized photochemical method. *J Org Chem*. 1983, *48(20)*, 3477-83.

398   6S-Hydroperoxy-9,10-seco-4,7E,10(19)-cholestatrien-3S-ol

9547435 $C_{27}H_{44}O_3$

(1S)-3-[(1S,2E)-2-[(1R,3aR,7aS)-7a-methyl-1-[(2R)-6-methylheptan-2-yl]-2,3,3a,5,6,7-hexahydro-1H-inden-4-ylidene]-1-hydroperoxy-ethyl]-4-methylidene-cyclohex-2-en-1-ol.
LMST03020253; 6S-hydroperoxy-9,10-seco-4,7E,10(19)-cholestatrien-3S-ol; 9547435; Formed as a minor product by the reaction of vitamin D_3 with singlet oxygen generated by a dye-sensitized photochemical method; λ_m = 232.5 nm (95% EtOH). *J Org Chem*. 1983, *48(20)*, 3477-83.

399   1α,18-Dihydroxyvitamin D_3

5283733 $C_{27}H_{44}O_3$

(1R,3S,5Z)-5-[(2E)-2-[(1R,3aR,7aR)-7a-(hydroxy-methyl)-1-[(2R)-6-methylheptan-2-yl]-2,3,3a,5,6,7-hexahydro-1H-inden-4-ylidene]ethylidene]-4-methylidene-cyclohexane-1,3-diol.
LMST03020254; 1α,18-dihydroxycholecalciferol; 9,10-seco-5Z,7E,10(19)-cholestatriene-1S,3R,18-triol; 5283733.

400   1α,24R-Dihydroxyvitamin D_3

57333-96-7 $C_{27}H_{44}O_3$

(1R,3S,5Z)-5-[(2E)-2-[(1R,3aR,7aS)-1-[(2R,5R)-5-hydroxy-6-methyl-heptan-2-yl]-7a-methyl-2,3,3a,5,6,7-hexahydro-1H-inden-4-ylidene]ethylidene]-4-methyl-idene-cyclohexane-1,3-diol.
LMST03020256; Tacalcitol; 9,10-Secocholesta-5Z,7E, 10(19)-triene-1α,3β,24R-triol; 1α,24R-Dihydroxy-vitamin D_3; Bonalfa; CCRIS 4211; Curatoderm; PRI 2191; TV 02; TV-02; Tacalcitol; Tacalcitolum; (+)-(5Z,7E,24R)-9,10-Secocholesta-5Z,7E,10(19)-triene-1α,3β,24R-triol; 1α,24(R)-Dihydroxyvitamin D_3; 9,10-Secocholesta-5,7,10(19)-triene-1α,3β,24R-triol; 5283734; 57333-96-7; 93129-94-3; Synthesized from 24-oxocholesterol. Compared to 1α,25-dihydroxy-vitamin D_3, is more active in HL-60 differentiation and nearly as active in its affinity for the chick intestinal receptor. Approved for topical use against psoriasis. Used topically in ointment form in a concentration of 4 μg/g, for the treatment of mild and/or moderate psoriasis (involvement of <20% of the surface of the skin); Solid; mp = 84-85° (mixture of 24R, 24S epimers); λ_m = 266 nm (ε = 17000). LD_{50} (dog sc) = 0.01 mg/kg, (mus iv) = 0.559 mg/kg, (mus orl) = 3.05 mg/kg, (mus sc) = 0.419 mg/kg, (rat iv) = 0.556 mg/kg, (rat orl) = 3.279 mg/kg, rat sc) = 0.1 mg/kg. *J Chem Soc Perkin 1*. 1975, 1421-4; *Chem Pharm Bull*. 1984, *32(10)*, 3866-72; *Tet Lett*. 1975, *26*, 2203-6; *J Biol Chem*. 1987, *262(29)*, 14164-71.

401 1α,24S-Dihydroxyvitamin D_3
5283735 $C_{27}H_{44}O_3$

(1R,3S,5Z)-5-[(2E)-2-[(1R,3aR,7aS)-1-[(2R,5S)-5-hydroxy-6-methyl-heptan-2-yl]-7a-methyl-2,3,3a,5,6,7-hexahydro-1H-inden-4-ylidene]ethylidene]-4-methylidene-cyclohexane-1,3-diol.
LMST03020257; 1α,24S-dihydroxyvitamin D_3; 1α,24S-dihydroxycholecalciferol; 1,24-Dihydroxyvitamin D_3; 1,24-Dihydroxycholecalciferol; 9,10-seco-5Z,7E,10 (19)-cholestatriene-1S,3R,24S-triol; 5283735; 6504811; Synthesized from 24-oxocholesterol. Equipotent with 1α,25-dihydroxyvitamin D_3 in HL-60 human promyelocytes differentiation; only 10% as active in its affinity for the chick intestinal receptor; Solid; mp = 84-85° (mixture of 24R, 24S epimers); $λ_m$ = 266 nm (ε = 17000). *J Chem Soc Perkin 1*. 1975, 1421-4; *Chem Pharm Bull*. 1984, *32(10)*, 3866-72; *Tet Lett*. 1975, *26*, 2203-6; *J Biol Chem*. 1987, *262(29)*, 14164-71.

402 (5E)-24R,25-Dihydroxy-[6,19,19-trideutero]vitamin D_3
9547371 $C_{27}H_{44}O_3$

(3R,6R)-6-[(1R,3aR,4E,7aS)-4-[(2E)-2-deuterio-2-[(5S)-5-hydroxy-2-methylidene-cyclohexylidene]ethylidene]-7a-methyl-2,3,3a,5,6,7-hexahydro-1H-inden-1-yl]-2-methyl-heptane-2,3-diol.
LMST03020126; (5E)-24R,25-dihydroxy-[6,19,19-trideutero]cholecalciferol; [6,19,19-trideutero]-9,10-seco-5E,7E,10(19)-cholestatriene-3S,24R,25-triol; 9547371; Synthesized from 24R,25-dihydroxyvitamin D_3 *via* proton-deuterium exchange at 6 and 19-positions of its sulfur dioxide adduct under basic conditions; $λ_m$ = 273 nm (EtOH). *Steroids*. 1989,

54(2), 145-57.

403 1α,25-Dihydroxy-previtamin D_3
57102-09-7 $C_{27}H_{44}O_3$

(1R,5S)-3-[(Z)-2-[(1R,3aR,7aS)-1-[(2R)-6-hydroxy-6-methyl-heptan-2-yl]-7a-methyl-1,2,3,3a,6,7-hexahydroinden-4-yl]ethenyl]-2-methyl-cyclohex-2-ene-1,5-diol.
1,25-Dihydroxy-previtamin D_3; 1α,25-Dihydroxy-previtamin D_3; 1,25-Dihydroxytachysterol; 9,10-Secocholesta-5(10),6Z,8-triene-1α,3β,25-triol; 6438507; 57102-09-7; Photoisomer of previtamin D_3. Intermediate in metabolism of vitamin D_3. *J Nutr Biochem*. 2000, *11(5)*, 267-72; *J Biol Chem*. 1993, *268(19)*, 13811-9.

404 1α,25-Dihydroxy-14-epiprevitamin D_3
5283738 $C_{27}H_{44}O_3$

(1S,5R)-3-[(Z)-2-[(1R,3aS,7aS)-1-[(2R)-6-hydroxy-6-methyl-heptan-2-yl]-7a-methyl-1,2,3,3a,6,7-hexahydroinden-4-yl]ethenyl]-2-methyl-cyclohex-2-ene-1,5-diol.
LMST03020261; (14R)-9,10-seco-5(10),6Z,8-cholestatriene-1S,3R,25-triol; 1α,25-dihydroxy-14-epiprevitamin D_3; 1α,25-dihydroxy-14-epiprecholecalciferol; 5283738.

405 18,25-Dihydroxyvitamin D_3
5283743 $C_{27}H_{44}O_3$

(6R)-6-[(1R,3aR,4E,7aR)-7a-(hydroxymethyl)-4-[(2Z)-2-[(5S)-5-hydroxy-2-methylidene-cyclohexylidene]ethyl-idene]-2,3,3a,5,6,7-hexahydro-1H-inden-1-yl]-2-methyl-heptan-2-ol.

LMST03020268; 18,25-dihydroxyvitamin D_3; 18,25-dihydroxycholecalciferol; 9,10-seco-5Z,7E,10(19)-cholestatriene-3S,18,25-triol; 5283743; Synthesized from des-AB vitamin D 8-ketone with A-ring phosphine oxide. *Tet Lett.* 1992, *33(11)*, 1503-6.

406 ***22,25-Dihydroxyvitamin D_3***

81446-12-0 $C_{27}H_{44}O_3$

(6S)-6-[(1R,3aR,4E,7aS)-4-[(2Z)-2-[(5R)-5-hydroxy-2-methylidene-cyclohexylidene]ethylidene]-7a-methyl-2,3,3a,5,6,7-hexahydro-1H-inden-1-yl]-2-methyl-heptane-2,5-diol.

LMST03020269; 22,25-Dihydroxyvitamin D_3; (22R)-22,25-Dihydroxy-vitamin D_3; (22S)-22,25-Dihydroxy-vitamin D_3; 22,25-(OH)2D_3; 9,10-Secocholesta-5Z,7E,10(19)-triene-3β, 22,25-triol; 6439950; 81446-12-0; *Chem Pharm Bull (Tokyo).* 1981, *29(8)*, 2254-60.

407 ***22R,25-Dihydroxyvitamin D_3***

5283744 $C_{27}H_{44}O_3$

(5R,6S)-6-[(1R,3aR,4E,7aS)-4-[(2Z)-2-[(5S)-5-hydroxy-2-methylidene-cyclohexylidene]ethylidene]-7a-methyl-2,3,3a,5,6,7-hexahydro-1H-inden-1-yl]-2-methyl-heptane-2,5-diol.

LMST03020269; 22R,25-dihydroxyvitamin D_3; 22R,25-dihydroxycholecalciferol; 9,10-seco-5Z,7E,10(19)-cholestatriene-3S,22R,25-triol; 5283744; λ_m = 263 nm (EtOH). *Chem Pharm Bull (Tokyo).* 1981, *29(8)*, 2254-60.

408 ***22S,25-Dihydroxyvitamin D_3***

5283745 $C_{27}H_{44}O_3$

(5S,6S)-6-[(1R,3aR,4E,7aS)-4-[(2Z)-2-[(5S)-5-hydroxy-2-methylidene-cyclohexylidene]ethylidene]-7a-methyl-2,3,3a,5,6,7-hexahydro-1H-inden-1-yl]-2-methyl-heptane-2,5-diol.

LMST03020270; 22S,25-dihydroxyvitamin D_3; 22S,25-dihydroxycholecalciferol; 9,10-seco-5Z,7E,10(19)-cholestatriene-3S,22S,25-triol; 5283745; λ_m = 263 nm (EtOH). *Chem Pharm Bull (Tokyo).* 1981, *29(8)*, 2254-60.

409 ***23R,25-Dihydroxyvitamin D_3***

5283746 $C_{27}H_{44}O_3$

(4R,6R)-6-[(1R,3aR,4E,7aS)-4-[(2Z)-2-[(5S)-5-hydroxy-2-methylidene-cyclohexylidene]ethylidene]-7a-methyl-2,3,3a,5,6,7-hexahydro-1H-inden-1-yl]-2-methyl-heptane-2,4-diol.

LMST03020271; 23R,25-dihydroxyvitamin D_3; 23R,25-dihydroxycholecalciferol; 9,10-seco-5Z,7E,10(19)-cholestatriene-3S,23R,25-triol; 5283746; Produced by incubating kidney homogenates from vitamin D supplemented chickens with 25-hydroxyvitamin D_3. Synthesized from 23,25-dihydroxyvitamin D_3; λ_m = 265 nm. *Biochemistry.* 1981, *20(13)*, 3875-9; *J Chem Soc Chem Comm.* 1981, 1157.

410 ***23S,25-Dihydroxyvitamin D_3***

77733-16-5 $C_{27}H_{44}O_3$

(4S,6R)-6-[(1R,3aR,4E,7aS)-4-[(2Z)-2-[(5S)-5-hydroxy-2-methylidene-cyclohexylidene]ethylidene]-7a-methyl-2,3,3a,5,6,7-hexahydro-1H-inden-1-yl]-2-methyl-heptane-2,4-diol.

LMST03020272; 23S,25-dihydroxyvitamin D_3; 23S,25-dihydroxycholecalciferol; 9,10-seco-5Z,7E,10(19)-cholestatriene-3S,23S,25-triol; 23,25-Dihydroxyvitamin D_3; 23,25-Dihydroxycholecalciferol; 5283747; 6438871; 77733-16-5; A major metabolite of 25-hydroxyvitamin D_3 and 1,25-dihydroxyvitamin D_3. Is further metabolized through the 26,23 lactone pathway. Produced by incubating kidney homogenates from vitamin D supplemented chickens with 25-hydroxyvitamin D_3. Synthesized from 23,25-dihydroxyvitamin D_3; λ_m = 265 nm. *Proc Natl Acad Sci U.S.A* 1981, *78(8),* 4805-8; *Biochemistry*. 1981, *20(13),* 3875-9; *J Chem Soc Chem Comm.* 1981, 1157.

411   ***24R,25-Dihydroxyvitamin D_3***
55721-11-4 $C_{27}H_{44}O_3$

(3R,6R)-6-[(1R,3aR,4E,7aS)-4-[(2Z)-2-[(5S)-5-hydroxy-2-methylidene-cyclohexylidene]ethylidene]-7a-methyl-2,3,3a,5,6,7-hexahydro-1H-inden-1-yl]-2-methyl-heptane-2,3-diol.

LMST03020273; Secalciferol; Osteo D; 24(R),25-Dihydroxyvitamin D_3; C07712; D00628; 5283748; 55721-11-4; Isolated from chicken kidney homogenates incubated with 24-hydroxyvitamin D_3. Synthesized from dinor-5,7-choladien-3β-ol or from 3β-acetoxy-27-nor-5-cholesten-25-one. Calcium regulator. *In vitro*, is oxidized to yield 25-hydroxy-24-oxovitamin D_3, then hydroxylated to give 23S,25-dihydroxy-24-oxovitamin D_3, and cleaved at the 23,24-bond to afford 23-hydroxy-24,25,26,27-tetranorvitamin D_3. All these metabolic transformations are catalyzed by

vitamin D_3 24-hydroxylase (CYP24). *In vivo*, 24R,25-dihdroxyvitamin D_3 has a long half-life (17 days in dog) and is metabolized similarly as found in the *in vitro* studies up to 23S,25-dihydroxy-24-oxovitamin D_3. However *in vivo* 23S,25-dihydroxy-24-oxovitamin D_3 is conjugated as it is produced in a form of 23β-glucuronide and excreted to bile; λ_m = 265 nm; LD_{50} (rat iv) > 0.2 mg/kg, (rat orl) > 10 mg/kg. *Biochemistry*. 1972, *11(23),* 4251-5; 1973, *12(24),* 4851-5; 1981, *20(6),* 1681-6; 1984, *23(16),* 3749-54; 1996, *35(25),* 8465-72; *Tet Lett.* 1980, *21(52),* 5027-8; *J Biol Chem.* 1983, *258(1),* 457-63; *Biochem Biophys Acta.* 1997, 1346(2), 147-57.

412   ***24S,25-Dihydroxyvitamin D_3***
5283749 $C_{27}H_{44}O_3$

(3S,6R)-6-[(1R,3aR,4E,7aS)-4-[(2Z)-2-[(5S)-5-hydroxy-2-methylidene-cyclohexylidene]ethylidene]-7a-methyl-2,3,3a,5,6,7-hexahydro-1H-inden-1-yl]-2-methyl-heptane-2,3-diol.

LMST03020274; 24S,25-dihydroxyvitamin D_3; 24S,25-dihydroxycholecalciferol; 9,10-seco-5Z,7E,10(19)-cholestatriene-3S,24S,25-triol; 5283749; Isolated from kidney homogenates from either vitamin D-supplemented or deficient chickens incubated with 25-hydroxy-24-oxovitamin D_3. Synthesized from either 3β-acetoxy-27-nor-5-cholesten-25-one or 5-cholen-24-oic acid. 25-hydroxy-24-oxovitamin D_3 is metabolized in kidney to 24S,25-dihydroxyvitamin D_3. *Arch Biochem Biophys.* 1975, *170(2),* 620-6; *Biochemistry*. 1973, *12(24),* 4851-5; *Proc Natl Acad Sci U.S.A.* 1985, *82(22),* 7485-9; *J Chem Soc Perkin Trans I.* 1983, 1401.

413   ***25R,26-Dihydroxyvitamin D_3***
9547439 $C_{27}H_{44}O_3$

(3R,6R)-6-[(1R,3aR,4E,7aS)-4-[(2Z)-2-[(5S)-5-hydroxy-2-methylidene-cyclohexylidene]ethylidene]-7a-methyl-2,3,3a,5,6,7-hexahydro-1H-inden-1-yl]-2-methyl-heptane-1,2-diol.
LMST03020275; 25R,26-dihydroxycholecalciferol; 9,10-seco-5Z,7E,10(19)-cholestatriene-3S,25R,26-triol; 9547439; Isolated from the plasma of pigs given large doses of vitamin D_3 and obtained as a mixture of (25S) and (25R) epimers either by incubating vitamin D supplemented chick kidney with 25-hydroxyvitamin D_3 or from rat serum. Synthesized from 3-hydroxy-27-nor-5-cholesten-25-one, 5-cholen-24-oic acid, or from stigmasterol-derived C(22) steroid-units. Effect on calcium transport and serum calcium of rats on a low calcium diet and stimulation of intestinal calcium transport by a epimeric mixture of synthetic 25,26-dihydroxyvitamin D_3 reported. Effect of a synthetic mixture of (25S)- and (25R)- dihydroxyvitamin D_3 on calcium mobilization from bone also noted; Solid; mp = 126-128°; $[\alpha]_D$ = +88° (c = 0.5 MeOH); λ_m = 265 nm (ε = 18000 EtOH). *Biochemistry*. 1970, *9(24)*, 4776-80; *Proc Natl Acad Sci U.S.A.* 1983, *80(17)*, 5286-8; *Steroids*. 1975, *25(2)*, 247-56; *J Chem Soc Perkin Trans I*. 1983, 1401; *J Amer Chem Soc.* 1981, *103(5)*, 1253-5.

414   *25S,26-Dihydroxyvitamin D_3*
29261-12-9 $C_{27}H_{44}O_3$

(2S,6R)-6-[(1R,3aR,4E,7aS)-4-[(2Z)-2-[(5S)-5-hydroxy-2-methylidene-cyclohexylidene]ethylidene]-7a-methyl-2,3,3a,5,6,7-hexahydro-1H-inden-1-yl]-2-methyl-heptane-1,2-diol.
LMST03020276; 25S,26-Dihydroxyvitamin D_3; AIDS-109016; AIDS-109016; 25S,26-dihydroxyvitamin D_3; 25S,26-dihydroxycholecalciferol; 9,10-Secocholesta-5Z,7E,10-triene-3S,25S,26-triol; 9,10-seco-5Z,7E,10(19)-cholestatriene-3S,25S,26-triol; 123705; 5283750; 5364803; 29261-12-9; Isolated from the plasma of pigs given large doses of vitamin D_3. Synthesized from 3-hydroxy-27-nor-5-cholesten-25-one or 5-cholen-24-oic acid or stigmasterol-derived C(22) steroids. Effects calcium transport and serum calcium of rats on a low calcium diet. Stimulates intestinal calcium transport and effects calcium mobilization from bone at dose levels of 0.25 and 2.5 µg. Significantly prolongs the survival time of mice with Lewis lung carcinoma; exhibits an antimetastatic effect on Lewis lung carcinoma, and also had an analgesic effect in mice with Lewis lung carcinoma; Solid; mp = 140-142°; $[\alpha]_D$ = +91° (c = 0.5 MeOH); λ_m = 265 nm (ε = 18000 EtOH). *Biochemistry*. 1970, *9(24)*, 4776-80; *Proc Natl*

Acad Sci U.S.A. 1983, *80(17)*, 5286-8; *Steroids*. 1975, *25(2)*, 247-56; *J Chem. Soc. Perkin Trans I*. 1983, 1401; *Steroids*. 1974, *24(4)*, 463-76; *J Amer Chem Soc.* 1981, *103(5)*, 1253-5.

415   *1α,25-Dihydroxy-2-methylene-19-nor-vitamin D_3*
9547653 $C_{27}H_{44}O_3$

(1R,3R)-5-[(2E)-2-[(1R,3aR,7aS)-1-[(2R)-6-hydroxy-6-methyl-heptan-2-yl]-7a-methyl-2,3,3a,5,6,7-hexahydro-1H-inden-4-ylidene]ethylidene]-2-methylidene-cyclohexane-1,3-diol.
LMST03020552; 2-methylene-19-nor-9,10-seco-5,7E-cholestadiene-1R,3R,25-triol; 1α,25-dihydroxy-2-methylene-19-norcholecalciferol; 2-methylene-19-nor-(20S)-1α,25(OH)(2)D(3); 2MD; 9547653; A highly potent analog of 1,25-dihydroxyvitamin D_3 whose actions are mediated through the vitamin D receptor (VDR). Induces bone formation both *in vitro and in vivo*. Synthesized by combination of Grundmann type 8-ketone with an A-ring synthon obtained from (-)-quinic acid. Compared to 1α,25-dihydroxyvitamin $D_{3'}$ affinity for the porcine intestinal vitamin D receptor is 6%, differentiation of HL-60 cells is 100%, intestinal calcium transport is 0% and bone calcium mobilization is >200%; λ_m = 243.5, 252, 262.5 nm (EtOH). *Biochemistry*. 2004, *43(14)*, 4101-10; *Proc Natl Acad Sci U.S.A.* 2002, *99(21)*, 13487-91; *J Biol Chem*. 2003, *278(34)*, 31756-65; *J Med Chem*. 1998, *41(23)*, 4662-74.

416   *1α,25-Dihydroxy-2-methylene-19-nor-20-epivitamin D_3*
5289549 $C_{27}H_{44}O_3$

(1R,3R)-5-[(2E)-2-[(1R,3aR,7aS)-1-[(2S)-6-hydroxy-6-methyl-heptan-2-yl]-7a-methyl-2,3,3a,5,6,7-hexa-hydro-1H-inden-4-ylidene]ethylidene]-2-methylidene-cyclohexane-1,3-diol.
LMST03020557;   1α,25-dihydroxy-2-methylene-19-nor-20-epicholecalciferol;   (20S)-2-methylene-19-nor-9,10-seco-5,7E-cholestadiene-1R,3R,25-triol;   VDZ;   5289549; Synthesized by combination of Grundmann type 8-ketone with an A-ring synthon obtained from (-)-quinic acid. Compared to 1α,25-dihydroxyvitamin D$_3$, affinity for the porcine intestinal vitamin D receptor is 77%, HL-60 cell differentiation is 2500%, intestinal calcium transport is 0% and bone calcium mobilization is >200%; λ$_m$ = 243.5, 252, 262.5 nm (EtOH). *J Med Chem.* 1998, *41(23)*, 4662-74.

417   24,25-Dihydroxyvitamin D$_3$
486588   C$_{27}$H$_{44}$O$_3$

(3R,6R)-6-[(1R,3aR,7aS)-4-[2-[(5S)-5-hydroxy-2-methyl-idene-cyclohexylidene]ethylidene]-7a-methyl-2,3,3a,5,6,7-hexahydro-1H-inden-1-yl]-2-methyl-heptane-2,3-diol.

24,25-Dihydroxyvitamin   D$_3$;   CBiol_001818;
KBioGR_000129;   KBioSS_000129;   KBio2_000129;
KBio2_002697;   KBio2_005265;   KBio3_000257;
KBio3_000258;   LMST03020274;   Bio1_000104;
Bio1_000593; 1590; 486588; *J Clin Endocrinol Metab.* 1980, *50(4)*, 773-5; *Chemistry.* 2002, *8(12)*, 2747-52; USP 7166585.

418   1α,25-Dihydroxy-19-Nor-vitamin D$_2$
131918-61-1   C$_{27}$H$_{44}$O$_3$

(1R,3R)-5-[2-[(1R,3aR,7aS)-1-[(2S,5S)-6-hydroxy-5,6-dimethyl-hept-3-en-2-yl]-7a-methyl-2,3,3a,5,6,7-hexahydro-1H-inden-4-ylidene]ethylidene]cyclo-hexane-1,3-diol.

Compound 49510; C08127; D00930; 19-Nor-9,10-secoergosta-5,7E,22E-triene-1α,3β,25-triol;   19-Nor-1,25-(OH)2D$_2$;   19-Nor-1α,25-dihydroxyvitamin D$_2$;
Paricalcitol; Zemplar; 1,25-Dihydroxyvitamin D$_2$;
77996; 4636600; 5281104; 131918-61-1; A vitamin D receptor (VDR) activator. Offers protection against cardiovascular disease in some patients. Approved for use in secondary hyperparathyroidism and used in treatment of hyperparathyroidism secondary to chronic kidney   disease.   Also   used   in   treatment   of osteodystrophy for the reduction of parathyroid hormone levels. *Expert Opin Pharmacother.* 2006, *7(5)*, 617-21; *Drugs.* 2005, *65(4)*, 559-76.

419   1α,2α,25-Trihydroxyvitamin D$_3$
59783-84-5   C$_{27}$H$_{44}$O$_3$

(1S,2R,3S,5Z)-5-[(2E)-2-[(1R,3aR,7aS)-7a-methyl-1-[(2R)-6-methylheptan-2-yl]-2,3,3a,5,6,7-hexahydro-1H-inden-4-ylidene]ethylidene]-4-methylidene-cyclohexane-1,2,3-triol.
1α,2α,25-Trihydroxycholecalciferol;   9,10-Secochol-esta-5Z,7E,10(19)-triene-1α,2α,3β-triol;   6441889;
59783-84-5. *J Org Chem.* 1991, *56(14)*, 4339-41; USP 6482812.

420   9,10-Secocholesta-5,7,10(19)-triene-3β,21,25-triol
5372266   C$_{27}$H$_{44}$O$_3$

2-[(4E)-4-[(2E)-2-(5-hydroxy-2-methylidene-cyclohexylidene)ethylidene]-7a-methyl-2,3,3a,5,6,7-hexahydro-1H-inden-1-yl]-6-methyl-heptane-1,6-diol.
9,10-Secocholesta-5,7,10(19)-triene-3β,21,25-triol;
9,10-Secocholesta-5Z,7E,10(19)-triene-3β,21,25-triol;
5372266.

421 **24R,25-Dihydroxyvitamin D$_3$**

55700-58-8 $C_{27}H_{44}O_3$

(3S,6R)-6-[(1R,3aR,4E,7aS)-4-[(2Z)-2-[(5R)-5-hydroxy-2-methylidene-cyclohexylidene]ethylidene]-7a-methyl-2,3,3a,5,6,7-hexahydro-1H-inden-1-yl]-2-methyl-heptane-2,3-diol.
24R,25-Dihydroxycholecalciferol; 9,10-Secocholesta-5Z,7E,10(19)-triene-3β,24R,25-triol; 9,10-Secocholesta-5Z,7E,10(19)-triene-3β,24R,25-triol; 6439679; 55700-58-8. *Biochemistry*. 1984, *23(21)*, 5041-8.

422 **24,25-Dihydroxyvitamin D$_3$**

40013-87-4 $C_{27}H_{44}O_3$

(6R)-6-[(1R,3aR,4E,7aS)-4-[(2Z)-2-[(5S)-5-hydroxy-2-methylidene-cyclohexylidene]ethylidene]-7a-methyl-2,3,3a,5,6,7-hexahydro-1H-inden-1-yl]-2-methyl-heptane-2,3-diol.
24,25-Dihydroxyvitamin D$_3$; 24,25-Dihydroxycholecalciferol; 24,25-Dihydroxyvitamin D; 24,25-Dihydroxyvitamin D$_3$; 24,25-Dihydroxyvitamine D$_3$; 9,10-Secocholesta-5Z,7E,10(19)-triene-3β,24,25-triol; EINECS 254-749-5; 5363181; 6434253; 40013-87-4; A physiologically active metabolite of Vitamin D, it is involved in the regulation of calcium metabolism, alkaline phosphatase activity, and enhances the calcemic effect of calcitriol. Used in treatment of uremic osteodystrophy. A highly active ^{3}H-labelled compound has been made. Formed by cytochrome P450 (CYP24) oxidation in the kidney of 25-hydroxyvitamin D$_3$. Involved in the regulation of calcium metabolism, alkaline phosphatase activity. *J Biol Chem*. 1983, *258(1)*, 457-463; *Biochemistry*. 1984, *23(21)*, 5041-8; *Chemistry*. 2002, *8(12)*, 2747-52; *Bioorg Med Chem*. 2001, *9(12)*, 3123-8; *Steroids*.

1989, *54(2)*, 145-57.

423 **4,25-Dihydroxyvitamin D$_3$**

6540551 $C_{27}H_{44}O_3$

(1S,3E)-3-[(2E)-2-[1-(6-hydroxy-6-methyl-heptan-2-yl)-7a-methyl-2,3,3a,5,6,7-hexahydro-1H-inden-4-ylidene]ethylidene]-4-methylidene-cyclohexane-1,2-diol.
4,25-dihydroxycholecalciferol; 6540551.

424 **(6R)-Vitamin D$_3$ 6,19-sulfur dioxide adduct**

9547432 $C_{27}H_{44}O_3S$

(3S,9R)-9-[(E)-[(1R,3aR,7aS)-7a-methyl-1-[(2R)-6-methylheptan-2-yl]-2,3,3a,5,6,7-hexahydro-1H-inden-4-ylidene]methyl]-8,8-dioxo-8-thiabicyclo[4.3.0]non-10-en-3-ol.
LMST03020250; 6R,19-epithio-9,10-seco-5(10),7E-cholestadien-3S-ol S,S-dioxide; (6R)-cholecalciferol 6,19-sulfur dioxide adduct; 9547432; Formed from vitamin D$_3$ by treatment with liquid sulfur dioxide. *J Org Chem*. 1983, *48(20)*, 3483-8.

425 **(6S)-Vitamin D$_3$ 6,19-sulfur dioxide adduct**

9547433 $C_{27}H_{44}O_3S$

(3S,9S)-9-[(E)-[(1R,3aR,7aS)-7a-methyl-1-[(2R)-6-methylheptan-2-yl]-2,3,3a,5,6,7-hexahydro-1H-inden-4-ylidene]methyl]-8,8-dioxo-8-thiabicyclo[4.3.0]non-10-en-3-ol.
LMST03020251;   6S,19-epithio-9,10-seco-5(10),7E-cholestadien-3S-ol S,S-dioxide; cholecalciferol 6S,19-sulfur dioxide adduct; 9547433; Formed from vitamin D₃ by treatment with liquid sulfur dioxide. *J Org Chem.* 1983, *48(20)*, 3483-8.

426   *1α,25-Dihydroxy-24a-homo-22-thiavitamin D₃*
9547437 $C_{27}H_{44}O_3S$

(1R,3S,5Z)-5-[(2E)-2-[(1S,3aR,7aR)-1-[(1S)-1-(4-hydroxy-4-methyl-pentyl)sulfanylethyl]-7a-methyl-2,3,3a,5,6,7-hexahydro-1H-inden-4-ylidene]ethylidene]-4-methylidene-cyclohexane-1,3-diol.
LMST03020263;   24a-homo-9,10-seco-22-thia-5Z,7E,10(19)-cholestatriene-1S,3R,25-triol; 1α,25-dihydroxy-24a-homo-22-thiacholecalciferol; 9547437; Synthesized from dehydroepiandrosterone *via* the 5,7-diene by a photochemical method. *Bioorg Med Chem Lett.* 1995, *5(3)*, 279-82.

427   *1α,25-Dihydroxy-24a-homo-22-thia-20-epivitamin D₃*
9547438 $C_{27}H_{44}O_3S$

(1R,3S,5Z)-5-[(2E)-2-[(1S,3aR,7aR)-1-[(1R)-1-(4-hydroxy-4-methyl-pentyl)sulfanylethyl]-7a-methyl-2,3,3a,5,6,7-hexahydro-1H-inden-4-ylidene]ethylidene]-4-methylidene-cyclohexane-1,3-diol.
LMST03020264;   (20R)-24a-homo-9,10-seco-22-thia-5Z,7E,10(19)-cholestatriene-1S,3R,25-triol;   1α,25-dihydroxy-24a-homo-22-thia-20-epicholecalciferol; 9547438; Synthesized from dehydroepiandrosterone *via* the 5,7-diene by a photochemical method. *Bioorg Med Chem Lett.* 1995, *5(3)*, 279-82.

428   *19-Nor-1,24,25-trihydroxyvitamin D₂*
                                           $C_{27}H_{44}O_4$

19-Nor-1,24,25-trihydroxyergocalciferol; Metabolite in keratinocytes of 19-nor-1α,25-dihydroxyvitamin D₂. *Arch Biochem Biophys.* 2001, *387(2)*, 297-306.

429   *6R,19-Epidioxy-1α-hydroxy-6,19-dihydrovitamin D₃*
9547440 $C_{27}H_{44}O_4$

(5R,8R,10S)-5-[(E)-[(1R,3aR,7aS)-7a-methyl-1-[(2R)-6-methylheptan-2-yl]-2,3,3a,5,6,7-hexahydro-1H-inden-

4-ylidene]methyl]-3,4-dioxabicyclo[4.4.0]dec-11-ene-8,10-diol.
LMST03020277; 6R,19-epidioxy-9,10-seco-5(10),7E-cholestadiene-1S,3R-diol; 6R,19-epidioxy-1α-hydroxy-6,19-dihydrocholecalciferol; 9547440; Formed as a major product of the reaction of 1α-hydroxyvitamin D$_3$ with singlet oxygen generated by a dye-sensitized photochemical method. Affinity for the vitamin D receptor in HL-60 cells is < 1/10000 as active as 1α,25-dihydroxyvitamin D$_3$. In differentiation of HL-60 cells is > 1/1000 as active as 1α,25-dihydroxyvitamin D$_3$. *J Med Chem*. 1985, **28(9)**, 1148-53.

430   6S,19-Epidioxy-1α-hydroxy-6,19-dihydrovitamin D$_3$
9547441   C$_{27}$H$_{44}$O$_4$

(5S,8R,10S)-5-[(E)-[(1R,3aR,7aS)-7a-methyl-1-[(2R)-6-methylheptan-2-yl]-2,3,3a,5,6,7-hexahydro-1H-inden-4-ylidene]methyl]-3,4-dioxabicyclo[4.4.0]dec-11-ene-8,10-diol.
LMST03020278; 6S,19-epidioxy-9,10-seco-5(10),7E-cholestadiene-1S,3R-diol; 6S,19-epidioxy-1α-hydroxy-6,19-dihydrocholecalciferol; 9547441; Formed as a major product of the reaction of 1α-hydroxyvitamin D$_3$ with singlet oxygen generated by a dye-sensitized photochemical method. Affinity for the vitamin D receptor in HL-60 cells is < 1/10000 as active as 1α,25-dihydroxyvitamin D$_3$. In differentiation of HL-60 cells is > 1/1000 as active as 1α,25-dihydroxyvitamin D$_3$; Solid; mp = 135-136°. *J Med Chem*. 1985, **28(9)**, 1148-53.

431   6R,19-Epidioxy-25-hydroxy-6,19-dihydrovitamin D$_3$
9547444   C$_{27}$H$_{44}$O$_4$

(3S,10R)-10-[(E)-[(1R,3aR,7aS)-1-[(2R)-6-hydroxy-6-methyl-heptan-2-yl]-7a-methyl-2,3,3a,5,6,7-hexa-hydro-1H-inden-4-ylidene]methyl]-8,9-dioxabicyclo[4.4.0]dec-11-en-3-ol.
LMST03020281; 6R,19-epidioxy-9,10-seco-5(10),7E-cholestadiene-3S,25-diol; 6R,19-epidioxy-25-hydroxy-6,19-dihydrocholecalciferol; 9547444; A major product of the reaction of 25-hydroxyvitamin D$_3$ with singlet oxygen generated by a dye-sensitized photochemical method. Affinity for the vitamin D receptor in HL-60 cells is < 1/3000 as active as that of 1α,25-dihydroxyvitamin D$_3$. In differentiation of HL-60 cells is > 1/120 as active as 1α,25-dihydroxyvitamin D$_3$; Crystals; mp = 137-138°. *J Med Chem*. 1985, **28(9)**, 1148-53.

432   6S,19-Epidioxy-25-hydroxy-6,19-dihydrovitamin D$_3$
9547445   C$_{27}$H$_{44}$O$_4$

(3S,10S)-10-[(E)-[(1R,3aR,7aS)-1-[(2R)-6-hydroxy-6-methyl-heptan-2-yl]-7a-methyl-2,3,3a,5,6,7-hexa-hydro-1H-inden-4-ylidene]methyl]-8,9-dioxabicyclo[4.4.0]dec-11-en-3-ol.
LMST03020282; 6S,19-epidioxy-9,10-seco-5(10),7E-cholestadiene-3S,25-diol; 6S,19-epidioxy-25-hydroxy-6,19-dihydrocholecalciferol; 9547445; A major product of the reaction of 25-hydroxyvitamin D$_3$ with singlet oxygen generated by a dye-sensitized photochemical method. Affinity for the vitamin D receptor in HL-60 cells is < 1/3000 as active as that of 1α,25-dihydroxyvitamin D$_3$. In differentiation of HL-60 cells is > 1/130 as active as 1α,25-dihydroxyvitamin D$_3$; Crystals; mp = 148-150°. *J Med Chem*. 1985, **28(9)**, 1148-53.

433   1α,25-Dihydroxy-24a-homo-22-oxa-vitamin D$_3$
9547446   C$_{27}$H$_{44}$O$_4$

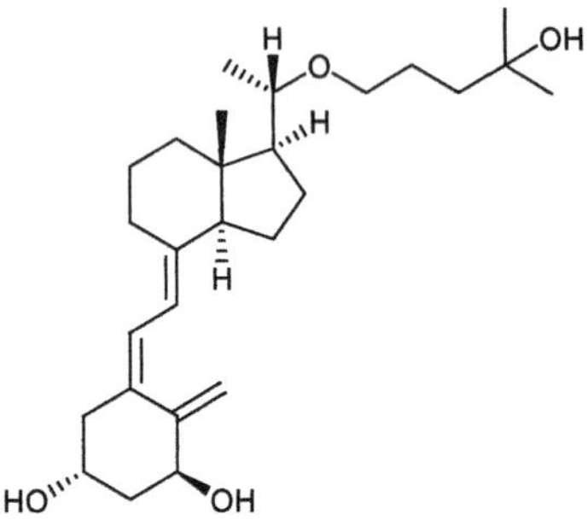

(1R,3S,5Z)-5-[(2E)-2-[(1S,3aR,7aR)-1-[(1S)-1-(4-hydroxy-4-methyl-pentoxy)ethyl]-7a-methyl-2,3,3a,5,6,7-hexahydro-1H-inden-4-ylidene]ethylidene]-4-methylidene-cyclohexane-1,3-diol.
LMST03020283; 24a-homo-22-oxa-9,10-seco-5Z,7E,10(19)-cholestatriene-1S,3R,25-triol; 1α,25-dihydroxy-24a-homo-22-oxacholecalciferol; 9547446; Synthesized photochemically from dehydroepiandrosterone. Induces differentiation of human myeloid HL-60 leukemia cells into macrophases *in vitro* ten times more effectively than 1α,25-dihydroxyvitamin D$_3$. Binds to the chick embryonic intestinal 1α,25-dihydroxyvitamin D$_3$ receptor 9% as well as 1α,25-dihydroxyvitamin D$_3$; λ$_m$ = 263 nm (EtOH). *Chem Pharm Bull (Tokyo)*. 1992, *40(6)*, 1494-9.

434   *1α,25-Dihydroxy-24a-homo-22-oxa-20-epivitamin D$_3$*
9547447   C$_{27}$H$_{44}$O$_4$

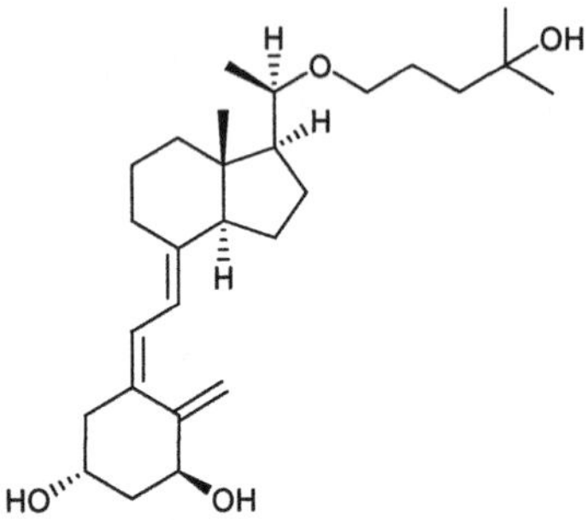

(1R,3S,5Z)-5-[(2E)-2-[(1S,3aR,7aR)-1-[(1R)-1-(4-hydroxy-4-methyl-pentoxy)ethyl]-7a-methyl-2,3,3a,5,6,7-hexahydro-1H-inden-4-ylidene]ethylidene]-4-methylidene-cyclohexane-1,3-diol.
LMST03020284; (20R)-24a-homo-22-oxa-9,10-seco-5Z,7E,10(19)-cholestatrien-1S,3R,25-triol; 1α,25-dihydroxy-24a-homo-22-oxa-20-epicholecalciferol; 9547447; Compared to 1α,25-dihydroxyvitamin D$_3$, inhibition of U937 cell (human histiocytic lymphoma cell line) proliferation is 8200:, induction of U937 cell differentiation is 117600%, binding to the 1α,25-dihydroxyvitamin D$_3$ receptor from rachitic chicken intestine is 9%, calcemic activity determined by the increase in urinary calcium excretion in rats is 270%. *Biochem Pharmacol*. 1991, *42(8)*, 1569-75.

435   *1α,25-Dihydroxy-10,19-methano-23-oxavitamin D$_3$*
9547448   C$_{27}$H$_{44}$O$_4$

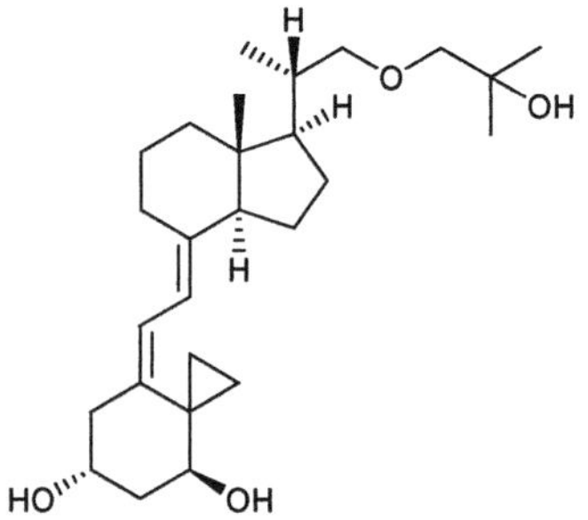

(4S,6R,8Z)-8-[(2E)-2-[(1R,3aR,7aS)-1-[(2S)-1-(2-hydroxy-2-methyl-propoxy)propan-2-yl]-7a-methyl-2,3,3a,5,6,7-hexahydro-1H-inden-4-ylidene]ethylidene]spiro[2.5]octane-4,6-diol.
LMST03020285; 10,19-methano-23-oxa-9,10-seco-5Z,7E-cholestadiene-1S,3R,25-triol; 1α,25-dihydroxy-10,19-methano-23-oxacholecalciferol; 9547448; Synthesized from protected (5E)-1α,22-dihydroxy-23,24,25,26,27-pentanorvitamin D$_3$. *Tet Lett*. 1991, *32(38)*, 5073-6.

436   *1α,2,25-Trihydroxyvitamin D$_3$*
9547449   C$_{27}$H$_{44}$O$_4$

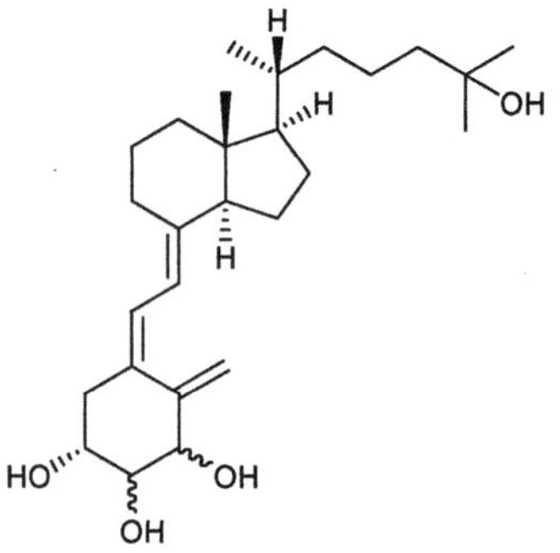

(1R,5Z)-5-[(2E)-2-[(1R,3aR,7aS)-1-[(2R)-6-hydroxy-6-methyl-heptan-2-yl]-7a-methyl-2,3,3a,5,6,7-hexahydro-1H-inden-4-ylidene]ethylidene]-4-methylidene-cyclohexane-1,2,3-triol.
LMST03020286; 1α,2,25-trihydroxycholecalciferol; 9,10-seco-5Z,7E,10(19)-cholestatriene-1,2,3R,25-tetrol; 9547449. *J Org Chem*. 1991, *56(14)*, 4339-41; USP 6482812.

437   *1α,11α,25-Trihydroxyvitamin D$_3$*
5283751   C$_{27}$H$_{44}$O$_4$

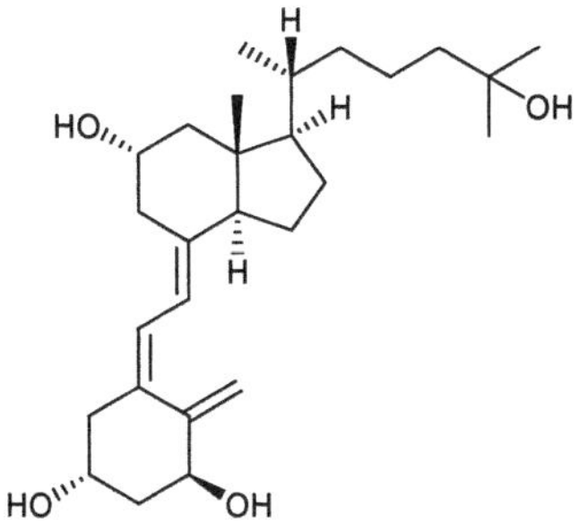

(1R,3S,5Z)-5-[(2E)-2-[(1R,3aR,6R,7aS)-6-hydroxy-1-[(2R)-6-hydroxy-6-methyl-heptan-2-yl]-7a-methyl-2,3,3a,5,6,7-hexahydro-1H-inden-4-ylidene]ethyl-idene]-4-methylidene-cyclohexane-1,3-diol.
LMST03020287; 9,10-seco-5Z,7E,10(19)-cholestatri-ene-1S,3R,11S,25-tetrol; 1α,11α,25-trihydroxyvitamin D₃; 1α,11α,25-trihydroxycholecalciferol; 5283751.

438   *1α,11β,25-Trihydroxyvitamin D₃*
5283752   C₂₇H₄₄O₄

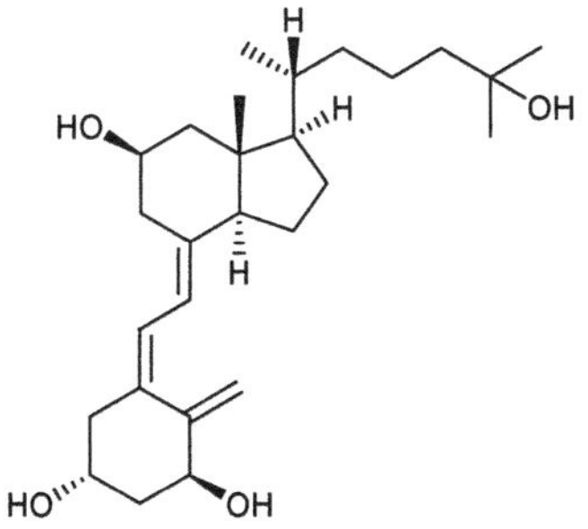

(1R,3S,5Z)-5-[(2E)-2-[(1R,3aR,6S,7aS)-6-hydroxy-1-[(2R)-6-hydroxy-6-methyl-heptan-2-yl]-7a-methyl-2,3,3a,5,6,7-hexahydro-1H-inden-4-ylidene]ethyl-idene]-4-methylidene-cyclohexane-1,3-diol.
LMST03020288; 9,10-seco-5Z,7E,10(19)-cholestatri-ene-1S,3R,11R,25-tetrol; 1α,11β,25-trihydroxyvitamin D₃; 1α,11β,25-trihydroxycholecalciferol; 5283752.

439   *1α,18,25-Trihydroxyvitamin D₃*
5283753   C₂₇H₄₄O₄

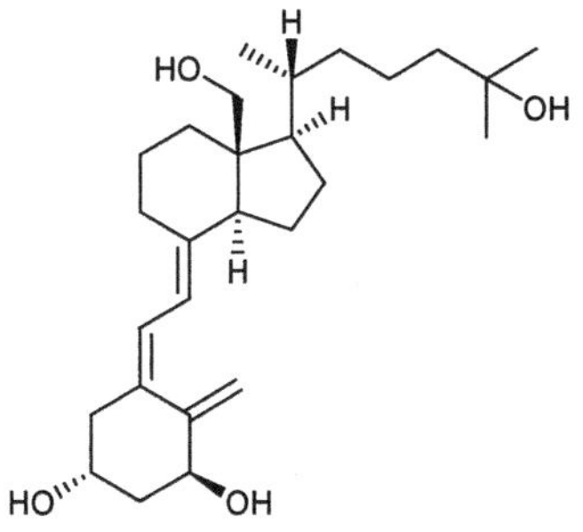

(1R,3S,5Z)-5-[(2E)-2-[(1R,3aR,7aR)-7a-(hydroxy-methyl)-1-[(2R)-6-hydroxy-6-methyl-heptan-2-yl]-2,3,3a,5,6,7-hexahydro-1H-inden-4-

ylidene]ethylidene]-4-methylidene-cyclohexane-1,3-diol.
LMST03020289; 9,10-seco-5Z,7E,10(19)-cholestatri-ene-1S,3R,18,25-tetrol; 1α,18,25-trihydroxyvitamin D₃; 1α,18,25-trihydroxycholecalciferol; 5283753; Prepared by Pd-catalyzed coupling of des-AB 8-triflate with 5(10)-en-6-yne A-ring fragment as the key step. *Tet Lett.* 1992, *33(11)*, 1503-6.

440   *1α,20S,25-Trihydroxyvitamin D₃*
5283754   C₂₇H₄₄O₄

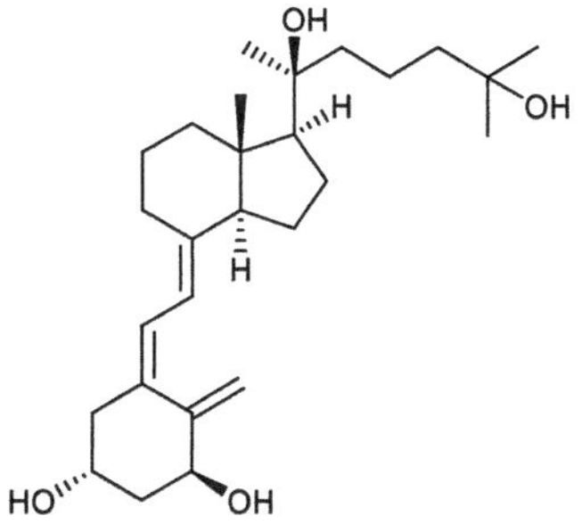

(2S)-2-[(1S,3aR,4E,7aR)-4-[(2Z)-2-[(3S,5R)-3,5-di-hydroxy-2-methylidene-cyclohexylidene]ethylidene]-7a-methyl-2,3,3a,5,6,7-hexahydro-1H-inden-1-yl]-6-methyl-heptane-2,6-diol.
LMST03020290; 1α,20S,25-trihydroxyvitamin D₃; 1α,20S,25-trihydroxycholecalciferol; 9,10-seco-5Z,7E,10(19)-cholestatriene-1S,3R,20S,25-tetrol; 5283754; Prepared from a (5E)-1α-hydroxy-20-keto-22,23,24,25,26,27-hexanorvitamin D₃ derivative. Compared to 1α,25-dihydroxyvitamin D₃, inhibition of U937 cell (human histiocytic lymphoma cell line) proliferation is 50%, binding to the 1,25-(OH)2D3 receptor from rachitic chicken intestine is 0.006%, calcemic activity determined by the increase in urinary calcium excretion in rats is 0.6%. Proceedings of the Ninth Workshop on Vitamin D, Orlando, Florida (USA) May 28-June 2. Synthesis and Biological Activity of 20-Hydroxylated Vitamin D Analogues. (Norman, A. W., Bouillon, R., Thomasset, M., eds), pp95-96, Walter de Gruyter Berlin New York (1994).

441   *(22R)-1α,22,25-Trihydroxy-20-epivitamin D₃*
5283755   C₂₇H₄₄O₄

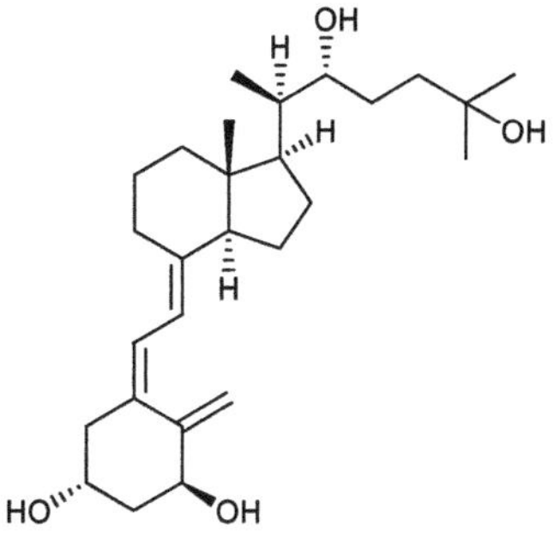

(5R,6R)-6-[(1R,3aR,4E,7aS)-4-[(2Z)-2-[(3S,5R)-3,5-di-hydroxy-2-methylidene-cyclohexylidene]ethylidene]-7a-methyl-2,3,3a,5,6,7-hexahydro-1H-inden-1-yl]-2-methyl-heptane-2,5-diol.

LMST03020291; (20R)-9,10-seco-5Z,7E,10(19)-chol-estatriene-1S,3R,22R,25-tetrol; (22R)-1α,22,25-tri-hydroxy-20-epivitamin D_3; (22R)-1α,22,25-trihydroxy-20-epicholecalciferol; 5283755; Synthesized from protected (5E)-1α-hydroxy-22-oxo-23,24,25,26,27-pentanor-20-epivitamin D_3. Is 10% as potent as 1α,25-dihydroxyvitamin D_3 in induction of differentiation of U937 cells and inhibition of proliferation. Proceedings of the Ninth Workshop on Vitamin D, Orlando, Florida (USA) May 28-June 2. Synthesis and Biological Activity of 20-Hydroxylated Vitamin D Analogues. (Norman, A. W., Bouillon, R., Thomasset, M., eds), pp95-96, Walter de Gruyter Berlin New York (1994).

442   **1,23,25-Trihydroxy Vitamin D_3**
86701-33-9 $C_{27}H_{44}O_4$

(4S,6R)-6-[(1R,3aR,4E,7aS)-4-[(2Z)-2-[(3R,5S)-3,5-di-hydroxy-2-methylidene-cyclohexylidene]ethylidene]-7a-methyl-2,3,3a,5,6,7-hexahydro-1H-inden-1-yl]-2-methyl-heptane-2,4-diol.

LMST03020388; 9,10-Secocholesta-5Z,7E,10(19)-tri-ene-1α,3β,23S,25-tetrol; 1,23,25-Trihydroxycholecal-ciferol; 1,23,25-Tri-hydroxyvitamin D_3; 6439793; 86701-33-9; Metabolite of 1α,25-dihydroxyvitamin D_3. Shows no intestinal calcium absorptive or bone calcium resorptive activity in vitamin D deficient rats. Less potent than 1,25-dihydroxycholecalciferol in the chick intestinal 1,25-dihydroxycholecalciferol receptor assay. **Biochemistry.** 1984, **23(17)**, 3973-9; **Biochem J.** 1983, **214(1)**, 261-4.

443   **1α,23R,25-Trihydroxyvitamin D_3**
5283795 $C_{27}H_{44}O_4$

(4R,6R)-6-[(1R,3aR,4E,7aS)-4-[(2Z)-2-[(3S,5R)-3,5-di-hydroxy-2-methylidene-cyclohexylidene]ethylidene]-7a-methyl-2,3,3a,5,6,7-hexahydro-1H-inden-1-yl]-2-methyl-heptane-2,4-diol.

LMST03020388; 1α,23,25-trihydroxycholecalciferol; 9,10-seco-5Z,7E,10(19)-cholestatriene-1S,3R,23R,25-tetrol; 5283795; Has a low affinity for the vitamin D receptor and fails to induce cell differentiation. Compared to 1α,25-dihydroxyvitamin D_3, inhibition of proliferation of human keratinocytes in culture is 100%, intestinal calcium absorption in rat is 6%, bone calcium mobilization in rat is 0%, competitive binding to rat intestinal vitamin D receptor is 1% and differentiation of human leukemia cells (HL-60) is 50%. **J Nutr Biochem.** 1993, **4(1)**, 49-57; **Mol Endocrinology.** 2000, **14**(11), 1788-96.

444   **1α,23S,25-Trihydroxyvitamin D_3**
9547450 $C_{27}H_{44}O_4$

(4S,6R)-6-[(1R,3aR,4E,7aS)-4-[(2Z)-2-[(3S,5R)-3,5-di-hydroxy-2-methylidene-cyclohexylidene]ethylidene]-7a-methyl-2,3,3a,5,6,7-hexahydro-1H-inden-1-yl]-2-methyl-heptane-2,4-diol.

LMST03020292; 1α,23S,25-trihydroxycholecalciferol; 9,10-seco-5Z,7E,10(19)-cholestatriene-1S,3R,23S,25-tetrol; 9547450; Compared to 1α,25-dihydoxyvitamin D_3, inhibition of proliferation of human keratinocytes in culture is 100%, intestinal calcium absorption in rat is 0%, bone calcium mobilization in rat is 0% and differentiation of human leukemia cells (HL-60) is 14%. **J Nutr Biochem.** 1993, **4(1)**, 49-57.

445   **1,24,25-Trihydroxyvitamin D_3**
50648-94-7 $C_{27}H_{44}O_4$

(6R)-6-[(1R,3aR,4E,7aS)-4-[(2Z)-2-[(5S)-3,5-dihydroxy-2-methylidene-cyclohexylidene]ethylidene]-7a-methyl-2,3,3a,5,6,7-hexahydro-1H-inden-1-yl]-2-methyl-

heptane-2,3-diol.
1,24,25-Trihydroxyvitamin D_3; 1,24,25-Trihydroxy-cholecalciferol; 9,10-Secocholesta-5Z,7E,10(19)-triene-1,3β,24,25-tetrol; LMST03020294; 6438336; 50649-94-7; A vitamin D_3 metabolite. Exhibits synergism with vitamin D_3. *Biochemistry.* 1984, *23(7),* 1473-8; *Chemistry.* 2002, *8(12),* 2747-52.

446   *1α,24,25-Trihydroxyvitamin D_3*
4525692 $C_{27}H_{44}O_4$

6-[4-[2-(3,5-dihydroxy-2-methylidene-cyclohexyl-idene)-ethylidene]-7a-methyl-2,3,3a,5,6,7-hexahydro-1H-inden-1-yl]-2-methyl-heptane-2,3-diol.
1,24,25-Trihydroxyvitamin D_3; 4525692; Metabolite of 1α,25-dihydroxyvitamin D_3. Formed in the kidney as a metabolite of 25-hydroxy-24-oxovitamin D_3. *Bio-chemistry.* 1984, *23(7),* 1473-8; *J Biol Chem.* 1983, *258(1),* 457-63; *Chemistry.* 2002, *8(12),* 2747-52.

447   *1α,24S,25-Trihydroxyvitamin D_3*
5283756 $C_{27}H_{44}O_4$

(3S,6R)-6-[(1R,3aR,4E,7aS)-4-[(2Z)-2-[(3S,5R)-3,5-di-hydroxy-2-methylidene-cyclohexylidene]ethylidene]-7a-methyl-2,3,3a,5,6,7-hexahydro-1H-inden-1-yl]-2-methyl-heptane-2,3-diol.
LMST03020294; 1α,24S,25-trihydroxycholecalciferol; 9,10-seco-5Z,7E,10(19)-cholestatriene-1S,3R,24S,25-tetrol; 5283756. *Chemistry.* 2002, *8(12),* 2747-52.

448   *1α,24R,25-Trihydroxyvitamin D_3*
56142-94-0 $C_{27}H_{44}O_4$

(3R,6R)-6-[(1R,3aR,4E,7aS)-4-[(2Z)-2-[(3S,5R)-3,5-di-hydroxy-2-methylidene-cyclohexylidene]ethylidene]-7a-methyl-2,3,3a,5,6,7-hexahydro-1H-inden-1-yl]-2-methyl-heptane-2,3-diol.
(1α,3β,5Z,7E,24R)-9,10-Secocholesta-5,7,10(19)-triene-1,3,24,25-tetrol; (24R)-24-hydroxycalcitriol; 1α,24R,25-Trihydroxyvitamin D_3; LMST03020294; Ro 21-7729; 9,10-Seco-cholesta-5Z,7E,10(19)-triene-1α,3β,24R,25-tetrol; 6446280; 56142-94-0; Increases the antrachitic activity of 1α,25-dihydroxy vitamin D_3; LD_{50} (rat sc) = 812 mg/kg. *Biochem J.* 1984, *219(3),* 713-7; *Biochemistry.* 1984, *23(7),* 1473-8; *Chemistry.* 2002, *8(12),* 2747-52.

449   *1,25,26-Trihydroxyvitamin D_3*
77372-59-9 $C_{27}H_{44}O_4$

(2S,6R)-6-[(1R,3aR,4E,7aS)-4-[(2Z)-2-[(3R,5S)-3,5-di-hydroxy-2-methylidene-cyclohexylidene]ethylidene]-7a-methyl-2,3,3a,5,6,7-hexahydro-1H-inden-1-yl]-2-methyl-heptane-1,2-diol.
1,25,26-OhD$_3$; 9,10-Secocholesta-5Z,7E,10(19)-triene-1α,3β,25S,26-tetrol; 1,25,26-Trihydroxyvitamin D_3; 1α,25,26-Trihydroxyvitamin D_3; 6444050; 77372-59-9; Metabolite of vitamin D_3. *Biochemistry.* 1981, *20(21),* 6230-5; *Arch Biochem Biophys.* 1981, *10(1),* 104-9; *Biochem Biophys Res Commun.* 1981, *99(1),* 302-7.

450   *1α,25S,26-Trihydroxyvitamin D_3*
5283758 $C_{27}H_{44}O_4$

(2S,6R)-6-[(1R,3aR,4E,7aS)-4-[(2Z)-2-[(3S,5R)-3,5-di-
hydroxy-2-methylidene-cyclohexylidene]ethylidene]-
7a-methyl-2,3,3a,5,6,7-hexahydro-1H-inden-1-yl]-2-
methyl-heptane-1,2-diol.
LMST03020296; 1α,25S,26-trihydroxyvitamin D_3; 1α,-
25S,26-trihydroxycholecalciferol; 9,10-seco-5Z,7E,10
(19)-cholestatriene-1S,3R,25S,26-tetrol; 5283758; Synthesized from C(17) steroid and C(6)-chiral side chain
synthons or *via* Wittig-Horner coupling of CD ring 8-
ketone with A ring phosphine oxide. Compared to
1α,25-dihydroxyvitamin D_3, vitamin D receptor
binding (chick intestine) is 9%, inhibition of cell (HL-
60) proliferation is 50%, induction cell (HL-60)
differentiation is 100%, and [45]Ca retention in kidney in
rats is <10%; Crystals; mp = 163-164°; $[\alpha]_D^{25}$ = +58.8° (c
= 0.5 MeOH); λ_m = 265 nm (ε = 17080 EtOH).
Helvetica Chimica Acta. 1981, *64(7)*, 2138-41;
Tetrahedron. 1984, 40(12), 2283-96.

451   (5E)-1α,25R,26-Trihydroxyvitamin D_3
5283757 $C_{27}H_{44}O_4$

(2R,6R)-6-[(1R,3aR,4E,7aS)-4-[(2E)-2-[(3S,5R)-3,5-di-
hydroxy-2-methylidene-cyclohexylidene]ethylidene]-
7a-methyl-2,3,3a,5,6,7-hexahydro-1H-inden-1-yl]-2-
methyl-heptane-1,2-diol.
LMST03020295;   9,10-seco-5E,7E,10(19)-cholestatri-
ene-1S,3R,25R,26-tetrol;   (5E)-1α,25R,26-trihydroxy-
vitamin D_3; (5E)-1α,25R,26-trihydroxycholecalciferol;
5283757; Affinity for the chick intestinal receptor is
1/8 that of 1α,25-dihydroxyvitamin D_3. *Endocrinology.*
1990, *127(2),* 695-701.

452   22,24,25-Trihydroxyvitamin D_3
5283759 $C_{27}H_{44}O_4$

(6S)-6-[(1R,3aR,4E,7aS)-4-[(2Z)-2-[(5S)-5-hydroxy-2-
methylidene-cyclohexylidene]ethylidene]-7a-methyl-
2,3,3a,5,6,7-hexahydro-1H-inden-1-yl]-2-methyl-
heptane-2,3,5-triol.
LMST03020297;   22,24,25-trihydroxycholecalciferol;
9,10-seco-5Z,7E,10(19)-cholestatriene-3S,22,24,25-
tetrol; 5283759; λ_m = 265 nm (EtOH). *Chem Pharm
Bull (Tokyo).* 1987, *35(3),* 970-9.

453   23,24,25-Trihydroxyvitamin D_3
80463-20-3 $C_{27}H_{44}O_4$

(6R)-6-[(1R,3aR,4E,7aS)-4-[(2Z)-2-[(5R)-5-hydroxy-2-
methylidene-cyclohexylidene]ethylidene]-7a-methyl-
2,3,3a,5,6,7-hexahydro-1H-inden-1-yl]-2-methyl-
heptane-2,3,4-triol.
LMST03020298;   23S,24,25-Trihydroxyvitamin D_3;
9,10-Secocholesta-5Z, 7E,10(19)-triene-3β,23S,24,25-
tetrol; 6439569; 80463-20-3; a vitamin D_3 metabolite,
formed in the kidney. *J Biol Chem.* 1983, *258(1),* 457-
63.

454   23,24,25-Trihydroxyvitamin D_3
5283760 $C_{27}H_{44}O_4$

(4S,6R)-6-[(1R,3aR,4E,7aS)-4-[(2Z)-2-[(5S)-5-hydroxy-2-methylidene-cyclohexylidene]ethylidene]-7a-methyl-2,3,3a,5,6,7-hexahydro-1H-inden-1-yl]-2-methyl-heptane-2,3,4-triol.

LMST03020298; 23S,24,25-trihydroxyvitamin D_3; 23S,24,25-trihydroxycholecalciferol; 9,10-seco-5Z,7E,10(19)-cholestatriene-3S,23S,24,25-tetrol; 5283760; *In vivo* metabolite of vitamin D_3. Isolated from chicken kidney homogenate incubated with 25-hydroxy-24-oxovitamin D_3 and from blood plasma of chicks given large doses of vitamin D_3; λ_m = 265 nm. *Biochemistry.* 1981, *20(26)*, 7385-91; *J Biol Chem.* 1983, *258(1)*, 457-63.

455   23,25,26-Trihydroxyvitamin D_3
81515-15-3 $C_{27}H_{44}O_4$

(6R)-6-[(1R,3aR,4E,7aS)-4-[(2Z)-2-[(5R)-5-hydroxy-2-methylidene-cyclohexylidene]ethylidene]-7a-methyl-2,3,3a,5,6,7-hexahydro-1H-inden-1-yl]-2-methyl-heptane-1,2,4-triol.

23,25,26-Trihydroxyvitamin D_3; 23,25,26-Trihydroxycholecalciferol; 9,10-Secocholesta-5Z,7E,10(19)-triene-3β,23,25,26-tetrol; LMST03020299; 6439591; 81515-15-3; Produces 25-hydroxycholecalciferol-26,23-lactone when incub-ated in chick kidney homogenates. Is metabolized to 23S,25R-dihydroxyvitamin D_3 and the 26,23-lactone. *Arch Biochem Biophys.* 1982, *217(1)*, 264-72; *Biochem J.* 1982, *206(1)*, 173-6.

456   23S,25R,26-Trihydroxyvitamin D_3
5283761 $C_{27}H_{44}O_4$

(2R,4S,6R)-6-[(1R,3aR,4E,7aS)-4-[(2Z)-2-[(5S)-5-hydroxy-2-methylidene-cyclohexylidene]ethylidene]-7a-methyl-2,3,3a,5,6,7-hexahydro-1H-inden-1-yl]-2-methyl-heptane-1,2,4-triol.

LMST03020299; 9,10-seco-5Z,7E,10(19)-cholestatriene-3S,23S,25R,26-tetrol; 23S,25R,26-trihydroxyvitamin D_3; 23S,25R,26-trihydroxycholecalciferol; 5283761; Isolated from chick kidney homogenates incubated with 23S,25-dihydroxyvitamin D_3. Synthesized from the triazolinedione adduct of (3S,23S,25R)-3,25-dihydroxy-5,7-cholestadiene 26,23-lactone by reduction with $LiAlH_4$ followed by photochemical and thermal isomerization. Metabolized to (23S,25R)-25-hydroxyvitamin D_3 26,23-lactone *via* (23S,25R)-225-hydroxyvitamin D_3 26,23-lactol by incubation with vitamin D supplemented chick kidney homogenates; λ_m = 265 nm (95% EtOH). *Biochem J.* 1982, *206(1)*, 173-6; *Chem Pharm Bull (Tokyo).* 1981, *29*, 2393; *J Org Chem.* 1982, *47*(24), 4770-2.

457   24,25,26-Trihydroxyvitamin D_3
76355-23-2 $C_{27}H_{44}O_4$

(6R)-6-[(1R,3aR,4E,7aS)-4-[(2Z)-2-[(5R)-5-hydroxy-2-methylidene-cyclohexylidene]ethylidene]-7a-methyl-2,3,3a,5,6,7-hexahydro-1H-inden-1-yl]-2-methyl-heptane-1,2,3-triol.

24,25,26-Trihydroxyvitamin D_3; 9,10-Secocholesta-5Z,7E,10(19)-triene-3β,24,25,26-tetrol; LMST03020300; 6444044; 76355-23-2; A vitamin D_3 metabolite. *Biochemistry.* 1981, *20(26)*, 7385-91.

458   24,25,26-Trihydroxyvitamin D_3
5283762 $C_{27}H_{44}O_4$

(6R)-6-[(1R,3aR,4E,7aS)-4-[(2Z)-2-[(5S)-5-hydroxy-2-methylidene-cyclohexylidene]ethylidene]-7a-methyl-2,3,3a,5,6,7-hexahydro-1H-inden-1-yl]-2-methyl-heptane-1,2,3-triol.
LMST03020300; 24,25,26-trihydroxyvitamin D_3; 24,25,26-trihydroxycholecalciferol; 9,10-seco-5Z,7E,10(19)-cholestatriene-3S,24,25,26-tetrol; 5283762; *In vivo* metabolite of vitamin D_3. Isolated from blood plasma of chicks given large doses of vitamin D_3; λ_m = 265 nm. *Biochemistry.* 1981, *20(26),* 7385-91.

459 ***1α,24,25-Trihydroxyvitamin D_3***
4525692 $\qquad$ $C_{27}H_{44}O_4$

6-[4-[2-(3,5-dihydroxy-2-methylidene-cyclohexyl-idene)ethylidene]-7a-methyl-2,3,3a,5,6,7-hexahydro-1H-inden-1-yl]-2-methyl-heptane-2,3-diol.
1,24,25-Trihydroxyvitamin D_3; 4525692; Metabolite of 1α,25-dihydroxyvitamin D_3. Formed in the kidney as a metabolite of 25-hydroxy-24-oxovitamin D_3. *Biochemistry.* 1984, *23(7),* 1473-8; *J Biol Chem.* 1983, *258(1),* 457-63.

460 ***1,24,25-Trihydroxyvitamin D_3***
50648-94-7 $\qquad$ $C_{27}H_{44}O_4$

(6R)-6-[(1R,3aR,4E,7aS)-4-[(2Z)-2-[(5S)-3,5-dihydroxy-2-methylidene-cyclohexylidene]ethylidene]-7a-methyl-2,3,3a,5,6,7-hexahydro-1H-inden-1-yl]-2-methyl-heptane-2,3-diol.
1,24,25-Trihydroxyvitamin D_3; 1,24,25-Trihydroxy-cholecalciferol; 9,10-Secocholesta-5Z,7E,10(19)-triene-1,3β,24,25-tetrol; 6438336; 50649-94-7; A vitamin D_3 metabolite. Exhibits synergism with vitamin D_3. *Biochem J.* 1984, *219(3),* 713-7; *Biochemistry.* 1984, *23(7),* 1473-8; *Arch Biochem Biophys.* 1982, *213(1),* 163-8.

461 ***23,25,26-Trihydroxyvitamin D_3***
81515-15-3 $\qquad$ $C_{27}H_{44}O_4$

(6R)-6-[(1R,3aR,4E,7aS)-4-[(2Z)-2-[(5R)-5-hydroxy-2-methylidene-cyclohexylidene]ethylidene]-7a-methyl-2,3,3a,5,6,7-hexahydro-1H-inden-1-yl]-2-methyl-heptane-1,2,4-triol.
23,25,26-Trihydroxyvitamin D_3; 23,25,26-Trihydroxy-cholecalciferol; 9,10-Secocholesta-5Z,7E,10(19)-triene-3β,23,25,26-tetrol; 6439591; 81515-15-3; Produces 25-hydroxycholecalciferol-26,23-lactone when incubated in chick kidney homogenates. Is metabolized to 23S,25R-dihydroxyvitamin D_3 and the 26,23-lactone. *Arch Biochem Biophys.* 1982, *217(1),* 264-72; *Biochem J.* 1982, *206(1),* 173-6.

462 ***19,25-Dihydroxy-6,19-dihydro-6,19-epoxyvitamin D_3***
96999-67-6 $\qquad$ $C_{27}H_{44}O_4$

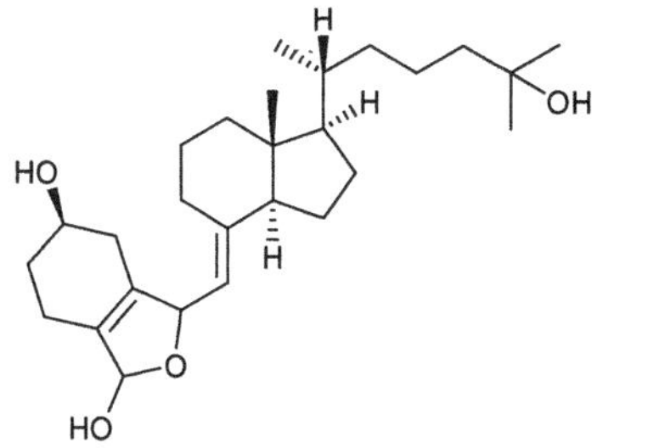

(3S,5R)-3-[(E)-[(1R,3aR,7aS)-1-[(2R)-6-hydroxy-6-methyl-heptan-2-yl]-7a-methyl-2,3,3a,5,6,7-hexahydro-1H-inden-4-ylidene]methyl]-1,3,4,5,6,7-hexahydroisobenzofuran-1,5-diol.

19-25-Dde-vitamin D_3; 19,25-Dihydroxy-6,19-dihydro-6,19-epoxyvitamin D_3; 6,10-Secocholesta-5(10),7E-diene-3β,19,25-triol, 6S,19-epoxy-; 19,25-hydroxy-6,19-epoxyvitamin D_3; 6439195; 96999-67-6; Has differentiating activity and bone-resorbing activity. *J Med Chem.* 1985, *28(9),* 1153-1158.

463 **2β-Methoxy-1α, 26-dihydroxy-27-nor-vitamin D_3**

$C_{27}H_{44}O_4$

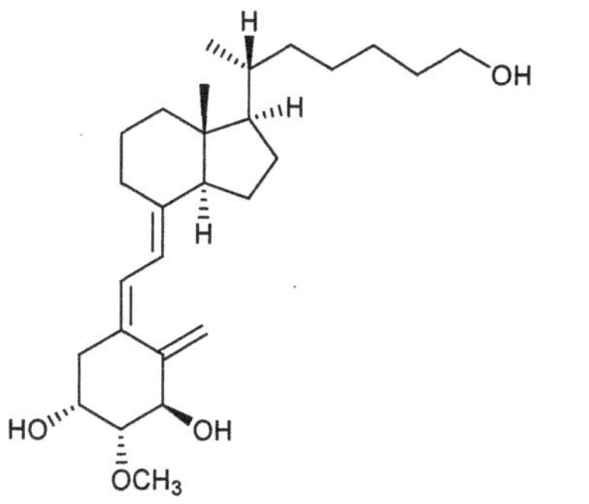

(5Z,7E)-2β-methoxy-9,10-seco-27-norcholesta-5,7,10(19)-triene-1α,3α,26-triol.

2β-methoxy-1α, 26-dihydroxy-27-nor-vitamin D_3; Synthetic analog of vitamin D_3. Has lower affinity than 1,25-dihydroxyvitamin D_3 for the vitamin D receptor; Colorless foam; λ_m = 265 nm (ε 16200, MeOH). *Steroids.* 1996, *61(10),* 598-608; *Endrocrinology.* 1991, *128(4),* 1687-92.

464 **25-Hydroxyvitamin D_3 6R,19-sulfur dioxide adduct**

9547442 $C_{27}H_{44}O_4S$

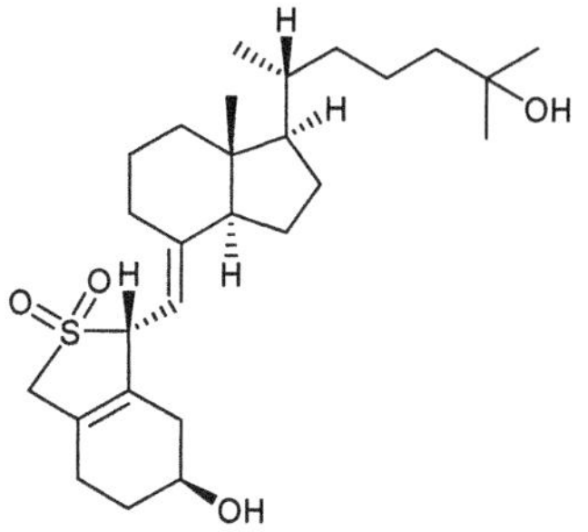

(3S,9R)-9-[(E)-[(1R,3aR,7aS)-1-[(2R)-6-hydroxy-6-methyl-heptan-2-yl]-7a-methyl-2,3,3a,5,6,7-hexa-hydro-1H-inden-4-ylidene]methyl]-8,8-dioxo-8-thia-bicyclo[4.3.0]non-10-en-3-ol.

LMST03020279; 6R,19-epithio-9,10-seco-5(10),7E-cholestadiene-3S,25-diol S,S-dioxide; 25-hydroxy-cholecalciferol 6R,19-sulfur dioxide adduct; 9547442; Synthesized from 25-hydroxyvitamin D_3 by treatment with liquid SO_2. *Chem Lett.* 1979, 583; *J Org Chem.* 1983, *48(20),* 3483-8.

465 **25-Hydroxyvitamin D_3 6S,19-sulfur dioxide adduct**

9547443 $C_{27}H_{44}O_4S$

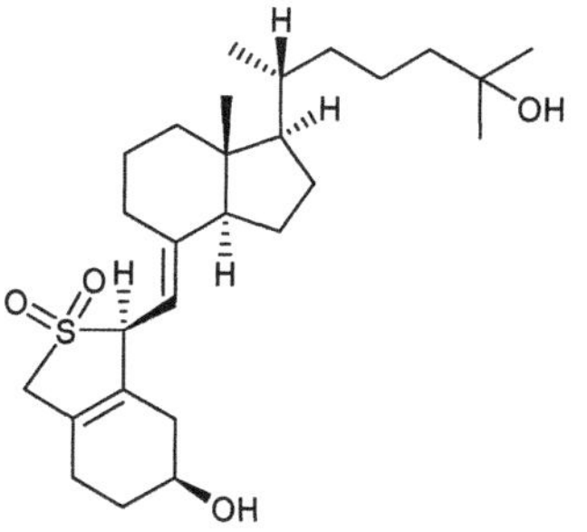

(3S,9S)-9-[(E)-[(1R,3aR,7aS)-1-[(2R)-6-hydroxy-6-methyl-heptan-2-yl]-7a-methyl-2,3,3a,5,6,7-hexa-hydro-1H-inden-4-ylidene]methyl]-8,8-dioxo-8-thia-bicyclo[4.3.0]non-10-en-3-ol.

LMST03020280; 6S,19-epithio-9,10-seco-5(10),7E-cholestadiene-3S,25-diol S,S-dioxide; 25-hydroxy-cholecalciferol 6S,19-sulfur dioxide adduct; 9547443; Synthesized from 25-hydroxyvitamin D_3 by treatment with liquid SO_2. *Chem Let.* 1979, 583-6; *J Org Chem.* 1983, *48(20),* 3483-8

466 **7-Dehydrocholesterol-3-sulfate ester**

10529-44-9 $C_{27}H_{44}O_4S$

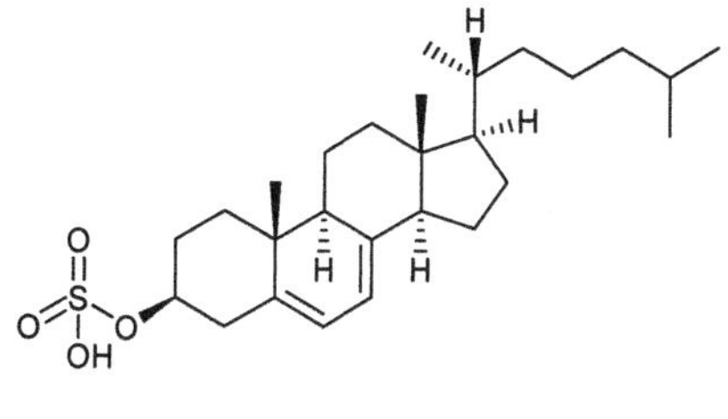

(3S,9R,10R,13S,14R,17R)-10,13-dimethyl-17-[(2R)-6-methylheptan-2-yl]-3-sulfooxy-2,3,4,9,11,12,14,15,16,17-decahydro-1H-cyclopenta[a]phenanthrene.

7-Dehydrocholesterol-3-sulfate ester; 7-DH-3-SE; Cholesta-5,7-dien-3β-ol, hydrogen sulfate; 3080672; 10529-44-9; Found in human and rat skin, it can serve as a precursor of cholecalciferol sulfate and 25-hydroxy-cholecalciferol 3-sulfate. *J Steroid Biochem.* 1987, *26(3)*, 369-73.

467 *Vitamin D₃ sulfoconjugate*

10529-43-8 $C_{27}H_{44}O_4S$

$$O=S=O$$
$$OH$$

(1R,3aR,4E,7aS)-7a-methyl-1-[(2R)-6-methylheptan-2-yl]-4-[(2Z)-2-[(5R)-2-methylidene-5-sulfooxy-cyclo-hexylidene]ethylidene]-2,3,3a,5,6,7-hexahydro-1H-indene.

Vitamin D₃ sulfoconjugate; Cholecalciferol sulfate; Vitamin D₃ sulfate; Vitamin D₃ sulfoconjugate; 9,10-Secocholesta-5Z,7E,10(19)-trien-3β-ol, hydrogen sulfate; 6440809; 10529-43-8; Increases active calcium transport in the duodenum and is also able to mobilize calcium from bone and soft tissue but has only 0.1% of the activity of vitamin D₃. *J Steroid Biochem.* 1987, *26(3)*, 369-73; *J Biol Chem.* 1981, *256(2)*, 823-6; *256(11)*, 5536-9.

468 *19-Nor-1,24,25,26-tetrahydroxyvitamin D₂*

$C_{27}H_{44}O_5$

Metabolite in keratinocytes of 19-nor-1α,25-dihydroxyvitamin D₂. *Arch Biochem Biophys.* 2001, *387(2)*, 297-306.

469 *6,19-Epidioxy-1α,24R-dihydroxy-6,19-dihydrovitamin D₃*

9547451 $C_{27}H_{44}O_5$

8R,10S)-5-[(E)-[(1R,3aR,7aS)-1-[(2R,4R)-4-hydroxy-6-methyl-heptan-2-yl]-7a-methyl-2,3,3a,5,6,7-hexa-hydro-1H-inden-4-ylidene]methyl]-3,4-dioxabicyclo[4.4.0]dec-11-ene-8,10-diol.

LMST03020301; 6,19-epidioxy-9,10-seco-5(10),7E-cholestadiene-1S,3R,24R-triol; 6,19-epidioxy-1α,24R-dihydroxy-6,19-dihydrocholecalciferol; 9547451; The less polar epimer, compared to 1α,25-dihydroxy-vitamin D₃, binding affinity for the vitamin D receptor in HL-60 cells is 1/800, differentiation of HL-60 cells is 1/150 as active as 1,25-(OH)2D3. For the more polar epimer, binding affinity for the vitamin D receptor in HL-60 cells is 1/800 as active and in differentiation of HL-60 cells is 1/200 as active. *J Med Chem*. 1985, *28(9)*, 1148-53.

470 *6R,19-Epidioxy-1α,25-dihydroxy-6,19-dihydrovitamin D₃*

9547452 $C_{27}H_{44}O_5$

(8R,10S)-5-[(E)-[(1R,3aR,7aS)-1-[(2R)-6-hydroxy-6-methyl-heptan-2-yl]-7a-methyl-2,3,3a,5,6,7-hexa-hydro-1H-inden-4-ylidene]methyl]-3,4-dioxabicyclo[4.4.0]dec-11-ene-8,10-diol.

LMST03020302; 6,19-epidioxy-9,10-seco-5(10),7E-cholestadiene-1S,3R,25-triol; 6R,19-epidioxy-1α,25-dihydroxy-6,19-dihydrovitamin D₃; 6R,19-epidioxy-1α,25-dihydroxy-6,19-dihydrocholecalciferol; 9547452; The less polar epimer, compared to 1α,25-dihydroxyvitamin D₃, binding affinity for the vitamin D receptor in HL-60 cells is 1/800, differentiation of HL-60 cells is 1/130 as active as 1,25-(OH)2D3. For the more polar epimer, binding affinity for the vitamin D receptor in HL-60 cells is 1/800 as active and in differentiation of HL-60 cells is 1/200 as active. *J Med Chem*. 1985, **28(9)**, 1148-53.

471   1α,23R,25S,26-Tetrahydroxyvitamin D$_3$

100634-18-2 $C_{27}H_{44}O_5$

(2R,4S,6R)-6-[(1R,3aR,4Z,7aS)-4-[(2Z)-2-[(3S,5R)-3,5-dihydroxy-2-methylidene-cyclohexylidene]ethylidene]-7a-methyl-2,3,3a,5,6,7-hexahydro-1H-inden-1-yl]-2-methyl-heptane-1,2,4-triol.
1,23,25,26-Tetrahydroxyvitamin D$_3$; 1α,23,25,26-Tetrahydroxyvitamin D$_3$; 6438879; 100643-18-2; Stimulates intestinal calcium absorption but is less active than 1,25-dihydroxyvitamin D$_3$. *Arch Biochem Biophys*. 1987, **254(1)**, 188-95.

472   1α,23S,25R,26-Tetrahydroxyvitamin D$_3$

5283763 $C_{27}H_{44}O_5$

(2R,4S,6R)-6-[(1R,3aR,4E,7aS)-4-[(2Z)-2-[(3S,5R)-3,5-dihydroxy-2-methylidene-cyclohexylidene]ethylidene]-7a-methyl-2,3,3a,5,6,7-hexahydro-1H-inden-1-yl]-2-methyl-heptane-1,2,4-triol.
LMST03020306; 9,10-seco-5Z,7E,10(19)-cholestatriene-1S,3R,23S,25R,26-pentol; 1α,23S,25R,26-tetrahydroxyvitamin D$_3$; 1α,23S,25R,26-tetrahydroxycholecalciferol; 5283763; Affinity for the chick intestinal receptor is 1/216 that of 1α,25-dihydroxyvitamin D$_3$. *Endocrinology* 1990, **127(2)**, 695-701.

473   24R,25-Dihydroxyvitamin D$_3$ 6RS,19-sulfur dioxide adduct

9547455 $C_{27}H_{44}O_5S$

(3R,6R)-6-[(1R,3aR,4E,7aS)-4-[[(3S)-3-hydroxy-8,8-dioxo-8-thiabicyclo[4.3.0]non-10-en-9-yl]methylidene]-7a-methyl-2,3,3a,5,6,7-hexahydro-1H-inden-1-yl]-2-methyl-heptane-2,3-diol.
LMST03020305; 24R,25-dihydroxycholecalciferol 6RS, 19-sulfur dioxide adduct; 6RS,19-epithio-9,10-seco-5(10),7E-cholestadiene-3S,24R,25-triol   S,S-dioxide; 9547455; Prepared from 24R,25-dihydroxyvitamin D$_3$ by treatment with liquid SO$_2$. *Steroids*. 1989, **54(2)**, 145-57.

474   1α,25-Dihydroxyvitamin D$_3$ 6R,19-sulfur dioxide adduct

9547453 $C_{27}H_{44}O_5S$

(2S,4R,7R)-7-[(E)-[(1R,3aR,7aS)-1-[(2R)-6-hydroxy-6-methyl-heptan-2-yl]-7a-methyl-2,3,3a,5,6,7-hexahydro-1H-inden-4-ylidene]methyl]-8,8-dioxo-8-thiabicyclo[4.3.0]non-10-ene-2,4-diol.
LMST03020303; 1α,25-dihydroxycholecalciferol 6R, 19-sulfur dioxide adduct; 6R,19-epithio-9,10-seco-5(10),7E-cholestadiene-1S,3R,25-triol   S,S-dioxide; 9547453; Prepared from the 1α,25-dihydroxyvitamin D$_3$ derivative by treatment with liquid SO$_2$. *Chem Lett*. 1979, 583; *J Org Chem*. 1983, **48(20)**, 3483-8.

475   1α,25-Dihydroxyvitamin D$_3$ 6S,19-sulfur dioxide adduct

9547454 $C_{27}H_{44}O_5S$

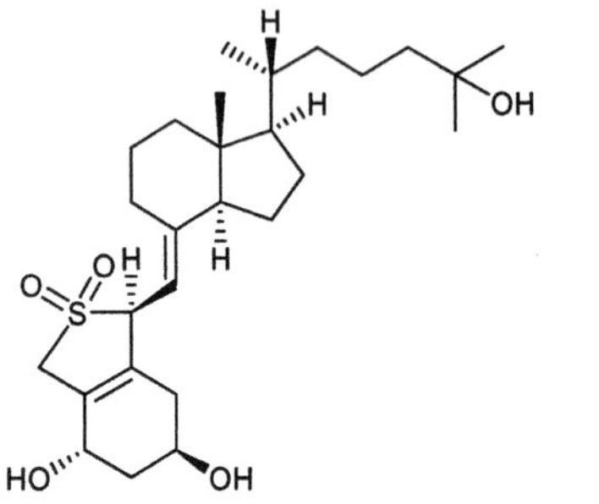

(3R,5S,9S)-9-[(E)-[(1R,3aR,7aS)-1-[(2R)-6-hydroxy-6-methyl-heptan-2-yl]-7a-methyl-2,3,3a,5,6,7-hexa-hydro-1H-inden-4-ylidene]methyl]-8,8-dioxo-8-thia-bicyclo[4.3.0]non-10-ene-3,5-diol.

LMST03020304; 1α,25-dihydroxycholecalciferol 6S, 19-sulfur dioxide adduct; 6S,19-epithio-9,10-seco-5(10),7E-cholestadiene-1S,3R,25-triol   S,S-dioxide; 9547454; Prepared from 1α,25-dihydroxyvitamin D₃ derivative by treatment with liquid SO₂. *Chem Lett.* 1979, 583; *J Org Chem.* 1983, *48(20)*, 3483-8.

476   *25-Hydroxyvitamin D₃ 3-sulfate ester*
99447-30-0   C₂₇H₄₄O₅S

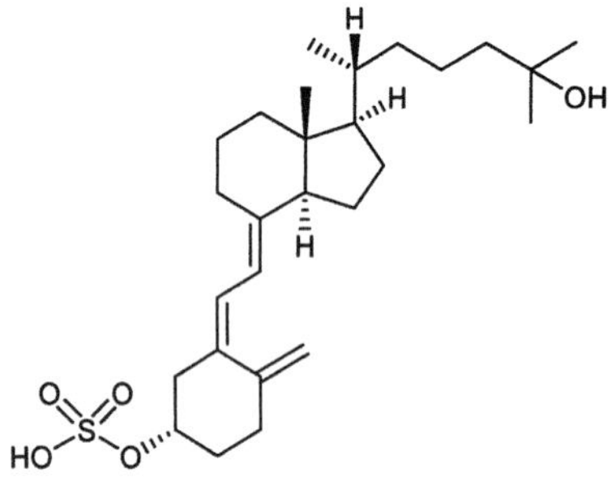

(1R,3aS,5Z,7aR)-1-[(2R)-6-hydroxy-6-methyl-heptan-2-yl]-7a-methyl-5-[(2Z)-2-[(5S)-2-methylidene-5-sulfo-oxy-cyclohexylidene]ethylidene]-2,3,3a,4,6,7-hexa-hydro-1H-indene.

25-Hydroxyvitamin D₃ 3-sulfate ester; 25-Hydroxy-cholecalciferol 3-sulfate; Calcifediol-3-sulfate; 9,10-Secocholesta-5Z,7E,10(19)-triene-3,25-diol, 3β-(hydro-gen sulfate); 6438836; 99447-30-0; A major circulating form of vitamin D in man; decreases sharply upon renal failure. *FEBS Lett.* 1985, *191(2)*, 171-5.

477   *3β-Thiovitamin D₃*
98353-78-7   C₂₇H₄₄S

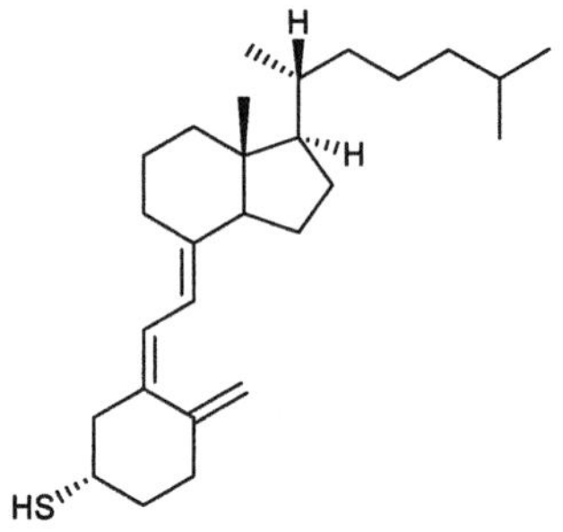

9,10-Secocholesta-5Z,7E,10(19)-trien-3β-thiol.

3-Thiovitamin   D₃;   3β-Thiocholecalciferol;   9,10-Secocholesta-5Z,7E,10(19)-triene-3β-thiol;   98353-78-7; Synthesized from 7-dehydrocholesterol. A weak vitamin D agonist; does not increase the synthesis of calcium binding protein or block the activity of 1,25-dihydroxyvitamin D₃ in chick embryoinc duodenum. *J Steroid Biochem.* 1985, *23(1)*, 81-5.

478   *(5E)-10S,19-Dihydrovitamin D₃*
9547456   C₂₇H₄₆O

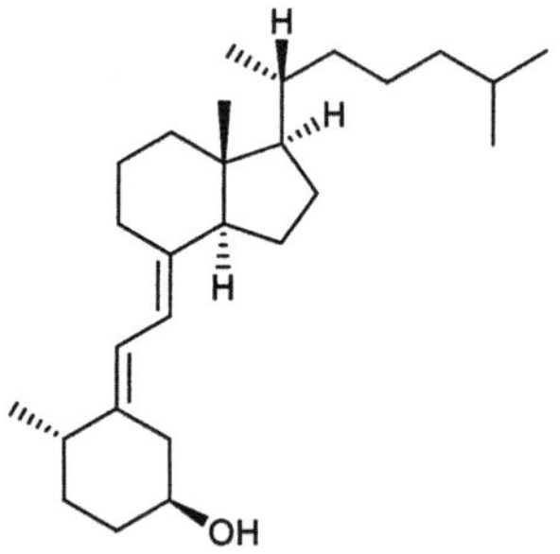

(1S,3E,4S)-3-[(2E)-2-[(1R,3aR,7aS)-7a-methyl-1-[(2R)-6-methylheptan-2-yl]-2,3,3a,5,6,7-hexahydro-1H-inden-4-ylidene]ethylidene]-4-methyl-cyclohexan-1-ol.

LMST03020307; 9,10S-seco-5E,7E-cholestatrien-3S-ol; dihydrotachysterol₃;   (5E)-10S,19-dihydrocholecalc-iferol; 9574456. *Science.* 1974, *186(4168)*, 1038-40.

479   *6,19-Dihydrovitamin D₃*
5823764   C₂₇H₄₆O

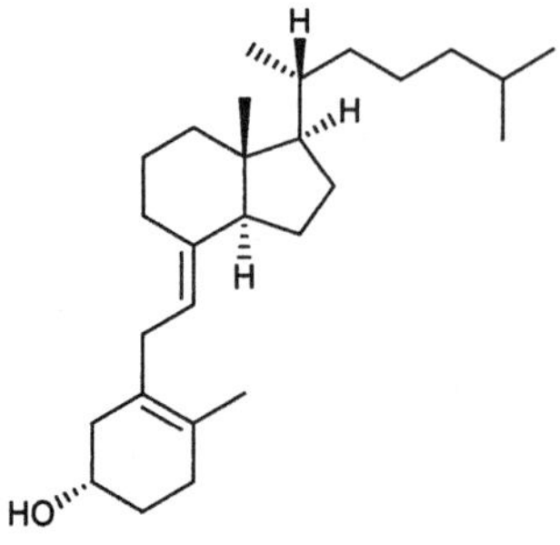

(1S)-3-[(2E)-2-[(1R,3aR,7aS)-7a-methyl-1-[(2R)-6-methylheptan-2-yl]-2,3,3a,5,6,7-hexahydro-1H-inden-4-ylidene]ethyl]-4-methyl-cyclohex-3-en-1-ol.

LMST03020308; 9,10-seco-5(10),7E-cholestadien-3S-ol; toxisterol-3 R1; 6,19-dihydrocholecalciferol; 5283764; Obtained as one of the overirradiation products of 7-dehydrocholesterol in EtOH or MeOH; No absorption maximum >210 nm. *J R Neth Chem Soc.* 1977, *96*, 104.

480   *3-Dihydro-25-hydroxytachysterol₃*

5370881 $C_{27}H_{46}O_2$

6-[(4E)-4-[(2E)-2-(5-hydroxy-2-methyl-cyclohexyl-idene)ethylidene]-7a-methyl-2,3,3a,5,6,7-hexahydro-1H-inden-1-yl]-2-methyl-heptan-2-ol.
3-Dihydro-25-hydroxytachysterol; 9,10-Secocholesta-5E,7E-diene-3β,25-diol; NIST 1401674646; 5370881.

481   *(3S,7E)-9,10-Secocholesta-5(10),7-diene-3,25-diol*

446421 $C_{27}H_{46}O_2$

(6R)-6-[(1R,3aR,4E,7aS)-4-[2-[(5S)-5-hydroxy-2-methyl-1-cyclohexenyl]ethylidene]-7a-methyl-2,3,3a,5,6,7-hexahydro-1H-inden-1-yl]-2-methyl-heptan-2-ol.
VDY; 446421.

482   *19-Hydroxy-10S,19-dihydrovitamin D₃*

62077-06-9 $C_{27}H_{46}O_2$

(1R,3Z,4R)-3-[(2E)-2-[(1R,3aR,7aS)-7a-methyl-1-[(2R)-6-methylheptan-2-yl]-2,3,3a,5,6,7-hexahydro-1H-inden-4-ylidene]ethylidene]-4-(hydroxymethyl)cyclo-hexan-1-ol.

Hdhv D₃; 19-Hydroxy-10S,19-dihydrovitamin D₃; 9,10α-Secocholesta-5Z,7E-diene-3β,19-diol; 6454339; 62077-06-9; Inhibits rat liver vitamin D₃-25-hydroxylase. In rats treated with this compound, intestinal calcium transport is normal, but bone calcium is not mobilized. *Biochemistry.* 1980, *19(23)*, 5335-9.

483   *24-Hydroxycholesterol*

474-73-7 $C_{27}H_{46}O_2$

(3S,8S,9S,10R,13R,14S,17R)-17-[(2R,5S)-5-hydroxy-6-methyl-heptan-2-yl]-10,13-dimethyl-2,3,4,7,8,9,11,12,14,15,16,17-dodecahydro-1H-cyclopenta[a]phen-anthren-3-ol.
LMST01010019; Cerebrosterol; 24-Hydroxycholest-erol; 24S-Hydroxycholesterol; Cholest-5-ene-3,24-diol; Cholest-5-en-3β,24S-diol; C13550; 121948; 474-73-7; 27460-27-1; Metabolite of cholesterol. *J Lipid Res.* 1978, *19(2)*, 191-6; 1985, *26(2)*, 230-40.

484   *25-Hydroxydihydrotachysterol₃*

25631-39-4 $C_{27}H_{46}O_2$

(6R)-6-[(1R,3aR,4E,7aS)-4-[(2Z)-2-[(2R,5R)-5-hydroxy-2-methyl-cyclohexylidene]ethylidene]-7a-methyl-

2,3,3a,5,6,7-hexahydro-1H-inden-1-yl]-2-methyl-heptan-2-ol.
25-Hydroxydihydrotachysterol(3); 9,10-Secocholesta-5,7-diene-3β,25-diol; 6443307; 25631-39-4; *J Steroid Biochem.* 1985, *23(2)*, 223-9.

485 1-Methyl-1,25-dihydroxy-4-nor-2,3-secovitamin D₃

9547457 $C_{27}H_{46}O_3$

(2E)-2-[(2E)-2-[(1R,3aR,7aS)-1-[(2R)-6-hydroxy-6-methyl-heptan-2-yl]-7a-methyl-2,3,3a,5,6,7-hexa-hydro-1H-inden-4-ylidene]ethylidene]-4-methyl-3-methylidene-pentane-1,4-diol.
LMST03020309; 1-methyl-A-nor-(2,3)-(9,10)-diseco-5E,7E,10(19)-cholestatriene-1,3,25-triol; 1-methyl-1,25-dihydroxy-4-nor-2,3-secocholecalciferol; 9547457.

486 1α,25-Dihydroxy-2α-methyl-19-norvitamin D₃

5289548 $C_{27}H_{46}O_3$

(1R,2R,3R)-5-[(2E)-2-[(1R,3aR,7aS)-1-[(2R)-6-hydroxy-6-methyl-heptan-2-yl]-7a-methyl-2,3,3a,5,6,7-hexa-hydro-1H-inden-4-ylidene]ethylidene]-2-methyl-cyclohexane-1,3-diol.
LMST03020553; 2AM20R; 2α-methyl-19-nor-1,25-di-hydroxyvitamin D₃; 2R-methyl-19-nor-9,10-seco-5,7E-cholestadiene-1R,3R,25-triol; 1α,25-dihydroxy-2α-methyl-19-norvitamin D₃; 1α,25-dihydroxy-2α-methyl-19-norcholecalciferol; 5-{2-[1-(5-hydroxy-1,5-di-methylhexyl)-7A-methyl-octahydro-inden-4-ethylid-ene]-2-methylcyclohexane-1,3-diol; VD2; 5289548; Synthesized by combination of Grundmann type 8-ketone with an A-ring synthon obtained from (-)-quinic acid. Binds to the vitamin D receptor but with only 20% of the affinity of 1α,25-dihydroxyvitamin D₃.

Compared to 1α,25-dihydroxyvitamin D₃, affinity for the porcine intestinal vitamin D receptor is 22%, HL-60 cell differentiation is 5000%, intestinal calcium transport and bone calcium mobilization are both slightly less active. *Biochemistry.* 2004, *43(14)*, 4101-10; *J Med Chem.* 1998, *41(23)*, 4662-74; USP 6306844.

487 1α,25-Dihydroxy-2β-methyl-19-nor-vitamin D₃

5289548 $C_{27}H_{46}O_3$

(1R,2S,3R)-5-[(2E)-2-[(1R,3aR,7aS)-1-[(2R)-6-hydroxy-6-methyl-heptan-2-yl]-7a-methyl-2,3,3a,5,6,7-hexa-hydro-1H-inden-4-ylidene]ethylidene]-2-methyl-cyclohexane-1,3-diol.
LMST03020554; 1α,25-dihydroxy-2β-methyl-19-nor-vitamin D₃; 1α,25-dihydroxy-2β-methyl-19-norchole-calciferol; 2S-methyl-19-nor-9,10-seco-5,7E-cholesta-diene-1R,3R,25-triol; 5289548; Synthesized by combination of Grundmann type 8-ketone with an A-ring synthon obtained from (-)-quinic acid. Compared to 1α,25-dihydroxyvitamin D₃, affinity for the porcine intestinal vitamin D receptor is 2.5%, HL-60 cell differentiation is 50%, intestinal calcium transport and bone calcium mobilization are both 0%. *J Med Chem.* 1998, *41(23)*, 4662-74.

488 1α,25-Dihydroxy-2α-methyl-19-nor-20-epivitamin D₃

9547655 $C_{27}H_{46}O_3$

(1R,2S,3R)-5-[(2E)-2-[(1R,3aR,7aS)-1-[(2S)-6-hydroxy-6-methyl-heptan-2-yl]-7a-methyl-2,3,3a,5,6,7-hexa-hydro-1H-inden-4-ylidene]ethylidene]-2-methyl-cyclohexane-1,3-diol.
LMST03020558; 1α,25-dihydroxy-2α-methyl-19-nor-

20-epicholecalciferol; (20S)-2S-methyl-19-nor-9,10-seco-5,7-cholestadiene-1R,3R,25-triol; 9547655; Synthesized by combination of Grundmann type 8-ketone with an A-ring synthon obtained from (-)-quinic acid. Compared to 1α,25-dihydroxyvitamin D$_3$, affinity for the porcine intestinal vitamin D receptor is 23%, HL-60 cell differentiation is 5000%, intestinal calcium transport is >200%, bone calcium mobilization is >200%. *J Med Chem*. 1998, *41(23)*, 4662-74.

489   1α,25-Dihydroxy-2β-methyl-19-nor-20-epivitamin D$_3$
9547655   C$_{27}$H$_{46}$O$_3$

(1R,2R,3R)-5-[(2E)-2-[(1R,3aR,7aS)-1-[(2S)-6-hydroxy-6-methyl-heptan-2-yl]-7a-methyl-2,3,3a,5,6,7-hexa-hydro-1H-inden-4-ylidene]ethylidene]-2-methyl-cyclohexane-1,3-diol.
LMST03020559; 1α,25-dihydroxy-2β-methyl-19-nor-20-epicholecalciferol; (20S)-2R-methyl-19-nor-9,10-seco-5,7-cholestadiene-1R,3R,25-triol; 9547655; Synthesized by combination of Grundmann type 8-ketone with an A-ring synthon obtained from (-)-quinic acid. Compared to 1α,25-dihydroxyvitamin D$_3$, affinity for the porcine intestinal vitamin D receptor is 18%, HL-60 cell differentiation is 600%, intestinal calcium transport is 0%, bone calcium mobilization is 0%. *J Med Chem*. 1998, *41(23)*, 4662-74.

490   25-Hydroxyvitamin D$_3$ hydrate
63283-36-3   C$_{27}$H$_{46}$O$_3$

(6R)-6-[(1R,3aR,4E,7aS)-4-[(2Z)-2-[(5S)-5-hydroxy-2-methylidene-cyclohexylidene]ethylidene]-7a-methyl-2,3,3a,5,6,7-hexahydro-1H-inden-1-yl]-2-methyl-heptan-2-ol hydrate.
Calcifediol; Calderol; Calderol; Calcifediol; D00122; Calcifediol hydrate; 9,10-Secocholesta-5Z,7E,10(19)-

triene-3β,25-diol, monohydrate; 5282368; 6441383; 63283-36-3; The major circulating metabolite of vitamin D$_3$ (calciferol). It is produced in the liver and is the best indicator of the body's vitamin D stores. It is effective in the treatment of rickets and osteomalacia, both in azotemic and non-azotemic patients. Calcifediol also has mineralizing properties. Used therapeutically as a vitamin D replacement. *Chirality*. 1999, *11(9)*, 701-6; *Arch Biochem Biophys*. 1998, *355(1)*, 77-83.

491   1,25-Dihydroxydihydrotachysterol$_3$
65878-49-1   C$_{27}$H$_{46}$O$_3$

(1R,3S,4R,5E)-5-[(2E)-2-[(1R,3aR,7aS)-1-[(2R)-6-hydroxy-6-methyl-heptan-2-yl]-7a-methyl-2,3,3a,5,6,7-hexahydro-1H-inden-4-ylidene]ethylidene]-4-methyl-cyclohexane-1,3-diol.
T3-Ha; T3-Hb; 1α,25-(OH)2DHT(3); 1β,25-(OH)2-DHT(3); 1,25-Dihydroxydihydrotachysterol(3); 9,10α-Secocholesta-5E,7E-diene-1α,3β,25-triol; 6438585; 65878-49-1; *In vivo* metabolite of dihydrotachysterol. Binds only very poorly to the mammalian vitamin D receptor with an affinity much less than that of 1α,25-dihydroxyvitamin D$_3$. *Biochem Pharmacol*. 1992, *43(9)*, 1893-905.

492   1,25-Dihydroxycholesterol
50392-32-0   C$_{27}$H$_{46}$O$_3$

Cholest-5-ene-1α,3β,25-triol.
1α,25-Dihydroxycholesterol; Cholest-5-ene-1α,3β,25-triol; 191124; 50392-32-0.

493   1α,25-Dihydroxy-2α-methyl-19-nor-vitamin D$_3$
9547655   C$_{27}$H$_{46}$O$_3$

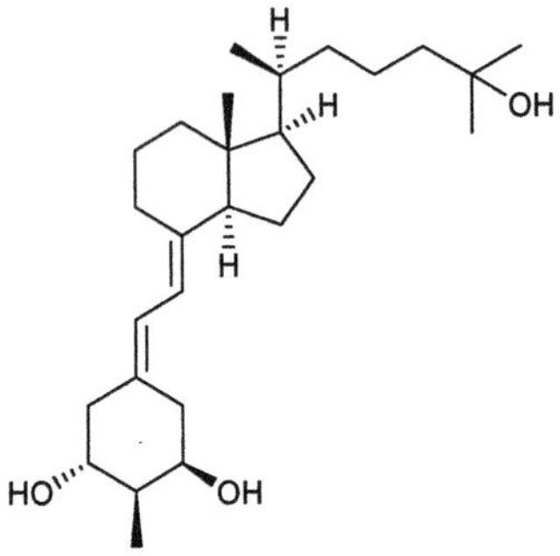

(1R,2S,3R)-5-[(2E)-2-[(1R,3aR,7aS)-1-[(2S)-6-hydroxy-6-methyl-heptan-2-yl]-7a-methyl-2,3,3a,5,6,7-hexahydro-1H-inden-4-ylidene]ethylidene]-2-methyl-cyclohexane-1,3-diol.
LMST03020555; 1α,25-dihydroxy-2α-methyl-19-norcholecalciferolcholecalciferol; 2S-ethyl-19-nor-9,10-seco-5,7E-cholestadiene-1R,3R,25-triol; 9547655. *J Med Chem.* 1998, *41(23)*, 4662-74.

494   1α,25-Dihydroxy-2β-methyl-19-nor-vitamin D₃

9547655 $C_{27}H_{46}O_3$

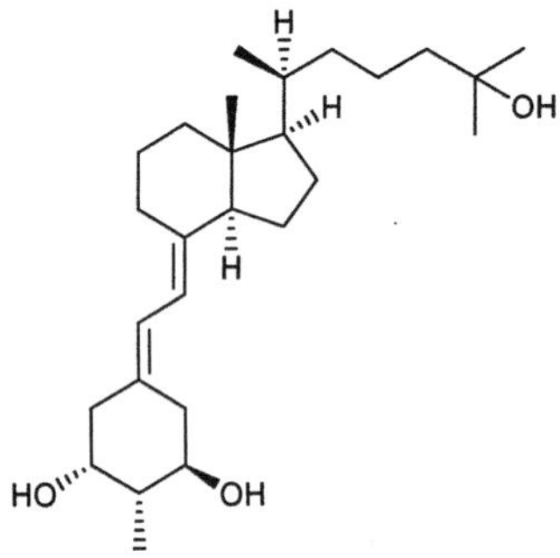

(1R,2R,3R)-5-[(2E)-2-[(1R,3aR,7aS)-1-[(2S)-6-hydroxy-6-methyl-heptan-2-yl]-7a-methyl-2,3,3a,5,6,7-hexahydro-1H-inden-4-ylidene]ethylidene]-2-methyl-cyclohexane-1,3-diol.
LMST03020556; 1α,25-dihydroxy-2β-hydroxymethyl-19-norcholecalciferolcholecalciferol; 2R-hydroxymethyl-19-nor-9,10-seco-5,7E-cholestadiene-1R,3R,25-triol; 9547655; Synthesized by combination of Grundmann type 8-ketone with an A-ring synthon obtained from (-)-quinic acid. Compared to 1α,25-dihydroxyvitamin D₃, affinity for the porcine intestinal vitamin D receptor is 85%, HL-60 cell differentiation is 4%. *J Med Chem.* 1998, *41(23)*, 4662-74.

495   1α,25-Dihydroxy-2α-hydroxymethyl-19-nor-20-epivitamin D₃

9547656 $C_{27}H_{46}O_4$

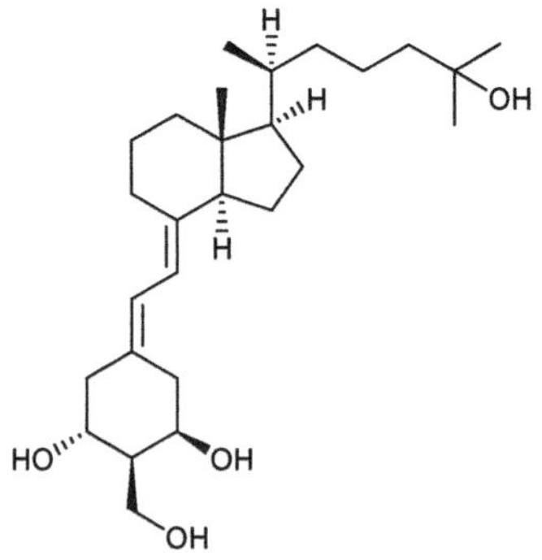

(1R,2R,3R)-5-[(2E)-2-[(1R,3aR,7aS)-1-[(2S)-6-hydroxy-6-methyl-heptan-2-yl]-7a-methyl-2,3,3a,5,6,7-hexahydro-1H-inden-4-ylidene]ethylidene]-2-(hydroxymethyl)cyclohexane-1,3-diol.
LMST03020560; (20S)-2R-hydroxymethyl-19-nor-9,10-seco-5,7E-cholestadiene-1R,3R,25-triol; 1α,25-dihydroxy-2α-hydroxymethyl-19-nor-20-epicholecalciferol; 9547656; Synthesized by combination of Grundmann type 8-ketone with an A-ring synthon obtained from (-)-quinic acid. Compared to 1α,25-dihydroxyvitamin D₃, affinity for the porcine intestinal vitamin D receptor is 88%, HL-60 cell differentiation is 200%, intestinal calcium transport is 0%, bone calcium mobilization is 0%. *J Med Chem.* 1998, *41(23)*, 4662-74.

496   1α,25-Dihydroxy-2β-hydroxymethyl-19-nor-20-epivitamin D₃

9547656 $C_{27}H_{46}O_4$

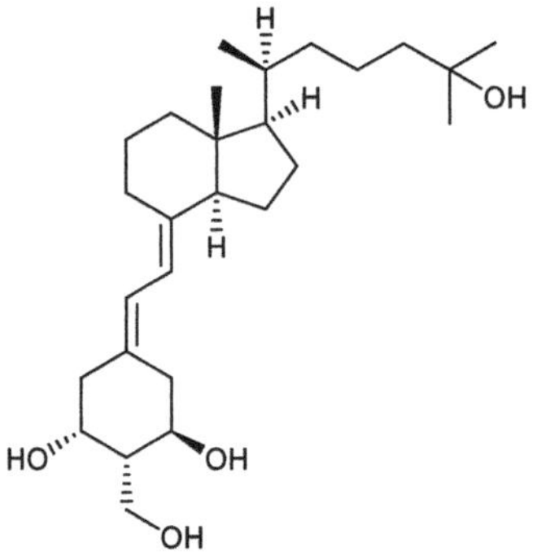

(1R,2S,3R)-5-[(2E)-2-[(1R,3aR,7aS)-1-[(2S)-6-hydroxy-6-methyl-heptan-2-yl]-7a-methyl-2,3,3a,5,6,7-hexahydro-1H-inden-4-ylidene]ethylidene]-2-(hydroxymethyl)cyclohexane-1,3-diol.
LMST03020561; (20S)-2S-hydroxymethyl-19-nor-9,10-seco-5,7E-cholestadiene-1R,3R,25-triol; 1α,25-dihydroxy-2β-hydroxymethyl-19-nor-20-epicholecalciferol; 9547656. *J Med Chem.* 1998, *41(23)*, 4662-74.

497   1α,19,25-Trihydroxy-10R,19-dihydro-vitamin D₃

5283799 $C_{27}H_{46}O_4$

(1R,3S,4R,5Z)-5-[(2E)-2-[(1R,3aR,7aS)-1-[(2R)-6-hydroxy-6-methyl-heptan-2-yl]-7a-methyl-2,3,3a,5,6,7-hexahydro-1H-inden-4-ylidene]ethylidene]-4-(hydroxymethyl)cyclohexane-1,3-diol.
LMST03020598; 9,10R-seco-5Z,7E-cholestadiene-1S,3R,19,25-tetrol; 1α,19,25-trihydroxy-10R,19-dihydrocholecalciferol; 5283799; Prepared by hydroboration of 1α,25-dihydroxyvitamin D_3 followed by oxidation to give a mixture of epimers at C(10); λ_m = 243, 251, 261 nm (EtOH). *J Med Chem.* 1998, *41(23)*, 4662-74.

498   *1α,19,25-Trihydroxy-10S,19-dihydro-vitamin D_3*
5283798 $C_{27}H_{46}O_4$

(1R,3S,4S,5Z)-5-[(2E)-2-[(1R,3aR,7aS)-1-[(2R)-6-hydroxy-6-methyl-heptan-2-yl]-7a-methyl-2,3,3a,5,6,7-hexahydro-1H-inden-4-ylidene]ethylidene]-4-(hydroxymethyl)cyclohexane-1,3-diol.
LMST03020597; (10S)-9,10-seco-5Z,7E-cholestadiene-1S,3R,19,25-tetrol; (10S)-1α,19,25-trihydroxy-10,19-dihydrocholecalciferol; 5283798; Prepared by hydroboration of 1α,25-dihydroxyvitamin D_3 followed by oxidation to give a mixture of epimers at C(10); λ_m = 243, 251.5, 261 nm (EtOH). *J Med Chem.* 1998, *41(23)*, 4662-74.

499   *9,11-Seco-3,6,11-trihydroxycholest-7-en-9-one*
143625-39-2 $C_{27}H_{46}O_4$

(8aS)-4,6-dihydroxy-2-[(2S,3R)-2-(2-hydroxyethyl)-2-methyl-3-[(2R)-6-methylheptan-2-yl]cyclopentyl]-8a-methyl-4,4a,5,6,7,8-hexahydronaphthalen-1-one.
9,11-Seco-3,6,11-trihydroxycholest-7-en-9-one;   9,11-Seco-3β,6α,11-trihydroxy-5α-cholest-7-en-9-one;   9,11-Sthceo; 132570; 143625-39-2; Isolated from marine sponge *Spongia officinalis*. Structure elucidated by analysis of NMR spectral data and compound synthesized. *Steroids.* 1992, *57(7)*, 344-7.

500   *2,3-Secocholestane-2,3-dioic acid*
1178-00-3 $C_{27}H_{46}O_4$

2,3-Seco-5α-cholestan-2,3-dioic acid.
Seco-CDA; 2,3-Secocholestan-2,3-dioic acid; 2,3-Seco-5α-cholestan-2,3-dioic acid; 2,3-Secocholestane-2,3-dioic acid; 3080581; 1178-00-3; Alters the phase transition temperatures in lipid bilayers. May promote endocytic drug delivery. *Biochim Biophys Acta.* 1988, 940(1), 85-92.

501   *26,26,26,27,27,27-Hexafluoro-25-hydroxyvitamin D_2*
9547224 $C_{28}H_{38}F_6O_2$

(E,3S,6S)-6-[(1R,3aR,4E,7aS)-4-[(2Z)-2-[(5S)-5-hydroxy-2-methylidene-cyclohexylidene]ethylidene]-7a-methyl-2,3,3a,5,6,7-hexahydro-1H-inden-1-yl]-1,1,1-trifluoro-3-methyl-2-(trifluoromethyl)hept-4-en-2-ol.
LMST03010008; (5E,7E,22E)-(3S)-26,26,26,27,27,27-hexafluoro-9,10-seco-5,7,10(19),22-ergostatetraene-3,25-diol; 26,26,26,27,27,27-hexafluoro-25-hydroxyvitamin D$_2$; 26,26,26,27,27,27-hexafluoro-25-hydroxyergocalciferol; 9547224; Promotes HL-60 cell differentiation, but less effectively than 1α,25-dihydroxyvitamin D$_3$. *J Biol Chem.* 1987, *262(27)*, 12939-44.

502   *26,26,26,27,27,27-Hexafluoro-25-hydroxyvitamin D_2*

9547225 $C_{28}H_{38}F_6O_2$

(E,6S)-6-[(1R,3aR,4E,7aS)-4-[(2Z)-2-[(5S)-5-hydroxy-2-methylidene-cyclohexylidene]ethylidene]-7a-methyl-2,3,3a,5,6,7-hexahydro-1H-inden-1-yl]-1,1,1-trifluoro-3-methyl-2-(trifluoromethyl)hept-4-en-2-ol.
LMST03010009; 26,26,26,27,27,27-hexafluoro-9,10-seco-5Z,7E,10(19),22E-ergostatetraene-3S,25-diol; 26,26,26,27,27,27-hexafluoro-25-hydroxyergo-calciferol; 9547225; Both C24 epimers synthesized. *Tet Lett.* 1988, *29(2)*, 227-30.

503   *1α-Hydroxy-22-(3-hydroxyphenyl)-23, 24,25,26,27-pentanorvitamin D₃*
9547458 $C_{28}H_{38}O_3$

(1R,3S,5Z)-5-[(2E)-2-[(1R,3aR,7aS)-1-[(2R)-1-(3-hydroxyphenyl)propan-2-yl]-7a-methyl-2,3,3a,5,6,7-hexahydro-1H-inden-4-ylidene]ethylidene]-4-methylidene-cyclohexane-1,3-diol.
LMST03020310; 22-(3-hydroxyphenyl)-23,24-dinor-9,10-seco-5Z,7E,10(19)-cholatriene-1S,3R-diol; 1α-hydroxy-22-(3-hydroxyphenyl)-23,24,25,26,27-penta-norcholecalciferol; 9547458; Synthetic vitamin D₃ derivative. Compared to 1α,25-dihydroxyvitamin D₃, intestinal calcium absorption is 0.28%, bone calcium mobilization is 1.0%, affinity for chick intestinal receptor, HL-60 cell receptor and serum vitamin D binding protein are 28%, 26% and 980%, respectively, inhibition of 1α-hydroxylase activity is 97% and differentiation of HL-60 cells is 60%. Arocalciferols: A New Class of Side-Chain Analogs of 1,25-(OH)2 D3. In Vitamin D Gene Regulation Structure-Function Analysis and Clinical Application (Norman, A.W., Bouillon R. and Thomasset, M., eds), pp 165-166, Walter de Gruyter, Berlin/New York (1991).

504   *1α-Hydroxy-22-(4-hydroxyphenyl)-23, 24,25,26,27-pentanorvitamin D₃*

9547459 $C_{28}H_{38}O_3$

(1R,3S,5Z)-5-[(2E)-2-[(1R,3aR,7aS)-1-[(2R)-1-(4-hydroxyphenyl)propan-2-yl]-7a-methyl-2,3,3a,5,6,7-hexahydro-1H-inden-4-ylidene]ethylidene]-4-methylidene-cyclohexane-1,3-diol.
LMST03020311; 22-(4-hydroxyphenyl)-23,24-dinor-9,10-seco-5Z,7E,10(19)-cholatriene-1S,3R-diol; 1α-hydroxy-22-(4-hydroxyphenyl)-23,24,25,26,27-penta-norcholecalciferol; 9547459; Synthetic vitamin D₃ derivative. Compared to 1α,25-dihydroxyvitamin D₃, intestinal calcium absorption is 0.04%, bone calcium mobilization is 0.08%, affinity for chick intestinal receptor, HL-60 cell receptor and serum vitamin D binding protein are 5%, 8% and 1980%, respectively, inhibition of 1α-hydroxylase activity is 104% and differentiation of HL-60 cells is 15%. Arocalciferols: A New Class of Side-Chain Analogs of 1,25-(OH)2 D3. In Vitamin D Gene Regulation Structure-Function Analysis and Clinical Application (Norman, A.W., Bouillon R. and Thomasset, M., eds), pp 165-166, Walter de Gruyter, Berlin/New York (1991).

505   *22(S)-24-Homo-26,26,26,27,27,27-hexa-fluoro-1α,22,25-trihydroxyvitamin D₃*
107793-48-6 $C_{28}H_{40}F_6O_4$

22(S)-24-Homo-26,26,26,27,27,27-hexafluoro-1,22,25-trihydroxyvitamin D₃; DD 003; 22(S)-24-Homo-26,26,26,27,27,27-hexafluoro-1α,22,25-trihydroxy-vitamin D₃; 107793-48-6. *Cancer Res.* 1994, *54(19)*, 5148-53.

506   *1α,26b-Dihydroxy-22E,23,24E,25,26E, 26a-hexadehydro-26a,26b-dihomo-27-nor-vitamin D₃*
9547460 $C_{28}H_{40}O_3$

(1R,3S,5Z)-5-[(2E)-2-[(1R,3aR,7aS)-1-[(2S,3E,5E,7E)-9-hydroxynona-3,5,7-trien-2-yl]-7a-methyl-2,3,3a,5,6,7-hexahydro-1H-inden-4-ylidene]ethylidene]-4-methylidene-cyclohexane-1,3-diol.
LMST03020312; 1α,26b-dihydroxy-22E,23,24E,25,26E,26a-hexadehydro-26a,26b-dihomo-27-norchole-calciferol; 26a,26b-di-homo-27-nor-9,10-seco-5Z,7E,10(19),22E,24E,26E(26a)-cholestahexaene-1,3,26b-triol; 9547460; Synthesized from protected (5E)-1α-hydroxy-22-oxo-23,24,25,26,27-pentanorvitamin D_3.

Compared to 1α,25-dihydroxyvitamin D_3, inhibition of proliferation of U937 cells is < 80% and induction of differentiation of U937 cells is 100%. Gene Regulation, Structure-Function Analysis and Clinical Application. Proceedings of the Eighth Workshop on Vitamin D Paris, France July 5-10. Synthesis and Biological Activity of 1α-Hydroxylated Vitamin D_3 Analogues with Hydroxylated Side Chains, Multi-Homologated in The 24- or 24,26,27-Positions. (Norman, A. W., Bouillon, R., Thomasset, M., eds), pp159-160, Walter de Gruyter Berlin (1991).

507 *(24RS)-28,28,28-Trifluoro-25-hydroxy-vitamin D_2*
9547226 $C_{28}H_{41}F_3O_2$

(E,6S)-6-[(1R,3aR,4E,7aS)-4-[(2Z)-2-[(5S)-5-hydroxy-2-methylidene-cyclohexylidene]ethylidene]-7a-methyl-2,3,3a,5,6,7-hexahydro-1H-inden-1-yl]-2-methyl-3-(trifluoromethyl)hept-4-en-2-ol.
LMST03010010; (24RS)-28,28,28-trifluoro-25-hydroxy-vitamin D_2; (24RS)-28,28,28-trifluoro-25-hydroxyergo-calciferol; (24RS)-28,28,28-trifluoro-9,10-seco-5Z,7E,10(19),22E-ergostatetraene-3S,25-diol; 9547226; Solid; λ_m = 265 nm. *Tet Lett.* 1988, *29(2)*, 227-30.

508 *26,26,26,27,27-Pentafluoro-1-hydroxy-27-methoxyvitamin D_3*
119839-97-3 $C_{28}H_{41}F_5O_3$

(1R,5Z)-5-[(2E)-2-[(7aS)-1-[(2R)-6-(difluoro-methoxy-methyl)-7,7,7-trifluoro-heptan-2-yl]-7a-methyl-2,3,3a,5,6,7-hexahydro-1H-inden-4-ylidene]ethylidene]-4-methylidene-cyclohexane-1,3-diol.
Pfhmv-D_3; 26,26,26,27,27-pentafluoro-1-hydroxy-27-methoxyvitamin D_3; 6439039; 119839-97-3; *Chem Pharm Bull (Tokyo)*. 1988, *36(10)*, 4144-7.

509 *Calicoferol D*
5283765 $C_{28}H_{42}O_2$

(1R,3aS,7aR)-1-[(E,2S)-3,6-dimethylhept-3-en-2-yl]-4-[2-(5-hydroxy-2-methyl-phenyl)ethyl]-7a-methyl-2,3,3a,4,6,7-hexahydro-1H-inden-5-one.
LMST03020313; Calicoferol D; (8S)-3-hydroxy-22-methyl-9,10-seco-1,3,5(10),22E-cholestatetraen-9-one; 5283765; Isolated with calicoferol C and E from an undescribed gorgonian of the genus *Muricella*. Structure determined by a combination of spectroscopic methods. Exhibits potent antiviral activity, brine-shrimp lethality and potent activity against Herpes simplex viruses I and II and polio virus. *J Nat Prod*. 1995, *58(8)*, 1291-5; *Nat Prod Rep.* 1997, *14(2)*, 259-302.

510 *1α,25-Dihydroxy-22E,23,24E,24a-tetradehydro-24a-homovitamin D_3*
9547461 $C_{28}H_{42}O_3$

(1R,3S,5Z)-5-[(2E)-2-[(1R,3aR,7aS)-1-[(2S,3E,5E)-7-hydroxy-7-methyl-octa-3,5-dien-2-yl]-7a-methyl-2,3,3a,5,6,7-hexahydro-1H-inden-4-ylidene]ethylidene]-4-methylidene-cyclohexane-1,3-diol. LMST03020314; 1α,25-dihydroxy-22E,23,24E,24a-tetradehydro-24a-homocholecalciferol; 24a-homo-9,10-seco-5Z,7E,10(19),22E,24E-cholestapentaene-1S,3R,25-triol; 9547461; Synthesized from protected (5E)-1α-hydroxy-22-oxo-23,24,25,26,27-pentanorvitamin D$_3$. Compared to 1α,25-dihydroxyvitamin D$_3$, inhibition of proliferation of U937 cells is 140%, induction of differentiation of U937 cells is 600% and calciuric activity is 70%. Gene Regulation, Structure-Function Analysis and Clinical Application. Proceedings of the Eighth Workshop on Vitamin D Paris, France July 5-10. Synthesis and Biological Activity of 1α-Hydroxylated Vitamin D$_2$ Analogues with Hydroxylated Side Chains, Multi-Homologated in The 24- or 24,26,27-Positions. (Norman, A. W., Bouillon, R., Thomasset, M., eds), pp159-160, Walter de Gruyter Berlin (1991).

511   *25-Hydroxy-1α-hydroxymethyl-23,23, 24,24-tetradehydrovitamin D$_3$*
9547671   C$_{28}$H$_{42}$O$_3$

(6R)-6-[(1R,3aR,4E,7aS)-4-[(2Z)-2-[(3S,5S)-5-hydroxy-3-(hydroxymethyl)-2-methylidene-cyclohexylidene]ethylidene]-7a-methyl-2,3,3a,5,6,7-hexahydro-1H-inden-1-yl]-2-methyl-hept-3-yn-2-ol. LMST03020576; 1S-hydroxymethyl-9,10-seco-5Z,7E,10(19)-cholestatrien-23-yne-3S,25-diol; 25-hydroxy-1α-hydroxymethyl-23,23,24,24-tetradehydrocholecalciferol; 9547671.

512   *25-Hydroxy-1β-hydroxymethyl-23,23,24,24-tetradehydro-3-epivitamin D$_3$*
9547672   C$_{28}$H$_{42}$O$_3$

(6R)-6-[(1R,3aR,4E,7aS)-4-[(2Z)-2-[(3R,5R)-5-hydroxy-3-(hydroxymethyl)-2-methylidene-cyclohexylidene]-ethylidene]-7a-methyl-2,3,3a,5,6,7-hexahydro-1H-inden-1-yl]-2-methyl-hept-3-yn-2-ol. LMST03020577;   (5Z,7E)-(1R,3R)-1R-hydroxymethyl-9,10-seco-5Z,7E,10(19)-cholestatrien-23-yne-3R,25-diol; 25-hydroxy-1β-hydroxymethyl-23,23,24,24-tetradehydro-3-epicholecalciferol; 9547672.

513   *1-Methyl,1,23-dihydroxy-22-ene-25,26-cyclopropylvitamin D$_3$*
2523   C$_{28}$H$_{42}$O$_3$

5-[2-[1-(5-cyclopropyl-5-hydroxy-pent-3-en-2-yl)-7a-methyl-2,3,3a,5,6,7-hexahydro-1H-inden-4-ylidene]ethylidene]-1-methyl-6-methylidene-cyclohexane-1,3-diol.

514   *1α,22S,25-Trihydroxy-23,24-tetra-dehydro-24a-homo-20-epivitamin D$_3$*
9547462   C$_{28}$H$_{42}$O$_4$

(6S,7R)-7-[(1R,3aR,4E,7aS)-4-[(2Z)-2-[(3S,5R)-3,5-dihydroxy-2-methylidene-cyclohexylidene]ethylidene]-

7a-methyl-2,3,3a,5,6,7-hexahydro-1H-inden-1-yl]-2-methyl-oct-4-yne-2,6-diol.
LMST03020315; 1α,22S,25-trihydroxy-23,24-tetradehydro-24a-homo-20-epicholecalciferol; (20R)-24a-homo-9,10-seco-5Z,7E,10(19)-cholestatrien-23-yne-1S,3R,22S,25-tetrol; 9547462; Synthesized from protected (5E)-1α-hydroxy-22-oxo-23,24,25,26,27-pentanor-20-epivitamin D_3. Compared to 1α,25-dihydroxyvitamin D_3, induction of differentiation of U937 cells is 10%, inhibition of proliferation is 17%, and VDR (rachitic chicken intestinal receptor) binding affinity is 0.007%. Vitamin D. A Pluripotent Steroid Hormone: Structural Studies, Molecular Endocrinology and Clinical Applications. Proceedings of the Ninth Workshop on Vitamin D, Orlando, Florida (USA) May 28-June 2. Chemistry and Biology of Highly Active 22-Oxy Analogs of 20-Epi Calcitriol with very low Binding Affinity to the Vitamin D Receptor. (Norman, A. W., Bouillon, R., Thomasset, M., eds), pp85-86, Walter de Gruyter Berlin New York (1994).

515 **1α-Hydroxy-18-(4-hydroxy-4-methyl-2-pentynyloxy)-23,24,25,26,27-pentanorvitamin D_3**
9547686 $C_{28}H_{42}O_4$

(1R,3S,5Z)-5-[(2E)-2-[(1R,3aR,7aR)-7a-[(4-hydroxy-4-methyl-pent-2-ynoxy)methyl]-1-propan-2-yl-2,3,3a,5,6,7-hexahydro-1H-inden-4-ylidene]ethylidene]-4-methylidene-cyclohexane-1,3-diol.
LMST03020591; 18-(4-hydroxy-4-methyl-2-pentynyloxy)-23,24-dinor-9,10-seco-5Z,7E,10(19)-cholatriene-1S,3R-diol; 1α-hydroxy-18-(4-hydroxy-4-methyl-2-pentynyloxy)-23,24,25,26,27-pentanorcholecalciferol; 9547686.

516 **(24R)-24-Fluoro-1α,25-dihydroxyvitamin D_2**
9547227 $C_{28}H_{43}FO_3$

(1R,3S,5Z)-5-[(2E)-2-[(1R,3aR,7aS)-1-[(E,2S,5R)-5-fluoro-6-hydroxy-5,6-dimethyl-hept-3-en-2-yl]-7a-methyl-2,3,3a,5,6,7-hexahydro-1H-inden-4-ylidene]ethylidene]-4-methylidene-cyclohexane-1,3-diol.
LMST03010011; (24R)-24-fluoro-1α,25-dihydroxyergocalciferol; 9547227; As active as 1α,25-dihydroxyvitamin D_3 and 1α,25-dihydroxyvitamin D_2 in both *in vivo and in vitro* bone calcium mobilization tests; $[\alpha]_D$ = +0.4° (c = 0.01 in EtOH). *Chem Pharm Bull (Tokyo)*. 1992 **40**, 2932.

517 **(24S)-24-Fluoro-1α,25-dihydroxyvitamin D_2**
9547228 $C_{28}H_{43}FO_3$

(1R,3S,5Z)-5-[(2E)-2-[(1R,3aR,7aS)-1-[(E,2S,5S)-5-fluoro-6-hydroxy-5,6-dimethyl-hept-3-en-2-yl]-7a-methyl-2,3,3a,5,6,7-hexahydro-1H-inden-4-ylidene]ethylidene]-4-methylidene-cyclohexane-1,3-diol.
LMST03010012; (24S)-24-fluoro-1α,25-dihydroxyvitamin D_2; (24S)-24-fluoro-1α,25-dihydroxyergocalciferol; (5Z,7E,22E)-(1S,3R,24S)-24-fluoro-9,10-seco-5,7,10(19),22-ergostatetraene-1,3,25-triol; 9547228; Shows weak response in both *in vivo and in vitro* bone calcium mobilization test and stimulates intestinal calcium transport poorly; $[\alpha]_D$ = +1.6° (c = 0.03 in EtOH); λ_m = 265 nm (ε = 18900, EtOH). *Chem Pharm Bull (Tokyo)*. 1992, **40**, 2932.

518 **Biodinamine vitamin D_2**
115586-24-8 $C_{28}H_{44}AsNa_3O_5$

(1S.3Z)-3-[(2E)-2-[(1R,3aR,7aS)-1-[(E,2S,5R)-5,6-di-methylhept-3-en-2-yl]-7a-methyl-2,3,3a,5,6,7-hexa-hydro-1H-inden-4-ylidene]ethylidene]-4-methylidene-cyclohexan-1-ol, mixt. with arsenic acid (H_3AsO_4) sodium salt.

Biodinamine vitamin D_2; 9,10-Secoergosta-5Z,7E, 10(9),22E-tetraen-3β-ol, (3β,5Z,7E,22E), mixt. with arsenic acid (H_3AsO_4) sodium salt; 6449838; 115586-24-8.

519   *24,24-Difluoro-1α,25-dihydroxy-24a-homovitamin D_3*

9547463 $C_{28}H_{44}F_2O_3$

(1R,3S,5Z)-5-[(2E)-2-[(1R,3aR,7aS)-1-[(2R)-5,5-difluoro-7-hydroxy-7-methyl-octan-2-yl]-7a-methyl-2,3,3a,5, 6,7-hexahydro-1H-inden-4-ylidene]ethylidene]-4-methylidene-cyclohexane-1,3-diol.
LMST03020316; 24,24-difluoro-24a-homo-9,10-seco-5Z,7E,10(19)-cholestatrien-1S,3R,25-triol; 24,24-di-fluoro-1α,25-dihydroxy-24a-homocholecalciferol; 9547463; Synthesized from the 1α-hydroxy-5-cholen-24-ol derivative. Compared to 1α,25-dihydroxyvitamin D_3, binding affinity for chick intestinal receptor is 28%, binding affinity for rat serum vitamin D binding protein is 15%. Potency in [45]Ca release from neonatal mouse parietal bones in culture is significantly higher and potency in the formation of osteoclast-like cells is 100 times of that of 1α,25-dihydroxyvitamin D_3. Potency in bone calcium mobilization in vitamin D deficient rats is significantly lower and potency in intestinal Ca transport response *in situ* is similar; $λ_m$ = 265 nm; *Chem Pharm Bull (Tokyo)*. 1987, *35(10)*, 4362-5; *Calcif Tissue Int.* 1993, *53(5)*, 318-23.

520   *24,24-Difluoro-24-homo-1,25-di-hydroxyvitamin D_3*

115540-42-6 $C_{28}H_{44}F_2O_3$

(1R,3S,5E)-5-[(2E)-2-[(3aR,7aS)-1-(6,6-difluoro-7-hydroxy-7-methyl-octan-2-yl)-7a-methyl-2,3,3a,5,6,7-hexahydro-1H-inden-4-ylidene]ethylidene]-4-methyl-idene-cyclohexane-1,3-diol.
24,24-DFHDC; 24aF2-Homo-1,25(OH)2D$_3$; 24,24-Di-fluoro-24-homo-1,25-dihydroxyvitamin D_3; 24a-Homo-24,24-difluoro-1α,25-dihydroxyvitamin D_3; 24,24-Difluoro-24-homo-1α,25-dihydroxycholecalc-iferol; 1,3-Cyclohexanediol, 5-((1-(5,5-difluoro-6-hydroxy-1,6-dimethylheptyl)octahydro-7a-methyl-4H-inden-4-ylidene)ethylidene)-4-methylene-, (1R-(1α(R*).3aβ,4E(1R*,3S*,5Z),7aα))-; 6439248; 115540-42-6; No more active in Ca-regulation than 1α,25-dihydroxy vitamin D_3. Fluoro substitution at 24-position of 1,25(OH)2D$_3$ and the elongation of the side chain of 1,25(OH)2D$_3$ does not intensify Ca-regulating activity. *Chem Pharm Bull (Tokyo)*. 1987, *35(10)*, 4362-5; *Calcif Tissue Int.* 1993, *53(5)*, 318-23.

521   *Ergosterol*

57-87-4   3694 $C_{28}H_{44}O$

(3S,9R,10R,13S,14R,17R)-17-[(E,2S,5R)-5,6-dimethyl hept-3-en-2-yl]-10,13-dimethyl-2,3,4,9,11,12,14,15, 16,17-decahydro-1H-cyclopenta[a]phenanthren-3-ol. AI3-18876; CCRIS 7220; Ergosta-5,7,22-trien-3β-ol; Ergosterin; Ergosterol; HSDB 395; 24-Methylcholesta-5,7,22-trien-3β-ol; 24R-Methylcholesta-5,7,2E-trien-3β-ol; 24α-Methyl-22E-dehydrocholesterol; Provitamin D_2; EINECS 200-352-7; 444679; 57-87-4; 18844-74-1; 37571-51-0; Inhibits the growth of human breast cancer cells *in vitro*; Solid, mp = 170°. *Lipids*. 1995, *30(3)*, 227-30.

522   *Lumisterol*

474-69-1   5623 $C_{28}H_{44}O$

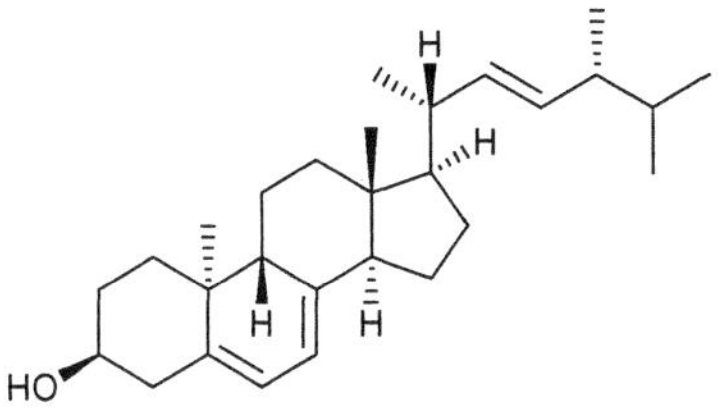

(3S,9S,10S,13S,14R,17R)-17-[(E,2S,5R)-5,6-dimethyl-hept-3-en-2-yl]-10,13-dimethyl-2,3,4,9,11,12,14,15,16,17-decahydro-1H-cyclopenta[a]phenanthren-3-ol.
Lumisterol; Lumisterol 3; 9β,10α-Ergosta-5,7,22-trien-3β-ol; EINECS 207-487-0; 6436872; 474-69-1; Prepared by irradiation of ergosterol. Converted by irradiation to previtamin D$_3$; Needles, mp = 118°; $[\alpha]_D^{19}$ = +191.5°, $[\alpha]_{546}^{19}$ - +235.4° (c = 2, Me$_2$CO); λ_m = 265, 280 nm; soluble in organic solvents, insoluble in H$_2$O. *Quart Rev Biol.*1960, *35(2)*, 162.

523 *Tachysterol$_2$*
115-61-7   9115 $C_{28}H_{44}O$

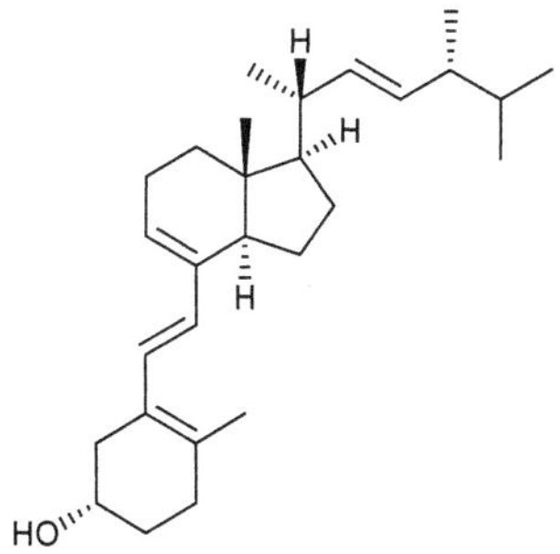

3-[(E)-2-[1-[(E)-5,6-dimethylhept-3-en-2-yl]-7a-methyl-1,2,3,3a,6,7-hexahydroinden-4-yl]ethenyl]-4-methyl-cyclohex-3-en-1-ol.
Tachysterol; 9,10-Secoergosta-5(10),6E,8,22E-tetraen-3R-ol; EINECS 204-096-7; LMST03010016; 6436868; 115-61-7; Formed photochemically from lumisterol or ergosterol; Liquid; $[\alpha]_{546}^{18}$ = -86.3° (petroleum ether); λ_m = 280 nm; soluble in organic solvents, insoluble in H$_2$O. *Physiol Rev.* 1973, *53(2)*, 327-72.

524 *24-Epivitamin D$_2$*
6713937 $C_{28}H_{44}O$

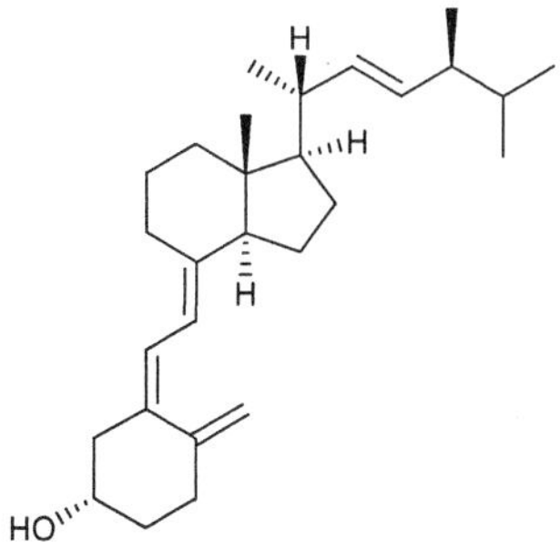

(3R,5Z,7E,20S,22E)-9,10-secoergosta-5,7,10,22-tetraen-3-ol.
Prestwick0_000420; Prestwick1_000420; SPBio_002-319; 6713937.

525 *Tachysterol$_2$*
9547230 $C_{28}H_{44}O$

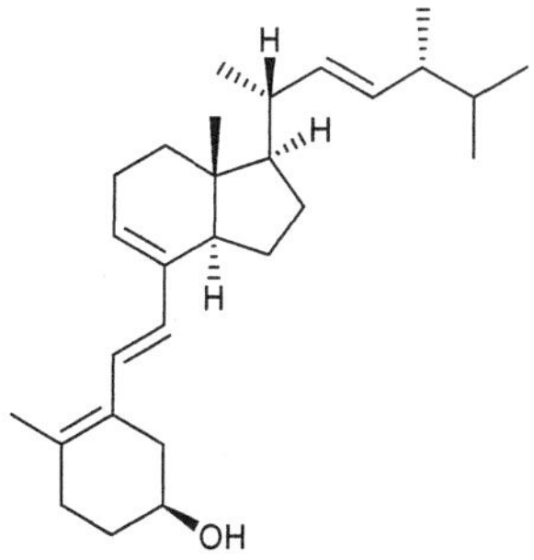

(1S)-3-[(E)-2-[(1R,3aR,7aS)-1-[(E,2S,5R)-5,6-dimethylhept-3-en-2-yl]-7a-methyl-1,2,3,3a,6,7-hexahydroinden-4-yl]ethenyl]-4-methyl-cyclohex-3-en-1-ol.
LMST03010016; (6E,22E)-(3S)-9,10-seco-5(10),6,8,22-ergostatetraen-3-ol; 9547230; Formed from ergosterol by ultraviolet light irradiation; Solid; λ_m = 281 nm (ε = 246000); $[\alpha]_D$ = -70°. *Chem Ber.* 1956, *89*, 2273.

526 *Isotachysterol$_2$*
9547231 $C_{28}H_{44}O$

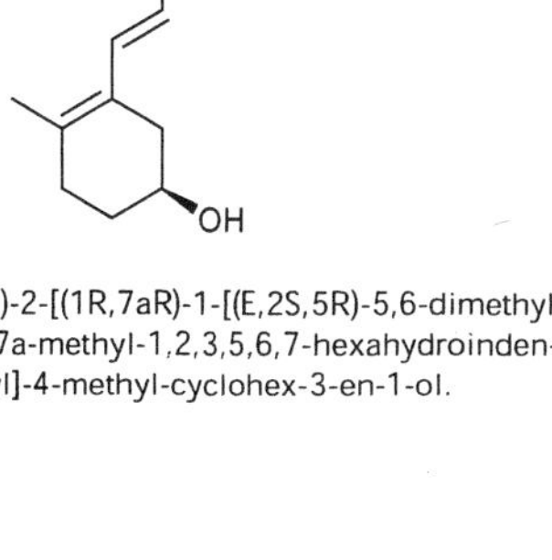

(1S)-3-[(E)-2-[(1R,7aR)-1-[(E,2S,5R)-5,6-dimethylhept-3-en-2-yl]-7a-methyl-1,2,3,5,6,7-hexahydroinden-4-yl]ethenyl]-4-methyl-cyclohex-3-en-1-ol.

LMST03010017;   (6E,22E)-(3S)-9,10-seco-5(10),6,8 (14),22-ergostatetraen-3-ol;   9547231;   treatment of vitamin D_2 in benzene with BF_3-Et_2O; Solid; λ_m = 280, 290, 302 nm, (ε = 31300, 40800, 30650); $[\alpha]_D$ = -71° (CHCl$_3$). **Chem Ber.** 1956, **89**, 2273.

527   *5,6-cis-Isovitamin D_2*
9547232 $C_{28}H_{44}O$

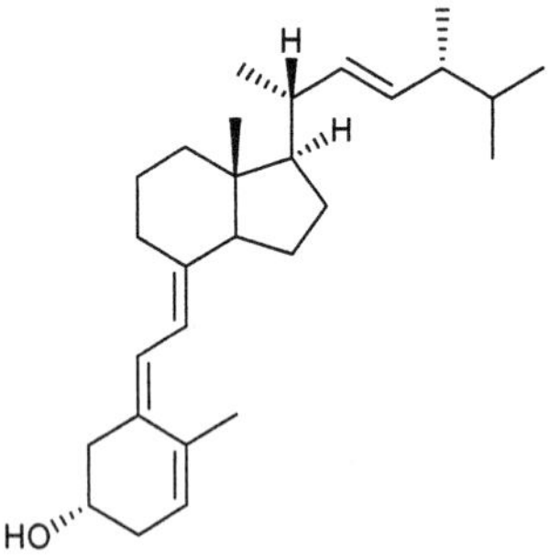

(1S,5Z)-5-[(2E)-2-[(1R,3aR,7aS)-1-[(E,2S,5R)-5,6-dimethylhept-3-en-2-yl]-7a-methyl-2,3,3a,5,6,7-hexahydro-1H-inden-4-ylidene]ethylidene]-4-methyl-cyclohex-3-en-1-ol.
LMST03010018; (5Z,7E,22E)-(3S)-9,10-seco-1(10),5,7, 22-ergostatetraen-3-ol; (5Z)-isovitamin D_2; (5Z)-iso-ergocalciferol; 5,6-cis-isoergocalcifero; 9547232; Prepared from vitamin D_2 by treatment with CaHPO$_4$; Solid; λ_m = 276, 286.5, 298 (log ε = 4.51, 4.58, 4.42, iPrOH). **Yakugaku Zasshi.** 1969, **89(7)**, 919-24.

528   *(5E)-Isoergocalciferol*
9547233 $C_{28}H_{44}O$

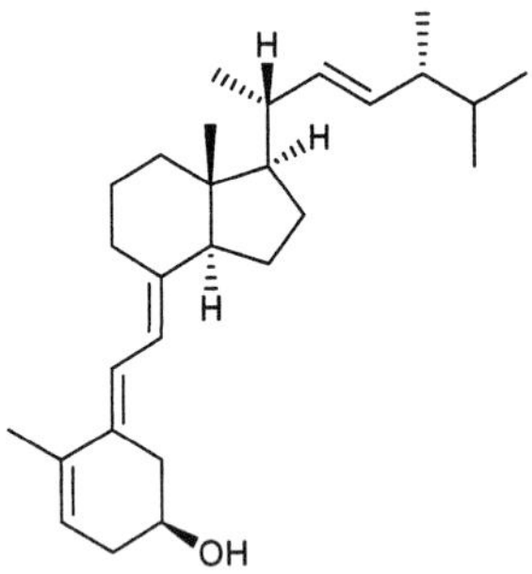

(1S,5E)-5-[(2E)-2-[(1R,3aR,7aS)-1-[(E,2S,5R)-5,6-dimethylhept-3-en-2-yl]-7a-methyl-2,3,3a,5,6,7-hexa-hydro-1H-inden-4-ylidene]ethylidene]-4-methyl-cyclohex-3-en-1-ol.
LMST03010019; (5E)-isovitamin D_2; (5E,7E,22E)-(3S)-9,10-seco-1(10),5,7,22-ergostatetraen 3-ol; 9547233; Formed by treatment of vitamin D_2 with BF_3-etherate in benzene or by pyrolytic dehydration of (5E)-10-hydroxy-10,19-dihydrovitamin D_2; Crystals; mp = 108-110°; λ_m = 278, 288, 300 nm (ε = 33600, 41800, 30500); $[\alpha]_D$ = +108° (CHCl$_3$). **Chem Ber.** 1954, **87**, 1; 1956, **89**, 2273.

529   *Suprasterol II*
42763-68-8 $C_{28}H_{44}O$

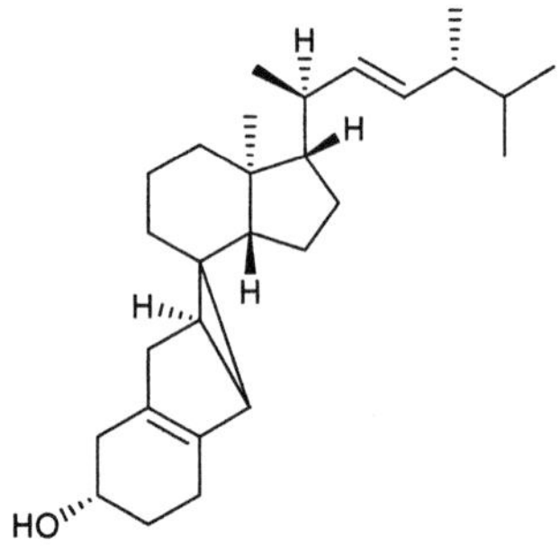

7α,19α:8R,19-Dicyclo-9,10-secoergosta-5(10),22E-dien-2α-ol.
Suprasterol II; Suprasterol$_2$ II; 7α,19α:8R,19-Dicyclo-9,10-secoergosta-5(10),22E-dien-2α-ol; 6443532; 42763-68-8; Formed by irradiation of vitamin D_3. **Quart Rev Biol.** 1960, **35(2)**, 162; **J Nutr Sci Vitaminol.** 1977, **23(4)**, 291-8.

530   *Previtamin D_2*
9547229 $C_{28}H_{44}O$

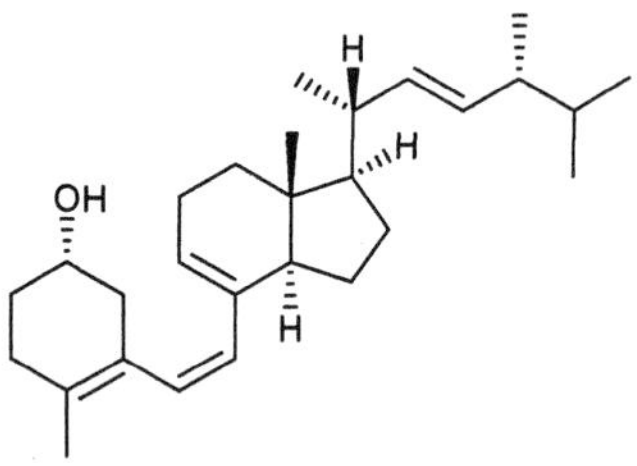

(1S)-3-[(Z)-2-[(1R,3aR,7aS)-1-[(E,2S,5R)-5,6-dimethylhept-3-en-2-yl]-7a-methyl-1,2,3,3a,6,7-hexahydroinden-4-yl]ethenyl]-4-methyl-cyclohex-3-en-1-ol.
LMST03010015; preergocalciferol; (6Z,24E)-(3S)-9,10-seco-5(10),6,8,22-ergostatetraen-3-ol; 9547229; Formed from ergosterol by ultraviolet light irradiation; Crystals; mp = 101-102°; λ_m = 262 nm (ε = 9000, EtOH); $[\alpha]_D$ = +30° (C$_6$H$_6$). **Chem Ber.** 1956, **89**, 2273.

531   *Vitamin D_2*
50-14-6   10156 $C_{28}H_{44}O$

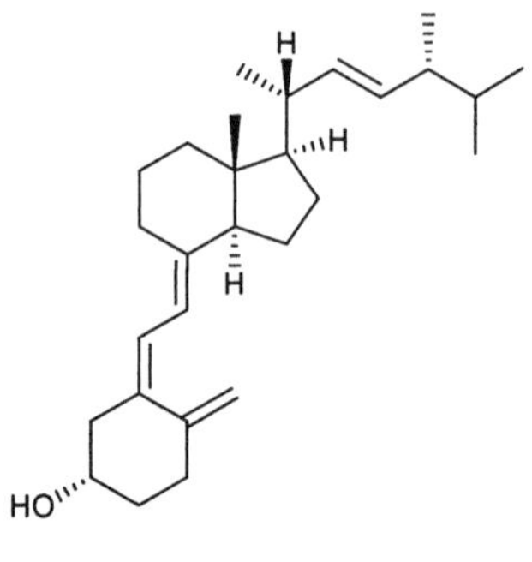

(1S,3Z)-3-[(2E)-2-[(1R,3aR,7aS)-1-[(E,2S,5R)-5,6-dimethylhept-3-en-2-yl]-7a-methyl-2,3,3a,5,6,7-hexahydro-1H-inden-4-ylidene]ethylidene]-4-methylidene-cyclohexan-1-ol.
LMST03010001; Activated ergosterol; D-Arthin; Buco-D; Calciferol; Calciferolum; Calciferon 2; Condocaps; Condol; Crystallina; Cyclohexanol, 4-methylene-3-(2-(tetrahydro-7a-methyl-1-; Daral; Davitamon D; Davitin; De-rat concentrate; Decaps; Dee-Roual; Dee-Ronal; Dee-Osterol; Dee-Ron; Deltalin; Deratol; Detalup; Diactol; Divit urto; Doral; Drisdol; EINECS 200-014-9; Ergocalciferol; Ergocalciferolo; DivK1c_000805; Ergocalciferolum; Ergorone; Ertron; Fortodyl; Geltabs; Geltabs Vitamin D; Haliver; HI-Deratol; HSDB 819; Hyperkil; Infron; Irradiated ergosta-5,7,22-trien-3β-ol; KBioGR_001169; KBioSS_001291; KBio1_000805; KBio2_001291; Metadee; Mina D2; Mulsiferol; Mykostin; NCGC00016213-01; Novovitamin-D; NSC 62792; Oleovitamin D2; Ostelin; Radiostol; Radsterin; Rodine C; Rodinec; 9,10-Secoergosta-5,7,10(19),22-tetraen-3-ol; 9,10-Seco(5Z,7E,22E)-5,7,10(19),22-ergostatetraen-3-ol; Spectrum_000811; Spectrum2_000126; Spectrum3_000417; Spectrum4_000535; SPBio_000172; ZINC01691365; 50-14-6; 3249; 5751; 5280793; 5284358; 5315257; 5353610; 5356615; 5702050; 5702762; 6432478; 6540731; 6604177; 6708745; 6861541; 7067801; Used as an agricultural chemical, a therapeutic agent, rodenticide and antirachitic vitamin. Used to treat rickets and osteomalacia; LD_{lo} (cat orl) = 5 mg/kg, (dog im) = 5 mg/kg, (dog ip) = 10 mg/kg, (dog iv) = 5 mg/kg, (dog orl) = 4 mg/kg, (gpg orl) = 40 mg/kg, LD_{50} (duck orl) > 2 mg/kg, (mus orl) = 23.7 mg/kg, (rat orl) = 10 mg/kg, TD_{lo} (wmn orl) = 12.6 mg/kg. *Chem Ber.* 1956, **89**, 2273; *Biochemistry.* 1969, **8(9)**, 3515-20; 1975, **14(6)**, 1250-6; *J Chem Soc Perkin I*, 1978, 590.

532   ***(22Z)-Vitamin D_2***
6604608 $C_{28}H_{44}O$

(1R,3Z)-3-[(2E)-2-[(1S,3aS,7aR)-1-[(Z,2R,5S)-5,6-dimethylhept-3-en-2-yl]-7a-methyl-2,3,3a,5,6,7-hexahydro-1H-inden-4-ylidene]ethylidene]-4-methylidene-cyclohexan-1-ol.
TNP00097; NCGC00017219-01; 6604608.

533   ***5,6-trans-Vitamin D_2***
6536972 $C_{28}H_{44}O$

(1S,3E)-3-[(2E)-2-[(1R,3aR,7aS)-1-[(E,2S,5R)-5,6-dimethylhept-3-en-2-yl]-7a-methyl-2,3,3a,5,6,7-hexahydro-1H-inden-4-ylidene]ethylidene]-4-methylidene-cyclohexan-1-ol.
LMST03010014; (5E,7E,22E)-(3S)-9,10-seco-5,7,10(19),22-ergostatetraen-3-ol; (5E)-vitamin D_2; (5E)-ergocalciferol; (5E)-ercalciol; 6536972; Prepared by treatment of vitamin D_2 in petroleum ether with I_2; Crystals; mp = 99-101°; λ_m = 273 nm (ε = 2400, EtOH); $[\alpha]_D$ = +223° (C_6H_6). *Chem Ber.* 1956, **89**, 2273.

534   ***Vitamin D_1***
520-91-2

$C_{28}H_{44}O.C_{28}H_{44}O$
(1R,3Z)-3-[(2E)-2-[(1R,3aR,7aS)-1-[(E,2S,5R)-5,6-dimethylhept-3-en-2-yl]-7a-methyl-2,3,3a,5,6,7-hexahydro-1H-inden-4-ylidene]ethylidene]-4-methylidene-cyclohexan-1-ol; (3S,9S,10S,13S,14R,17R)-17-[(E,2S,5R)-5,6-dimethylhept-3-en-2-yl]-10,13-dimethyl-2,3,4,9,11,12,14,15,16,17-decahydro-1H-cyclopenta[a]phenanthren-3-ol.

Vitamin D_1; 9,10-Secoergosta-5,7,10(19),22E-tetraen-3β-ol, compd. with (3β,9β,10α,22E)-Ergosta-5,7,22-trien-3-ol (1:1); 6452575; 11048-08-1; 53776-52-6; EINECS 10077; A 1:1 mixture of lumisterol and vitamin D_2; Solid, mp = 124-125°; dec, 180°; soluble in Me_2CO (3.7 g/100 ml), petroleum ether (2.4 g/100 ml), MeOH (20 g/100 ml); λ_m = 265 nm (ε = 1.56); $[\alpha]_D^{20}$ = +140.5° (Me_2CO c = 0.445), +140.5° (EtOH c = 0.445), +127° ($CHCl_3$ c = 0.445). A vitamin that includes both cholecalciferols and ergocalciferols, which have the common effect of preventing or curing rickets in animals. It can also be viewed as a hormone since it can be formed in skin by action of UV radiation upon the precursors, 7-dehydrocholesterol and ergosterol, and acts on vitamin D receptors to regulate calcium, thus opposing parathyroid hormone.

Ann Rev Biochem. 1976, **45**, 631-6.

535   ***25R-Hydroxy-26-methyl-22E,23-di-***
       dehydrovitamin D₃

5283766 $C_{28}H_{44}O_2$

(E,3R,7S)-7-[(1R,3aR,4E,7aS)-4-[(2Z)-2-[(5S)-5-hydroxy-
2-methylidene-cyclohexylidene]ethylidene]-7a-methyl-
2,3,3a,5,6,7-hexahydro-1H-inden-1-yl]-3-methyl-oct-
5-en-3-ol.
LMST03020317;   26-methyl-9,10-seco-5Z,7E,10(19),
22E-cholestatetraene-3S,25R-diol;   25R-hydroxy-26-
methyl-22E,23-didehydrocholecalciferol;   5293766;
Synthesized from a 25-hydroxy-5,22-cholestadien-26-
oic acid derivative;   λ$_m$ = 264 nm (EtOH). **J Steroid
Biochem**. 1990, **35(6)**, 655-64.

536   ***25S-Hydroxy-26-methyl-22E,23-di-***
       dehydrovitamin D₃

5283767 $C_{28}H_{44}O_2$

(E,3S,7S)-7-[(1R,3aR,4E,7aS)-4-[(2Z)-2-[(5S)-5-hydroxy-
2-methylidene-cyclohexylidene]ethylidene]-7a-methyl-
2,3,3a,5,6,7-hexahydro-1H-inden-1-yl]-3-methyl-oct-
5-en-3-ol.
LMST03020318;   26-methyl-9,10-seco-5Z,7E,10(19),
22E-cholestatetraene-3S,25S-diol;   25S-hydroxy-26-
methyl-22E,23-didehydrocholecalciferol;   5283767;
Synthesized from a 25-hydroxy-5,22-cholestadien-26-
oic acid derivative **via** methylation of the 25,26-
epoxide as key step; λ$_m$ = 264 nm (EtOH). **J Steroid
Biochem**. 1990, **35(6)**, 655-64.

537   ***9,10-Secoergosta-5,7,10(19),22-tetraene-***
       3β,25-diol

5372246 $C_{28}H_{44}O_2$

(E)-6-[(4E)-4-[(2E)-2-(5-hydroxy-2-methylidene-cyclo-
hexylidene)ethylidene]-7a-methyl-2,3,3a,5,6,7-hexa-
hydro-1H-inden-1-yl]-2,3-dimethyl-hept-4-en-2-ol.
9,10-Secoergosta-5,7,10(19),22-tetraene-3β,25-diol;
9,10-Secoergosta-5Z,7E,10(19),22E-tetraene-3,25-diol;
5372246.

538   ***Doxercalciferol***

54573-75-0 $C_{28}H_{44}O_2$

(1R,3S)-5-[2-[(1R,3aR,7aS)-1-[(2S,5R)-5,6-dimethyl-
hept-3-en-2-yl]-7a-methyl-2,3,3a,5,6,7-hexahydro-1H-
inden-4-ylidene]ethylidene]-4-methylidene-cyclo-
hexane-1,3-diol.
1α-Hydroxyergocalciferol; Doxercalciferol; 1-Hydroxy-
ergocalciferol; 1α-Hydroxyvitamin D₂; 1α-Hydroxy-
ergocalciferol; BRN 4716774; Hectorol; TSA 840;
9,10-Secoergosta-5Z,7E,10(19),22E-tetraene-1α,3β-
diol; 4479094; 5281107; 6438325; 54573-75-0; A
metabolite of 1α-hydroxyvitamin D₂ in human liver
cells. A vitamin D prodrug, biologically active in
growth hormone and chloramphenicol acetyltransferase reporter gene expression systems **in vitro**, but
binds poorly to rat vitamin D-binding globulin. Used
systemically to treat osteoporosis. **Blood Purif**. 2002,
20(1), 109-12; **Nephrol Dial Transplant**. 1996, **11
Suppl 3**, 153-7; **Biochem J**. 1995, **310(Pt 1)**, 233-41.

539   ***1α-Hydroxy-3-epivitamin D₂***

6850801 $C_{28}H_{44}O_2$

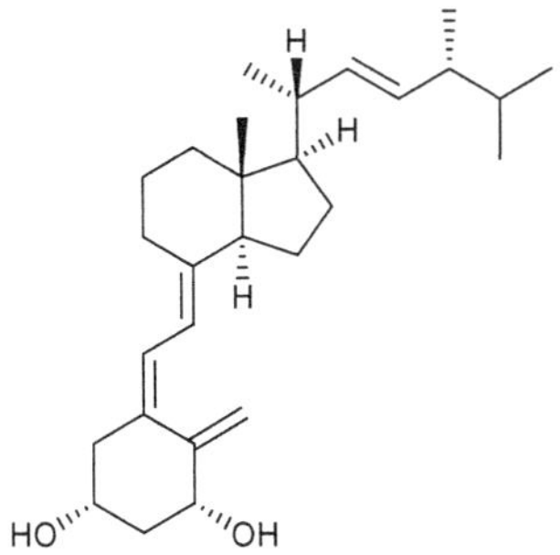

(1S,3R,5Z)-5-[(2E)-2-[(1R,3aR,7aS)-1-[(E,2S,5R)-5,6-dimethylhept-3-en-2-yl]-7a-methyl-2,3,3a,5,6,7-hexahydro-1H-inden-4-ylidene]ethylidene]-4-methylidene-cyclohexane-1,3-diol; LS-144720; 1α-Hydroxy-3-epiergocalciferol; 6850801. *J Org Chem.* 1978, *43(4)*, 574-80.

540   *1α-Hydroxyvitamin D₂*

5281107 $C_{28}H_{44}O_2$

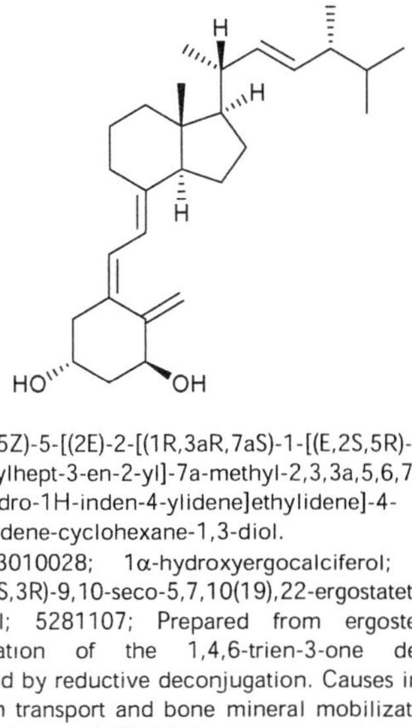

(1R,3S,5Z)-5-[(2E)-2-[(1R,3aR,7aS)-1-[(E,2S,5R)-5,6-dimethylhept-3-en-2-yl]-7a-methyl-2,3,3a,5,6,7-hexahydro-1H-inden-4-ylidene]ethylidene]-4-methylidene-cyclohexane-1,3-diol.
LMST03010028; 1α-hydroxyergocalciferol; (5Z,7E,22E)-(1S,3R)-9,10-seco-5,7,10(19),22-ergostatetraene-1,3-diol; 5281107; Prepared from ergosterol by epoxidation of the 1,4,6-trien-3-one derivative followed by reductive deconjugation. Causes intestinal calcium transport and bone mineral mobilization and differentiation of HL-60 human promyelocytes. One order of magnitude less effective than 1α,25-dihydroxyvitamin D₃. *Science.* 1974, *186(4168)*, 1038-40; *Steroids.* 1977, *30(5)*, 671-7.

541   *20-Hydroxyvitamin D₂*

$C_{28}H_{44}O_2$

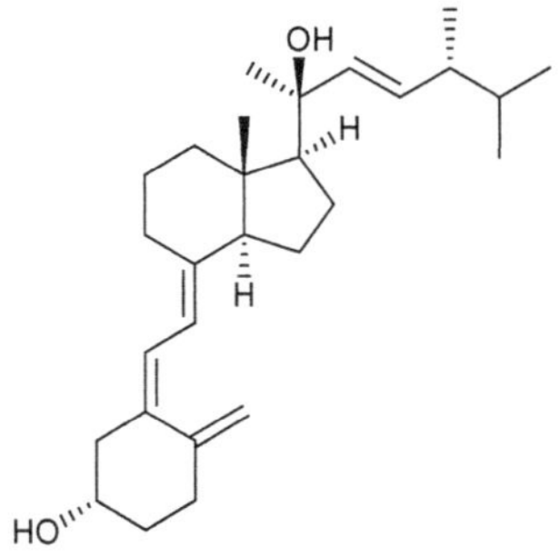

(1S,3Z)-3-[(2E)-2-[(1R,3aR,7aS)-1-[(E,2S,5R)-2-hydroxy-5,6-dimethylhept-3-en-2-yl]-7a-methyl-2,3,3a,5,6,7-hexahydro-1H-inden-4-ylidene]ethylidene]-4-methylidene-cyclohexan-1-ol.
20-hydroxyergocalciferol; Formed in the Cyt450-mediated metabolism of vitamin D₂. Inhibits DNA synthesis in keratinocytes. *FEBS J.* 2006, *273(13)*, 2891-901.

542   *24-Hydroxyvitamin D₂*

58050-56-9 $C_{28}H_{44}O_2$

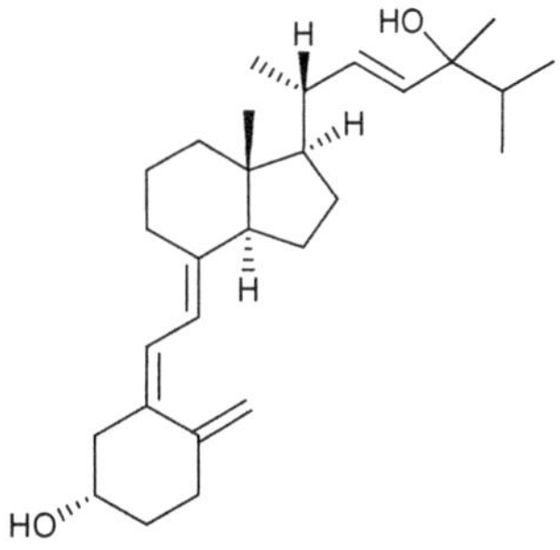

(E,6S)-6-[(1R,3aR,4E,7aS)-4-[(2Z)-2-[(5S)-5-hydroxy-2-methylidene-cyclohexylidene]ethylidene]-7a-methyl-2,3,3a,5,6,7-hexahydro-1H-inden-1-yl]-2,3-dimethyl-hept-4-en-3-ol.
LMST03010029; 24-Hydroxyergocalciferol; 24-hydroxyvitamin D₂; 24-hydroxyergocalciferol; (3S)-9,10-seco-5Z,7E,10(19),22E-ergostatetraene-3β,24-diol; 9,10-Secoergosta-5Z,7E,10(19),22E-tetraene-3β,24-diol; 6443813; 58050-56-9; Isolated from rat blood. Metabolite of Vitamin D₂. Part of a possible activation pathway for vitamin D₂ and vitamin D₃. *Arch Biochem Biophys.* 1980, *202(2)*, 450-7.

543   *25-Hydroxyvitamin D₂*

21343-40-8 $C_{28}H_{44}O_2$

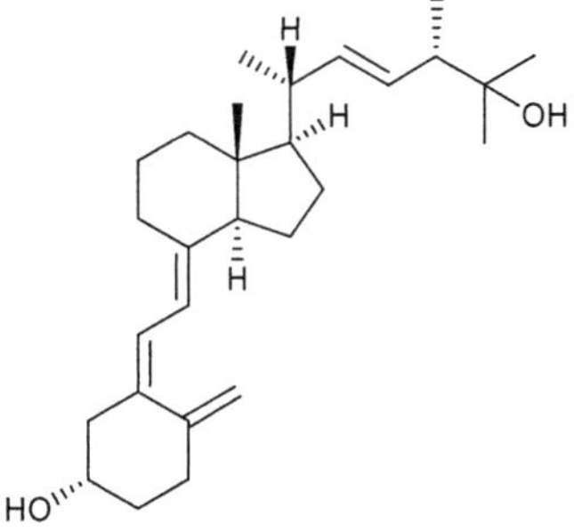

(E,3S,6S)-6-[(1R,3aR,4E,7aS)-4-[(2Z)-2-[(5S)-5-hydroxy-2-methylidene-cyclohexylidene]ethylidene]-7a-methyl-2,3,3a,5,6,7-hexahydro-1H-inden-1-yl]-2,3-dimethyl-hept-4-en-2-ol.
LMST03010030; 25-Hydroxyvitamin D; 25-Hydroxy-ergocalciferol; 25-hydroxyvitamin D₂; 25-hydroxy-ergocalciferol; (5Z,7E,22E)-(3S)-9,10-seco-5Z,7E,10(19),22E-ergostatetraene-3β,25-diol; 9,10-Secoergosta-

5Z,7E,10(19),22E-tetraene-3,25-diol; 5710148; 21343-34-0; 21343-40-8; 29864-49-1; 50351-34-3; Biologically active metabolite of vitamin D_2 which is more active in curing rickets than its parent. The compound is believed to attach to the same receptor as vitamin D_2 and 25-hydroxyvitamin D_3; Solid; λ_m = 265nm (ε = 17950); $[\alpha]_D^{25}$ = +56.8° (c = 0.2, EtOH). *J Org Chem.* 1984, *49(12)*, 2148-51; *J Biol Chem.* 1987, *262(29)*, 14164-71.

544   *1α,25-Dihydroxy-20-epivitamin D_2*
9547235 $C_{28}H_{44}O_2$

(1R,3S,5Z)-5-[(2E)-2-[(1R,3aR,7aS)-1-[(E,2R,5R)-5,6-dimethylhept-3-en-2-yl]-7a-methyl-2,3,3a,5,6,7-hexahydro-1H-inden-4-ylidene]ethylidene]-4-methylidene-cyclohexane-1,3-diol.
LMST03010032; (20S)-9,10-seco-5Z,7E,10(19),22E-ergostatetraene-1S,3R,25-triol; 1α,25-dihydroxy-20-epiergocalciferol; 9547235; Inhibits proliferation of U937 cells, 40 times as effective as 1α,25-dihydroxcyvitamin D_3; induces differentiation of U937 cells, (twice as effective); calciuric effects on normal rats, 80% that of 1α,25-dihydroxyvitamin D_3. Proceedings of the Eighth Workshop on Vitamin D Paris, France July 5-10. The 20-Epi Modification in The Vitamin D Series : Selective Enhancement of 'Non-Classical' Receptpr-Mediated Effects. (Norman, A. W., Bouillon, R., Thomasset, M., eds), pp163-164, Walter de Gruyter Berlin New York.

545   *25-Hydroxy-24-epivitamin D_2*
9547234 $C_{28}H_{44}O_2$

(E,3R,6S)-6-[(1R,3aR,4E,7aS)-4-[(2Z)-2-[(5S)-5-hydroxy-2-methylidene-cyclohexylidene]ethylidene]-7a-methyl-2,3,3a,5,6,7-hexahydro-1H-inden-1-yl]-2,3-dimethyl-hept-4-en-2-ol.

LMST03010031; 24-epi-25-hydroxyvitamin D_2; 24-epi-25-hydroxyergocalciferol; (24R)-9,10-seco-5Z,7E,10(19),22-ergostatetraene-3S,25-diol; 9547234; Promotes HL-60 human promyelocytes differentiation but is ten times less effective than 1α,25-dihydroxyvitamin D_3; $[\alpha]_D$ = +50.7 (c = 0.2, EtOH); λ_m = 265nm (ε = 17300). *J Org Chem.* 1983, *49(12)*, 2148-51; *J Biol. Chem.* 1987, *262(29)*, 14164-71.

546   *25-Hydroxy-24-methyl-23,24-didehydro-vitamin D_3*
9547464 $C_{28}H_{44}O_2$

(E,6R)-6-[(1R,3aR,4E,7aS)-4-[(2Z)-2-[(5S)-5-hydroxy-2-methylidene-cyclohexylidene]ethylidene]-7a-methyl-2,3,3a,5,6,7-hexahydro-1H-inden-1-yl]-2,3-dimethyl-hept-3-en-2-ol.
LMST03020319; 9,10-seco-5Z,7E,10(19),23E-ergosta-tetraene-3S,25-diol; 25-hydroxy-24-methyl-23,24-di-dehydrocholecalciferol; 9547464; *Bioorg Chem.* 1987, *15*, 152-66; *J Biol Chem.* 1987, *262(29)*, 14164-71.

547   *1α,25-Dihydroxy-26-methyl-22E,23-di-dehydrovitamin D_3*
5283768 $C_{28}H_{44}O_3$

(1R,3S,5Z)-5-[(2E)-2-[(1R,3aR,7aS)-1-[(E,2S)-6-hydroxy-6-methyl-oct-3-en-2-yl]-7a-methyl-2,3,3a,5,6,7-hexa-hydro-1H-inden-4-ylidene]ethylidene]-4-methylidene-cyclohexane-1,3-diol.
LMST03020320; LMST03020325; (22E)-1α,25-di-hydroxy-26-methyl-22,23-didehydrovitamin D_3; (22E)-1α,25-dihydroxy-26-methyl-22,23-didehydrochole-calciferol; 26-methyl-9,10-seco-5Z,7E,10(19),22E-cholestatetraene-1S,3R,25-triol; 5283768; Synthesized from a 1α-hydroxylated C(23) steroid precursor by introduction of the desired side chain and the double bond at C(7), and finally, photochemical and thermal

isomerization. Formed by *in vitro* incubation of 24-epi-25-hydroxyvitamin D_2 with chicken kidney homogenate. Affinity for chick intestinal receptor and bone calcium mobilization in hypocalcemic rat are the same as for $1\alpha,25$-dihydroxyvitamin D_3; λ_m = 265 nm. *Biochemistry*. 1986, *25(19)*, 5512-18.

548   *$1\alpha,25S$-Dihydroxy-26-methyl-22E,23-didehydrovitamin D_3*

5283770

$C_{28}H_{44}O_3$

(1R,3S,5Z)-5-[(2E)-2-[(1R,3aR,7aS)-1-[(E,2S,6S)-6-hydroxy-6-methyl-oct-3-en-2-yl]-7a-methyl-2,3,3a,5,6,7-hexahydro-1H-inden-4-ylidene]ethylidene]-4-methylidene-cyclohexane-1,3-diol.
LMST03020324; $1\alpha,25S$-dihydroxy-26-methyl-22E,23-didehydrovitamin D_3; $1\alpha,25S$-dihydroxy-26-methyl-22E,23-didehydrocholecalciferol; 26-methyl-9,10-seco-5Z,7E,10(19),22E-cholestatetraene-1S,3R,25S-triol; 5283770; Synthesized from 25S-hydroxy-26-methyl-22E,23-didehydrovitamin D_3 by enzymatic 1α-hydroxylation using vitamin D-deficient chick kidney homogenates. Compared to $1\alpha,25$-dihydroxyvitamin D_3, affinity for chick intestinal vitamin D receptor is 150%, intestinal calcium transport is 210%, bone calcium mobilization is 160%. Induction of differentiation (NBT reduction) of HL-60 cells is compared with the effect of $1\alpha,25$-dihydroxyvitamin D_3 (value in parentheses) : 37.6% (18.1%) at 10^{-9} M and 76.5% (56.4%) at 10^{-8} M. *J Steroid Biochem*. 1990, *35(6)*, 655-64; *Arch Biochem Biophys*. 1990, *276(2)*, 310-16.

549   *$1\alpha,25$-Dihydroxy-18-methylidenevitamin D_3*

9547465

$C_{28}H_{44}O_3$

(1R,3S,5Z)-5-[(2E)-2-[(1R,3aR,7aR)-7a-ethenyl-1-[(2R)-6-hydroxy-6-methyl-heptan-2-yl]-2,3,3a,5,6,7-hexahydro-1H-inden-4-ylidene]ethylidene]-4-methylidene-cyclohexane-1,3-diol.
LMST03020321; 18-methylidene-9,10-seco-5Z,7E,10(19)-cholestatriene-1S,3R,25-triol; $1\alpha,25$-dihydroxy-18-methylidenecholecalciferol; 9547465; Synthesized from Inhoffen-Lythgoe diol (CD-ring plus 20-22 side chain) and A-ring enyne *via* Pd-catalyzed coupling.
Compared to $1\alpha,25$-dihydroxyvitamin D_3, affinity for calf thymus receptor and vitamin D binding protein are 25% and 100%, differentiation of HL-60 cells is 50%, intestinal calcium transport activity, evaluated using Caco-2 intestinal cancer cell line, is much less than that of $1\alpha,25$-dihydroxyvitamin D_3. *Bioorg Med Chem Lett*. 1993, *3*, 1855-8

550   *$1\alpha,25$-Dihydroxy-22E,23-didehydro-24a-homovitamin D_3*

9547466

$C_{28}H_{44}O_3$

(1R,3S,5Z)-5-[(2E)-2-[(1R,3aR,7aS)-1-[(E,2S)-7-hydroxy-7-methyl-oct-3-en-2-yl]-7a-methyl-2,3,3a,5,6,7-hexahydro-1H-inden-4-ylidene]ethylidene]-4-methylidene-cyclohexane-1,3-diol.
LMST03020322; $1\alpha,25$-dihydroxy-22E,23-didehydro-24a-homocholecalciferol; 24a-homo-9,10-seco-5Z,7E,10(19),22E-cholestatetraene-1S,3R,25-triol; 9547466; λ_m = 265 nm (EtOH). *Chem Pharm Bull (Tokyo)*. 1986, *34(11)*, 4508-15.

551   *25-Hydroxy-1α-hydroxymethyl-16,17-didehydrovitamin D_3*

9547673

$C_{28}H_{44}O_3$

(6R)-6-[(3aR,4E,7aS)-4-[(2Z)-2-[(3S,5S)-5-hydroxy-3-(hydroxymethyl)-2-methylidene-cyclohexylidene]-ethylidene]-7a-methyl-3a,5,6,7-tetrahydro-3H-inden-1-

yl]-2-methyl-heptan-2-ol.
LMST03020578; 1S-hydroxymethyl-9,10-seco-5Z,7E,
10(19),16-cholestatetraene-3S,25-diol; 25-hydroxy-1α-
hydroxymethyl-16,17-didehydrocholecalciferol;
9547673.

552   25-Hydroxy-1β-hydroxymethyl-16,17-di-
dehydro-3-epivitamin D₃
9547674 $C_{28}H_{44}O_3$

(6R)-6-[(3aR,4E,7aS)-4-[(2Z)-2-[(3R,5R)-5-hydroxy-3-
(hydroxymethyl)-2-methylidene-cyclohexylidene]-
ethylidene]-7a-methyl-3a,5,6,7-tetrahydro-3H-inden-1-
yl]-2-methyl-heptan-2-ol.
LMST03020579; 25-hydroxy-1β-hydroxymethyl-16,17-
didehydro-3-epicholecalciferol;   1R-hydroxymethyl-
9,10-seco-5Z,7E,10(19),16-cholestatetraene-3R,25-
diol; 9547674.

553   1,25-Dihydroxyvitamin D₂
55248-15-2 $C_{28}H_{44}O_3$

(1R,5Z)-5-[(2E)-2-[(1R,3aR,7aS)-1-[(E,2S,5S)-6-hydroxy-
5,6-dimethyl-hept-3-en-2-yl]-7a-methyl-2,3,3a,5,6,7-
hexahydro-1H-inden-4-ylidene]ethylidene]-4-methyl-
idene-cyclohexane-1,3-diol.
LMST03010041; 1,25-Dihydroxyergocalciferol; 1,25-
(OH)2D2; 1,25-Dihydroxyvitamin D₂; 1α,25-
Dihydroxyvitamin D₂; 9,10-Segoergosta-5Z,7E,10(19),
22E-tetraene-1,3β,25-triol;   6437855;   55248-15-2;
Induces leukemia cell differentiation. In intestinal
calcium transport, mineralization of bone, mobilization
of bone calcium, and elevation of plasma inorganic
phosphorus of rachitic rats it has activity similar to that
of 1,25-dihydroxyvitamin D₃. *Biochim Biophys Acta.*
2006, *1761(2)*, 221-34; 1991, *1091(2)*, 188-92; *Am J
Physiol.* 1988, *254(4 Pt 1)*, E402-6.

554   1α,25-Dihydroxyvitamin D₂
9547243 $C_{28}H_{44}O_3$

(1R,3S,5Z)-5-[(2E)-2-[(1R,3aR,7aS)-1-[(E,2S,5S)-6-
hydroxy-5,6-dimethyl-hept-3-en-2-yl]-7a-methyl-
2,3,3a,5,6,7-hexahydro-1H-inden-4-ylidene]ethyl-
idene]-4-methylidene-cyclohexane-1,3-diol.
LMST03010040;   1,25(OH)2D2;   1α,25-dihydroxy-
ergocalciferol;   9,10-seco-5Z,7E,10(19),22E-ergosta-
tetraene-1S,3R,25-triol; 9547243; Synthesized by SeO₂
oxidation of the 3,5-cyclovitamin D derivative. Has
antirachitic activity. Causes differentiation in HL-60
human promyelocytes;   λₘ = 265 nm (EtOH).
Biochemistry 1975, *14(6)*, 1250-6.

555   1α,25-Dihydroxy-24-epivitamin D₂
9547244 $C_{28}H_{44}O_3$

(1R,3S,5Z)-5-[(2E)-2-[(1R,3aR,7aS)-1-[(E,2S,5R)-6-
hydroxy-5,6-dimethyl-hept-3-en-2-yl]-7a-methyl-
2,3,3a,5,6,7-hexahydro-1H-inden-4-ylidene]ethyl-
idene]-4-methylidene-cyclohexane-1,3-diol.
LMST03010041;   (24R)-9,10-seco-5Z,7E,10(19),22E-
ergostatetraene-1S,3R,25-triol;   1α,25-dihydroxy-24-
epivitamin D₂; 1α,25-dihydroxy-24-epiergocalciferol;
9547244; Synthesized by SeO₂ oxidation of the 3,5-
cyclovitamin D derivative. Has antirachitic activity.
Causes differentiation in HL-60 human promyelocytes;
λₘ = 265.5 nm (EtOH). *Bioorg Chem.* 1985, *13(2)*,
158-69.

556   5(E)-1α,25-Dihydroxyvitamin D₂
9547245 $C_{28}H_{44}O_3$

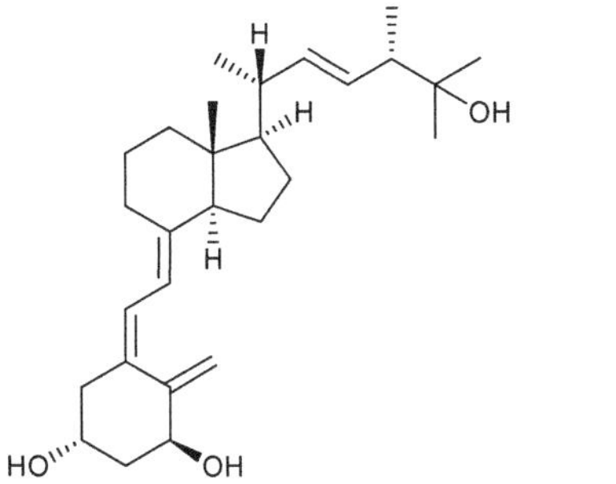

(1R,3S,5E)-5-[(2E)-2-[(1R,3aR,7aS)-1-[(E,2S,5S)-6-hydroxy-5,6-dimethyl-hept-3-en-2-yl]-7a-methyl-2,3,3a,5,6,7-hexahydro-1H-inden-4-ylidene]ethylidene]-4-methylidene-cyclohexane-1,3-diol.
LMST03010042; 9,10-seco-5E,7E,10(19),22E-ergostatetraene-1S,3R,25-triol; 5(E)-1α,25-dihydroxyergocalciferol; 9547245; Synthesized by SeO_2 oxidation of the 3,5-cyclovitamin D derivative. Has antirachitic activity. Causes differentiation in HL-60 human promyelocytes; λ_m = 273.5 nm (EtOH). *Bioorg Chem*. 1985, *13(2)*, 158-69.

557   *5(E)-1α,25-Dihydroxy-24-epivitamin D₂;*
9547246 $C_{28}H_{44}O_3$

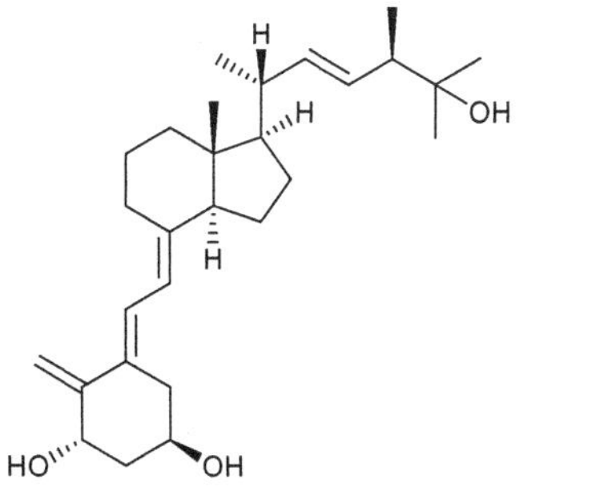

(1R,3S,5E)-5-[(2E)-2-[(1R,3aR,7aS)-1-[(E,2S,5R)-6-hydroxy-5,6-dimethyl-hept-3-en-2-yl]-7a-methyl-2,3,3a,5,6,7-hexahydro-1H-inden-4-ylidene]ethylidene]-4-methylidene-cyclohexane-1,3-diol.
LMST03010043; (24R)-9,10-seco-5E,7E,10(19),22E-ergostatetraene-1S,3R,25-triol; 5(E)-1α,25-dihydroxy-24-epiergocalciferol; 9547246; Synthesized by SeO_2 oxidation of the 3,5-cyclovitamin D derivative. Has antirachitic activity. Causes differentiation in HL-60 human promyelocytes; λ_m = 273.5 nm (EtOH). *Bioorg Chem*. 1985, *13(2)*, 158-69.

558   *1β,25-Dihydroxyvitamin D₂*
9547247 $C_{28}H_{44}O_3$

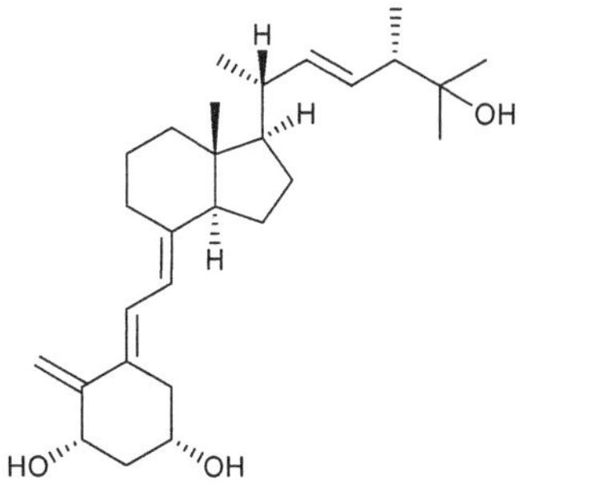

(1R,3R,5Z)-5-[(2E)-2-[(1R,3aR,7aS)-1-[(E,2S,5S)-6-hydroxy-5,6-dimethyl-hept-3-en-2-yl]-7a-methyl-2,3,3a,5,6,7-hexahydro-1H-inden-4-ylidene]ethylidene]-4-methylidene-cyclohexane-1,3-diol.
LMST03010044; 1β,25-dihydroxyergocalciferol; 9,10-seco-5Z,7E,10(19),22E-ergostatetraene-1R,3R,25-triol; 9547247; Synthesized by SeO_2 oxidation of the 3,5-cyclovitamin D derivative. Has antirachitic activity. Causes differentiation in HL-60 human promyelocytes; λ_m = 263.5 nm (EtOH). *Bioorg Chem*. 1985, *13(2)*, 158-69.

559   *1β,25-Dihydroxy-24-epivitamin D₂*
9547248 $C_{28}H_{44}O_3$

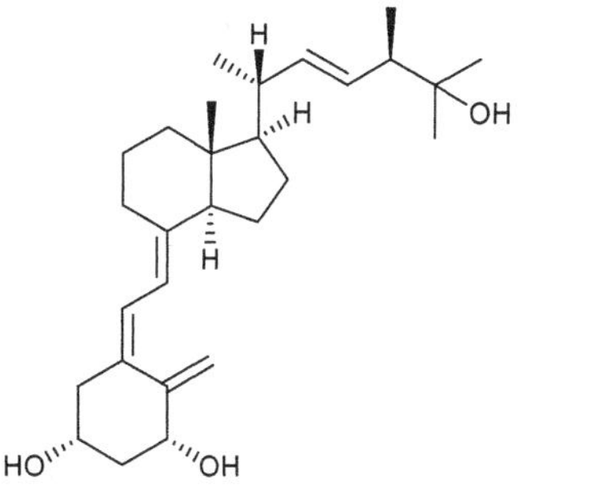

(1R,3R,5Z)-5-[(2E)-2-[(1R,3aR,7aS)-1-[(E,2S,5R)-6-hydroxy-5,6-dimethyl-hept-3-en-2-yl]-7a-methyl-2,3,3a,5,6,7-hexahydro-1H-inden-4-ylidene]ethylidene]-4-methylidene-cyclohexane-1,3-diol.
LMST03010045; (24R)-9,10-seco-5Z,7E,10(19),22E-ergostatetraene-1R,3R,25-triol; 1β,25-dihydroxy-24-epiergocalciferol; 9547248; Synthesized by SeO_2 oxidation of the 3,5-cyclovitamin D derivative. Has antirachitic activity. Causes differentiation in HL-60 human promyelocytes; λ_m = 263.5 nm (EtOH). *Bioorg Chem*. 1985, *13(2)*, 158-69.

560   *5(E)-1β,25-Dihydroxyvitamin D₂*
9547249 $C_{28}H_{44}O_3$

135

(1R,3R,5E)-5-[(2E)-2-[(1R,3aR,7aS)-1-[(E,2S,5S)-6-hydroxy-5,6-dimethyl-hept-3-en-2-yl]-7a-methyl-2,3,3a,5,6,7-hexahydro-1H-inden-4-ylidene]ethyl-idene]-4-methylidene-cyclohexane-1,3-diol.
LMST03010046; 1β,25-dihydroxyergocalciferol; 9,10-seco-5E,7E,10(19),22E-ergostatetraene-1R,3R,25-triol; 9547249; Synthesized by SeO₂ oxidation of the 3,5-cyclovitamin D derivative. Has antirachitic activity. Causes differentiation in HL-60 human promyelocytes; λ_m = 270 nm (EtOH). **Bioorg Chem**. 1985, **13(2)**, 158-69.

561   ***5(E)-1β,25-Dihydroxy-24-epivitamin D₂***
9547250   C₂₈H₄₄O₃

(1R,3R,5E)-5-[(2E)-2-[(1R,3aR,7aS)-1-[(E,2S,5R)-6-hydroxy-5,6-dimethyl-hept-3-en-2-yl]-7a-methyl-2,3,3a,5,6,7-hexahydro-1H-inden-4-ylidene]ethyl-idene]-4-methylidene-cyclohexane-1,3-diol.
LMST03010047; (24R)-9,10-seco-5E,7E,10(19),22E-ergostatetraene-1R,3R,25-triol; 5(E)-1β,25-dihydroxy-24-epiergocalciferol; 9547250; Synthesized by SeO₂ oxidation of the 3,5-cyclovitamin D derivative. Has antirachitic activity. Causes differentiation in HL-60 human promyelocytes; λ_m = 270 nm (EtOH). **Bioorg Chem**. 1985, **13(2)**, 158-69.

562   ***Carcinomedin***
108387-51-5   C₂₈H₄₄O₃

(3Z,5S)-3-[(2E)-2-[(1S,3aR,7aS)-1-[(2R,5S)-6-hydroxy-5,6-dimethyl-heptan-2-yl]-7a-methyl-2,3,3a,5,6,7-hexahydro-1H-inden-4-ylidene]ethylidene]-5-hydroxy-2-methylidene-cyclohexan-1-one.
Carcinomedin; 1-Keto-24-methylcalcifediol; 1-Ceto-24-methyl-calcifediol; 1-Keto-24-methyl-25-hydroxy-cholecalciferol; 9,10-Secoergosta-5Z,7E,10(19)-trien-1-one, 3β,25-dihydroxy-; 6439174; 108387-51-5; Serum levels of carcinomedin, an abnormal cholecalciferol derivative, may be useful as an indicator of the progression of cancer. **Int J Vitam Nutr Res**. 1987, **57(4)**, 19-23; 367-73; 1988, **58(4)**, 381-6.

563   ***1α,24S-Dihydroxyvitamin D₂***
9547261   C₂₈H₄₄O₃

(1R,3S,5Z)-5-[(2E)-2-[(1R,3aR,7aS)-5-hydroxy-5,6-dimethyl-hept-3-en-2-yl]-7a-methyl-2,3,3a,5,6,7-hexahydro-1H-inden-4-ylidene]ethyl-idene]-4-methylidene-cyclohexane-1,3-diol.
LMST03010062; 1α,24S-dihydroxyergocalciferol; 9,10-seco-5Z,7E,10(19),22E-ergostatetraene-1S,3R,24S-triol; 9547261. **Bioorg Med Chem Lett**. 1994, **4(12)**, 1523-6.

564   ***24,25-Dihydroxyvitamin D₂***
71183-99-8   C₂₈H₄₄O₃

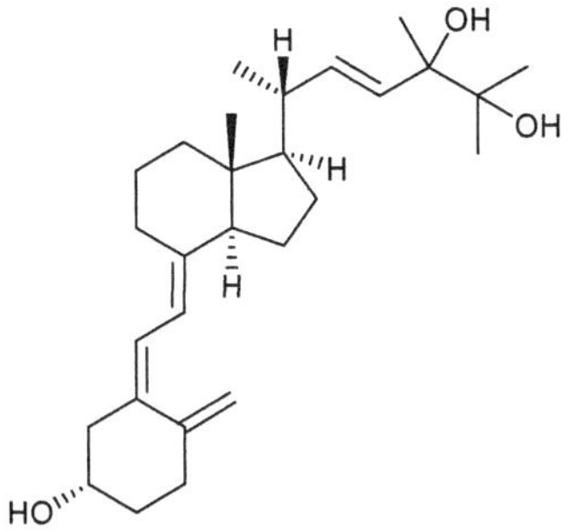

(E,3R,6S)-6-[(1R,3aR,4E,7aS)-4-[(2Z)-2-[(5R)-5-hydroxy-2-methylidene-cyclohexylidene]ethylidene]-7a-methyl-2,3,3a,5,6,7-hexahydro-1H-inden-1-yl]-2,3-dimethyl-hept-4-ene-2,3-diol.

24,25-Dihydroxyvitamin D$_2$; 24,25-Dihydroxyergo-calciferol; 9,10-Secoergosta-5Z,7E,10(19),22E-tetraene-3β,24,25-triol; LMST03010048; 6438393; 71183-99-8; Normally present in human serum, but markedly decreased in patients undergoing anticonvulsant therapy. *Arch Biochem Biophys.* 1980, *202(2),* 450-7; *Ann Clin Res.* 1981, *13(1),* 26-33; *Chem Pharm Bull (Tokyo).* 1987, *35(3),* 970-9.

565   *24,25-Dihydroxy-24-epivitamin D$_2$*
5314030   C$_{28}$H$_{44}$O$_3$

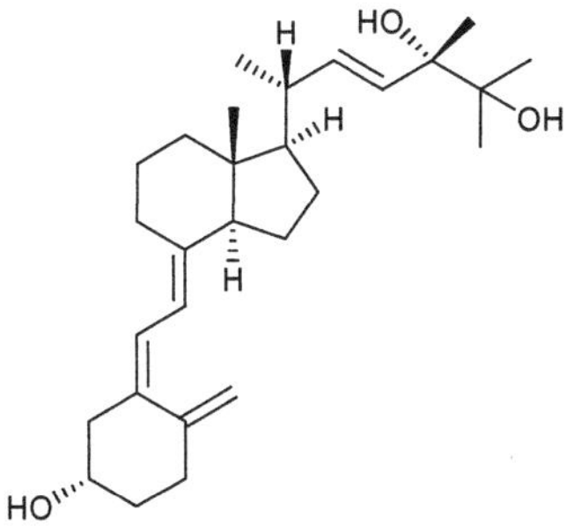

(E,3S,6S)-6-[(1R,3aR,4E,7aS)-4-[(2Z)-2-[(5S)-5-hydroxy-2-methylidene-cyclohexylidene]ethylidene]-7a-methyl-2,3,3a,5,6,7-hexahydro-1H-inden-1-yl]-2,3-dimethyl-hept-4-ene-2,3-diol.

LMST03010049; 24,25-dihydroxy-24-epivitamin D$_2$; 24,25-dihydroxy-24-epiergocalciferol; 24S-methyl-9,10-seco-5Z,7E,10(19),22E-cholestatetraene-3S,24S,25-triol; 5314030; λ$_m$ = 265 nm (EtOH). *Chem Pharm Bull (Tokyo).* 1987, *35(3),* 970-9.

566   *24R,25-Dihydroxyvitamin D$_2$*
9547251   C$_{28}$H$_{44}$O$_3$

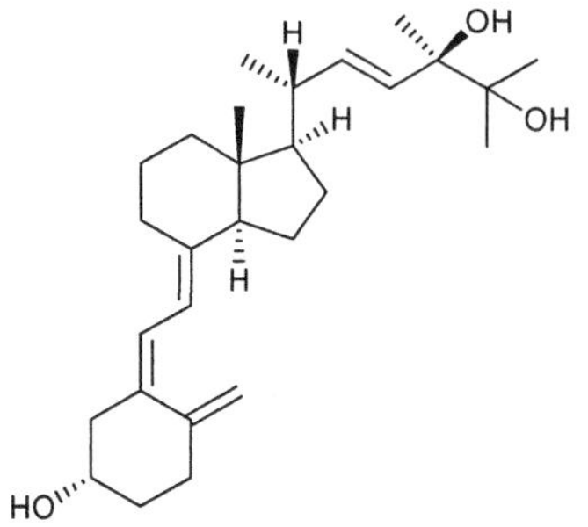

(E,3R,6S)-6-[(1R,3aR,4E,7aS)-4-[(2Z)-2-[(5S)-5-hydroxy-2-methylidene-cyclohexylidene]ethylidene]-7a-methyl-2,3,3a,5,6,7-hexahydro-1H-inden-1-yl]-2,3-dimethyl-hept-4-ene-2,3-diol.

LMST03010048; 24R,25-dihydroxyergocalciferol; 9,10-seco-5Z,7E,10(19),22E-ergostatetraene-3S,24R,25-triol; 9547251; Asymmetric synthesis and biological formation reported; λ$_m$ = 265 nm (EtOH). *Arch Biochem Biophys* 1990, *202,* 450; *Biochem.* 1979, *18*(6), 1094-1101; *Chem Pharm Bull (Tokyo).* 1987, *35(3),* 970-9.

567   *22-Dehydro-1,25-dihydroxy-24-homo-vitamin D$_3$*
103732-08-7   C$_{28}$H$_{44}$O$_3$

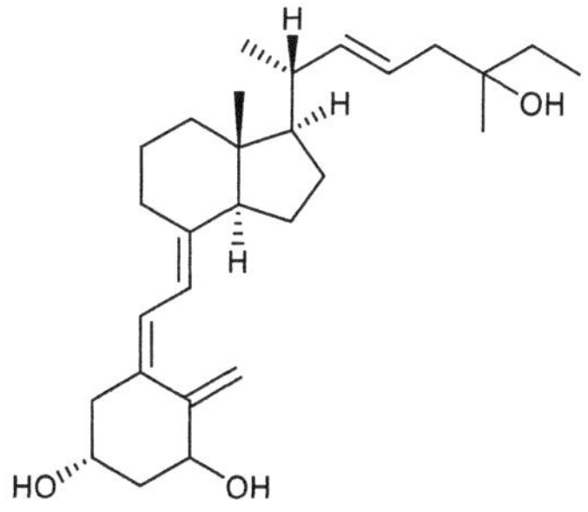

(1S,3R,5Z)-5-[(2E)-2-[(1R,3aR,7aS)-1-[(E,2S)-6-hydroxy-6-methyl-oct-3-en-2-yl]-7a-methyl-2,3,3a,5,6,7-hexa-hydro-1H-inden-4-ylidene]ethylidene]-4-methylidene-cyclohexane-1,3-diol.

22-Dehydro-1,25-dihydroxy-24-homovitamin D$_3$; 1,25(OH)2-22-Dehydro-monohomo-vitamin D$_3$; 1,25-Dihydroxy-22-dehydro-26-homovitamin D$_3$; Dhdh-vitamin D$_3$; Dhmh-calcitriol; 27-Nor-9,10-secocholesta-5Z,7E,10(19),22E-tetraene-1α,3β,25-triol, 25-ethyl-; LMST03020320; 6438976 103732-08-7; Inhibits IL-6 and IL-8 production in human fibroblast cell lines. *Arch Biochem Biophys.* 1990, *276(2),* 310-6.

568   *1α,25-Dihydroxy-25-dehydro-26-methylvitamin D$_3$*
123000-44-2   C$_{28}$H$_{44}$O$_3$

(1S,3R,5Z)-5-[(2E)-2-[(3aR,7aS)-1-[(E,2S,6S)-6-hydroxy-
6-methyl-oct-3-en-2-yl]-7a-methyl-2,3,3a,5,6,7-hexa-
hydro-1H-inden-4-ylidene]ethylidene]-4-methylidene-
cyclohexane-1,3-diol.
DDMVD; LMST03020323; 25-Dehydro-1,25-di-
hydroxy-26-methyl-vitamin D_3; 22E-Dehydro-1α,25R-
dihydroxy-26-methylvitamin D_3; 22E-Dehydro-1α,25S-
dihydroxy-26-methylvitamin D_3; 25-Ethyl-27-nor-9,10-
secocholesta-5Z,7E,10(19),22E-tetraene-1α,3β,25S-
triol; 6439141; 123000-44-2; Slightly more active than
1,25-dihydroxyvitamin D_3 in intestinal calcium
transport and bone calcium mobilization. *J Steroid
Biochem.* 1990, *35(6)*, 655-64.

569   ***24R,26-Dihydroxyvitamin D2***
9547252 $C_{28}H_{44}O_3$

(E,3S,6S)-6-[(1R,3aR,4E,7aS)-4-[(2Z)-2-[(5S)-5-hydroxy-
2-methylidene-cyclohexylidene]ethylidene]-7a-methyl-
2,3,3a,5,6,7-hexahydro-1H-inden-1-yl]-2,3-dimethyl-
hept-4-ene-1,3-diol.
LMST03010050;   24R,26-dihydroxyergocalciferol;
9,10-seco-5Z,7E,10(19),22E-ergostatetraene-
3S,24R,26-triol; 9547252; Displaces [^{3}H]-25-OHD$_3$
from the rat plasma vitamin D binding protein.
Biochemistry 1988, *27(15)*, 5785-90.

570   ***1,24S-Dihydroxyvitamin D2***
124043-51-2 $C_{28}H_{44}O_3$

(1R,5Z)-5-[(2E)-2-[(1R,7aS)-1-[(E,2S,5S)-5-hydroxy-5,6-
dimethyl-hept-3-en-2-yl]-7a-methyl-2,3,3a,5,6,7-hexa-
hydro-1H-inden-4-ylidene]ethylidene]-4-methylidene-
cyclohexane-1,3-diol.
LMST03010062; 1,24-Dihydroxyvitamin D_2; 1,24S-
Dihydroxyvitamin D_2; 9,10-secoergosta-5Z,7E,10(19),
22E-tetraene-1,3,   24ε-triol;   9,10-Secoergosta-5Z,7E,
10(19),22E-tetraene-1α,3β,24ε-triol;   6439163;
124043-51-2; A metabolite of vitamin D_2 with low
calcemic activity *in vivo*, inhibits the growth of
myeloma, breast and prostate cancer cells. Formed
from 1α-hydroxyvitamin D_2 in human liver cells.
Biologically   active   in   growth   hormone   and
chloramphenicol   acetyltransferase   reporter   gene
expression systems *in vitro*, but binds poorly to rat
vitamin D-binding globulin. *Biochem J.* 1995, *310(Pt
1)*, 233-41; *Anticancer Res.* 2005, *25(1α)*, 235-41.

571   ***4,25-Dihydroxyvitamin D2***
                                               $C_{28}H_{44}O_3$

$4,25(OH)_2D_2$; Formed in rats intoxicated with vitamin
D_2. Identified by UV and mass spectroscopy; λ_m = 265
nm (iPrOH). *J Steroid Biochem Mol Biol.* 1999, *71(1-
2)*, 63-70.

572   ***17,20-Dihydroxyvitamin D2***
                                               $C_{28}H_{44}O_3$

Formed in the Cyt450-mediated metabolism of vitamin D₂. Inhibits DNA synthesis in keratinocytes. **FEBS J.** 2006, **273(13)**, 2891-901.

573 *1,25-Dihydroxylumisterol₃*

$C_{28}H_{44}O_3$

9β,10α-Ergosta-5,7,22-trien-1, 3β,25-triol; An agonist for the membrane vitamin D receptor. **FASEB J.** 2002, **16(13)**, 1808-10; **Circ Res.** 2002, **91(1)**, 17-24.

574 *(22Z)-1α,25-Dihydroxy-20-epivitamin D₂*
9547236 $C_{28}H_{44}O_3$

(1R,3S,5Z)-5-[(2E)-2-[(1R,3aR,7aS)-1-[(Z,2R,5R)-6-hydroxy-5,6-dimethyl-hept-3-en-2-yl]-7a-methyl-2,3,3a,5,6,7-hexahydro-1H-inden-4-ylidene]ethylidene]-4-methylidene-cyclohexane-1,3-diol. LMST03010033; (20S)-9,10-seco-5Z,7E,10(19),22Z-ergostatetraene-1S,3R,25-triol; (22Z)-1α,25-dihydroxy-20-epiergocalciferol; 9547236; Compared to 1α,25-dihydroxyvitamin D₃, three times more effective in proliferation of U937 cells, similar induction of differentiation of U937 cells. Proceedings of the Eighth Workshop on Vitamin D Paris, France July 5-10. The 20-Epi Modification in The Vitamin D Series : Selective Enhancement of 'Non-Classical' Receptpr-Mediated Effects. (Norman, A. W., Bouillon, R., Thomasset, M.,

eds), pp163-164, Walter de Gruyter Berlin New York.

575 *(6R)-6,19-Epidioxy-6,19-dihydrovitamin D₂*
9547239 $C_{28}H_{44}O_3$

(3S,10R)-10-[(E)-[(1R,3aR,7aS)-1-[(E,2S,5R)-5,6-dimethylhept-3-en-2-yl]-7a-methyl-2,3,3a,5,6,7-hexahydro-1H-inden-4-ylidene]methyl]-8,9-dioxabicyclo[4.4.0]dec-11-en-3-ol. LMST03010036; 6R,19-epidioxy-9,10-seco-5(10),7E,22E-ergostatrien-3S-ol; (6R)-6,19-epidioxy-6,19-dihydroergocalciferol; 9547239; A major product in the reaction of vitamin D₂ with singlet oxygen generated by the dye-sensitized photochemical method. **Tet Lett.** 1975, **49(16)**, 4317-20; **J Org Chem.** 1983, **48(20)**, 3477-83.

576 *6R-Hydroperoxy-9,10-seco-4,7E,10(19),22E-ergostatetraen-3S-ol*
9547241 $C_{28}H_{44}O_3$

(1S)-3-[(1R,2E)-2-[(1R,3aR,7aS)-1-[(E,2S,5R)-5,6-dimethylhept-3-en-2-yl]-7a-methyl-2,3,3a,5,6,7-hexahydro-1H-inden-4-ylidene]-1-hydroperoxy-ethyl]-4-methylidene-cyclohex-2-en-1-ol. LMST03010038; 6R-hydroperoxy-9,10-seco-4,7E,10(19),22E-ergostatetraen-3S-ol; 9547241; A minor product by the reaction of vitamin D₂ with singlet oxygen generated by dye-sensitized photochemical method; λ_m = 235 nm (95% EtOH). **J Org Chem.** 1983, **48(20)**, 3477-83.

577 *6S-Hydroperoxy-9,10-seco-4,7E,10(19),22E-ergostatetraen-3S-ol*
9547242 $C_{28}H_{44}O_3$

(1S)-3-[(1S,2E)-2-[(1R,3aR,7aS)-1-[(E,2S,5R)-5,6-dimethylhept-3-en-2-yl]-7a-methyl-2,3,3a,5,6,7-hexahydro-1H-inden-4-ylidene]-1-hydroperoxy-ethyl]-4-methylidene-cyclohex-2-en-1-ol.
LMST03010039; 6S-hydroperoxy-9,10-seco-4,7E,10 (19),22E-ergostatetraen-3S-ol; 9547242; A minor product by the reaction of vitamin D_2 with singlet oxygen generated by dye-sensitized photochemical method; λ_m = 234 nm (95% EtOH). *J Org Chem.* 1983, *48(20)*, 3477-83.

578   **(6R)-Vitamin D_2 6,19-sulfur dioxide adduct**
9547237 $C_{28}H_{44}O_3S$

(3S,9R)-9-[(E)-[(1R,3aR,7aS)-1-[(E,2S,5R)-5,6-dimethylhept-3-en-2-yl]-7a-methyl-2,3,3a,5,6,7-hexahydro-1H-inden-4-ylidene]methyl]-8,8-dioxo-8-thiabicyclo[4.3.0]non-10-en-3-ol.
LMST03010034; 6R,19-epithio-9,10-seco-5(10),7E,22-ergostatrien-3S-ol S,S-dioxide; (6R)-ergocalciferol 6,19-sulfur dioxide adduct; 9547237; Prepared from vitamin D_2 by treatment with liquid SO_2. *Chem Lett.* 1979, 583.

579   **(6S)-Vitamin D_2 6,19-sulfur dioxide adduct**
9547238 $C_{28}H_{44}O_3S$ ·

(3S,9S)-9-[(E)-[(1R,3aR,7aS)-1-[(E,2S,5R)-5,6-dimethylhept-3-en-2-yl]-7a-methyl-2,3,3a,5,6,7-hexahydro-1H-inden-4-ylidene]methyl]-8,8-dioxo-8I^{6}-thiabicyclo[4.3.0]non-10-en-3-ol.
LMST03010035; 6S,19-epithio-9,10-seco-5(10),7E,22-ergostatrien-3S-ol S,S-dioxide; ergocalciferol 6S,19-sulfur dioxide adduct; 9547238; Prepared from vitamin D_2 by treatment with liquid SO_2. *Chem Lett.* 1979, 583.

580   **1,24,25-Trihydroxyvitamin D_2**
100496-04-6 $C_{28}H_{44}O_4$

(E,3R,6S)-6-[(1R,3aR,4Z,7aS)-4-[(2Z)-2-[(3S,5R)-3,5-dihydroxy-2-methylidene-cyclohexylidene]ethylidene]-7a-methyl-2,3,3a,5,6,7-hexahydro-1H-inden-1-yl]-2,3-dimethyl-hept-4-ene-2,3-diol.
1,24,25-Trihydroxyergocalciferol; 9,10-Secoergosta-5Z,7E,10(19),22E-tetraene-1α,3β,24,25-tetrol; 1,24,25-Trihydroxyvitamin D_2; 6438872; 100496-04-6; A metabolite of vitamin D_2, formed by hydroxylation at C24 of 1,25-dihydroxyvitamin D_2. Has bone-resorbing activity. *Biochemistry.* 1986, *25(18)*, 5328-36.

581   **1α,24R,25-Trihydroxyvitamin D_2**
9547253 $C_{28}H_{44}O_4$

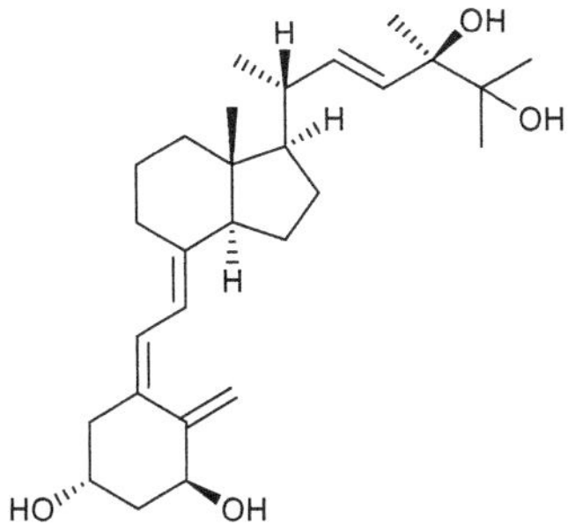

(E,3R,6S)-6-[(1R,3aR,4E,7aS)-4-[(2Z)-2-[(3S,5R)-3,5-dihydroxy-2-methylidene-cyclohexylidene]ethylidene]-7a-methyl-2,3,3a,5,6,7-hexahydro-1H-inden-1-yl]-2,3-dimethyl-hept-4-ene-2,3-diol.
LMST03010051; 1α,24R,25-trihydroxyergocalciferol; 9,10-seco-5Z,7E,10(19),22E-ergostatetraene-1S,3R, 24R,25-tetrol; 9547253; λ_m = 265 nm. *Biochemistry*. 1986, *25(18)*, 5328-36.

582   *1α,25S,26-Trihydroxyvitamin D$_2$*
9547254 $C_{28}H_{44}O_4$

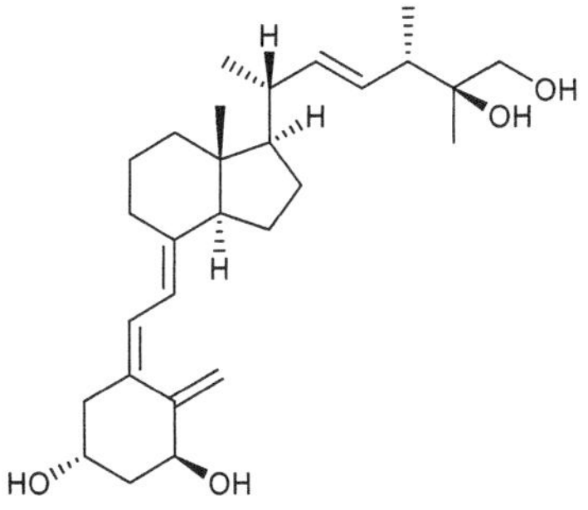

(E,2S,3S,6S)-6-[(1R,3aR,4E,7aS)-4-[(2Z)-2-[(3S,5R)-3,5-dihydroxy-2-methylidene-cyclohexylidene]ethylidene]-7a-methyl-2,3,3a,5,6,7-hexahydro-1H-inden-1-yl]-2,3-dimethyl-hept-4-ene-1,2-diol.
LMST03010052; 1α,25S,26-trihydroxyergocalciferol; 9,10-seco-5Z,7E,10(19),22E-ergostatetraene-1S,3R, 25S,26-tetrol; 9547254; *J Biol Chem*. 1986, *261(20)*, 9250-6.

583   *1α,25,28-Trihydroxyvitamin D$_2$*
104870-37-3 $C_{28}H_{44}O_4$

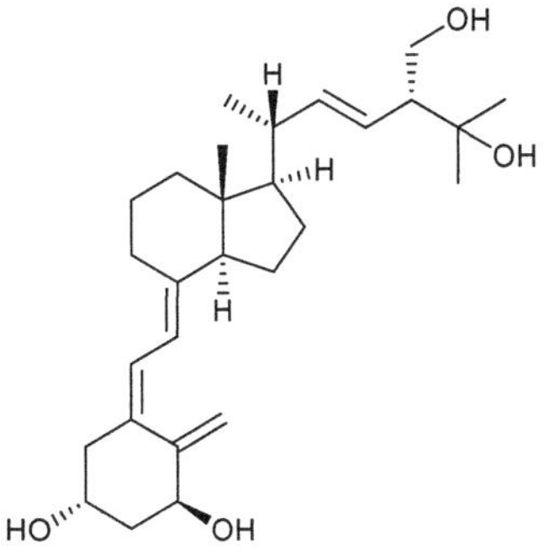

(2R)-2-[(E,3S)-3-[(1R,3aR,4E,7aS)-4-[(2Z)-2-[(3S,5R)-3,5-dihydroxy-2-methylidene-cyclohexylidene]ethylidene]-7a-methyl-2,3,3a,5,6,7-hexahydro-1H-inden-1-yl]but-1-enyl]-3-methyl-butane-1,3-diol.
LMST03010053; 1α,25,28-trihydroxyvitamin D$_2$; 1α, 25,28-trihydroxyergocalciferol; 9,10-seco-5Z,7E,10 (19),22E-ergostatetraene-1S,3R,25,28-tetrol; 9,10-Seco-ergosta-5Z,7E,10(19),22E-tetraen-1α,3β,25,28-tetrol; 6442094; 104870-37-3; Has no effect on intestinal calcium absorption, bone calcium mobilization, or intestinal calbindin-D9K protein and mRNA. *J Biol Chem*. 1986, *261(20)*, 9250-6.

584   *24,25,26-Trihydroxyvitamin D$_2$*
123992-85-8 $C_{28}H_{44}O_4$

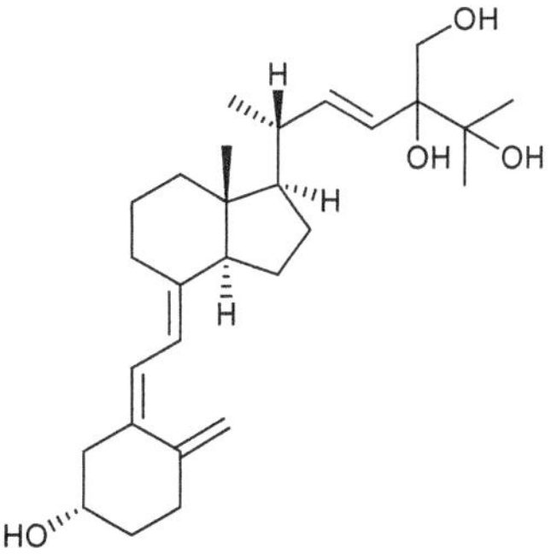

(E,3R,6S)-6-[(1R,4E,7aS)-4-[(2Z)-2-[(5S)-5-hydroxy-2-methylidene-cyclohexylidene]ethylidene]-7a-methyl-2,3,3a,5,6,7-hexahydro-1H-inden-1-yl]-2,3-dimethyl-hept-4-ene-1,2,3-triol.
24,25,26-Trihydroxyvitamin D$_2$; 9,10-Secoergosta-5Z, 7E,10(19),22E-tetraene-3β,24χ,25,26-tetrol; 24,25,26-(OH)3-Vitamin D$_2$; 9,10-Secoergosta-5Z,7E,10(19), 22E-tetraene-3β,24χ,25,26-tetrol; 6439162; 123992-85-8; Metabolite of 25-hydroxyvitamin D$_2$ in the isolated perfused rat kidney. *Biochemistry*. 1990, *29(4)*, 943-9.

585   *24,25,28-Trihydroxyvitamin D$_2$*
123992-86-9 $C_{28}H_{44}O_4$

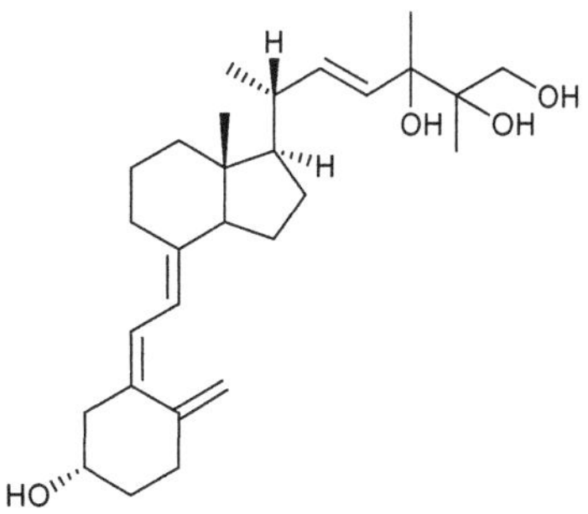

2-[(E,3S)-3-[(1R,3aR,4E,7aS)-4-[(2Z)-2-[(5S)-5-hydroxy-2-methylidene-cyclohexylidene]ethylidene]-7a-methyl-2,3,3a,5,6,7-hexahydro-1H-inden-1-yl]but-1-enyl]-3-methyl-butane-1,2,3-triol.

24,25.28-Trihydroxyvitamin D_2; 24,25,28-(OH)3-Vitamin D_2; 9,10-Secoergosta-5Z,7E,10(19),22E-tetra-ene-3,24xi,25,28-tetrol; 6449938; 123992-86-9; Metabolite of 25-hydroxyvitamin D_2 in the isolated perfused rat kidney. **Biochemistry.** 1990, **29(4)**, 943-9.

586 **1,25,28-Trihydroxyvitamin D_2**

6506519 $C_{28}H_{44}O_4$

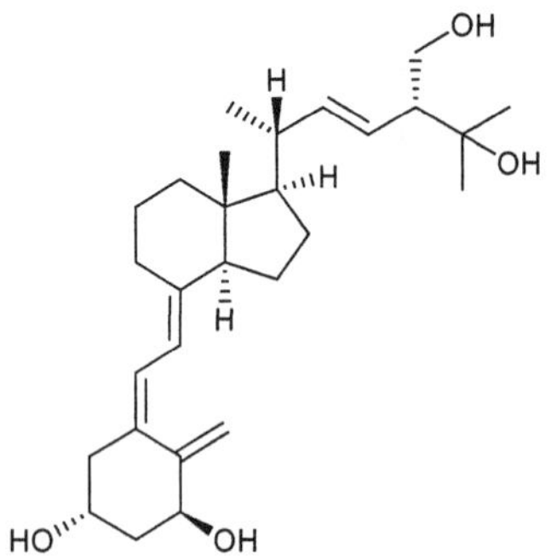

(2R)-2-[(E,3S)-3-[(1R,3aR,4E,7aS)-4-[(2E)-2-[(3S,5R)-3,5-dihydroxy-2-methylidene-cyclohexylidene]ethyl-idene]-7a-methyl-2,3,3a,5,6,7-hexahydro-1H-inden-1-yl]but-1-enyl]-3-methyl-butane-1,3-diol. 6506519. **J Bone Miner Res**. 1993, **8(12)**, 1483-90.

587 **22-Ene-25-oxavitamin D**

$C_{28}H_{44}O_4$

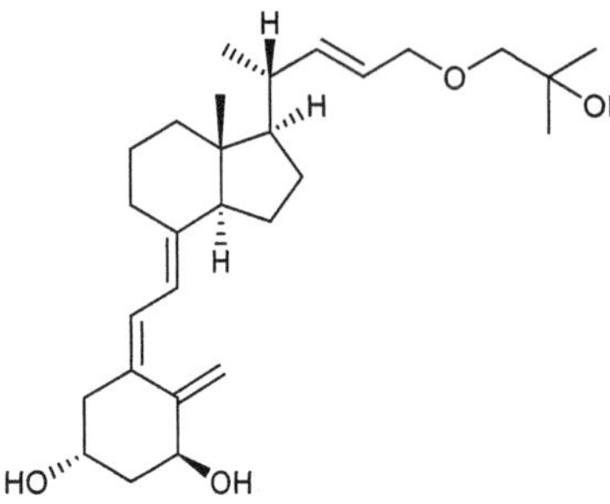

ZK156979; ZK 156979; ZK-156979; 26-Nor-25-oxaD3; An immunosuppressive agent with little calcemic activity. Inhibits Th1 cytokines. A novel low calcemic vitamin D analogue, may be useful in connection with T-cell-mediated diseases. **Eur J Clin Invest.** 205, **35(5)**, 343-9.

588 **Vitamin D_2 sulfate**

1784-46-9 $C_{28}H_{44}O_4S$

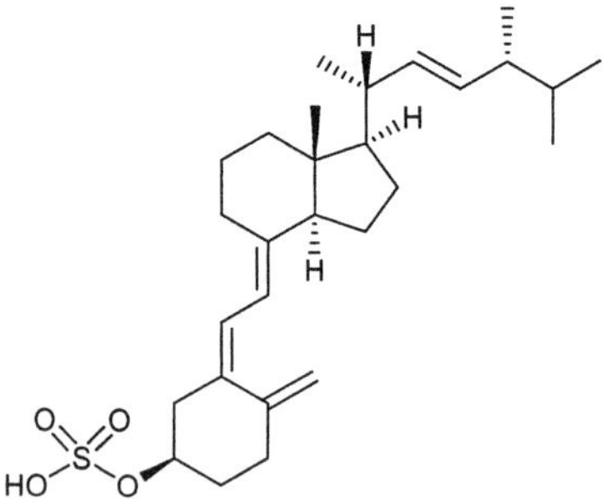

(1S,3Z)-3-[(2E)-2-[(1R,3aR,7aS)-1-[(E,2S,5R)-5,6-dimethylhept-3-en-2-yl]-7a-methyl-2,3,3a,5,6,7-hexahydro-1H-inden-4-ylidene]ethylidene]-4-methylidene-cyclohexan-1-ol(1R,3aR,4E,7aS)-1-[(E,2S,5R)-5,6-dimethylhept-3-en-2-yl]-7a-methyl-4-[(2Z)-2-[(5R)-2-methylidene-5-sulfooxy-cyclohexyl-idene]ethylidene]-2,3,3a,5,6,7-hexahydro-1H-indene.

Vitamin D_2 sulfate; Vitamin D_2 3β sulfate; Ergocalciferol hydrogen sulfate; 9,10-Secoergosta-5Z,7E,10(19),22E-tetraen-3β-ol, hydrogen sulfate; 6441761; 1784-46-9; 3308-52-9; 6609-90-1. **J Steroid Biochem.** 1982, **17(5)**, 495-502.

589 **1α,24,25,28-Tetrahydroxyvitamin D_2**

103305-10-8 $C_{28}H_{44}O_5$

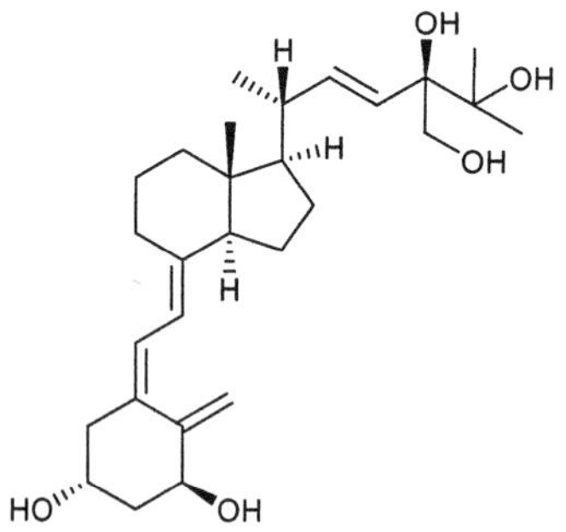

(2S)-2-[(E,3S)-3-[(1R,3aR,4E,7aS)-4-[(2Z)-2-[(3S,5R)-3,5-dihydroxy-2-methylidene-cyclohexylidene]ethylidene]-7a-methyl-2,3,3a,5,6,7-hexahydro-1H-inden-1-yl]but-1-enyl]-3-methyl-butane-1,2,3-triol.
LMST03010055; 9,10-seco-5Z,7E,10(19),22E-ergosta-tetraene-1S,3R,24,25,28-pentol; 1α,24,25,28-tetra-hydroxyergocalciferol; 1,24,25,28-Tetrahydroxy-vitamin D_2; 1,24,25,28-Tetrahydroxyergocalciferol; 6438960; 9547256; 103305-10-8; Metabolite of 1,25-dihydroxyvitamin D_2 in rat kidney. Formed by successive hydroxylation of 1,25-dihydroxyvitamin D_2 and 1,24,25-trihydroxyvitamin D_2; λ_m = 265 nm. **Biochemistry.** 1986, **25(18)**, 5328-36.

590 **1,24,25,26-Tetrahydroxyvitamin D_2**

103305-11-9 $C_{28}H_{44}O_5$

(E,3R,6S)-6-[(1R,3aR,4Z,7aS)-4-[(2Z)-2-[(3S,5R)-3,5-dihydroxy-2-methylidene-cyclohexylidene]ethylidene]-7a-methyl-2,3,3a,5,6,7-hexahydro-1H-inden-1-yl]-2,3-dimethyl-hept-4-ene-1,2,3-triol.
LMST03010054; 1,24,25,26-Tetrahydroxyvitamin D$_2$; 1,24,25,26-Tetra-hydroxyergocalciferol; 6438961; 103305-11-9; Metab-olite in rat kidney of 1,25-dihydroxyvitamin D$_2$. *Biochemistry.* 1986, *25(18)*, 5328-36.

591   24-Nor-9,11-seco-11-acetoxy-3,6-dihydroxycholest-7,22-dien-9-one

147879-65-0   C$_{28}$H$_{44}$O$_5$

6α,9,11-Secocholesta-7,22E-dien-9-one, 11-(acetyloxy)-3β,6α-dihydroxy-.
NSADC, 24-Nor-9,11-seco-11-acetoxy-3,6-dihydroxy-cholest-7,22-dien-9-one; Isolated from soft coral *Gersemia fruticosa*. Is a growth inhibitor. Cytotoxic against human leukemia K562, human cervical cancer HeLa and Ehrlich ascites tumor cells *in vitro*. *Steroids.* 1994, *59(4)*, 274-81.

592   1α,24R,25,26-Tetrahydroxyvitamin D$_2$

9547255   C$_{28}$H$_{44}$O$_5$

(E,3R,6S)-6-[(1R,3aR,4E,7aS)-4-[(2Z)-2-[(3S,5R)-3,5-dihydroxy-2-methylidene-cyclohexylidene]ethylidene]-7a-methyl-2,3,3a,5,6,7-hexahydro-1H-inden-1-yl]-2,3-dimethyl-hept-4-ene-1,2,3-triol.
LMST03010054; 1α,24R,25,26-tetrahydroxyergocal-ciferol; 9,10-seco-5Z,7E,10(19),22E-ergostatetraene-1S,3R,24R,25,26-pentol; 9547255; λ_m = 265 nm. *Biochemistry.* 1986, *25(18)*, 5328-36.

593   11α-(Chloromethyl)-1α,25-dihydroxy-vitamin D$_3$

9547468   C$_{28}$H$_{45}$ClO$_3$

(1R,3S,5Z)-5-[(2E)-2-[(1R,3aR,6R,7aS)-6-(chloro-methyl)-1-[(2R)-6-hydroxy-6-methyl-heptan-2-yl]-7a-methyl-2,3,3a,5,6,7-hexahydro-1H-inden-4-ylidene]ethylidene]-4-methylidene-cyclohexane-1,3-diol.
LMST03020327; 11S-(chloromethyl)-9,10-seco-5Z,7E,10(19)-cholestatriene-1S,3R,25-triol; 11α-(chloro-methyl)-1α,25-dihydroxycholecalciferol; 9547468; Synthesized by Horner coupling of 11-substituted CD-ring ketone with the A-ring phosphine oxide. Vitamin D Gene Regulation Structure-Function Analysis and Clinical Application (Norman, A.W., Bouillon R. and Thomasset, M., eds), pp 165-166, Walter de Gruyter, Berlin/New York (1991).

594   25-Fluoro-1α,24R-dihydroxy-24R-methylvitamin D$_3$

9547469   C$_{28}$H$_{45}$FO$_3$

(1R,3S,5Z)-5-[(2E)-2-[(1R,3aR,7aS)-1-[(2R,5R)-6-fluoro-5-hydroxy-5,6-dimethyl-heptan-2-yl]-7a-methyl-2,3,3a,5,6,7-hexahydro-1H-inden-4-ylidene]ethylidene]-4-methylidene-cyclohexane-1,3-diol.
LMST03020328;   25-fluoro-1α,24-dihydroxy-24R-methylcholecalciferol;   25-fluoro-24-methyl-9,10-seco-5Z,7E,10(19)-cholestatrien-1S,3R,24R-triol; 9547469.

595   *11α-(Fluoromethyl)-1α,25-dihydroxy-vitamin D₃*
9547470 $C_{28}H_{45}FO_3$

(1R,3S,5Z)-5-[(2E)-2-[(1R,3aR,6R,7aS)-6-(fluoromethyl)-1-[(2R)-6-hydroxy-6-methyl-heptan-2-yl]-7a-methyl-2,3,3a,5,6,7-hexahydro-1H-inden-4-ylidene]ethylidene]-4-methylidene-cyclohexane-1,3-diol.
LMST03020329;   11S-(fluoromethyl)-9,10-seco-5Z,7E,10(19)-cholestatriene-1S,3R,25-triol;   11α-(fluoromethyl)-1α,25-dihydroxycholecalciferol;   9547470;
Synthetic analog of vitamin D₃. Biological activity reported. *J Biol Chem.* 1992, *267(5)*, 3044-51.

596   *6-Methylvitamin D₃*
5283771 $C_{28}H_{46}O$

(1S,3Z)-3-[(1E)-1-[(1R,3aR,7aS)-7a-methyl-1-[(2R)-6-methylheptan-2-yl]-2,3,3a,5,6,7-hexahydro-1H-inden-4-ylidene]propan-2-ylidene]-4-methylidene-cyclo-hexan-1-ol.
LMST03020331;   6-methylcholecalciferol;   6-methyl-9,10-seco-5Z,7E,10(19)-cholestatrien-3S-ol;   5283771;
Synthesized from vitamin D₃ by methylation of its sulfur dioxide adduct; λ_m = 240 nm (hexane). *Tet Lett.* 1981, *22(39)*, 3085-8.

597   *6-Methylprevitamin D₃*
5283772 $C_{28}H_{46}O$

(1S)-3-[(E)-1-[(1R,3aR,7aS)-7a-methyl-1-[(2R)-6-methylheptan-2-yl]-1,2,3,3a,6,7-hexahydroinden-4-yl]prop-1-en-2-yl]-4-methyl-cyclohex-3-en-1-ol.
LMST03020333;   6-methylprecholecalciferol;   6E-methyl-9,10-seco-5(10),6,8-cholestatrien-3S-ol;
5283772; Synthesized from vitamin D₃ by methylation of its sulfur dioxide adduct; λ_m = 248 nm (95% EtOH). *Tet Lett.* 1981, *22(39)*, 3085-8.

598   *(10E)-19-Methylvitamin D₃*
5283773 $C_{28}H_{46}O$

(1S,3Z,4E)-3-[(2E)-2-[(1R,3aR,7aS)-7a-methyl-1-[(2R)-6-methylheptan-2-yl]-2,3,3a,5,6,7-hexahydro-1H-inden-4-ylidene]ethylidene]-4-ethylidene-cyclohexan-1-ol.
LMST03020334;   (10E)-19-methylcholecalciferol;   19-methyl-9,10-seco-5Z,7E,10E(19)-cholestatrien-3S-ol;
5283773; Synthesized from vitamin D₃ by methylation of its sulfur dioxide adduct; λ_m = 268 nm (95% EtOH). *Tet Lett.* 1981, *22(39)*, 3085-8.

599   *(5E,10E)-19-Methylvitamin D₃*
5283774 $C_{28}H_{46}O$

(1S,3E,4E)-3-[(2E)-2-[(1R,3aR,7aS)-7a-methyl-1-[(2R)-6-methylheptan-2-yl]-2,3,3a,5,6,7-hexahydro-1H-inden-4-ylidene]ethylidene]-4-ethylidene-cyclohexan-1-ol. LMST03020335; (5E,10E)-19-methylcholecalciferol; 19-methyl-9,10-seco-5E,7E,10E(19)-cholestatrien-3S-ol; 5283774; Synthesized from vitamin D_3 by methylation of its sulfur dioxide adduct; λ_m = 264 nm (95% EtOH). *Tet Lett.* 1981, *22(39)*, 3085-8.

600 *Vitamin D₄*
511-28-4 $C_{28}H_{46}O$

(1S,3Z)-3-[(2E)-2-[(1R,7aS)-1-(5,6-dimethylheptan-2-yl)-7a-methyl-2,3,3a,5,6,7-hexahydro-1H-inden-4-ylidene]ethylidene]-4-methylidene-cyclohexan-1-ol. LMST03030001; EINECS 208-127-5; 9,10-Secoergosta-5Z,7E,10(19)-trien-3β-ol; 9,10-Secocholesta-5,7,10(9)-trien-3-ol, 24-methyl-; 5460703; 6450185; 511-28-4; Vitamin mediating intestinal calcium absorption and bone calcium metabolism. Isolated from fish liver oils. Approximately as effective as vitamin D_2 in humans, but 50-100 times more effective than vitamin D_2 in chicks. Has 75% of the antirachitic effect of vitamin D_3; Needles, mp = 84-85°; $[\alpha]_D^{20}$ = +84.8° (c = 1.6 Me₂CO), +51.9° (c = 1.6 CHCl₃); λ_M = 264.5 nm ($E_{1\,cm}^{1\%}$ = 450-490); soluble in most organic solvents. *Hoppe-Seyler's Z Physiol. Chem.* 1937, *247*, 185.

601 *Provitamin D₄*
9547699 $C_{28}H_{46}O$

(3S,10R,13S,14R,17R)-17-[(2R,5S)-5,6-dimethylheptan-2-yl]-10,13-dimethyl-2,3,4,9,11,12,14,15,16,17-decahydro-1H-cyclopenta[a]phenanthren-3-ol. LMST03030002; Ergosta-5,7-dien-3β-ol; 9574699.

602 *Vitamin D₇*
5460703 $C_{28}H_{46}O$

(1S,3Z)-3-[(2E)-2-[(1R,3aR,7aS)-1-[(2R,5S)-5,6-dimethylheptan-2-yl]-7a-methyl-2,3,3a,5,6,7-hexahydro-1H-inden-4-ylidene]ethylidene]-4-methylidene-cyclohexan-1-ol. LMST03060001; Vitamin D_4; (24S)-methylcalciol; 22,23-dihydroercalciol; 22,23-dihydroergocalciferol; CHEBI:33237; 9,10-seco-5Z,7E,10(19)-ergostatrien-3S-ol; 9,10-secoergosta-5Z,7E,10(19)-trien-3S-ol; 5460703; Has 10% of the antirachitic effect of vitamin D_3. *J Am Chem Soc.* 1942, *64(8)*, 1900-2.

603 *Provitamin D₇*
9547704 $C_{28}H_{46}O$

(3S,10R,13S,14R,17R)-17-[(2R,5R)-5,6-dimethylheptan-2-yl]-10,13-dimethyl-2,3,4,9,11,12,14,15,16,17-decahydro-1H-cyclopenta[a]phenanthren-3-ol. LMST03060002; Campesta-5,7-trien-3β-ol; 24R-methyl-5,7-cholestadien-3S-ol; 9547704. *J Amer Chem Soc.* 1982, *104(21)*, 5780-1; USP 7211172.

604 *Dihydrotachysterol₂*
67-96-9 $C_{28}H_{46}O$

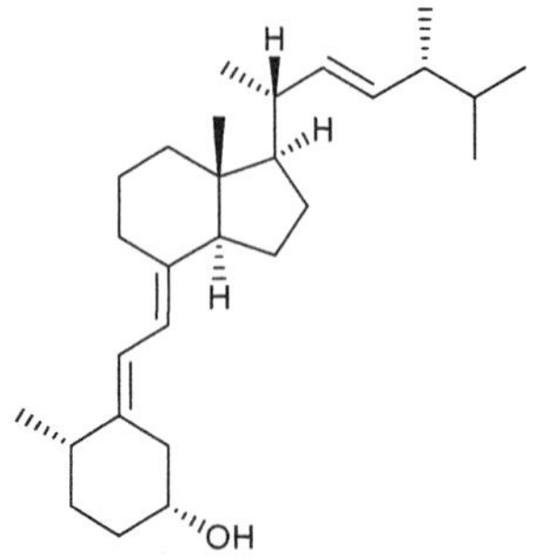

(3Z)-3-[(2E)-2-[1-[(E)-5,6-dimethylhept-3-en-2-yl]-7a-methyl-2,3,3a,5,6,7-hexahydro-1H-inden-4-ylidene]ethylidene]-4-methyl-cyclohexan-1-ol.
Dihydrotachysterol; 9,10α-Secoergosta-5E,7E,22E-trien-3β-ol; 9,10-Secoergosta-5,7,22-trien-3β-ol; A.T. 10; AT 10; Anti-tetany substance 10; Antitanil; Calcamine; DHT2; Dht₂; Dichystrolum; Dihidrotaquisterol; Dihydral; Dihydrotachysterol; Dihydrotachysterol₂; Dihydrotachysterolum; Diidrotachisterolo; Dygratyl; EINECS 200-672-7; HSDB 3314; Hytakerol; Parterol; Tachysterol, dihydro-; Tachysterol2, dihydro-; Tachystin; 9,10α-Secoergosta-5E,7E,22E-trien-3β-ol; Dihydrotachysterol; Tachysterol, dihydro-; 5281010; 11953890; 67-96-9; A vitamin D prodrug.A calcium active steroid not dependent upon kidney metabolism. Used to treat renal failure and osteoporosis; Needles, mp = 125-127°; $[\alpha]_D^{22}$ = +97.5° (CHCl₃); λ_m = 242, 251, 261 nm (E$_{1\ cm}^{1\%}$ 870, 1010, 650); soluble in organic solvents, insoluble in H₂O; LD₅₀ (mus ip) = 104 mg/kg, (mus orl) = 288 mg/kg. *J Biol Chem*. 1993, *268(1)*, 282-92; *J Org Chem*. 1988, *53(26)*, 6094-9; USP 6017908

605   *9,10S-Secoergosta-5E,7E,22E-trien-3R-ol*
6708477   C₂₈H₄₆O

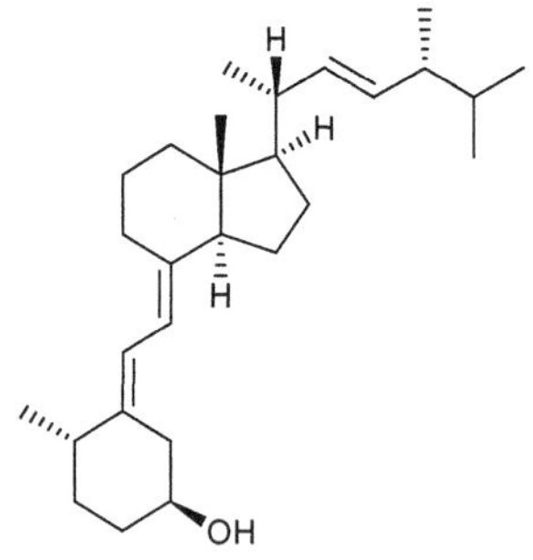

(1S,4S)-3-[2-[(3aR,7aS)-1-[(2S,5R)-5,6-dimethylhept-3-en-2-yl]-7a-methyl-2,3,3a,5,6,7-hexahydro-1H-inden-4-ylidene]ethylidene]-4-methyl-cyclohexan-1-ol.
DivK1c_000869; KBio1_000869; NINDS_000869; 6708477.

606   *1α-Hydroxy-3α-methyl-3-deoxyvitamin D₃*
54473-74-4   C₂₈H₄₆O

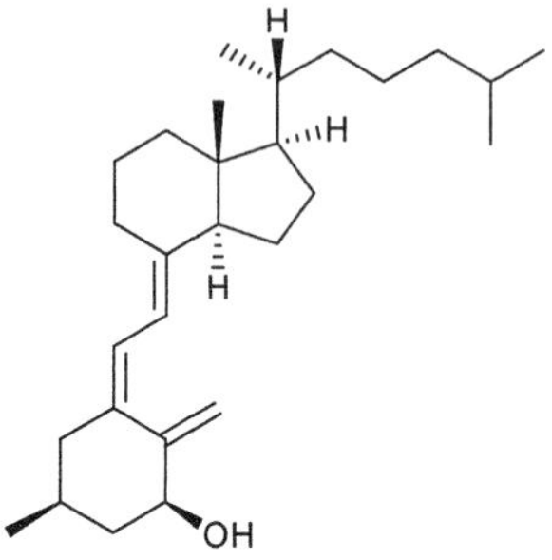

(6R)-6-[(1R,3aR,4E,7aS)-4-[(2Z)-2-[(3S,5S)-5-methyl-2-methylidene-cyclohexylidene]ethylidene]-7a-methyl-2,3,3a,5,6,7-hexahydro-1H-inden-1-yl]-2-methyl-heptan-2-ol.
3S-methyl-9,10-seco-5Z,7E,10(19)-cholestatriene-1S-ol; 1α-hydroxy-3α-methyl-3-deoxyvitamin D₃; 1α-hydroxy-3α-methyl-3-deoxycholeciferol; 54473-74-4; Synthetic vitamin D analog. Elicits intestinal calcium absorption and bone calcium mobilization in the chick. *J Org Chem*. 1978, *43(4)*, 574-80.

607   *Dihydrotachysterol₂*
5311071   C₂₈H₄₆O

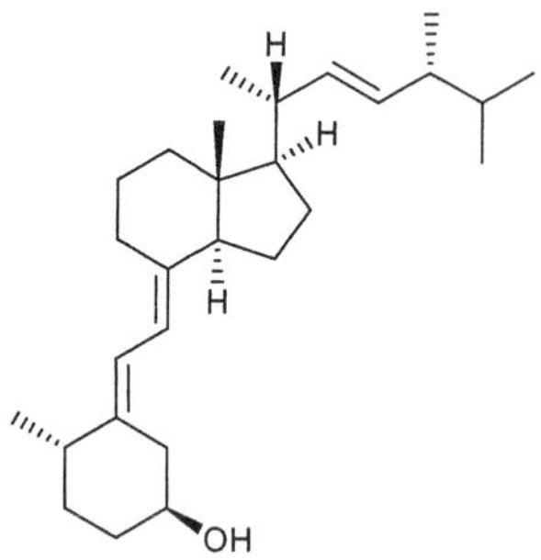

(1S,3E,4S)-3-[(2E)-2-[(1R,3aR,7aS)-1-[(E,2S,5R)-5,6-dimethylhept-3-en-2-yl]-7a-methyl-2,3,3a,5,6,7-hexahydro-1H-inden-4-ylidene]ethylidene]-4-methyl-cyclohexan-1-ol.
LMST03010056; 9,10S-seco-5E,7E,22E-ergostatrien-3S-ol; dihydrotachysterol-2; (5E)-(10S)-10,19-dihydro-vitamin D₂; (5E)-(10S)-10,19-dihydroergocalciferol; 5311071; Crystals; mp = 125-127°; $[\alpha]_D^{22}$ = +97.5° (CHCl₃); λ_m = 242, 251, 261 nm (E$_{1\ cm}^{1\%}$ = 870, 1010, 650). *Physiol*. 1939, *260*, 119; *J Org Chem*. 1988, *53(26)*, 6094-9; 1997, *42(21)*, 3325-30; USP 6017908

608   *Campesta-7,22E-dien-3β-ol*
5283669   C₂₈H₄₆O

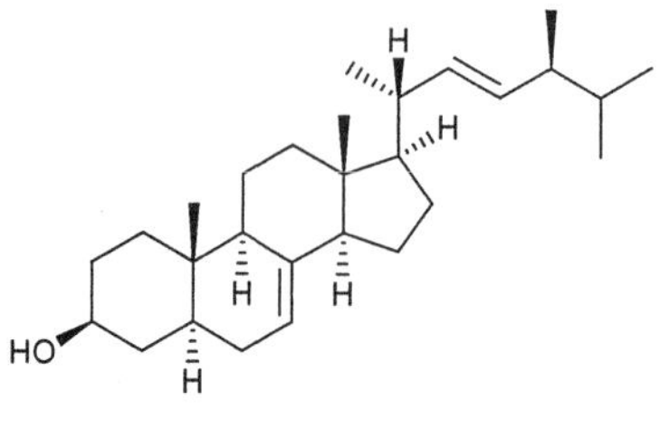

(3S,5S,9R,10S,13R,14R,17R)-17-[(E,2S,5S)-5,6-di-methylhept-3-en-2-yl]-10,13-dimethyl-2,3,4,5,6,9,11,12,14,15,16,17-dodecahydro-1H-cyclopenta[a]phen-anthren-3-ol.
Stellasterol; LMST01030117; 24S-methyl-5α-cholesta-7,22E-dien-3β-ol; 5283669; Isolated from Porifera, ***Axinella cannabina; Asteroidea, Asterias amurensis, A.rubens,*** and various plants of the ***Cucurbitaceae***; Crystals; mp = 159-160°. *J Chem Soc Perkin Trans I*, 1983, 147; *Lipids.* 1986, *21,* 39-47; *Phytochemistry.* 1983, *22(5),* 1300-1; 1991, *30(11),* 3621-4; *Tetrahedron.* 1973, *29(9),* 1193-4; *J Org Chem.* 1981, *46(8),* 1726-8; *Biochem J.* 1973, *135(3),* 443-55.

609   5,6-Dihydroergosterol
5283628 $C_{28}H_{46}O$

(3S,5S,9R,10S,13R,14R,17R)-17-[(E,2S,5R)-5,6-dimethylhept-3-en-2-yl]-10,13-dimethyl-2,3,4,5,6,9,11,12.14,15,16,17-dodecahydro-1H-cyclopenta[a]phenanthren-3-ol.
24R-methyl-5α-cholesta-7,22E-dien-3β-ol; ergosta-7,22-dien-3β-ol; LMST01030094; 5283628; Isolated from yeast, Cleodendron plants, Cucurbitaceae plants, Porifera ***Axinella cannabina, Haliclona flavescens, chlorella: chlorella vulgaris, C.ellipsoidea*** metabolite, and from Basidiomycetes (***Astraeus hygrometricus)***; Crystals; mp = 174-175°. *Steroids.* 1975, *25(6),* 741-51; 1982, *39(6),* 675-80; 1989, *53(3-5),* 625-38; *J. Chem. Soc., Perkin Trans I.* 1983, 147-53; *Phytochemistry.* 1972, *11(12),* 3473-7; 1987, *26(8),* 2341-4; 1991, *30(11),* 3621-4; 1992, *31(5),* 1769-72; *Tetrahedron.* 1973, *29(9),* 1193-6; Lipids. 1986, *21,* 39-47.

610   (5E)-10S,19-Dihydro-3-epivitamin D_2
9547257 $C_{28}H_{46}O$

(1R,3E,4S)-3-[(2E)-2-[(1R,3aR,7aS)-1-[(E,2S,5R)-5,6-dimethylhept-3-en-2-yl]-7a-methyl-2,3,3a,5,6,7-hexahydro-1H-inden-4-ylidene]ethylidene]-4-methyl-cyclohexan-1-ol.

LMST03010057;   9,10S-seco-5E,7E,22E-ergostatrien-3R-ol; 3-epidihydrotachysterol-2; (5E)-10S,19-dihydro-3-epiergocalciferol;   9547257;   Prepared   from dihydrotachysterol by inverting the configuration at C(3); λ_m = 242, 251, 261 (ε = 34500, 40000, 26000). *J Org Chem.* 1977, *42(21),* 3325-30.

611   24a-Homo-25-hydroxy-5,6-trans-vitamin D_3
                                           $C_{28}H_{46}O_2$

Synthesized from 24-homo-25-hydroxyvitamin D_3. *J Med Chem.* 1978, *21(10),* 1025-9.

612   1α,25-Dihydroxy-3α-methyl-3-deoxy-vitamin D_3
5283775 $C_{28}H_{46}O_2$

(6R)-6-[(1R,3aR,4E,7aS)-4-[(2Z)-2-[(3S,5S)-3-hydroxy-5-methyl-2-methylidene-cyclohexylidene]ethylidene]-7a-methyl-2,3,3a,5,6,7-hexahydro-1H-inden-1-yl]-2-methyl-heptan-2-ol.
LMST03020336;   3S-methyl-9,10-seco-5Z,7E,10(19)-cholestatriene-1S,25-diol; 1α,25-dihydroxy-3α-methyl-3-deoxycholecalciferol; 5283775; Synthesized from 1α,25-dihydroxycholesterol 3-tosylate. Elicits intestinal calcium absorption and bone calcium mobilization in the chick; λ_m = 262 nm. *J Org Chem.* 1978, *43(4),* 574-80.

613   25-Hydroxydihydrotachysterol₂
98830-20-7 $C_{28}H_{46}O_2$

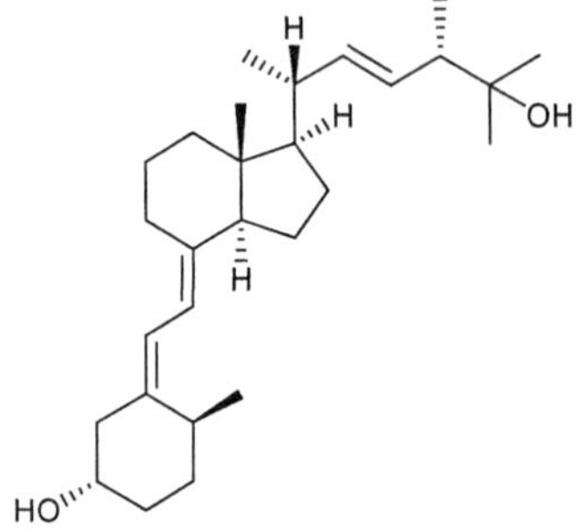

(E,3S,6S)-6-[(1R,3aR,4E,7aS)-4-[(2E)-2-[(2S,5S)-5-
hydroxy-2-methyl-cyclohexylidene]ethylidene]-7a-
methyl-2,3,3a,5,6,7-hexahydro-1H-inden-1-yl]-2,3-
dimethyl-hept-4-en-2-ol.
25-Dht(2); 25-Hydroxydihydrotachysterol(2); (10α,
22E)-9,10-Secoergosta-5E,7E,22-triene-3β,25-diol;
9,10-Secoergosta-5E,7E,22-triene-3β,25-diol,
(10α,22E)-; 6438813; 98830-20-7; A metabolite of
dihydrotachysterol(2), causes suppression of the
parathyroid glands. *J Steroid Biochem Mol Biol.* 1992,
43(4), 359-61.

614 *22-Hydroxy Vitamin D₄*

51504-03-1 $C_{28}H_{46}O_2$

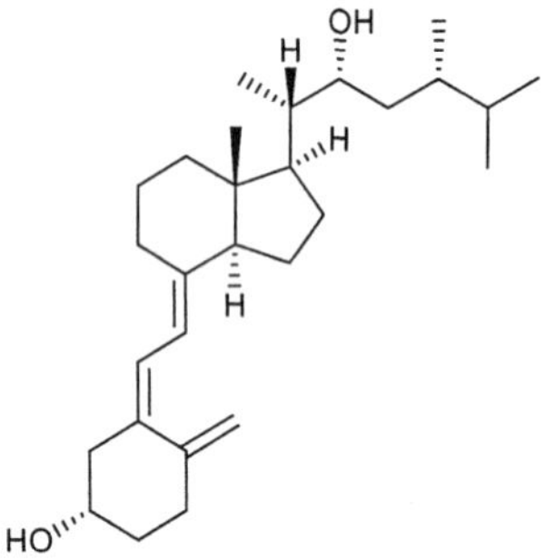

(2S,3S,5S)-2-[(1R,3aR,4E,7aS)-4-[(2Z)-2-[(5R)-5-
hydroxy-2-methylidene-cyclohexylidene]ethylidene]-
7a-methyl-2,3,3a,5,6,7-hexahydro-1H-inden-1-yl]-5,6-
dimethyl-heptan-3-ol.
22-Hydroxyvitamin D₄; 22S,23-Dihydro-22-hydroxy-
ergocalciferol; 9,10-secoergosta-5,7,10(19)-triene-3α,
22S-diol; 9,10-Secoergosta-5Z,7E,10(19)-triene-3β,
22S-diol; 6443542; 51504-03-1.

615 *1,25-Dihydroxyvitamin D₄*

$C_{28}H_{46}O_2$

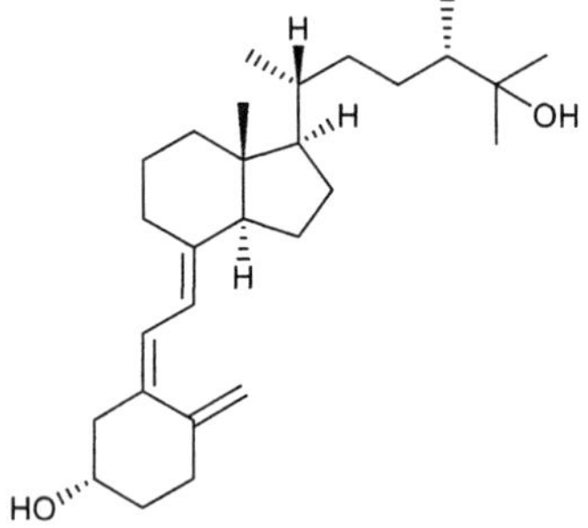

9,10-Secocholesta-5,7,10(19)-trien-3β,25-diol.
9,10-Secoergosta-5Z,7E,10(19)-trien-3β,25-diol; Metabolism parallels that of 1α,25-dihydroxyvitamin D₂
but not that of 1α,25-dihydroxyvitamin D₃. *Steroids.*
2001, *66(2)*, 93-7; *Biochim Biophys Acta.* 2002,
1583(2), 151-66.

616 *25-Hydroxy-3,3-dimethyl-3-deoxy-A-homo-2,4-dioxavitamin D₃*

135821-90-8 $C_{28}H_{46}O_3$

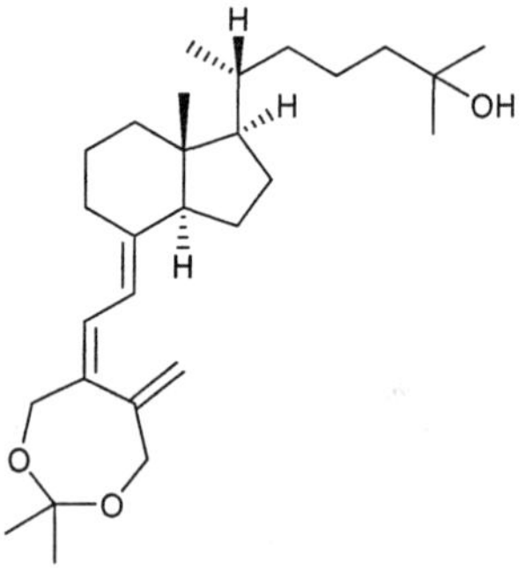

(6R)-6-[(1R,3aR,4E,7aS)-4-[(2E)-2-(2,2-dimethyl-6-
methylidene-1,3-dioxepan-5-ylidene)ethylidene]-7a-
methyl-2,3,3a,5,6,7-hexahydro-1H-inden-1-yl]-2-
methyl-heptan-2-ol.
LMST03020344; 3,3-dimethyl-A-homo-2,4-dioxa-9,10-
seco-5Z,7E,10(19)-cholestatrien-25-ol; 25-hydroxy-3,3-
dimethyl-3-deoxy-A-homo-2,4-dioxacholecalciferol; A-
homo-3-deoxy-3,3-dimethyl-2,4-dioxa-25-OH-D3;
6449866; 9547477; 135821-90-8; Fails to inhibit 25-
hydroxyvitamin D₃ 1α-hydroxylase in isolated mitochondria but induces 25-hydroxyvitamin D₃ metabolism in cultured chick kidney cells. Affinity for
chick intestinal receptor is 4% of that of 1α,25-
dihydroxyvitamin D₃. *J Steroid Biochem Mol Biol.*
1991, *38(6)*, 775-9.

617 *1α-Hydroxy-25-methoxyvitamin D₃*

9547471 $C_{28}H_{46}O_3$

(6R)-6-[(1R,3aR,4E,7aS)-4-[(2Z)-2-[(3R,5R)-5-hydroxy-3-(hydroxymethyl)-2-methylidene-cyclohexylidene] ethylidene]-7a-methyl-2,3,3a,5,6,7-hexahydro-1H-inden-1-yl]-2-methyl-heptan-2-ol.
LMST03020340; 1-Hbdv D3; 1-Hydroxymethyl-3-norhydroxy-3,25-dihydroxyvitamin D_3; 1β-(Hydroxymethyl)-9,10-secocholesta-5Z,7E,10(19)-triene-3α,25-diol; 1R-(hydroxymethyl)-9,10-seco-5Z,7E,10,(19)-cholestatriene-3R,25-diol; 1β-(Hydroxymethyl)-9,10-secocholesta-5Z,7E,10(19)-triene-3α,25-diol; 25-hydroxy-1R-(hydroxymethyl)cholecalciferol; 9,10-Secocholesta-5Z,7E,10(19)-triene-3α,25-diol, 1β-(hydroxymethyl)-; 6438753; 142508-68-7; Synthesized by combining enantiomerically pure 25-hydroxylated C,D-ring ketone with highly enantiomerically enriched A-ring phosphine oxide. Retains the antiproliferative activity of natural calcitriol in murine keratinocytes. Less than 0.1% as effective as 1α,25-dihydroxyvitamin D_2 for binding to the 1,25-(OH)2D3 receptor. Is equipotent at inhibiting growth of PE cells (murine keratinocyte cell line) and inhibiting the effect of TPA (12-O-tetradecanoylphorbol-13-acetate) on the activity of ornithinedecarboxylase; $[α]_D^{23}$ = +24° (c = 0.74 CH_2Cl_2); $λ_m$ = 265 nm (MeOH). *J Med Chem*. 1992, **35(17)**, 3280-7.

620   **25-Hydroxy-1S-(hydroxymethyl)vitamin D_3**
142508-67-6 $C_{28}H_{46}O_3$

(6R)-6-[(1R,3aR,4E,7aS)-4-[(2Z)-2-[(3S,5S)-5-hydroxy-3-(hydroxymethyl)-2-methylidene-cyclohexylidene]ethylidene]-7a-methyl-2,3,3a,5,6,7-hexahydro-1H-inden-1-yl]-2-methyl-heptan-2-ol.
LMST03020341; 1-Hmhv D3; 1S-(hydroxymethyl)-9,10-seco-5Z,7E,10(19)-cholestatriene-3S,25-diol; 25-hydroxy-1S-(hydroxymethyl)vitamin D_3; 25-hydroxy-1S-(hydroxymethyl)cholecalciferol; 9,10-Secocholesta-5Z,7E,10(19)-triene-3β,25-diol, 1α-(hydroxymethyl)-; 6438752; 142508-67-6; Synthesized by combining enantiomerically pure 25-hydroxylated C,D-ring ketone with highly enantiomerically enriched A-ring phosphine oxide. Retains the antiproliferative activity of natural calcitriol in murine keratinocytes but is less than 0.1% as effective as calcitriol for binding to the 1α,25-dihydroxyvitamin D_3 receptor and less than 0.1% as potent as calcitriol for calbindin-D28K induction in organ-cultured embryonic chick duodenum. Equipotent at inhibiting growth of PE cells (murine keratinocyte cell line) and inhibiting the effect of TPA (12-O-tetradecanoylphorbol-13-acetate) on the

(1R,3S,5Z)-5-[(2E)-2-[(1R,3aR,7aS)-1-[(2R)-6-methoxy-6-methyl-heptan-2-yl]-7a-methyl-2,3,3a,5,6,7-hexahydro-1H-inden-4-ylidene]ethylidene]-4-methylidene-cyclohexane-1,3-diol.
LMST03020330; 25-methoxy-9,10-seco-5Z,7E,10(19)-cholestatriene-1S,3R-diol; 1α-hydroxy-25-methoxy-cholecalciferol; 9547471; Compared to 1α,25-dihydroxyvitamin D_3 (ED_{50} = 1.0 x 10^{-8}), HL-60 human promyelocytes differentiation ED_{50} is 6.5 x 10^{-8} M. *J Biol Chem*. 1987, **262(27)**, 12939-44.

618   **25-Methyl-1α,26-dihydroxyvitamin D_3**
9547474 $C_{28}H_{46}O_3$

(1R,3S,5Z)-5-[(2E)-2-[(1R,3aR,7aS)-1-[(2R)-7-hydroxy-6,6-dimethyl-heptan-2-yl]-7a-methyl-2,3,3a,5,6,7-hexahydro-1H-inden-4-ylidene]ethylidene]-4-methylidene-cyclohexane-1,3-diol.
LMST03020339; 25-methyl-9,10-seco-5Z,7E,10(19)-cholestatriene-1S,3R,26-triol; 25-methyl-1α,26-dihydroxycholecalciferol; 9547474.

619   **25-Hydroxy-1R-(hydroxymethyl)vitamin D_3**
142508-68-7 $C_{28}H_{46}O_3$

activity of ornithinedecarboxylase (ODC); $[\alpha]_D^{25}$ = -64°
(c = 0.09 CH_2Cl_2); λ_m = 264 nm (MeOH). *J Med Chem.*
1992, *35(17)*, 3280-7.

621   *25-Hydroxy-22R-methoxyvitamin D₃*
9547475 $C_{28}H_{46}O_3$

(5R,6S)-6-[(1R,3aR,4E,7aS)-4-[(2Z)-2-[(5S)-5-hydroxy-2-
methylidene-cyclohexylidene]ethylidene]-7a-methyl-
2,3,3a,5,6,7-hexahydro-1H-inden-1-yl]-5-methoxy-2-
methyl-heptan-2-ol.
LMST03020342; 25-hydroxy-22R-methoxycholecalc-
iferol;   22R-methoxy-9,10-seco-5Z,7E,10(19)-cholesta
triene-3S,25-diol; 9547475; λ_m = 267 nm (EtOH).
Chem Pharm Bull (Tokyo). 1981, *29(8)*, 2254-60.

622   *25-Hydroxy-22S-methoxyvitamin D₃*
9547476 $C_{28}H_{46}O_3$

(5S,6S)-6-[(1R,3aR,4E,7aS)-4-[(2Z)-2-[(5S)-5-hydroxy-2-
methylidene-cyclohexylidene]ethylidene]-7a-methyl-
2,3,3a,5,6,7-hexahydro-1H-inden-1-yl]-5-methoxy-2-
methyl-heptan-2-ol.
LMST03020343; 25-hydroxy-22S-methoxycholecalc-
iferol;   22S-methoxy-9,10-seco-5Z,7E,10(19)-cholesta-
triene-3S,25-diol; 9547476; λ_m = 267 nm (EtOH).
Chem Pharm Bull (Tokyo). 1981, *29(8)*, 2254-60.

623   *1α,25-Dihydroxy-1β-methylvitamin D₃*
9547478 $C_{28}H_{46}O_3$

(1S,3R,5Z)-5-[(2E)-2-[(1R,3aR,7aS)-1-[(2R)-6-hydroxy-
6-methyl-heptan-2-yl]-7a-methyl-2,3,3a,5,6,7-hexa-
hydro-1H-inden-4-ylidene]ethylidene]-1-methyl-6-
methylidene-cyclohexane-1,3-diol.
LMST03020345;   1S-methyl-9,10-seco-5Z,7E,10(19)-
cholestatriene-1S,3R,25-triol;   1α,25-dihydroxy-1β-
methylcholecalciferol; 9547478; Synthesized from
1α,25-dihydroxyvitamin D₃ or from the bisnorchola-
5,7-diene-1α,3β,22-triol   N-phenyltriazolinedione
(PTAD) adduct. Binding affinity for calf thymus vitamin
D receptor is 1/150 that of 1α,25-dihydroxyvitamin D₃;
λ_m = 252, 264.6 nm (95% EtOH). *J Org Chem.* 1993,
58(7), 1895-9.

624   *1β,25-Dihydroxy-1α-methylvitamin D₃*
9547479 $C_{28}H_{46}O_3$

(1R,3R,5Z)-5-[(2E)-2-[(1R,3aR,7aS)-1-[(2R)-6-hydroxy-
6-methyl-heptan-2-yl]-7a-methyl-2,3,3a,5,6,7-hexa-
hydro-1H-inden-4-ylidene]ethylidene]-1-methyl-6-
methylidene-cyclohexane-1,3-diol.
LMST03020346;   1R-methyl-9,10-seco-5Z,7E,10(19)-
cholestatriene-1R,3R,25-triol;   1β,25-dihydroxy-1α-
methylcholecalciferol; 9547479; Synthesized from
1α,25-dihydroxyvitamin D₃. Binding affinity for calf
thymus vitamin D receptor is < 1/1000 that of 1α,25-
dihydroxyvitamin D₃; λ_m = 262 nm (95% EtOH). *J Org
Chem.* 1993, *58(7)*, 1895-9.

625   *1α,25-Dihydroxy-8(14)a-homovitamin D₃*
145459-22-9 $C_{28}H_{46}O_3$

(1R,3S,5Z)-5-[(2E)-2-[(1R,3aS,8aR)-1-[(2R)-6-hydroxy-6-methyl-heptan-2-yl]-8a-methyl-1,2,3,3a,4,6,7,8-octahydroazulen-5-ylidene]ethylidene]-4-methylidene-cyclohexane-1,3-diol.
LMST03020347; 8(14)a-Homocalcitriol; ZK 150123; 8(14)a-homo-9,10-seco-5Z,7E,10(19)-cholestatriene-1S,3R,25-triol; 8(14a)-Homo-9,10-secocholesta-5Z,7E,10(19)-triene-1S,3R,25-triol; C(14a)-Home-9,10-seco-cholesta-5Z,7E,10(19)-triene-1α,3β,24-triol; 1α,25-dihydroxy-8(14)a-homocholecalciferol; 6438777; 145459-22-9; A synthetic vitamin D agonist, prepared by Horner coupling of the 25-hydroxylated C-homo-CD-ring ketone with A-ring phosphine oxide. Binds to the pig intestinal receptor with an affinity slightly less than that of 1α,25-dihydroxyvitamin D$_3$ but shows the same potency in inducing HL-60 cell differentiation and inhibition of keratinocyte proliferation, and was found to be approximately 10-fold less potent than 1α,25-dihydroxyvitamin D$_3$ in inducing hypercalcemia and hypercalciuria after a single injection in normal rats. *Steroids.* 1992, *57(9)*, 447-52.

626   *2β-Methyl-1α,25-dihydroxyvitamin D$_3$*
5283776   C$_{28}$H$_{46}$O$_3$

(1R,2R,3S,5Z)-5-[(2E)-2-[(1R,3aR,7aS)-1-[(2R)-6-hydroxy-6-methyl-heptan-2-yl]-7a-methyl-2,3,3a,5,6,7-hexahydro-1H-inden-4-ylidene]ethylidene]-2-methyl-4-methylidene-cyclohexane-1,3-diol.
LMST03020348; 2R-methyl-9,10-seco-5Z,7E,10(19)-cholestatriene-1S,3R,25-triol; 2β-methyl-1α,25-di-hydroxycholecalciferol; 5283776; Synthetic vitamin D$_3$ analog. Compared to 1α,25-dihydroxyvitamin D$_3$, bovine thymus VDR binding is 13%, bone calcium mobilization is 2%, HL-60 cell differentiation is 10% and DBP binding is 79%. *Bioorg Med Chem Lett.* 1998, *8(2)*, 151-6.

627   *2α-Methyl-3-epi-1α,25-dihydroxyvitamin D$_3$*
5283777   C$_{28}$H$_{46}$O$_3$

(1S,2S,3S,5Z)-5-[(2E)-2-[(1R,3aR,7aS)-1-[(2R)-6-hydroxy-6-methyl-heptan-2-yl]-7a-methyl-2,3,3a,5,6,7-hexahydro-1H-inden-4-ylidene]ethylidene]-2-methyl-4-methylidene-cyclohexane-1,3-diol.
LMST03020349; 2S-methyl-9,10-seco-5Z,7E,10(19)-cholestatriene-1S,3S,25-triol; 2α-methyl-3-epi-1α,25-dihydroxycholecalciferol; 5283777; Synthetic vitamin D$_3$ analog. Compared to 1α,25-dihydroxyvitamin D$_3$, bovine thymus VDR binding is 4%, HL-60 cell differentiation is 13% and DBP binding is 45%. *Bioorg Med Chem Lett.* 1998, *8(2)*, 151-6.

628   *2β-Methyl-3-epi-1α,25-dihydroxyvitamin D$_3$*
5283778   C$_{28}$H$_{46}$O$_3$

(1S,2R,3S,5Z)-5-[(2E)-2-[(1R,3aR,7aS)-1-[(2R)-6-hydroxy-6-methyl-heptan-2-yl]-7a-methyl-2,3,3a,5,6,7-hexahydro-1H-inden-4-ylidene]ethylidene]-2-methyl-4-methylidene-cyclohexane-1,3-diol.
LMST03020350; 2R-methyl-9,10-seco-5Z,7E,10(19)-cholestatriene-1S,3S,25-triol; 2β-methyl-3-epi-1α,25-dihydroxycholecalciferol; 5283778; Synthetic vitamin D$_3$ analog. Compared to 1α,25-dihydroxyvitamin D$_3$, bovine thymus VDR binding is 0.3%, HL-60 cell differentiation is 1.5% and DBP binding is 21%. *Bioorg Med Chem Lett.* 1998, *8(2)*, 151-6; *J Med Chem.* 2000, 43(21), 4247-8; *Steroids.* 2001, *66(3-5)*, 277-85.

629   *2α-Methyl-1α,25-dihydroxyvitamin D$_3$*
5283797   C$_{28}$H$_{46}$O$_3$

(1R,2S,3S,5Z)-5-[(2E)-2-[(1R,3aR,7aS)-1-[(2R)-6-hydroxy-6-methyl-heptan-2-yl]-7a-methyl-2,3,3a,5,6,7-hexahydro-1H-inden-4-ylidene]ethylidene]-2-methyl-4-methylidene-cyclohexane-1,3-diol.
LMST03020540;   2α-Methyl-1α,25-dihydroxychole-calciferol; 2S-methyl-9,10-seco-5Z,7E,10(19)-cholesta-triene-1S,3R,25-triol;   2α-methyl-1α,25-dihydroxy-cholecalciferol;   5283797;   Synthetic   vitamin   D$_3$ derivative. Compared to 1α,25-dihydroxyvitamin D$_3$, bovine thymus VDR binding is 400%, bone calcium mobilization is 400%, HL-60 cell differentiation is 200%, and vitamin D receptor binding is 68%; λ$_m$ = 265 nm (EtOH). *Bioorg Med Chem Lett.* 1998, *8(2)*, 151-6; *J Med Chem.* 2000, *43(21)*, 4247-8; *Steroids.* 2001, *66(3-5)*, 277-85.

630   **2α-Methyl-1β,25-dihydroxyvitamin D$_3$**
5283779   C$_{28}$H$_{46}$O$_3$

(1R,2S,3R,5Z)-5-[(2E)-2-[(1R,3aR,7aS)-1-[(2R)-6-hydroxy-6-methyl-heptan-2-yl]-7a-methyl-2,3,3a,5,6,7-hexahydro-1H-inden-4-ylidene]ethylidene]-2-methyl-4-methylidene-cyclohexane-1,3-diol.
LMST03020351;   2S-methyl-9,10-seco-5Z,7E,10(19)-cholestatriene-1R,3R,25-triol;   2α-methyl-1β,25-di-hydroxycholecalciferol; 5283779; Synthetic vitamin D$_3$ analog. Compared to 1α,25-dihydroxyvitamin D$_3$, bovine thymus VDR binding is <0.1%, HL-60 cell differentiation is 1.0% and DBP binding is 200%. *Bioorg Med Chem Lett.* 1998, *8(2)*, 151-156; *J Med Chem.* 2000, *43(21)*, 4247-8; *Steroids.* 2001, *66(3-5)*, 277-85.

631   **2β-Methyl-1β,25-dihydroxyvitamin D$_3$**
5283780   C$_{28}$H$_{46}$O$_3$

(1R,2R,3R,5Z)-5-[(2E)-2-[(1R,3aR,7aS)-1-[(2R)-6-hydroxy-6-methyl-heptan-2-yl]-7a-methyl-2,3,3a,5,6,7-hexahydro-1H-inden-4-ylidene]ethylidene]-2-methyl-4-methylidene-cyclohexane-1,3-diol.
LMST03020352;   2R-methyl-9,10-seco-5Z,7E,10(19)-cholestatriene-1R,3R,25-triol;   2β-methyl-1β,25-di-hydroxycholecalciferol; 5283780; Synthetic vitamin D$_3$ analog. Compared to 1α,25-dihydroxyvitamin D$_3$, bovine thymus VDR binding is <0.1%, HL-60 cell differentiation is 1.5% and DBP binding is 1000%. *Bioorg Med Chem Lett.* 1998, *8(2)*, 151-6; *J Med Chem.* 2000, *43(21)*, 4247-8; *Steroids.* 2001, *66(3-5)*, 277-85.

632   **2α-Methyl-3-epi-1β,25-dihydroxyvitamin D$_3$**
5283781   C$_{28}$H$_{46}$O$_3$

(1S,2S,3R,5Z)-5-[(2E)-2-[(1R,3aR,7aS)-1-[(2R)-6-hydroxy-6-methyl-heptan-2-yl]-7a-methyl-2,3,3a,5,6,7-hexahydro-1H-inden-4-ylidene]ethylidene]-2-methyl-4-methylidene-cyclohexane-1,3-diol.
LMST03020353;   2S-methyl-9,10-seco-5Z,7E,10(19)-cholestatriene-1R,3S,25-triol;   2α-methyl-3-epi-1β,25-dihydroxycholecalciferol; 5283781; Synthetic vitamin D$_3$ analog. Compared to 1α,25-dihydroxyvitamin D$_3$, bovine thymus VDR binding is <0.1%, HL-60 cell differentiation is 0.5% and DBP binding is 1200%. *Bioorg Med Chem Lett.* 1998, *8(2)*, 151-6; *J Med Chem.* 2000, *43(21)*, 4247-8; *Steroids.* 2001, *66(3-5)*, 277-85.

633   **2β-Methyl-3-epi-1β,25-dihydroxyvitamin D$_3$**
5283782   C$_{28}$H$_{46}$O$_3$

(1S,2R,3R,5Z)-5-[(2E)-2-[(1R,3aR,7aS)-1-[(2R)-6-hydroxy-6-methyl-heptan-2-yl]-7a-methyl-2,3,3a,5,6,7-hexahydro-1H-inden-4-ylidene]ethylidene]-2-methyl-4-methylidene-cyclohexane-1,3-diol.
LMST03020354; 2R-methyl-9,10-seco-5Z,7E,10(19)-cholestatriene-1R,3S,25-triol; 2β-methyl-3-epi-1β,25-dihydroxycholecalciferol; 5283782; Synthetic vitamin D_3 analog. Compared to 1α,25-dihydroxyvitamin D_3, bovine thymus VDR binding is 0.8%, HL-60 cell differentiation is 3.0% and DBP binding is 1300%. *Bioorg Med Chem Lett.* 1998, *8(2)*, 151-6; *J Med Chem.* 2000, *43(21)*, 4247-8; *Steroids.* 2001, *66(3-5)*, 277-85.

634   *2α-Methyl-20-epi-1α,25-dihydroxy-vitamin D₃*

5283783 $C_{28}H_{46}O_3$

(1R,2S,3S,5Z)-5-[(2E)-2-[(1R,3aR,7aS)-1-[(2S)-6-hydroxy-6-methyl-heptan-2-yl]-7a-methyl-2,3,3a,5,6,7-hexahydro-1H-inden-4-ylidene]ethylidene]-2-methyl-4-methylidene-cyclohexane-1,3-diol.
LMST03020355; 2S-methyl-9,10-seco-5Z,7E,10(19)-cholestatriene-1S,3R,25-triol; 2α-methyl-20-epi-1α,25-dihydroxycholecalciferol; 95045; 5283783; Synthetic vitamin D_3 analog. Highly potent synthetic vitamin D derivative. Compared to 1α,25-dihydroxyvitamin D_3, bovine thymus VDR binding is 1200%, HL-60 cell differentiation is 59000%, bone calcum mobilization is 655% and DBP binding is <0.3%. Antagonist to 2α-methyl-1α,25-dihydroxyvitamin D_3. *Bioorg Med Chem Lett.* 1998, *8(2)*, 151-6; *J Med Chem.* 2000, *43(21)*, 4247-8; *Steroids.* 2001, *66(3-5)*, 277-85; *J Steroid Biochem Mol Biol.* 2005, *94(5)*, 469-79.

635   *2β-Methyl-20-epi-1α,25-dihydroxy-vitamin D₃*

5283784 $C_{28}H_{46}O_3$

(1R,2R,3S,5Z)-5-[(2E)-2-[(1R,3aR,7aS)-1-[(2S)-6-hydroxy-6-methyl-heptan-2-yl]-7a-methyl-2,3,3a,5,6,7-hexahydro-1H-inden-4-ylidene]ethylidene]-2-methyl-4-methylidene-cyclohexane-1,3-diol.
LMST03020356; (20S)-2R-methyl-9,10-seco-5Z,7E,10(19)-cholestatriene-1S,3R,25-triol; 2β-methyl-20-epi-1α,25-dihydroxycholecalciferol; 5283784; Synthetic vitamin D_3 analog. Compared to 1α,25-dihydroxyvitamin D_3, bovine thymus VDR binding is 160%, HL-60 cell differentiation is 2600%, bone calcum mobilization is 115% and DBP binding is <0.3%. *Bioorg Med Chem Lett.* 1998, *8(2)*, 151-6; *J Med Chem.* 2000, *43(21)*, 4247-8; *Steroids.* 2001, *66(3-5)*, 277-85.

636   *2α-Methyl-3-epi-20-epi-1α,25-di-hydroxyvitamin D₃*

5283785 $C_{28}H_{46}O_3$

(1S,2S,3S,5Z)-5-[(2E)-2-[(1R,3aR,7aS)-1-[(2S)-6-hydroxy-6-methyl-heptan-2-yl]-7a-methyl-2,3,3a,5,6,7-hexahydro-1H-inden-4-ylidene]ethylidene]-2-methyl-4-methylidene-cyclohexane-1,3-diol.
LMST03020357; (20S)-2S-methyl-9,10-seco-5Z,7E,10(19)-cholestatriene-1S,3S,25-triol; 2α-methyl-3-epi-20-epi-1α,25-dihydroxycholecalciferol; 5283785; Synthetic vitamin D_3 analog. Compared to 1α,25-dihydroxyvitamin D_3, bovine thymus VDR binding is 17%, HL-60 cell differentiation is 730%, bone calcum mobilization is 144% and DBP binding is <0.3%. Bovine thymus VDR binding is 17. *Bioorg Med Chem*

Lett. 1998, *8(2)*, 151-6; *J Med Chem.* 2000, *43(21)*, 4247-8; *Steroids.* 2001, *66(3-5)*, 277-85.

Med Chem. 2000, *43(21)*, 4247-8; *Steroids.* 2001, *66(3-5)*, 277-85.

637   **2β-Methyl-3-epi-20-epi-1α,25-dihydroxyvitamin D₃**
5283786 $C_{28}H_{46}O_3$

(1S,2R,3S,5Z)-5-[(2E)-2-[(1R,3aR,7aS)-1-[(2S)-6-hydroxy-6-methyl-heptan-2-yl]-7a-methyl-2,3,3a,5,6,7-hexahydro-1H-inden-4-ylidene]ethylidene]-2-methyl-4-methylidene-cyclohexane-1,3-diol.
LMST03020358;   (20S)-2R-methyl-9,10-seco-5Z,7E,10(19)-cholestatriene-1S,3S,25-triol;   2β-methyl-3-epi-20-epi-1α,25-dihydroxycholecalciferol;   5283786; Synthetic vitamin D₃ analog. Compared to 1α,25-dihydroxyvitamin D₃, bovine thymus VDR binding is <0.1%, HL-60 cell differentiation is 6%, and DBP binding is <0.3%. *Bioorg Med Chem Lett.* 1998, *8(2)*, 151-6; *J Med Chem.* 2000, *43(21)*, 4247-8; *Steroids.* 2001, *66(3-5)*, 277-85.

639   **2β-Methyl-20-epi-1β,25-dihydroxyvitamin D₃**
5283788 $C_{28}H_{46}O_3$

(1R,2R,3R,5Z)-5-[(2E)-2-[(1R,3aR,7aS)-1-[(2S)-6-hydroxy-6-methyl-heptan-2-yl]-7a-methyl-2,3,3a,5,6,7-hexahydro-1H-inden-4-ylidene]ethylidene]-2-methyl-4-methylidene-cyclohexane-1,3-diol.
LMST03020360;   (20S)-2R-methyl-9,10-seco-5Z,7E,10(19)-cholestatriene-1R,3R,25-triol;   2β-methyl-20-epi-1β,25-dihydroxycholecalciferol; 5283788; Synthetic vitamin D₃ analog. Compared to 1α,25-dihydroxyvitamin D₃, bovine thymus VDR binding is <0.1%, HL-60 cell differentiation is 1%, and DBP binding is <0.3%. *Bioorg Med Chem Lett.* 1998, *8(2)*, 151-6; *J Med Chem.* 2000, *43(21)*, 4247-8; *Steroids.* 2001, *66(3-5)*, 277-85.

638   **2α-Methyl-20-epi-1β,25-dihydroxyvitamin D₃**
5283787 $C_{28}H_{46}O_3$

(1R,2S,3R,5Z)-5-[(2E)-2-[(1R,3aR,7aS)-1-[(2S)-6-hydroxy-6-methyl-heptan-2-yl]-7a-methyl-2,3,3a,5,6,7-hexahydro-1H-inden-4-ylidene]ethylidene]-2-methyl-4-methylidene-cyclohexane-1,3-diol.
LMST03020359;   (20S)-2S-methyl-9,10-seco-5Z,7E,10(19)-cholestatriene-1R,3R,25-triol;   2α-methyl-20-epi-1β,25-dihydroxycholecalciferol; 5283787; Synthetic vitamin D₃ analog. Compared to 1α,25-dihydroxyvitamin D₃, bovine thymus VDR binding is <0.1%, HL-60 cell differentiation is 3%, and DBP binding is <0.3%. *Bioorg Med Chem Lett.* 1998, *8(2)*, 151-6; *J*

640   **2α-Methyl-3-epi-20-epi-1β,25-dihydroxyvitamin D₃**
5283789 $C_{28}H_{46}O_3$

(1S,2S,3R,5Z)-5-[(2E)-2-[(1R,3aR,7aS)-1-[(2S)-6-hydroxy-6-methyl-heptan-2-yl]-7a-methyl-2,3,3a,5,6,7-hexahydro-1H-inden-4-ylidene]ethylidene]-2-methyl-4-methylidenecyclohexane-1,3-diol.
LMST03020361;   (20S)-2S-methyl-9,10-seco-5Z,7E,10(19)-cholestatriene-1R,3S,25-triol; 2α-methyl-3-epi-20-epi-1β,25-dihydroxycholecalciferol;   5823789; Synthetic vitamin D₃ analog. Compared to 1α,25-dihydroxyvitamin D₃, bovine thymus VDR binding is <0.1%, and DBP binding is <0.3%. *Bioorg Med Chem Lett.* 1998, *8(16)*,

2145-8; *J Med Chem*. 2000, *43(21)*, 4247-8; *Steroids*. 2001, *66(3-5)*, 277-85.

641 *2β-Methyl-3-epi-20-epi-1β,25-dihydroxy-vitamin D₃*

5283790 $C_{28}H_{46}O_3$

(1S,2R,3R,5Z)-5-[(2E)-2-[(1R,3aR,7aS)-1-[(2S)-6-hydroxy-6-methyl-heptan-2-yl]-7a-methyl-2,3,3a,5,6,7-hexahydro-1H-inden-4-ylidene]ethylidene]-2-methyl-4-methylidene-cyclohexane-1,3-diol. LMST03020362; (20S)-2R-methyl-9,10-seco-5Z,7E,10(19)-cholestatriene-1R,3S,25-triol; 2β-methyl-3-epi-20-epi-1β,25-dihydroxycholecalciferol; 5283790; Synthetic vitamin D₃ analog. Compared to 1α,25-dihydroxyvitamin D₃, bovine thymus VDR binding is 7%, bone calcium mobilization is 19%, HL-60 cell differentiation is 190%, and DBP binding is <0.3%. ; *Bioorg Med Chem Lett*. 1998, *8(2)*, 151-6; *J Med Chem*. 2000, *43(21)*, 4247-8; *Steroids*. 2001, *66(3-5)*, 277-85.

642 *1α,25-Dihydroxy-11α-methylvitamin D₃*

5283791 $C_{28}H_{46}O_3$

(1R,3S,5Z)-5-[(2E)-2-[(1R,3aR,6R,7aS)-1-[(2R)-6-hydroxy-6-methyl-heptan-2-yl]-6,7a-dimethyl-2,3,3a,5,6,7-hexahydro-1H-inden-4-ylidene]ethylidene]-4-methylidene-cyclohexane-1,3-diol. LMST03020363; 11S-methyl-9,10-seco-5Z,7E,10(19)-cholestatriene-1S,3R,25-triol; 1α,25-dihydroxy-11α-methylcholecalciferol; 5283791; Synthetic vitamin D₃ analog. Compared to 1α,25-dihydroxyvitamin D₃, affinity for rat duodenum receptor, UMR 106 cell receptor and human vitamin D binding protein are 230%, 80% and 340%. Differentiation of HL-60 cells is 113%, inhibition of the proliferation of peripheral

blood mononuclear cells (PBMC) is 90%, bone resorption is 100% (3 days), 100% (6 days). In rachitic chicks: serum calcium is 40%, bone calcium is 46%, serum osteocalcin is 50% and calbindin D-28K is 60%. *J Biol Chem*. 1992, *267(5)*, 3044-51.

643 *1α,25-Dihydroxy-11β-methylvitamin D₃*

5283792 $C_{28}H_{46}O_3$

(1R,3S,5Z)-5-[(2E)-2-[(1R,3aR,6S,7aS)-1-[(2R)-6-hydroxy-6-methyl-heptan-2-yl]-6,7a-dimethyl-2,3,3a,5,6,7-hexahydro-1H-inden-4-ylidene]ethylidene]-4-methylidene-cyclohexane-1,3-diol. LMST03020364; 11R-methyl-9,10-seco-5Z,7E,10(19)-cholestatriene-1S,3R,25-triol; 1α,25-dihydroxy-11β-methylcholecalciferol; 5283792; Synthetic vitamin D₃ analog. Compared to 1α,25-dihydroxyvitamin D₃, affinity for rat duodenum receptor and human vitamin D binding protein are 37% and 86%, differentiation of HL-60 cells is 19%, inhibition of the proliferation of peripheral blood mononuclear cells (PBMC) is 1.4%, bone resorption is <1 (3 days), 1 (6 days), biological activity in rachitic chick: serum calcium is 2%, bone calcium is 5%, serum osteocalcin is 5% and calbindin D-28K is 3%. *J Biol Chem*. 1992, *267(5)*, 3044-51.

644 *1α,25-Dihydroxy-18-methylvitamin D₃*

9547480 $C_{28}H_{46}O_3$

(1R,3S,5Z)-5-[(2E)-2-[(1R,3aR,7aS)-7a-ethyl-1-[(2R)-6-hydroxy-6-methyl-heptan-2-yl]-2,3,3a,5,6,7-hexahydro-1H-inden-4-ylidene]ethylidene]-4-methylidene-cyclohexane-1,3-diol. LMST03020365; 18-methyl-9,10-seco-5Z,7E,10(19)-cholestatriene-1S,3R,25-triol; 1α,25-dihydroxy-18-methylcholecalciferol; 9547480; Synthetic vitamin D₃ analog. Compared to 1α,25-dihydroxyvitamin D₃, affinity for calf thymus receptor and vitamin D binding protein are 100% and 100%, differentiation of HL-60 cells is 50%, intestinal calcium transport activity evaluated using Caco-2 intestinal cancer cell line is 100%. *Bioorg MedChem Lett.* 1993, *3*, 1855-8.

645   *1α,25-Dihydroxy-24a-homovitamin D₃*

103656-40-2 $C_{28}H_{46}O_3$

(1R,3S,5Z)-5-[(2E)-2-[(1R,3aR,7aS)-1-[(2R)-7-hydroxy-7-methyl-octan-2-yl]-7a-methyl-2,3,3a,5,6,7-hexahydro-1H-inden-4-ylidene]ethylidene]-4-methylidene-cyclohexane-1,3-diol.
LMST03020366; 24a-homo-9,10-seco-5Z,7E,10(19)-cholestatriene-1S,3R,25-triol; 1α,25-dihydroxy-24a-homocholecalciferol; 9547481; 24-Homo-1,25-dihydroxyvitamin D₃; 24-Homocalcitriol; 6438975; 9547481; 103656-40-2; Synthesized from hydroxy-protected (5E)-23,24,25,26,27-pentanorvitamin D₃ 22-bis(methylseleno)acetal or from hydroxy-protected (5E)-22-tosyloxy-23,24,25,26,27-pentanorvitamin D₃. About ten times more active than 1α,25-dihydroxyvitamin D₃ in HL-60 human promyelocytes differentiation. Involved in bone resorption; λ_m = 265 nm (EtOH). *Chem Pharm Bull (Tokyo).* 1986, *34(11)*, 4508-15; *Biochemistry.* 1986, *25(19)*, 5512-8; *J Biol Chem.* 1987, *262(29)*, 14164-71.

646   *1α,25-Dihydroxy-24a-homo-20-epi-vitamin D₃*

9547482 $C_{28}H_{46}O_3$

(1R,3S,5Z)-5-[(2E)-2-[(1R,3aR,7aS)-1-[(2S)-7-hydroxy-7-methyl-octan-2-yl]-7a-methyl-2,3,3a,5,6,7-hexahydro-1H-inden-4-ylidene]ethylidene]-4-methylidene-cyclohexane-1,3-diol.
LMST03020368; (20S)-24a-homo-9,10-seco-5Z,7E,10(19)-cholestatrien-1S,3R,25-triol; 1α,25-dihydroxy-24a-homo-20-epicholecalciferol; 9547482; Synthes-

ized from protected (5E)-1α-hydroxy-22-p-toluenesulfonyloxy-23,24,25,26,27-pentanorvitamin D₃. Compared to 1α,25-dihydroxyvitamin D₃, inhibition of U937 cell (human histiocytic lymphoma cell line) proliferation is 31800%, induction of U937 cell differentiation is 20000%, binding to the 1α,25-dihydroxyvitamin D₃ receptor from rachitic chicken intestine is 110%, calcemic activity determined by the increase in urinary calcium excretion in rats is 500%, inhibitory effects on murine thymocyte activation is 38000%. *Biochem Pharmacol.* 1991, *42(8)*, 1569-75.

647   *1α,25-Dihydroxy-24R-methylvitamin D₃*

5283793 $C_{28}H_{46}O_3$

(1R,3S,5Z)-5-[(2E)-2-[(1R,3aR,7aS)-1-[(2R,5R)-6-hydroxy-5,6-dimethyl-heptan-2-yl]-7a-methyl-2,3,3a,5,6,7-hexahydro-1H-inden-4-ylidene]ethylidene]-4-methylidene-cyclohexane-1,3-diol.
LMST03020367; 24R-methyl-9,10-seco-5Z,7E,10(19)-cholestatriene-1S,3,25-triol; 1α,25-dihydroxy-24R-methylcholecalciferol; 5283793.

648   *1α,25-Dihydroxy-26-methylvitamin D₃*

105687-81-8 $C_{28}H_{46}O_3$

(1R,3S,5Z)-5-[(2E)-2-[(1R,3aR,7aS)-1-[(2R)-6-hydroxy-6-methyl-octan-2-yl]-7a-methyl-2,3,3a,5,6,7-hexahydro-1H-inden-4-ylidene]ethylidene]-4-methylidene-cyclohexane-1,3-diol.
LMST03020371; 26-methyl-9,10-seco-5Z,7E,10(19)-cholestatriene-1S,3R,25-triol; 1α,25-dihydroxy-26-methylcholecalciferol; 26-Homo-1,25-dihydroxyvitamin D₃; 26-Homocalcitriol; 27-Nor-9,7-secocholesta-5Z,7E,10(19)triene-1α,3β,25-triol,25-ethyl-; 5283594; 6439053; 105687-81-8; Less effective that 1α,25-dihydroxyvitamin D₃ by one order of magnitude in HL-60 human promyelocytes differentiation. Fails to

show a higher bone-resorbing activity than 1,25-dihydroxyvitamin D_3. *J Biol Chem.* 1987, *262(27),* 12939-44.

649 ***1α,25-Dihydroxy-26-methylvitamin D_3***

5283794 $C_{28}H_{46}O_3$

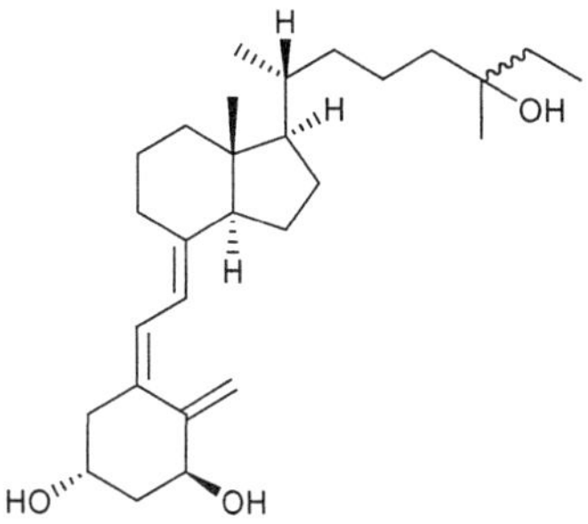

(1R,3S,5Z)-5-[(2E)-2-[(1R,3aR,7aS)-1-[(2R)-6-hydroxy-6-methyl-octan-2-yl]-7a-methyl-2,3,3a,5,6,7-hexa-hydro-1H-inden-4-ylidene]ethylidene]-4-methylidene-cyclohexane-1,3-diol.
LMST03020372; 26-methyl-9,10-seco-5Z,7E,10(19)-cholestatriene-1S,3R,25-triol; 1α,25-dihydroxy-26-methylcholecalciferol; 5283794; $λ_m$ = 265 nm (EtOH). *Chem Pharm Bull (Tokyo).* 1986, *34(11),* 4508-15.

650 ***2α-Methyl-20-epi-1α,25-dihydroxy-vitamin D_3***

5283783 $C_{28}H_{46}O_3$

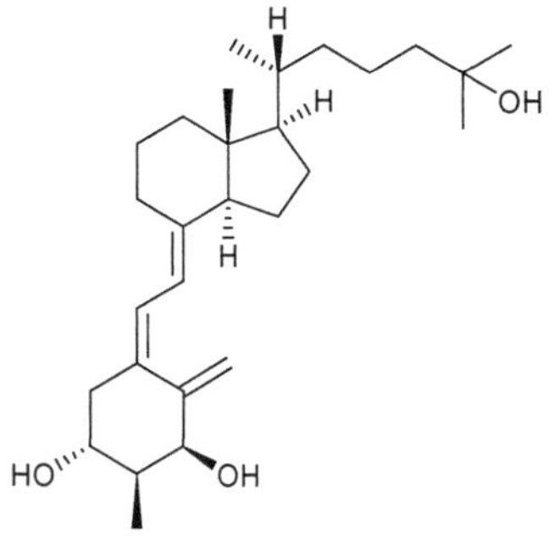

(5Z,7E)-(1S,2S,3R,20S)-2-methyl-9,10-seco-5,7,10(19)-cholestatriene-1,3,25-triol.
2α-Methyl-20-epi-1α,25-dihydroxyvitamin D_3; 2α-methyl-20-epi-1α,25-dihydroxycholecalciferol; LMST03020355; 5283783; Highly potent synthetic vitamin D derivative. Antagonist to 2α-methyl-1α,25-dihydroxyvitamin D_3. Compared to 1α,25-dihydroxy-vitamin D_3, binding to bovine thymus VDR binding is 1200%, bone calcium mobilization is 655%, HL-60 cell differentiation is 59000%, and DBP binding is < 0.3%. *J Steroid Biochem Mol Biol.* 2005, *94(5),* 469-79; *Bioorg Med Chem Lett.* 1998, *8(16),* 2145-8.

651 ***(7E,22E)-9,10-Secoergosta-7E,10,22E-triene-3,5,6-triol***

5366081 $C_{28}H_{46}O_3$

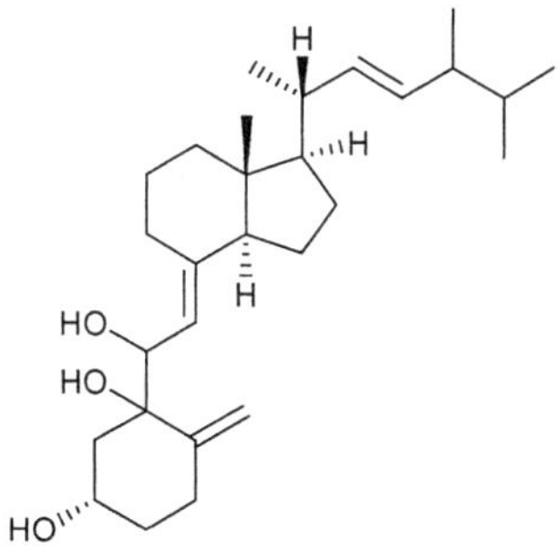

1-[(2E)-2-[1-[(E)-5,6-dimethylhept-3-en-2-yl]-7a-methyl-2,3,3a,5,6,7-hexahydro-1H-inden-4-ylidene]-1-hydroxy-ethyl]-6-methylidene-cyclohexane-1,3-diol.
(7E,22E)-9,10-Secoergosta-7E,10,22E-triene-3,5,6-triol; 9,10-Secoergosta-7,10(19),22-triene-3β,5,6-triol; 5366081.

652 ***20-Methyl-1α,25-dihydroxyvitamin D_3***

$C_{28}H_{46}O_3$

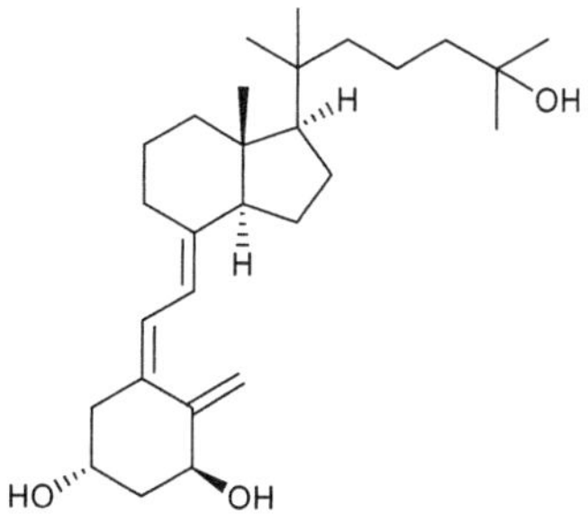

20-Methyl-1,25-dihydroxyvitamin D_3; Synthetic analog of vitamin D_3. Metabolized much like 1,25-dihydroxyvitamin D_3 but the efficiency of 23-hydroxylation is low compared with that of the natural hormone. *Biochem Pharmacol.* 2001, *61(7),* 893-902.

653 ***1α,25-Dihydroxy-2E-ethylidene-19-norvitamin D_3***

$C_{28}H_{46}O_3$

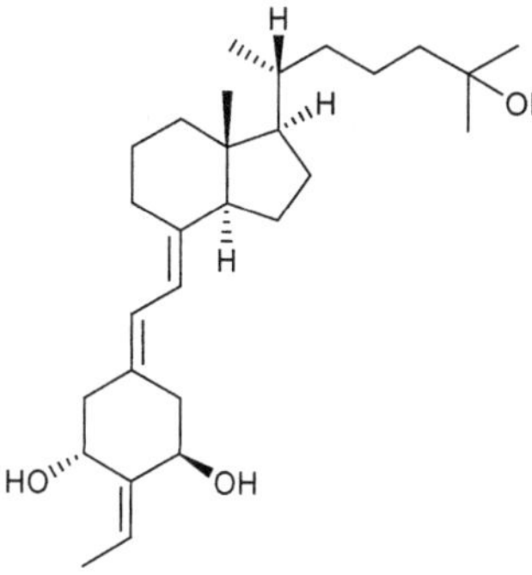

Synthetic vitamin D_3 analog. Has higher HL-60 differentiation activity and calcemic activity than 1,25-

dihydroxyvitamin D_3. *J Med Chem*. 2002, *45(16)*, 3366-80.

654   1α,25-Dihydroxy-2Z-ethylidene-19-norvitamin D_3

$C_{28}H_{46}O_3$

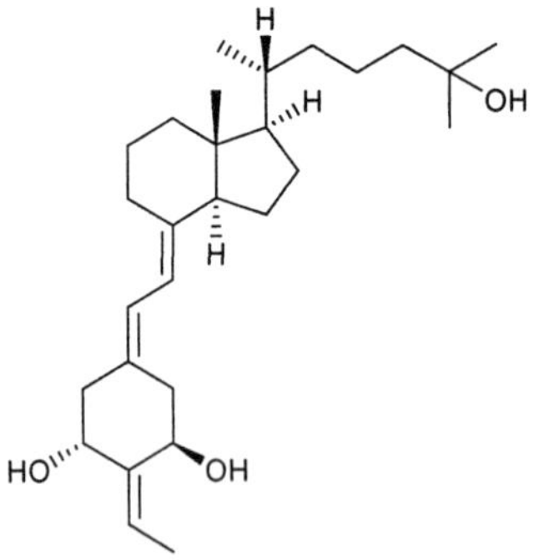

Synthetic vitamin D_3 analog. Has higher HL-60 differentiation activity and calcemic activity than 1,25-dihydroxyvitamin D_3. *J Med Chem*. 2002, *45(16)*, 3366-80.

655   20(17→12β)-abeo-1α,25-Dihydroxy-24-dihomo-21-norvitamin D_3

$C_{28}H_{46}O_3$

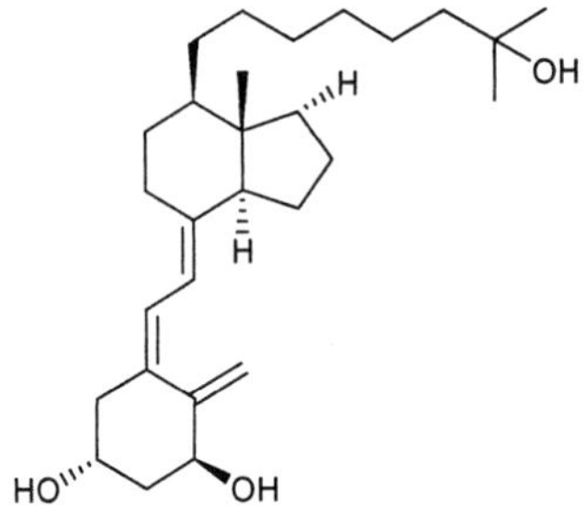

Has high affinity for the vitamin D receptor and transactivation activity about 20% that of 1α,25-dihydroxyvitamin D_3; White solid. *J Med Chem*. 2006, *49(5)*, 1509-16.

656   6R-Methylvitamin D_3 6,19-sulfur dioxide adduct

9547472 $C_{28}H_{46}O_3S$

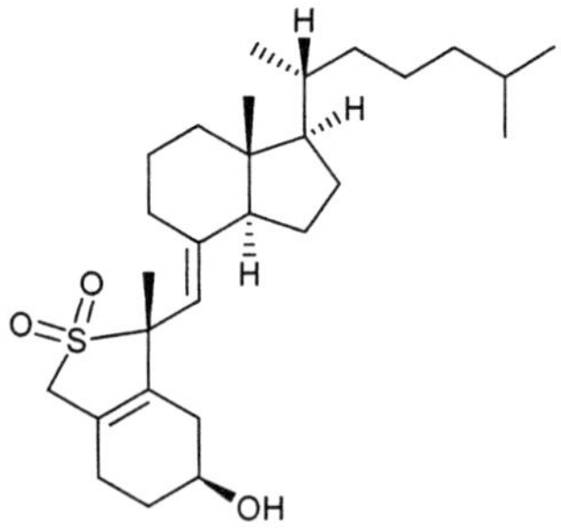

(3S,9R)-9-[(E)-[(1R,3aR,7aS)-7a-methyl-1-[(2R)-6-methylheptan-2-yl]-2,3,3a,5,6,7-hexahydro-1H-inden-4-ylidene]methyl]-9-methyl-8,8-dioxo-8-thiabicyclo[4.3.0]non-10-en-3-ol.
LMST03020337;   6R-methyl-6,19-epithio-9,10-seco-5(10),7E-cholestadien-3S-ol   S,S-dioxide;   6R-methyl-cholecalciferol 6,19-sulfur dioxide adduct; 9547472; Synthesized from vitamin D_3 SO_2 adduct by treatment with methyl iodide in the presence of NaH. *Tet Lett*. 1981, *22(39)*, 3085-8.

657   6S-Methylvitamin D_3 6,19-sulfur dioxide adduct

9547473 $C_{28}H_{46}O_3S$

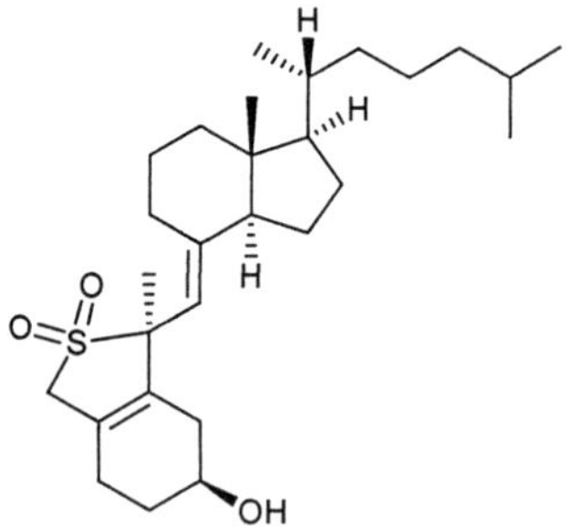

(3S,9S)-9-[(E)-[(1R,3aR,7aS)-7a-methyl-1-[(2R)-6-methylheptan-2-yl]-2,3,3a,5,6,7-hexahydro-1H-inden-4-ylidene]methyl]-9-methyl-8,8-dioxo-8-thiabicyclo[4.3.0]non-10-en-3-ol.
LMST03020338;   6S-methyl-6,19-epithio-9,10-seco-5(10),7E-cholestadien-3S-ol   S,S-dioxide;   6S-methyl-cholecalciferol 6,19-sulfur dioxide adduct; 9547473; Synthesized from vitamin D_3 SO_2 adduct by treatment with methyl iodide in the presence of NaH. *Tet Lett*. 1981, *22(39)*, 3085-8.

658   1α,25-Dihydroxy-24a,24b-dihomo-22-thia-20-epivitamin D_3

9547483 $C_{28}H_{46}O_3S$

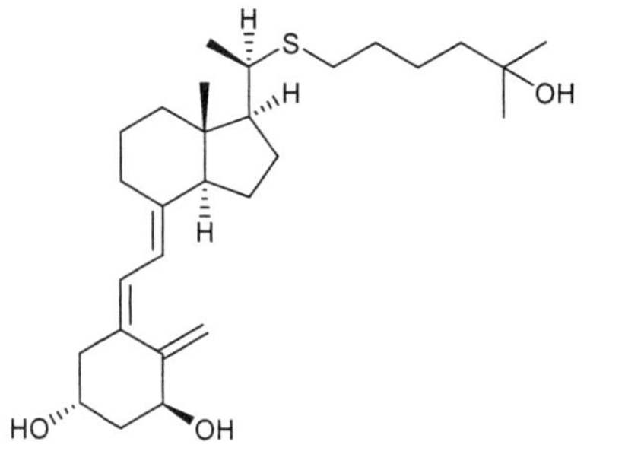

(1R,3S,5Z)-5-[(2E)-2-[(1S,3aR,7aR)-1-[(1R)-1-(5-hydroxy-5-methyl-hexyl)sulfanylethyl]-7a-methyl-2,3,3a,5,6,7-hexahydro-1H-inden-4-ylidene]ethylidene]-4-methylidene-cyclohexane-1,3-diol. LMST03020369; (20R)-24a,24b-dihomo-9,10-seco-22-thia-5Z,7E,10(19)-cholestatriene-1S,3R,25-triol; 1α,25-dihydroxy-24a,24b-dihomo-22-thia-20-epicholecalciferol; 9547483; Synthesized from dehydroepiandrosterone. *Bioorg Med Chem Lett.* 1995, *5(3)*, 279-82.

659 **1α,25-Dihydroxy-24a,24b-dihomo-22-thiavitamin D$_3$**

9547484 $C_{28}H_{46}O_3S$

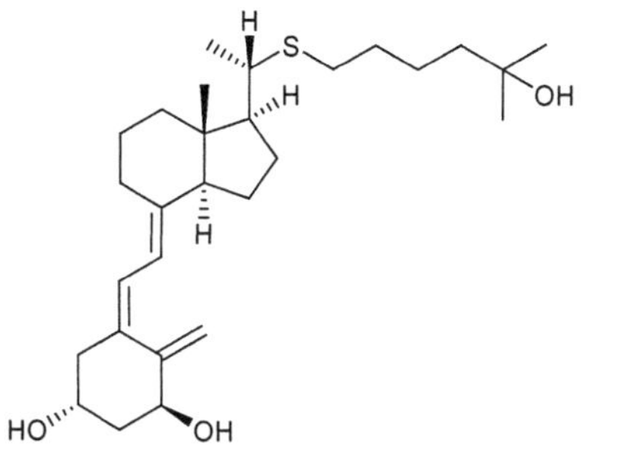

(1R,3S,5Z)-5-[(2E)-2-[(1S,3aR,7aR)-1-[(1S)-1-(5-hydroxy-5-methyl-hexyl)sulfanylethyl]-7a-methyl-2,3,3a,5,6,7-hexahydro-1H-inden-4-ylidene]ethylidene]-4-methylidene-cyclohexane-1,3-diol. LMST03020370; 24a,24b-dihomo-9,10-seco-22-thia-5Z,7E,10(19)-cholestatriene-1S,3R,25-triol; 1α,25-dihydroxy-24a,24b-dihomo-22-thiacholecalciferol; 9547484; Synthesized from dehydroepiandrosterone. *Bioorg Med Chem Lett.* 1995, *5(3)*, 279-82.

660 **1α,25-Dihydroxy-26,27-dimethyl-22-thiavitamin D$_3$**

9547485 $C_{28}H_{46}O_3S$

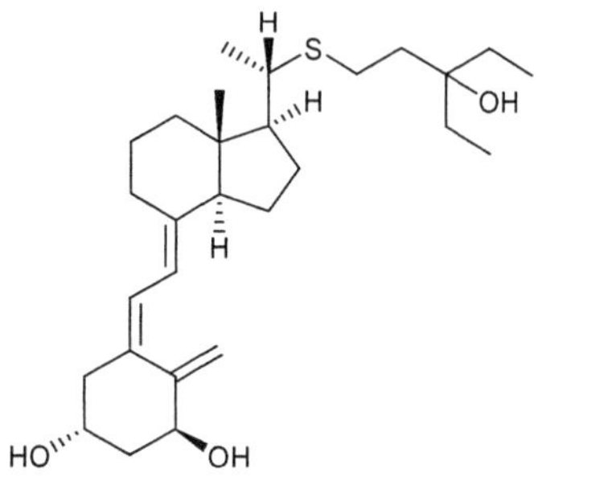

(1R,3S,5Z)-5-[(2E)-2-[(1S,3aR,7aR)-1-[(1S)-1-(3-ethyl-3-hydroxy-pentyl)sulfanylethyl]-7a-methyl-2,3,3a,5,6,7-hexahydro-1H-inden-4-ylidene]ethylidene]-4-methylidene-cyclohexane-1,3-diol. LMST03020373; 26,27-dimethyl-9,10-seco-22-thia-5Z,7E,10(19)-cholestatriene-1S,3R,25-triol; 1α,25-di-hydroxy-26,27-dimethyl-22-thiacholecalciferol; 9547485; Synthesized from dehydroepiandrosterone. *Bioorg Med Chem Lett.* 1995, *5(3)*, 279-82.

661 **1α,25-Dihydroxy-26,27-dimethyl-22-thia-20-epivitamin D$_3$**

9547486 $C_{28}H_{46}O_3S$

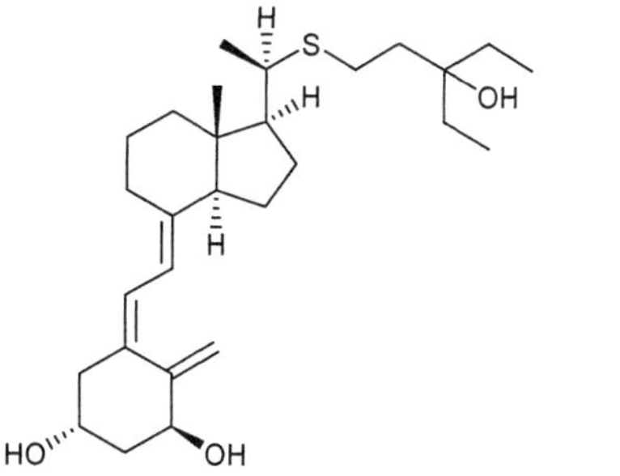

(1R,3S,5Z)-5-[(2E)-2-[(1S,3aR,7aR)-1-[(1R)-1-(3-ethyl-3-hydroxy-pentyl)sulfanylethyl]-7a-methyl-2,3,3a,5,6,7-hexahydro-1H-inden-4-ylidene]ethylidene]-4-methylidene-cyclohexane-1,3-diol. LMST03020374; (20R)-26,27-dimethyl-9,10-seco-22-thia-5Z.7E,10(19)-cholestatriene-1S,3R,25-triol; 1α,25-dihydroxy-26,27-dimethyl-22-thia-20-epicholecalc-iferol; 9547486; Synthesized from dehydroepiandro-sterone. *Bioorg Med Chem Lett.* 1995, *5(3)*, 279-82.

662 **1α,25-Dihydroxy-26,27-dimethyl-20,21-didehydro-23-oxavitamin D$_3$**

9547467 $C_{28}H_{46}O_4$

(1R,3S,5Z)-5-[(2E)-2-[(1R,3aR,7aS)-1-[(2S)-1-(2-ethyl-2-hydroxy-butoxy)propan-2-yl]-7a-methyl-2,3,3a,5,6,7-hexahydro-1H-inden-4-ylidene]ethylidene]-4-methylidene-cyclohexane-1,3-diol.
LMST03020326; 26,27-dimethyl-23-oxa-9,10-seco-5Z,7E,10(19),20-cholestatetraene-1S,3R,25-triol; 1α,25-dihydroxy-26,27-dimethyl-20,21-didehydro-23-oxacholecalciferol; 9547467; Compared to 1α,25-dihydroxyvitamin D₃, affinity for pig intestinal nuclear receptor is 33%, affinity for human vitamin D binding protein is < 0.02%, differentiation of HL-60 cells is 33%, and calciuric effect is 20%. Vitamin D. A Pluripotent Steroid Hormone: Structural Studies, Molecular Endocrinology and Clinical Applications. Proceedings of the Ninth Workshop on Vitamin D, Orlando, Florida (USA) May 28-June 2. Synthesis and Biological Activity of 23-Oxa Vitamin D Analogues. (Norman, A. W., Bouillon, R., Thomasset, M., eds), pp93-94, Walter de Gruyter Berlin New York (1994).

663   **1α,24R-Dihydroxy-26,27-dimethyl-22-oxavitamin D₃**
9547487 $C_{28}H_{46}O_4$

(1R,3S,5Z)-5-[(2E)-2-[(1S,3aR,7aR)-1-[(1S)-1-[(2R)-3-ethyl-2-hydroxy-pentoxy]ethyl]-7a-methyl-2,3,3a,5,6,7-hexahydro-1H-inden-4-ylidene]ethylidene]-4-methylidene-cyclohexane-1,3-diol.
LMST03020375; 1α,24R-dihydroxy-26,27-dimethyl-22-oxacholecalciferol; 26,27-dimethyl-22-oxa-9,10-seco-5Z,7E,10(19)-cholestatriene-1S,3R,24R-triol; 9547487; Synthesized from dehydroepiandrosterone. Induction of differentiation of human myeloid leukemia cells (HL-60) into macrophages **in vitro** estimated by superoxide anion generation compared to that of 1α,25-dihydroxyvitamin D₃; $[α]_D^{24}$ = +43.47° (c = 0.115 EtOH); $λ_m$ = 263 nm (EtOH). **Bioorg Med Chem** Lett. 1993, **3(9)**, 1845-8.

664   **1α,24S-Dihydroxy-26,27-dimethyl-22-oxavitamin D₃**
9547488 $C_{28}H_{46}O_4$

(1R,3S,5Z)-5-[(2E)-2-[(1S,3aR,7aR)-1-[(1S)-1-[(2S)-3-ethyl-2-hydroxy-pentoxy]ethyl]-7a-methyl-2,3,3a,5,6,7-hexahydro-1H-inden-4-ylidene]ethylidene]-4-methylidene-cyclohexane-1,3-diol.
LMST03020376; 1α,24S-dihydroxy-26,27-dimethyl-22-oxacholecalciferol; 26,27-dimethyl-22-oxa-9,10-seco-5Z,7E,10(19)-cholestatriene-1S,3R,24S-triol; 9547488; Synthesized from dehydroepiandrosterone. Induction of differentiation of human myeloid leukemia cells (HL-60) into macrophages **in vitro** estimated by superoxide anion generation compared to that of 1α,25-dihydroxyvitamin D₃; $[α]_D$ = +38.99° (c = 0.159 EtOH); $λ_m$ = 263 nm (EtOH). **Bioorg Med Chem Lett.** 1993, **3(9)**, 1845-8.

665   **1α,25-Dihydroxy-11α-methoxyvitamin D₃**
9547489 $C_{28}H_{46}O_4$

(1R,3S,5Z)-5-[(2E)-2-[(1R,3aR,6R,7aS)-1-[(2R)-6-hydroxy-6-methyl-heptan-2-yl]-6-methoxy-7a-methyl-2,3,3a,5,6,7-hexahydro-1H-inden-4-ylidene]ethylidene]-4-methylidene-cyclohexane-1,3-diol.
LMST03020377; 11S-methoxy-9,10-seco-5Z,7E,10(19)-cholestatriene-1S,3R,25-triol; 1α,25-dihydroxy-11α-methoxyvitamin D₃; 1α,25-dihydroxy-11α-methoxycholecalciferol; 9547489.

666   **1α,25-Dihydroxy-11β-methoxyvitamin D₃**
9547491 $C_{28}H_{46}O_4$

(1R,3S,5Z)-5-[(2E)-2-[(1R,3aR,6S,7aS)-1-[(2R)-6-
hydroxy-6-methyl-heptan-2-yl]-6-methoxy-7a-methyl-
2,3,3a,5,6,7-hexahydro-1H-inden-4-ylidene]
ethylidene]-4-methylidene-cyclohexane-1,3-diol.
LMST03020379; 11R-methoxy-9,10-seco-5Z,7E,10
(19)-cholestatriene-1S,3R,25-triol; 1α,25-dihydroxy-
11β-methoxycholecalciferol; 9547491.

667   1α,25-Dihydroxy-11α-(hydroxymethyl)-vitamin D₃

9547490 $C_{28}H_{46}O_4$

(1R,3S,5Z)-5-[(2E)-2-[(1R,3aR,6R,7aS)-6-(hydroxymethyl)-1-[(2R)-6-hydroxy-6-methyl-heptan-2-yl]-7a-
methyl-2,3,3a,5,6,7-hexahydro-1H-inden-4-ylidene]
ethylidene]-4-methylidene-cyclohexane-1,3-diol.
LMST03020378; 11S-(hydroxymethyl)-9,10-seco-5Z,
7E,10(19)-cholestatriene-1S,3R,25-triol; 1α,25-di-
hydroxy-11α-(hydroxymethyl)cholecalciferol;
9547490; Synthesized by Horner coupling of 25-
hydroxylated CD-ring ketone. Biological activity
reported. *J Biol Chem*. 1992, **267(5)**, 3044-51.

668   1α,25-Dihydroxy-20S-methoxyvitamin D₃

9547492 $C_{28}H_{46}O_4$

(1R,3S,5Z)-5-[(2E)-2-[(1S,3aR,7aR)-1-[(2S)-6-hydroxy-2-
methoxy-6-methyl-heptan-2-yl]-7a-methyl-2,3,3a,5,
6,7-hexahydro-1H-inden-4-ylidene]ethylidene]-4-
methylidene-cyclohexane-1,3-diol.
LMST03020380; 20S-methoxy-9,10-seco-5Z,7E,10
(19)-cholestatriene-1S,3R,25-triol; 1α,25-dihydroxy-
20S-methoxycholecalciferol; 9547492; Synthesized
from a (5E)-1α-hydroxy-20-keto-22,23,24,25,26,27-
hexanorvitamin D₃ derivative. Compared to 1α,25-
dihydroxyvitamin D₃, inhibition of U937 cell (human
histiocytic lymphoma cell line) proliferation is 900%,
binding to the 1α,25-dihydroxyvitamin D₃ receptor
from rachitic chicken intestine is 2%, calcemic activity
determined by the increase in urinary calcium
excretion in rats is 8%. Vitamin D. A Pluripotent
Steroid Hormone: Structural Studies, Molecular
Endocrinology and Clinical Applications. Proceedings
of the Ninth Workshop on Vitamin D, Orlando, Florida
(USA) May 28-June 2. Synthesis and Biological Activity
of 20-Hydroxylated Vitamin D Analogues. (Norman, A.
W., Bouillon, R., Thomasset, M., eds), pp95-96, Walter
de Gruyter Berlin New York (1994).

669   1α,25-Dihydroxy-24a,24b-dihomo-22-oxavitamin D₃

9547493 $C_{28}H_{46}O_4$

(1R,3S,5Z)-5-[(2E)-2-[(1S,3aR,7aR)-1-[(1S)-1-(5-
hydroxy-5-methyl-hexoxy)ethyl]-7a-methyl-
2,3,3a,5,6,7-hexahydro-1H-inden-4-ylidene]
ethylidene]-4-methylidene-cyclohexane-1,3-diol.
LMST03020381; 24a,24b-dihomo-22-oxa-9,10-seco-
5Z,7E,10(19)-cholestatriene-1S,3R,25-triol; 1α,25-di-
hydroxy-24a,24b-dihomo-22-oxacholecalciferol;
9547493; Synthesized from dehydroepiandrosterone.
About one order of magnitude more effective than
1α,25-dihydroxyvitamin D₃ in inducing differentiation
of human myeloid leukemia cells (HL-60) into
macrophages *in vitro* estimated by superoxide anion

generation; Colorless foam; λ_m = 263 nm (EtOH). **Chem Pharm Bull (Tokyo)**. 1992, **40(6)**, 1494-9.

670   *1,25-Dihydroxy-22-oxavitamin D$_3$*
132014-43-8 $C_{28}H_{46}O_4$

(1R,3S,5E)-5-[(2E)-2-[(3aR,7aR)-1-[(1R)-1-(5-hydroxy-5-methyl-hexoxy)ethyl]-7a-methyl-2,3,3a,5,6,7-hexa-hydro-1H-inden-4-ylidene]ethylidene]-4-methylidene-cyclohexane-1,3-diol.
KH 1049; KH-1049; 1,3-Cyclohexanediol, 4-methyl-ene-5-((octahydro-1-(1-((5-hydroxy-5-methylhexyl)-oxy)ethyl)-7a-methyl-4H-inden-4-ylidene)ethylidene)-, (1S-(1α(S*),3aβ,4E(1S*,3R*,5Z),7aα))-; 6439324; 132014-43-8. **J Steroid Biochem Mol Biol**. 1995, **55(3-4)**, 337-46.

671   *1α,25-Dihydroxy-24a,24b-dihomo-22-oxa-20-epivitamin D$_3$*
9547494 $C_{28}H_{46}O_4$

(1R,3S,5Z)-5-[(2E)-2-[(1S,3aR,7aR)-1-[(1R)-1-(5-hydroxy-5-methyl-hexoxy)ethyl]-7a-methyl-2,3,3a,5,6,7-hexahydro-1H-inden-4-ylidene]ethylidene]-4-methylidene-cyclohexane-1,3-diol.
LMST03020382; (20R)-24a,24b-dihomo-22-oxa-9,10-seco-5Z,7E,10(19)-cholestatrien-1S,3R,25-triol; 1α,25-dihydroxy-24a,24b-dihomo-22-oxa-20-epicholecal-ciferol; 9547494; Compared to 1α,25-dihydroxy-vitamin D$_3$, inhibition of U937 cell (human histiocytic lymphoma cell line) proliferation is 33300%, induction of U937 cell differentiation is 40000%, binding to the 1α,25-dihydroxyvitamin D$_3$ receptor from rachitic chicken intestine is 35%, and calcemic activity determined by the increase in urinary calcium excretion in rats is 140%. **Biochem Pharmacol**. 1991, **42(8)**, 1569-75.

672   *1α,25-Dihydroxy-26,27-dimethyl-22-oxavitamin D$_3$*
9547495 $C_{28}H_{46}O_4$

(1R,3S,5Z)-5-[(2E)-2-[(1S,3aR,7aR)-1-[(1S)-1-(3-ethyl-3-hydroxy-pentoxy)ethyl]-7a-methyl-2,3,3a,5,6,7-hexa-hydro-1H-inden-4-ylidene]ethylidene]-4-methylidene-cyclohexane-1,3-diol.
LMST03020383; 26,27-dimethyl-22-oxa-9,10-seco-5Z,7E,10(19)-cholestatriene-1S,3R,25-triol; 1α,25-di-hydroxy-26,27-dimethyl-22-oxacholecalciferol; 9547495; Synthesized from dehydroepiandrosterone.
Compared to 1α,25-dihydroxyvitamin D$_3$, induction of differentiation of human myeloid leukemia cells (HL-60) into macrophages **in vitro** estimated by superoxide anion generation is about one order of magnitude greater. Binding to the chick embryonic intestinal 1α,25-dihydroxyvitamin D$_3$ receptor is 50%; λ_m = 263 nm (EtOH). **Chem Pharm Bull (Tokyo)**. 1992, **40(6)**, 1494-9.

673   *1α,25-Dihydroxy-24a,24b-dihomo-23-oxavitamin D$_3$*
9547496 $C_{28}H_{46}O_4$

(1R,3S,5Z)-5-[(2E)-2-[(1R,3aR,7aS)-1-[(2S)-1-(4-hydroxy-4-methyl-pentoxy)propan-2-yl]-7a-methyl-2,3,3a,5,6,7-hexahydro-1H-inden-4-ylidene]ethylidene]-4-methylidene-cyclohexane-1,3-diol.
LMST03020384; 24a,24b-dihomo-23-oxa-9,10-seco-5Z,7E,10(19)-cholestatriene-1S,3R,25-triol; 1α,25-dihydroxy-24a,24b-dihomo-23-oxacholecalciferol; 9547496; Synthesized from protected (5E)-1α,22-dihydroxy-23,24,25,26,27-pentanorvitamin D$_3$. Compared to 1α,25-dihydroxyvitamin D$_3$, affinity for pig intestinal nuclear receptor is 50%, affinity for human vitamin D binding protein is 0.02%, differentiation of HL-60 cells is 100%, and calciuric effect is 1%. **Tet**

Lett. 1991, *32(38)*, 5073-6; Schering AG, Int. Pat. Appl. No. WO 94/00428.

674   *1α,25-Dihydroxy-24a,24b-dihomo-23-oxa-20-epivitamin D₃*
9547497   C₂₈H₄₆O₄

(1R,3S,5Z)-5-[(2E)-2-[(1R,3aR,7aS)-1-[(2R)-1-(4-hydroxy-4-methyl-pentoxy)propan-2-yl]-7a-methyl-2,3,3a,5,6,7-hexahydro-1H-inden-4-ylidene] ethylidene]-4-methylidene-cyclohexane-1,3-diol. LMST03020385; (20R)-24a,24b-dihomo-23-oxa-9,10-seco-5Z,7E,10(19)-cholestatriene-1S,3R,25-triol; 1α, 25-dihydroxy-24a,24b-dihomo-23-oxa-20-epicholecalciferol; 9548497; Synthesized from protected (5E)-1α-hydroxy-22-oxo-23,24,25,26,27-pentanorvitamin D₃.

Compared to 1α,25-dihydroxyvitamin D₃, inhibition of cell (U937) proliferation is 100000%, induction of cell (U937) differentiation is 200000%, inhibition of T-cell (murine MLR) activation is 8000%, vitamin D receptor (chick intestine) binding is 60%, and calciuric effect (urinary calcium excretion in rats) is 60%; λ_m = 264 nm (ε = 17400 EtOH). *Bioorg Med Chem Lett.* 1993, *3(9)*, 1809-14.

675   *1α,20S,25-Trihydroxy-24a-homovitamin D₃*
9547498   C₂₈H₄₆O₄

(2S)-2-[(1S,3aR,4E,7aR)-4-[(2Z)-2-[(3S,5R)-3,5-dihydroxy-2-methylidene-cyclohexylidene]ethylidene]-7a-methyl-2,3,3a,5,6,7-hexahydro-1H-inden-1-yl]-7-methyl-octane-2,7-diol. LMST03020387; 24a-homo-9,10-seco-5Z,7E,10(19)-cholestatriene-1S,3R,20S,25-tetrol; 1α,20S,25-tri-hydroxy-24a-homocholecalciferol; 9547498; Synthesized from protected (5E)-1α-hydroxy-22-oxo-23,24,25,26,27-pentanorvitamin D₃. Compared to

1α,25-dihydroxyvitamin D₃, inhibition of U937 cell (human histiocytic lymphoma cell line) proliferation is 70%, binding to the 1α,25-dihydroxyvitamin D₃ receptor from rachitic chicken intestine is 0.04%, calcemic activity was not determined. Vitamin D. A Pluripotent Steroid Hormone: Structural Studies, Molecular Endocrinology and Clinical Applications. Proceedings of the Ninth Workshop on Vitamin D, Orlando, Florida (USA) May 28-June 2. Synthesis and Biological Activity of 20-Hydroxylated Vitamin D Analogues. (Norman, A. W., Bouillon, R., Thomasset, M., eds), pp95-96, Walter de Gruyter Berlin New York (1994).

676   *1α,25-Trihydroxy-23R-methylvitamin D₃*
5283796   C₂₈H₄₆O₄

(4R,6R)-6-[(1R,3aR,4E,7aS)-4-[(2Z)-2-[(3S,5R)-3,5-dihydroxy-2-methylidene-cyclohexylidene]ethylidene]-7a-methyl-2,3,3a,5,6,7-hexahydro-1H-inden-1-yl]-2,4-dimethyl-heptane-2,4-diol. LMST03020389; 23R-methyl-9,10-seco-5Z,7E,10(19)-cholestatrien-1S,3R,23R,25-tetrol; 1α,23,25-trihydroxy-23-methylcholecalciferol; 5283796.

677   *1α-Hydroxy-18-(4-hydroxy-4-methyl-pentyloxy)-23,24,25,26,27-pentanorvitamin D₃*
9547681   C₂₈H₄₆O₄

(1R,3S,5Z)-5-[(2E)-2-[(1R,3aR,7aR)-7a-[(4-hydroxy-4-methyl-pentoxy)methyl]-1-propan-2-yl-2,3,3a,5,6,7-hexahydro-1H-inden-4-ylidene]ethylidene]-4-methylidene-cyclohexane-1,3-diol. LMST03020586; 18-(4-hydroxy-4-methylpentyloxy)-23,24-dinor-9,10-seco-5Z,7E,10(19)-cholatriene-1S,3R-diol; 1α-hydroxy-18-(4-hydroxy-4-methylpentyloxy)-23,24,25,26,27-pentanorcholecalciferol; 9547681.

678   *2β-Methoxy-1α,25-dihydroxyvitamin D$_3$*
9547696 $C_{28}H_{46}O_4$

(1R,2R,3R,5Z)-5-[(2E)-2-[(1R,3aR,7aS)-1-[(2R)-6-
hydroxy-6-methyl-heptan-2-yl]-7a-methyl-2,3,3a,5,6,7-
hexahydro-1H-inden-4-ylidene]ethylidene]-2-methoxy-
4-methylidene-cyclohexane-1,3-diol.
LMST03020603;   2β-methoxy-1α,25-dihydroxychole-
calciferol;   2R-methoxy-9,10-seco-5Z,7E,10(19)-chol-
estatriene-1R,3R,25-triol;   9547696;   Synthesized
photochemically from a C(22) steroid precursor.
Compared to 1α,25-dihydroxyvitamin D$_3$, affinity for
chicken intestine vitamin D receptor is 9%, affinity for
sheep serum vitamin D binding protein is 3500%,
inhibition of proliferation of C3H10T(BMP-4) cells is
11% (FCS-free), 1.5% (10% FCS); CAT assay
(osteocalcin) is 70%, inhibition of adipogenesis is
0.08%. *Steroids*. 1998, *63(12)*, 633-43.

679   *9,11-Seco-3,6,11-trihydroxy-24-methyl-
enecholest-7-en-9-one*
143625-40-5 $C_{28}H_{46}O_4$

(8aS)-4,6-dihydroxy-2-[(2S,3R)-2-(2-hydroxyethyl)-2-
methyl-3-[(2R)-6-methyl-5-methylidene-heptan-2-
yl]cyclopentyl]-8a-methyl-4,4a,5,6,7,8-
hexahydronaphthalen-1-one.
9,11-seco-3β,6α,11-trihydroxy-24-methylene-5α-chol-
est-7-en-9-one;   9,11-Symco;   9,11-Seco-3,6,11-tri-
hydroxy-24-methylenecholest-7-en-9-one;   5α,9,11-
Secoergosta-7,24(28)-dien-9-one, 3β,6α,11-trihydroxy-
; 132571; 143625-40-5; Isolated from the marine
sponge *Spongia officinalis*. Structure elucidated by
analysis of NMR spectral data. *Steroids*. 1992, *57(7)*,
344-7.

680   *1α-Hydroxy-20-epi-22-oxa-26-sulfone
3β-vitamin D$_3$*

$C_{28}H_{46}O_5S$

Non-calcemic but powerfully antiproliferative and
transcriptionally active *in vivo*; White solid; $[\alpha]^{25}$ = -
50.8° (c = 4.4 CHCl$_3$); λ_m = 263 nm (MeOH. ε =
13679). *J Med Chem*. 2000, *43(19)*, 3581-6.

681   *1β-Hydroxy-20-epi-22-oxa-26-sulfone
3α-vitamin D$_3$*

$C_{28}H_{46}O_5S$

Shows little biological activity *in vivo*; White solid;
$[\alpha]^{25}$ =-36.2° (c = 3.0 CHCl$_3$); λ_m = 263 nm (MeOH, ε
= 10319). *J Med Chem*. 2000, *43(19)*, 3581-6.

682   *Dihydrotachysterol$_2$*
22481-38-5 $C_{28}H_{48}O$

(3Z)-3-[(2E)-2-[1-(5,6-dimethylheptan-2-yl)-7a-methyl-
2,3,3a,5,6,7-hexahydro-1H-inden-4-ylidene]
ethylidene]-4-methyl-cyclohexan-1-ol.
Dihydrotachysterol$_2$; 9,10-Ergocholesta-5E,7E-dien-3β-
ol; 6450246; 22481-38-5. *Biochem Pharmacol*. 1992,
43(9), 1893-905.

683   *2,2-Dimethyl-1α,25-dihydroxyvitamin D₃*

$C_{28}H_{48}O_3$

Synthetic vitamin D derivative. Has 30% of the cell differentiating activity of calcitriol. *J Steroid Biochem Mol Biol*. 2004, **89-90(1-5)**, 89-92.

684   *1α-Hydroxy-22-(3-methylphenyl)-23,24, 25,26,27-pentanorvitamin D₃*

9547499 $C_{29}H_{40}O_2$

(1R,3S,5Z)-5-[(2E)-2-[(1R,3aR,7aS)-7a-methyl-1-[(2R)-1-(3-methylphenyl)propan-2-yl]-2,3,3a,5,6,7-hexahydro-1H-inden-4-ylidene]ethylidene]-4-methylidene-cyclohexane-1,3-diol.
LMST03020390;   22-(3-methylphenyl)-23,24-dinor-9,10-seco-5Z,7E,10(19)-cholatriene-1S,3R-diol;   1α-hydroxy-22-(3-methylphenyl)-23,24,25,26,27-pentanorcholecalciferol;   9547499;   Compared to 1α,25-dihydroxyvitamin D₃, intestinal calcium absorption is 0.3%, bone calcium mobilization is 0.06%, affinity for chick intestinal receptor and HL-60 cell receptor are 1.4% and 3.5%, inhibition of 1α-hydroxylase activity is 96% and differentiation of HL-60 cells is 15%. Arocalciferols: A New Class of Side-Chain Analogs of 1,25-(OH)2 D3. In Vitamin D Gene Regulation Structure-Function Analysis and Clinical Application (Norman, A.W., Bouillon R. and Thomasset, M., eds), pp 165-166, Walter de Gruyter, Berlin/New York (1991).

685   *1α,25-Dihydroxy-26,27-dimethyl-17Z,20,22,22,23,23-hexadehydrovitamin D₃*

9547500 $C_{29}H_{42}O_3$

(1R,3S,5Z)-5-[(2E)-2-[(1Z,3aR,7aS)-1-(6-ethyl-6-hydroxy-oct-3-yn-2-ylidene)-7a-methyl-2,3,3a,5,6,7-hexahydroinden-4-ylidene]ethylidene]-4-methylidene-cyclohexane-1,3-diol.
LMST03020391;   1α,25-dihydroxy-26,27-dimethyl-17Z,20,22,22,23,23-hexadehydrocholecalciferol;   26,27-dimethyl-9,10-seco-5Z,7E,10(19),17Z(20)-cholesta-tetraen-22-yne-1S,3R,25-triol;   9547500;   Synthesized from hydroxy-protected   (5E)-1α-hydroxy-20-oxo-22,23,24,25,26,27-hexanorvitamin D₃. Compared to 1α,25-dihydroxyvitamin D₃, inhibition of proliferation of U937 cells is 3200% and affinity for chicken intestinal vitamin D receptor is 7%. Vitamin D. A Pluripotent Steroid Hormone: Structural Studies, Molecular Endocrinology and Clinical Applications. Proceedings of the Ninth Workshop on Vitamin D, Orlando, Florida (USA) May 28-June 2. Chemistry and Biology of 22,23-Yne Analogs of Calcitriol. (Norman, A. W., Bouillon, R., Thomasset, M., eds), pp73-74, Walter de Gruyter Berlin New York (1994).

686   *1α,25-Dihydroxy-26,27-dimethyl-20,21,22,22,23,23-hexadehydrovitamin D₃*

9547501 $C_{29}H_{42}O_3$

(1R,3S,5Z)-5-[(2E)-2-[(1S,3aR,7aS)-1-(6-ethyl-6-hydroxy-oct-1-en-3-yn-2-yl)-7a-methyl-2,3,3a,5,6,7-hexahydro-1H-inden-4-ylidene]ethylidene]-4-methylidene-cyclohexane-1,3-diol.
LMST03020392;   26,27-dimethyl-9,10-seco-5Z,7E,10(19),20-cholestatetraen-22-yne-1S,3R,25-triol;   1α,25-dihydroxy-26,27-dimethyl-20,21,22,22,23,23-hexadehydrocholecalciferol;   9547501;   Synthesized from hydroxy-protected   (5E)-1α-hydroxy-20-oxo-22,23,24,25,26,27-hexanorvitamin D₃. Compared to 1α,25-dihydroxyvitamin D₃, inhibition of proliferation of U937 cells is 140% and affinity for chicken intestinal vitamin D receptor is 87%. Vitamin D. A Pluripotent

Steroid Hormone: Structural Studies, Molecular Endocrinology and Clinical Applications. Proceedings of the Ninth Workshop on Vitamin D, Orlando, Florida (USA) May 28-June 2. Chemistry and Biology of 22,23-Yne Analogs of Calcitriol. (Norman, A. W., Bouillon, R., Thomasset, M., eds), pp73-74, Walter de Gruyter Berlin New York (1994).

687   *24a,24b-Epoxy-23-tetradehydro-24a,*
24b-dihomo-1α,25-dihydroxyvitamin D$_3$
9547502 $C_{29}H_{42}O_4$

(1R,3S,5Z)-5-[(2E)-2-[(1R,3aR,7aS)-1-[(2R)-5-[3-(2-hydroxypropan-2-yl)oxiran-2-yl]pent-4-yn-2-yl]-7a-methyl-2,3,3a,5,6,7-hexahydro-1H-inden-4-ylidene]ethylidene]-4-methylidene-cyclohexane-1,3-diol. LMST03020393; 24a,24b-dihomo-24a,24b-epoxy-9,10-seco-5Z,7E,10(19)-cholestatriene-23-yne-1S,3R,25-triol; 24a,24b-epoxy-23-tetradehydro-24a,24b-di-homo-1α,25-dihydroxycholecalciferol; 9547502; Prepared by epoxidation of the corresponding olefinic precursor which was constructed from Inhoffen-Lythgoe diol. Compared to 1α,25-dihydroxyvitamin D$_3$, affinity for pig intestinal receptor and human vitamin D binding protein is 3%, inhibition of proliferation or differentiation induction of human promyeloid leukemia (HL-60) and osteosarcoma (MG-63) cells are 4% and 1%. *Steroids.* 1995, *60*(4), 324-32.

688   *1α-Fluoro-25-hydroxy-16,23E-diene-*
26,27-bishomo-20-epivitamin D$_3$
199798-84-0 $C_{29}H_{43}FO_2$

Cyclohexanol, 3-((2E)-((3aS,7aS)-1-((1S,3E)-5-ethyl-5-hydroxy-1-methyl-3-heptenyl)-3,3a,5,6,7,7a-hexahydro-7a-methyl-4H-inden-4-ylidene)ethylidene)-5-fluoro-4-methylene-, (1R,3Z,5S)-.
1α-Fluoro-25-hydroxy-16,23E-diene-26,27-bishomo-

20-epi-cholecalciferol; BXL 628; Ro 26-9228; Cyclohexanol, 3-((1-(5-ethyl-5-hydroxy-1-methyl-3-heptenyl)-3,3a,5,6,7,7a-hexahydro-7a-methyl-4H-inden-4-ylidene)ethylidene)-5-fluoro-4-methylene-, (3aS-(1(1R*,3E),3aα,4E(1S*,3Z,5R*),7aβ))-; Ro 26-9228; 199798-84-0; Is degraded to the previtamin form in acid solution.Cannot promote the accumulation of proteasome-sensitive transcriptionally active vitamin D receptor in Caco-2 cells, but can do so in hFOB cells. *J Pharm Sci.* 2003, *92(10)*, 1981-9; *Mol Endocrinology.* 2000, *18(4)*, 874-7.

689   *3-Deoxy-3-ene-4-vinyl-19-nor-25-*
hydroxy-vitamin D$_2$
                                           $C_{29}H_{44}O_2$

Synthesized as an example of A-ring functionalization of vitamin D$_2$. *Org Lett.* 2003, *5(5)*, 669-72.

690   *24a,24b-Dihomo-9,10-secocholesta-5,7,*
10(19),24a-tetraen-1α,3,25-triol
                                           $C_{29}H_{44}O_3$

(5Z,7E)-(1S,3R)-(24a)-24a,24b-dihomo-9,10-seco-cholesta-5,7,10(19),24a-tetraen-1,3,25-triol. 24a,24b-Dihomo-9,10-secocholesta-5,7,10(19),24a-tetraen-1,3,25-triol; PRI 1901; Synthetic derivative of 1,25-dihydroxyvitamin D$_3$. Binds to the vitamin D receptor 1000 times more effectively than its dihydro derivative, 24a,24b-dihomo-1α,25-dihydroxyvitamin D$_3$; Solid; mp = 139°; λ$_m$ = 263.8 nm. *Steroids.* 1997, *62(7)*, 546-53.

691   *1α,25-Dihydroxy-22E,23,24E,24a-*
tetradehydro-24a,24b-dihomovitamin D$_3$
9547504 $C_{29}H_{44}O_3$

(1R,3S,5Z)-5-[(2E)-2-[(1R,3aR,7aS)-1-[(2S,3E,5E)-8-hydroxy-8-methyl-nona-3,5-dien-2-yl]-7a-methyl-2,3,3a,5,6,7-hexahydro-1H-inden-4-ylidene]ethylidene]-4-methylidene-cyclohexane-1,3-diol.
LMST03020395; 1α,25-dihydroxy-22E,23,24E,24a-tetradehydro-24a,24b-dihomocholecalciferol; 24a,24b-dihomo-9,10-seco-5Z,7E,10(19),22E,24E-cholesta-pentaene-1S,3R,25-triol; 9547504; Synthesized from protected (5E)-1α-hydroxy-22-oxo-23,24,25,26,27-pentanorvitamin D_3. Compared to 1α,25-dihydroxy-vitamin D_3, inhibition of proliferation of U937 cells is 500%, induction of differentiation of U937 cells is 600%, and calciuric activity is 30%. Gene Regulation, Structure-Function Analysis and Clinical Application. Proceedings of the Eighth Workshop on Vitamin D Paris, France July 5-10. Synthesis and Biological Activity of 1α-Hydroxylated Vitamin D_3 Analogues with Hydroxylated Side Chains, Multi-Homologated in The 24- or 24,26,27-Positions. (Norman, A. W., Bouillon, R., Thomasset, M., eds), pp159-160, Walter de Gruyter Berlin (1991).

692 **1α,25-Dihydroxy-16,17-didehydro-20S-cyclopropyl-21-norvitamin D_3**
9547678 $C_{29}H_{44}O_3$

(1R,3S,5Z)-5-[(2E)-2-[(3aR,7aS)-1-[(1S)-1-cyclopropyl-5-hydroxy-5-methyl-hexyl]-7a-methyl-3a,5,6,7-tetrahydro-3H-inden-4-ylidene]ethylidene]-4-methylidene-cyclohexane-1,3-diol.
LMST03020583; 20S-cyclopropyl-1α,25-dihydroxy-16,17-didehydro-21-norcholecalciferol; 20S-cyclopropyl-21-nor-9,10-seco-5Z,7E,10(19),16-cholestatetraene-1S,3R,25-triol; 9547678.

693 **1α,25-Dihydroxy-26,27-dimethyl-22,22,23,23-tetradehydrovitamin D_3**

9547505 $C_{29}H_{44}O_2$

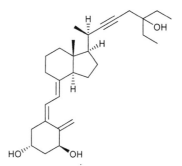

(1R,3S,5Z)-5-[(2E)-2-[(1R,3aR,7aS)-1-[(2S)-6-ethyl-6-hydroxy-oct-3-yn-2-yl]-7a-methyl-2,3,3a,5,6,7-hexahydro-1H-inden-4-ylidene]ethylidene]-4-methylidene-cyclohexane-1,3-diol.
LMST03020396; 26,27-dimethyl-9,10-seco-5Z,7E,10(19)-cholestatrien-22-yne-1S,3R,25-triol; 1α,25-dihydroxy-26,27-dimethyl-22,22,23,23-tetradehydro-cholecalciferol; 9547505; Synthesized from protected (5E)-1α-hydroxy-22-oxo-23,24,25,26,27-pentanorvitamin D_3. Compared to 1α,25-dihydroxyvitamin D_3, inhibition of proliferation of U937 cells is 350%, and affinity for chicken intestinal vitamin D receptor is 89%. Vitamin D. A Pluripotent Steroid Hormone: Structural Studies, Molecular Endocrinology and Clinical Applications. Proceedings of the Ninth Workshop on Vitamin D, Orlando, Florida (USA) May 28-June 2. Chemistry and Biology of 22,23-Yne Analogs of Calcitriol. (Norman, A. W., Bouillon, R., Thomasset, M., eds), pp73-74, Walter de Gruyter Berlin New York (1994).

694 **1α,25-Dihydroxy-26,27-dimethyl-22,22,23,23-tetradehydro-20-epivitamin D_3**
9547506 $C_{29}H_{44}O_3$

(1R,3S,5Z)-5-[(2E)-2-[(1R,3aR,7aS)-1-[(2R)-6-ethyl-6-hydroxy-oct-3-yn-2-yl]-7a-methyl-2,3,3a,5,6,7-hexahydro-1H-inden-4-ylidene]ethylidene]-4-methylidene-cyclohexane-1,3-diol.
LMST03020397; (20R)-26,27-dimethyl-9,10-seco-5Z,7E,10(19)-cholestatrien-22-yne-1S,3R,25-triol; 1α,25-dihydroxy-26,27-dimethyl-22,22,23,23-tetradehydro-20-epicholecalciferol; 9547506; Synthesized from protected (5E)-1α-hydroxy-22-oxo-23,24,25,26,27-pentanorvitamin D_3. Compared to 1α,25-dihydroxy-vitamin D_3, inhibition of proliferation of U937 cells is

390%, and affinity for chicken intestinal vitamin D receptor is 27%. Vitamin D. A Pluripotent Steroid Hormone: Structural Studies, Molecular Endocrinology and Clinical Applications. Proceedings of the Ninth Workshop on Vitamin D, Orlando, Florida (USA) May 28-June 2. Chemistry and Biology of 22,23-Yne Analogs of Calcitriol. (Norman, A. W., Bouillon, R., Thomasset, M., eds), pp73-74, Walter de Gruyter Berlin New York (1994).

695   *1α,25-Dihydroxy-11α-ethynyl-vitamin D₃*

9547503 $C_{29}H_{44}O_3$

(1R,3S,5Z)-5-[(2E)-2-[(1R,3aR,6R,7aS)-6-ethynyl-1-[(2R)-6-hydroxy-6-methyl-heptan-2-yl]-7a-methyl-2,3,3a,5,6,7-hexahydro-1H-inden-4-ylidene]ethylidene]-4-methylidene-cyclohexane-1,3-diol. LMST03020394; 11S-ethynyl-9,10-seco-5Z,7E,10(19)-cholestatriene-1S,3R,25-triol; 11α-ethynyl-1α,25-dihydroxycholecalciferol; 9547503.

696   *1α,22S,25-Trihydroxy-26,27-dimethyl-23,23,24,24-tetradehydrovitamin D₃*

9547507 $C_{29}H_{44}O_4$

(2S,3S)-2-[(1R,3aR,4E,7aS)-4-[(2Z)-2-[(3S,5R)-3,5-dihydroxy-2-methylidene-cyclohexylidene]ethylidene]-7a-methyl-2,3,3a,5,6,7-hexahydro-1H-inden-1-yl]-6-ethyl-oct-4-yne-3,6-diol. LMST03020398; 1α,22,25-trihydroxy-26,27-dimethyl-23,23,24,24-tetradehydrocholecalciferol; 26,27-dimethyl-9,10-seco-5Z,7E,10(19)-cholestatrien-23-yne-1S,3R,22S,25-tetrol; 9547507.

697   *1α,22R,25-Trihydroxy-23,24-tetra-dehydro-24a,24b-dihomo-20-epivitamin D₃*

9547508 $C_{29}H_{44}O_4$

(2R,3R)-2-[(1R,3aR,4E,7aS)-4-[(2Z)-2-[(3S,5R)-3,5-dihydroxy-2-methylidene-cyclohexylidene]ethylidene]-7a-methyl-2,3,3a,5,6,7-hexahydro-1H-inden-1-yl]-6-ethyl-oct-4-yne-3,6-diol. LMST03020399; 1α,22R,25-trihydroxy-23,24-tetra-dehydro-24a,24b-dihomo-20-epicholecalciferol; (20R)-24a,24b-dihomo-9,10-seco-5Z,7E,10(19)-cholestatrien-23-yne-1S,3R,22R,25-tetrol; 9547508; Synthesized from protected (5E)-1α-hydroxy-22-oxo-23,24,25,26,27-pentanorvitamin D₃. Compared to 1α,25-dihydroxyvitamin D₃, induction of differentiation of U937 cells is 0.02%, inhibition of proliferation, < 0.05%, VDR (rachitic chicken intestinal receptor) binding affinity is < 0.005%. Vitamin D. A Pluripotent Steroid Hormone: Structural Studies, Molecular Endocrinology and Clinical Applications. Proceedings of the Ninth Workshop on Vitamin D. Orlando, Florida (USA) May 28-June 2. Chemistry and Biology of 22,23-Yne Analogs of Calcitriol. (Norman, A. W., Bouillon, R., Thomasset, M., eds), pp73-74, Walter de Gruyter Berlin New York (1994).

698   *1α,22,25-Trihydroxy-23,24-tetra-dehydro-24a,24b-dihomo-20-epivitamin D₃*

9547509 $C_{29}H_{44}O_4$

(7S,8R)-8-[(1R,3aR,4E,7aS)-4-[(2Z)-2-[(3S,5R)-3,5-dihydroxy-2-methylidene-cyclohexylidene]ethylidene]-7a-methyl-2,3,3a,5,6,7-hexahydro-1H-inden-1-yl]-2-methyl-non-5-yne-2,7-diol. LMST03020400; 1α,22,25-trihydroxy-23,24-tetra-dehydro-24a,24b-dihomo-20-epicholecalciferol; (20R)-24a,24b-dihomo-9,10-seco-5Z,7E,10(19)-cholestatrien-23-yne-1S,3R,22S,25-tetrol; 9547509; Synthesized from protected (5E)-1α-hydroxy-22-oxo-23,24,25,26,27-pentanorvitamin D₃. Compared to 1α,25-

dihydroxyvitamin D_3, induction of differentiation of U937 cells is 10%, inhibition of proliferation, 12%, VDR (rachitic chicken intestinal receptor) binding affinity is 0.005%, and calciuric effect in normal rats is 0.3%. Vitamin D. A Pluripotent Steroid Hormone: Structural Studies, Molecular Endocrinology and Clinical Applications. Proceedings of the Ninth Workshop on Vitamin D, Orlando, Florida (USA) May 28-June 2. Chemistry and Biology of 22,23-Yne Analogs of Calcitriol. (Norman, A. W., Bouillon, R., Thomasset, M., eds), pp73-74, Walter de Gruyter Berlin New York (1994).

699   *1α,22R,25-Trihydroxy-26,27-dimethyl-23,24-tetradehydro-20-epivitamin D_3*

9547510 $C_{29}H_{44}O_4$

(7R,8R)-8-[(1R,3aR,4E,7aS)-4-[(2Z)-2-[(3S,5R)-3,5-dihydroxy-2-methylidene-cyclohexylidene]ethylidene]-7a-methyl-2,3,3a,5,6,7-hexahydro-1H-inden-1-yl]-2-methyl-non-5-yne-2,7-diol.
LMST03020401;   1α,22R,25-trihydroxy-26,27-dimethyl-23,24-tetradehydro-20-epicholecalciferol; (20R)-26,27-dimethyl-9,10-seco-5Z,7E,10(19)-cholestatrien-23-yne-1S,3R.22R,25-tetrol;   9547510; Prepared from protected (5E)-1α-hydroxy-22-oxo-23, 24,25,26,27-pentanor-20-epivitamin D_3. Compared to 1α,25-dihydroxyvitamin D_3, induction of differentiation of U937 cells is 100%, inhibition of proliferation is 110%, VDR (rachitic chicken intestinal receptor) binding affinity is < 0.3%. Vitamin D. A Pluripotent Steroid Hormone: Structural Studies, Molecular Endocrinology and Clinical Applications. Proceedings of the Ninth Workshop on Vitamin D, Orlando, Florida (USA) May 28-June 2. Chemistry and Biology of 22,23-Yne Analogs of Calcitriol. (Norman, A. W., Bouillon, R., Thomasset, M., eds), pp73-74, Walter de Gruyter Berlin New York (1994).

700   *1α-Hydroxy-18-(5-hydroxy-5-methyl-2-hexynyloxy)-23,24,25,26,27-pentanorvitamin D_3*

9547688 $C_{29}H_{44}O_4$

(1R,3S,5Z)-5-[(2E)-2-[(1R,3aR,7aR)-7a-[(5-hydroxy-5-methyl-hex-2-ynoxy)methyl]-1-propan-2-yl-2,3,3a,5,6,7-hexahydro-1H-inden-4-ylidene]ethylidene]-4-methylidene-cyclohexane-1,3-diol.
LMST03020593;   18-(5-hydroxy-5-methyl-2-hexynyl-oxy)-23,24-dinor-9,10-seco-5Z,7E,10(19)-cholatriene-1S,3R-diol;   1α-hydroxy-18-(5-hydroxy-5-methyl-2-hexynyloxy)-23,24,25,26,27-pentanorcholecalciferol; 9547688.

701   *1α-Hydroxy-18-(5-hydroxy-5-methyl-3-hexynyloxy)-23,24,25,26,27-pentanorvitamin D_3*

9547689 $C_{29}H_{44}O_4$

(1R,3S,5Z)-5-[(2E)-2-[(1R.3aR,7aR)-7a-[(5-hydroxy-5-methyl-hex-3-ynoxy)methyl]-1-propan-2-yl-2,3,3a,5,6,7-hexahydro-1H-inden-4-ylidene]ethylidene]-4-methylidene-cyclohexane-1,3-diol.
LMST03020594;   18-(5-hydroxy-5-methyl-3-hexynyl-oxy)-23,24-dinor-9,10-seco-5Z,7E,10(19)-cholatriene-1S,3R-diol;   1α-hydroxy-18-(5-hydroxy-5-methyl-3-hexynyloxy)-23,24,25,26,27-pentanorcholecalciferol; 9547689.

702   *25-Hydroxyvitamin D_3-bromoacetate*

                      $C_{29}H_{45}BrO_3$

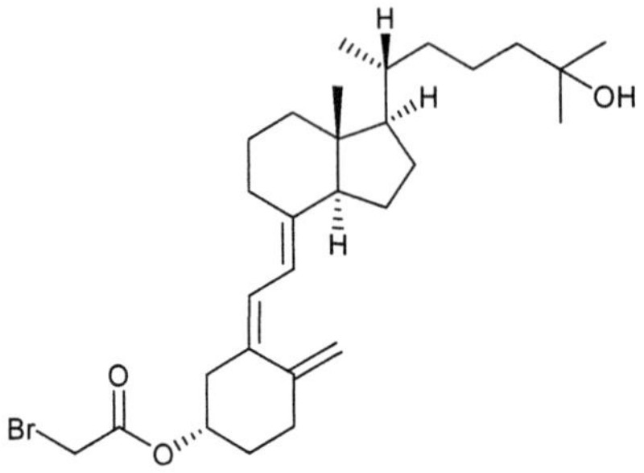

9,10-Secocholesta-5Z,7E,10(19)-triene-3β,25-diol 3-bromoacetate.

25-hydroxyvitamin D_3-3 β-bromoacetate; 25-OH-D_3-BE; Synthetic vitamin D_3 derivative. Covalently modifies the 25-hydroxyvitamin D_3-binding site in rat vitamin D binding protein. Interacts with Cys288 and Val418 in the vitamin D receptor protein. **Biochemistry**. 2000, **39(40)**, 12162-71; **Arch Biochem Biophys**. 1996, **333(1)**, 139-44.

703 **1,25-Dihydroxyvitamin D_3-bromoacetate**

$C_{29}H_{45}BrO_4$

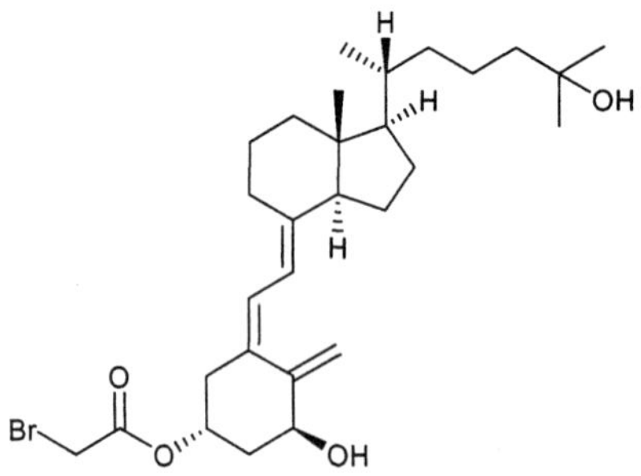

9,10-Secocholesta-5Z,7E,10(19)-triene-1α,3β,25-triol 3-bromoacetate.

1,25-dihydroxyvitamin D_3-3 β-bromoacetate; 1,25-(OH)2-D_3-BE; Synthetic vitamin D_3 derivative. Binds covalently to the vitamin D receptor protein. **J Cell Biochem**. 1996, **63(3)**, 302-10.

704 **Ecalcidene**

150337-94-3

$C_{29}H_{45}NO_3$

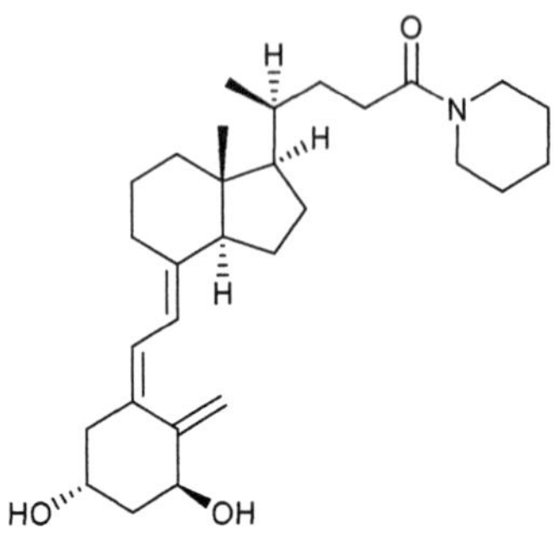

(1-[(20S)-1α,3β-dihydroxy-24-oxo-9,10-secochola-5Z,7E,10(19)-trien-24-yl]-piperidine).
(1-[(1α,3β,5Z,7E,20S)-1,3-dihydroxy-24-oxo-9,10-secochola-5,7,10(19)-trien-24-yl]-piperidine); piperidine, 1-[(1α,3β,5Z,7E,20S)-1,3-dihydroxy-24-oxo-9,10-secocholesta-5,7,10(19)-trien-24-yl]-Ecalcidine; 150337-94-3; A synthetic analog of 1-hydroxyvitamin D_3 under development for use in treatment of psoriasis. In solution is thermally and reversibly transformed to a pre-Vitamin D type isomer. **J Pharm Biomed Anal**. 2006, **40(4)**, 850-63.

705 **24,24-Difluoro-25-hydroxy-26,27-dimethylvitamin D_3**

106647-61-4

$C_{29}H_{46}F_2O_2$

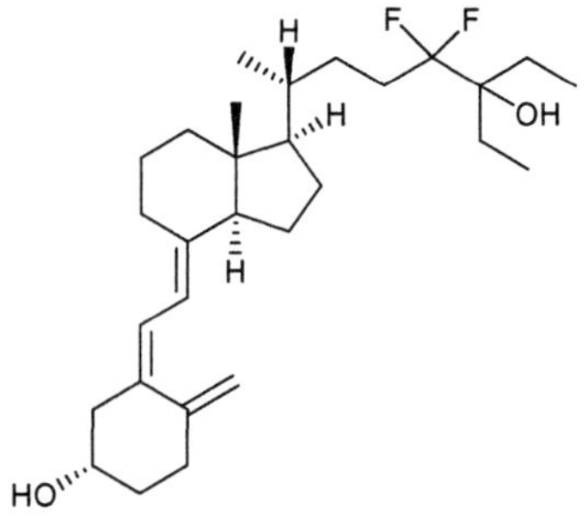

(7R)-7-[(1R,4E,7aS)-4-[(2Z)-2-[(5S)-5-hydroxy-2-methylidene-cyclohexylidene]ethylidene]-7a-methyl-2,3,3a,5,6,7-hexahydro-1H-inden-1-yl]-3-ethyl-4,4-difluoro-octan-3-ol.
Dfdm-calcifediol; 24,24-Difluoro-25-hydroxy-26,27-dimethylvitamin D_3; 6439944; 106647-61-4; A highly potent vitamin D analogue with bioactivity **in vivo** similar to that of 25-hydroxyvitamin D_3. Bound by vitamin D binding protein with an affinity slightly less than that of 25-hydroxyvitamin D_3. It is bound to the intestinal cytosol receptor for 1,25-dihydroxyvitamin D_3 with approximately the same affinity as that of 25-hydroxyvitamin D_3. In the organ-culture duodenum, it induced the synthesis of calcium binding protein with a potency approximately 1/20 that of 1,25-dihydroxyvitamin D_3. **J Med Chem**. 1990, **33(2)**, 480-90.

706 **24,24-Difluoro-1α,25-dihydroxy-26,27-dimethylvitamin D_3**

106647-71-6

$C_{29}H_{46}F_2O_3$

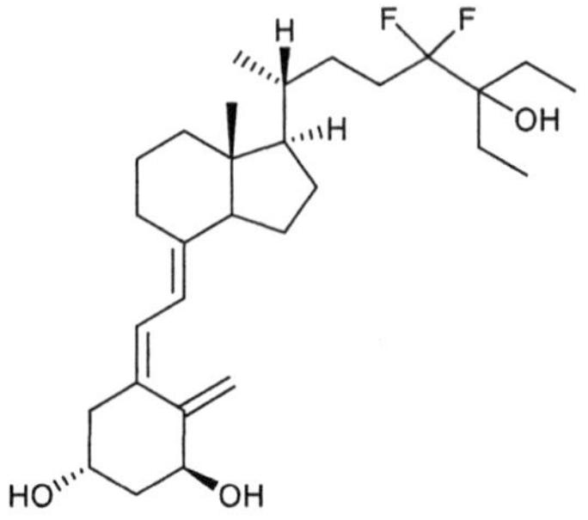

(1R,5Z)-5-[(2E)-2-[(1R,7aS)-1-[(2R)-6-ethyl-5,5-difluoro-6-hydroxy-octan-2-yl]-7a-methyl-2,3,3a,5,6,7-hexa-hydro-1H-inden-4-ylidene]ethylidene]-4-methylidene-cyclohexane-1,3-diol.
LMST03020403; 24,24-difluoro-26,27-dimethyl-9,10-seco-5Z,7E,10(19)-cholestatrien-1S,3R,25-triol; 24,24-difluoro-1α,25-dihydroxy-26,27-dimethylcholecalc-iferol; Dfdm-calcitriol; Dmdfdh-D$_3$; 24F2-1,25 (OH)2(Me)2D$_3$; 24,24-Difluoro-1,25-dihydroxy-26,27-dimethylvitamin D$_3$; 26,27-Dimethyl-24,24-difluoro-1,25-dihydroxyvitamin D$_3$; 26,27-Dimethyl-24,24-di-fluoro-1α,25-dihydroxyvitamin D$_3$; 6439159; 9547511; 106647-71-6; 123836-13-5; Fluoro substitution at 24-position of 1,25(OH)2D$_3$ and the elongation of side chain of 1,25(OH)2D$_3$ is found not to intensify Ca-regulating activity. Compared to 1α,25-dihydroxyvitamin D$_3$, binding affinity for chick intestinal receptor is 53% and binding affinity for vitamin D binding protein in rat serum is 4%. Potency in ^{45}Ca release from neonatal mouse parietal bones in culture is higher, potency in the formation of osteoclast-like cells is 100 times higher. Potency in bone calcium mobilization in vitamin D deficient rats is slightly lower and potency in intestinal Ca transport response *in situ* is similar to that of 1α,25-dihydroxyvitamin D$_3$. *J Med Chem*. 1990, *33(2)*, 480-90: *Calcif Tissue*. Int.,1993, *53(5)*, 318-23.

707 *Vitamin D$_6$*
9547702   C$_{29}$H$_{46}$O

(1S,3Z)-3-[(2E)-2-[(1R,3aR,7aS)-1-[(E,2S,5R)-5-ethyl-6-methyl-hept-3-en-2-yl]-7a-methyl-2,3,3a,5,6,7-hexahydro-1H-inden-4-ylidene]ethylidene]-4-methylidene-cyclohexan-1-ol.
LMST03050001; 9,10-seco-5Z,7E,10(19),22E-porifera-statetraen-3S-ol; 24R-ethyl-9,10-seco-5Z,7E,10(19),22E-cholestatetraen-3S-ol; 9547702; Has <1% of the antirachitic effect of vitamin D$_3$. *Hoppe-Seyler's Z Physiol. Chem*. 1936, *241*, 125.

708 *Provitamin D$_6$*
9547703   C$_{29}$H$_{46}$O

(3S,10R,13S,14R,17R)-17-[(E,2S,5R)-5-ethyl-6-methyl-hept-3-en-2-yl]-10,13-dimethyl-2,3,4,9,11,12,14,15,16,17-decahydro-1H-cyclopenta[a]phenanthren-3-ol.
LMST03050002; Poriferasta-5,7,22E-trien-3β-ol; 24R-ethyl-5,7,22E-cholestatrien-3S-ol; 9547703. *J Amer Chem Soc*. 1982, *104(21)*, 5780-1; USP 7211172.

709 *1α-Hydroxy-24,24-dimethyl-22-dehydrovitamin D$_3$*
104211-64-5   C$_{29}$H$_{46}$O$_2$

(1S,3R,5Z)-5-[(2E)-2-[(1R,3aR,7aS)-7a-methyl-1-[(E,2S)-5,5,6-trimethylhept-3-en-2-yl]-2,3,3a,5,6,7-hexahydro-1H-inden-4-ylidene]ethylidene]-4-methylidene-cyclohexane-1,3-diol.
LMST03010058; OH-Dmdh-D$_3$; 1α-Hydroxy-24,24-dimethyl-22-dehydrovitamin D$_3$; 6438985; 104211-64-5; *Chem Pharm Bull (Tokyo)*. 1985, *33(11)*, 4815-20.

710 *1α-Hydroxy-24-methylvitamin D$_2$*
9547258   C$_{29}$H$_{46}$O$_2$

(1R,3S,5Z)-5-[(2E)-2-[(1R.3aR,7aS)-7a-methyl-1-[(E,2S)-5,5,6-trimethylhept-3-en-2-yl]-2,3,3a,5,6,7-hexahydro-1H-inden-4-ylidene]ethylidene]-4-methylidene-cyclohexane-1,3-diol.
LMST03010058; 1α-hydroxy-24-methylergocalciferol;

24-methyl-9,10-seco-5Z,7E,10(19),22-ergostatetraene-1S,3R-diol; 9547258; Increases serum calcium and inorganic phosphorus concentration and decreases alkaline phosphatase activity in response to 1α-OH-24,24-Me2-D22-D3, 1α,25-(OH)2-24,24-Me2-D22-D3 and 1α-OH-D3; λ_m = 265 nm. **Chem Pharm Bull (Tokyo).** 1985, *33(11)*, 4815-20.

711 *25-Hydroxy[26,27-methyl-³H]vitamin D₃ 3β-(1,2-epoxypropyl)ether*

$C_{29}H_{46}O_3$

25-OH-D3-epoxide; Tritium labelled in the methyl groups at C26 and C27. Used as an affinity labeling reagent of human vitamin D binding protein. **Arch Biochem Biophys.**, 1995, 319(2), 504-7.

712 *18-Acetoxy-vitamin D₃*
9547512

$C_{29}H_{46}O_3$

[(3R,3aR,7E,7aR)-7-[(2Z)-2-[(5S)-5-hydroxy-2-methylidene-cyclohexylidene]ethylidene]-3-[(2R)-6-methylheptan-2-yl]-2,3,4,5,6,7a-hexahydro-1H-inden-3a-yl]methyl acetate.
LMST03020404; 18-acetoxy-cholecalciferol; 18-acetoxy-9,10-seco-5Z,7E,10(19)-cholestatrien-3S-ol; 9547512; Synthesized from upper part (CD ring ketone with side chain) and lower part (A-ring plus seco-B-ring part) phosphine oxide by Horner-Wittig coupling. Compared to 1α,25-dihydroxyvitamin D₃, intestinal calcium absorption is <0.03%, bone calcium mobilization is <0.2%, and affinity for the chick intestinal receptor is <0.001%. **J Org Chem.** 1992, *57(11)*, 3214-17.

713 *1α,25-Dihydroxy-11?-vinylvitamin D₃*
9547513

$C_{29}H_{46}O_3$

(1R,3S,5Z)-5-[(2E)-2-[(1R,3aR,6R,7aS)-6-ethenyl-1-[(2R)-6-hydroxy-6-methyl-heptan-2-yl]-7a-methyl-2,3,3a,5,6,7-hexahydro-1H-inden-4-ylidene]ethylidene]-4-methylidene-cyclohexane-1,3-diol.
LMST03020405, 11S-vinyl-9,10-seco-5Z,7E,10(19)-cholestatriene-1S,3R,25-triol; 1α,25-dihydroxy-11?-vinylcholecalciferol; 9547513; Synthesized by Horner coupling of 25-hydroxylated CD-ring ketone, which was constructed from Inhoffen-Lythgoe diol (vitamin D₂ ozonolysis product), with A-ring phosphine oxide. Biological activity reported. **J Biol Chem.** 1992, *267(5)*, 3044-51.

714 *1α,25-Dihydroxy-26,27-dimethyl-22E, 23-didehydrovitamin D₃*
134508-36-4

$C_{29}H_{46}O_3$

(1R,3S,5Z)-5-[(2E)-2-[(1R,3aR,7aS)-1-[(E,2S)-6-ethyl-6-hydroxy-oct-3-en-2-yl]-7a-methyl-2,3,3a,5,6,7-hexahydro-1H-inden-4-ylidene]ethylidene]-4-methylidene-cyclohexane-1,3-diol.
LMST03020406; 1α,25-dihydroxy-26,27-dimethyl-22, 23-didehydrochocalciferol; 26,27-dimethyl-9,10-seco-5Z,7E,10(19),22E-cholestatetraene-1S,3R,25-triol; 1,25-Dihydroxy-26,27-dihomo-22-ene-vitamin D₃; 1,25-Dde-D₃; 1,25-Dihydroxy-26,27-dihomo-22-ene-cholecalciferol; 6439377; 9547514; 134508-36-4; Has less calcemic activity and lower receptor binding affinity than 1α,25-dihydroxyvitamin D₃. Has a low affinity for vitamin D binding protein, >100 times less than that of 1α,25-dihydroxyvitamin D₃. **Endocrinology.** 1991, *128(4)*, 1687-92.

715 *1α,25-Dihydroxy-22,23-didehydro-24a,24b-dihomovitamin D₃*
154356-84-0

$C_{29}H_{46}O_3$

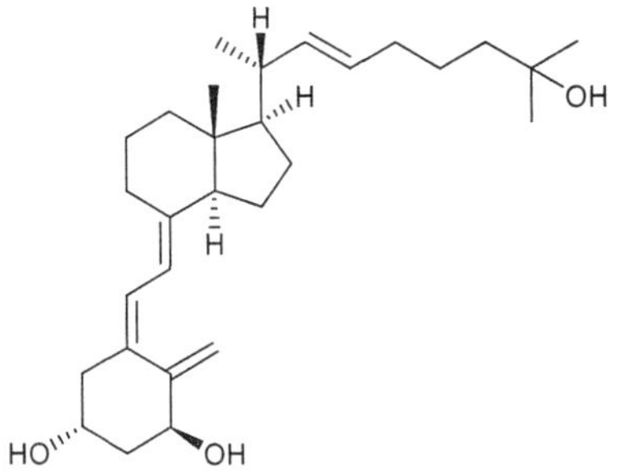

(1R,3S,5Z)-5-[(2E)-2-[(1R,3aR,7aS)-1-[(E,2S)-8-hydroxy-8-methyl-non-3-en-2-yl]-7a-methyl-2,3,3a,5,6,7-hexahydro-1H-inden-4-ylidene]ethylidene]-4-methylidene-cyclohexane-1,3-diol.
LMST03020407; 24a,24b-dihomo-9,10-seco-5Z,7E,10 (19),22E-cholestatetraene-1S,3R,25-triol; 1α,25-di-hydroxy-22,23-didehydro-24a,24b-dihomocholecalc-iferol; Dhbh-calcitriol; 1,25-(OH)2-22-Dehydro-di-homo-vitamin D$_3$; 22-Dehydro-1,25-dihydroxy-24-di-homovitamin D$_3$; 6441581; 9547515; 154356-84-0; Synthesized from 1α-hydroxylated pentanorvitamin D 22-carboxaldehyde and a C(7) side chain fragment. An interleukin-6 and 8 antagonist. Less active on calcium metabolism and less toxic than calcitriol. Ten times more active than 1α,25-dihydroxyvitamin D$_3$ in differentiating HL-60 cells, but approximately 1000 times less active in mobilizing skeletal calcium; λ_m = 264 nm (EtOH). *Biochemistry*. 1990, **29(1)**, 190-6.

716   **1α,25-Dihydroxy-26,27-ethanovitamin D$_3$**
9547516   C$_{29}$H$_{46}$O$_3$

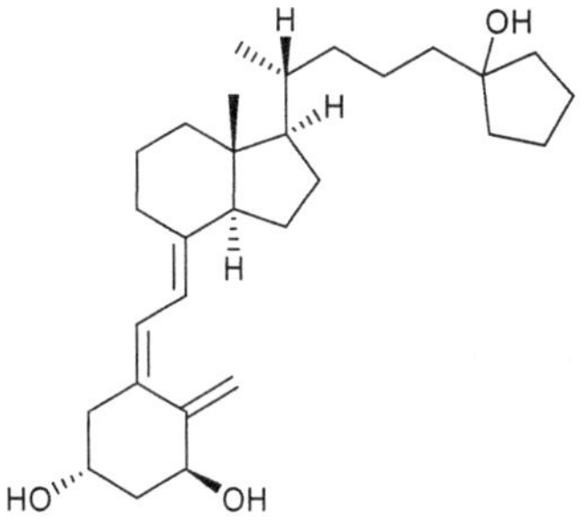

(1R,3S,5Z)-5-[(2E)-2-[(1R,3aR,7aS)-1-[(2R)-5-(1-hydroxycyclopentyl)pentan-2-yl]-7a-methyl-2,3,3a,5,6,7-hexahydro-1H-inden-4-ylidene]ethylidene]-4-methylidene-cyclohexane-1,3-diol.
LMST03020408;   26,27-ethano-9,10-seco-5Z,7E,10 (19)-cholestatriene-1S,3R,25-triol; 1α,25-dihydroxy-26,27-ethanocholecalciferol; 9547516; λ_m = 264 nm (iPrOH-hexane). *J Org Chem*. 1988, **53(15)**, 3450-7.

717   **1α,25-Dihydroxy-24E,24a-didehydro-24a,24b-dihomovitamin D$_3$**
9547690   C$_{29}$H$_{46}$O$_3$

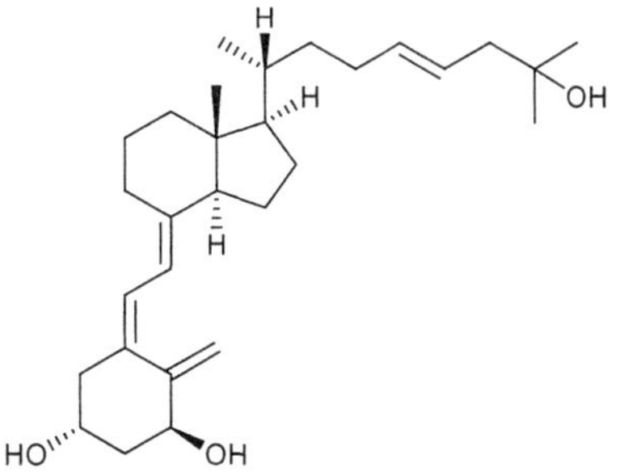

(1R,3S,5Z)-5-[(2E)-2-[(1R,3aR,7aS)-1-[(E,2R)-8-hydroxy-8-methyl-non-5-en-2-yl]-7a-methyl-2,3,3a,5,6,7-hexahydro-1H-inden-4-ylidene]ethylidene]-4-methylidene-cyclohexane-1,3-diol.
LMST03020595; 1α,25-dihydroxy-24E,24a-didehydro-24a,24b-dihomocholecalciferol; 24a,24b-dihomo-9, 10-seco-5Z,7E,10(19),24E-cholestatetraene-1S,3R,25-triol; 9547690.

718   **1α,25-Dihydroxy-24aE,24b-didehydro-24a,24b-dihomovitamin D$_3$**
9547691   C$_{29}$H$_{46}$O$_3$

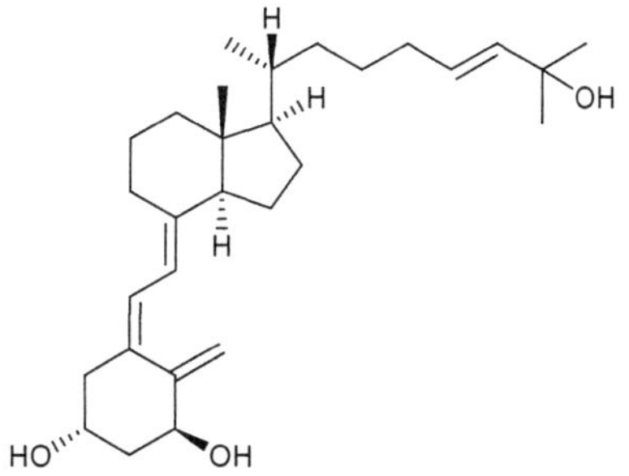

(1R,3S,5Z)-5-[(2E)-2-[(1R,3aR,7aS)-1-[(E,2R)-8-hydroxy-8-methyl-non-6-en-2-yl]-7a-methyl-2,3,3a,5,6,7-hexahydro-1H-inden-4-ylidene]ethylidene]-4-methylidene-cyclohexane-1,3-diol.
LMST03020596;   1α,25-dihydroxy-24aE,24b-di-dehydro-24a,24b-dihomocholecalciferol; 24a,24b-dihomo-9,10-seco-5Z,7E,10(19),24aE-cholestatetraene-1S,3R,25-triol; 9547691.

719   **1α,25-Dihydroxy-24-methylvitamin D$_2$**
104211-73-6   C$_{29}$H$_{46}$O$_3$

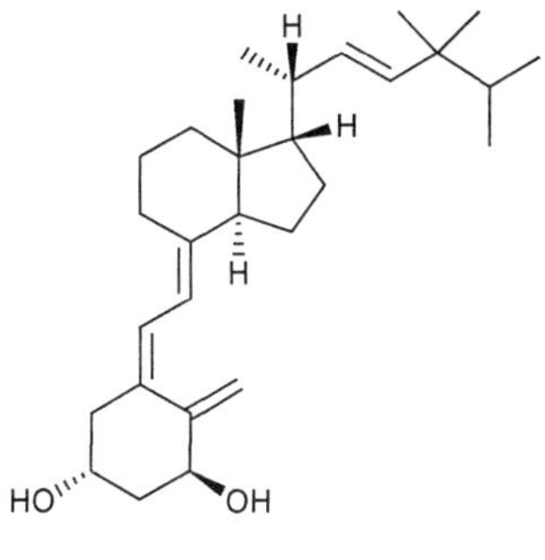

(1R,3S,5Z)-5-[(2E)-2-[(1R,3aR,7aS)-1-[(E,2S)-6-hydroxy-5,5,6-trimethyl-hept-3-en-2-yl]-7a-methyl-2,3,3a,5,6,7-hexahydro-1H-inden-4-ylidene]ethylidene]-4-methylidene-cyclohexane-1,3-diol.
LMST03010059; 1α,25-dihydroxy-24-methylvitamin D$_2$; 1α,25-dihydroxy-24-methylergocalciferol; 24-methyl-9,10-seco-5Z,7E,10(19),22E-ergostatetraene-1S,3R,25-triol; 1,25-Dihydroxy-24,24-dimethyl-22-dehydrovitamin D$_3$; Doh-DM-D$_3$; 6438986; 9547259; 104211-73-6; Increases serum calcium and inorganic phosphorus concentration and decreases alkaline phosphatase activity in response to 1α-OH-24,24-Me2-D22-D3, 1α,25-(OH)2-24,24-Me2-D22-D3 and 1α-OH-D3; λ_m = 265 nm. ***Chem Pharm Bull (Tokyo)***. 1985, ***33(11)***, 4815-20.

720   *18-Acetoxy-1α-hydroxyvitamin D$_3$*
9547517 $C_{29}H_{46}O_4$

[(3R,3aR,7E,7aR)-7-[(2Z)-2-[(3S,5R)-3,5-dihydroxy-2-methylidene-cyclohexylidene]ethylidene]-3-[(2R)-6-methylheptan-2-yl]-2,3,4,5,6,7a-hexahydro-1H-inden-3a-yl]methyl acetate.
LMST03020409; 18-acetoxy-9,10-seco-5Z,7E,10(19)-cholestatriene-1S,3R-diol; 18-acetoxy-1α-hydroxy-cholecalciferol; 9547517; Prepared from upper part (CD ring ketone with side chain) and lower part (A-ring plus seco-B-ring part) phosphine oxide by Horner-Wittig coupling. Compared to 1α,25-dihydroxyvitamin D$_3$, intestinal calcium absorption is <0.05%, bone calcium mobilization is <0.3%, affinity for the chick intestinal receptor is 0.017 0.010%. ***J Org Chem***. 1992, ***57(11)***, 3214-7.

721   *18-Acetoxy-25-hydroxyvitamin D$_3$*
9547518 $C_{29}H_{46}O_4$

[(3R,3aR,7E,7aR)-3-[(2R)-6-hydroxy-6-methyl-heptan-2-yl]-7-[(2Z)-2-[(5S)-5-hydroxy-2-methylidene-cyclohexylidene]ethylidene]-2,3,4,5,6,7a-hexahydro-1H-inden-3a-yl]methyl acetate.
LMST03020410; 18-acetoxy-25-hydroxycholecalc-iferol; 18-acetoxy-9,10-seco-5Z,7E,10(19)-cholestatri-ene-3S,25-diol; 9547518; Synthesized from upper part (CD ring ketone with side chain) and lower part (A-ring plus seco-B-ring part) phosphine oxide by Horner-Wittig coupling. Compared to 1α,25-dihydroxyvitamin D$_3$, intestinal calcium absorption is <0.05%, bone calcium mobilization is <0.3%, and affinity to the chick intestinal receptor is <0.01%. ***J Org Chem***. 1992, ***57(11)***, 3214-7.

722   *1α,25-Dihydroxy-11α-[(1R)-oxiranyl]-vitamin D$_3$*
9547519 $C_{29}H_{46}O_4$

(1R,3S,5Z)-5-[(2E)-2-[(1R,3aR,6R,7aS)-1-[(2R)-6-hydroxy-6-methyl-heptan-2-yl]-7a-methyl-6-(oxiran-2-yl)-2,3,3a,5,6,7-hexahydro-1H-inden-4-ylidene]ethylidene]-4-methylidene-cyclohexane-1,3-diol.
LMST03020411; 11S-[(1R)-oxiranyl]-9,10-seco-5Z,7E,10(19)-cholestatriene-1S,3R,25-triol; 1α,25-dihydroxy-11α-[(1R)-oxiranyl]cholecalciferol; 9547519; Compared to 1α,25-dihydroxyvitamin D$_3$, affinity for rat intestinal receptor and human vitamin D binding protein are 2% and 80% and differentiation of HL-60 cells is <1%. ***Endocr Rev***. 1995, ***16(2)***, 200-57.

723   *1α,25-Dihydroxy-11α-[(1S)-oxiranyl]-vitamin D$_3$*
9547520 $C_{29}H_{46}O_4$

(1R,3S,5Z)-5-[(2E)-2-[(1R,3aR,6S,7aS)-1-[(2R)-6-hydr-oxy-6-methyl-heptan-2-yl]-7a-methyl-6-(oxiran-2-yl)-2,3,3a,5,6,7-hexahydro-1H-inden-4-ylidene]ethylidene]-4-methylidene-cyclohexane-1,3-diol.

LMST03020412; 11S-[(1S)-oxiranyl]-9,10-seco-5Z,7E, 10(19)-cholestriene-1S,3R,25-triol; 1α,25-dihydroxy-11α-[(1S)-oxiranyl]cholecalciferol; 9547520; Compared to 1α,25-dihydroxyvitamin D_3, affinity for rat intestinal receptor and human vitamin D binding protein are 1% and 62% and differentiation of HL-60 cells is <1%. *Endocr Rev*. 1995, *16(2)*, 200-57.

724   1α,25-Dihydroxy-26,27-dimethyl-20,21-methano-23-oxavitamin D_3
9547520 $C_{29}H_{46}O_4$

(1R,3S,5Z)-5-[(2E)-2-[(1S,3aR,7aS)-1-[1-[(2-ethyl-2-hydroxy-butoxy)methyl]cyclopropyl]-7a-methyl-2,3,3a,5,6,7-hexahydro-1H-inden-4-ylidene]ethylidene]-4-methylidene-cyclohexane-1,3-diol.
LMST03020413; 26,27-dimethyl-20,21-methano-23-oxa-9,10-seco-5Z,7E,10(19)-cholestatriene-1S,3R,25-triol; 1α,25-dihydroxy-26,27-dimethyl-20,21-methano-23-oxavitamin D_3; 1α,25-dihydroxy-26,27-dimethyl-20,21-methano-23-oxacholecalciferol; 9547520a;

Compared to 1α,25-dihydroxyvitamin D_3, affinity for pig intestinal nuclear receptor is 50%, affinity for human vitamin D binding protein is <0.02%, differentiation of HL-60 cells is 20% and calciuric effect is 100%. Vitamin D. A Pluripotent Steroid Hormone: Structural Studies, Molecular Endocrinology and Clinical Applications. Proceedings of the Ninth Workshop on Vitamin D, Orlando, Florida (USA) May 28-June 2. Synthesis and Biological Activity of 23-Oxa Vitamin D Analogues. (Norman, A. W., Bouillon, R., Thomasset, M., eds), pp93-94, Walter de Gruyter Berlin New York (1994).

725   18-Acetoxy-1α,25-dihydroxyvitamin D_3
9547521 $C_{29}H_{46}O_5$

[(3R,3aR,7E,7aR)-7-[(2Z)-2-[(3S,5R)-3,5-dihydroxy-2-methylidene-cyclohexylidene]ethylidene]-3-[(2R)-6-hydroxy-6-methyl-heptan-2-yl]-2,3,4,5,6,7a-hexahydro-1H-inden-3a-yl]methyl acetate.
LMST03020414; 18-acetoxy-9,10-seco-5Z,7E,10(19)-cholestriene-1S,3R,25-triol; 18-acetoxy-1α,25-dihydroxycholecalciferol; 9547521; Synthesized from upper part (CD ring ketone with side chain) and lower part (A-ring plus seco-B-ring part) phosphine oxide by Horner-Wittig coupling. Compared to 1α,25-dihydroxyvitamin D_3, intestinal calcium absorption is <0.15%, bone calcium mobilization is <1.0% and affinity to the chick intestinal receptor is 0.036 0.021%. *J Org Chem*. 1992, *57(11)*, 3214-7.

726   3β-Fluoro-9,10-secostigmasta-5Z,7E,10 (19)-triene
76026-39-6 $C_{29}H_{47}F$

(3R,6R)-6-[(1R,3aR,4E,7aS)-4-[(2Z)-2-[(5R)-5-fluoro-2-methylidene-cyclohexylidene]ethylidene]-7a-methyl-2,3,3a,5,6,7-hexahydro-1H-inden-1-yl]-3-ethyl-2-methyl-heptane.
9,10-Secostigmasta-5Z,7E,10(19)-triene, 3β-fluoro-; 6440542; 76026-39-6.

727   *Vitamin D_5*
9547700 $C_{29}H_{48}O$

(1S,3Z)-3-[(2E)-2-[(1R,3aR,7aS)-1-[(2R,5S)-5-ethyl-6-methyl-heptan-2-yl]-7a-methyl-2,3,3a,5,6,7-hexahydro-1H-inden-4-ylidene]ethylidene]-4-methylidene-cyclohexan-1-ol.
LMST03040002; 9,10-seco-5Z,7E,10(19)-poriferasta-trien-3S-ol; 24S-ethyl-9,10-seco-5Z,7E,10(19)-cholestatrien-3S-ol; 9547700; Synthesized and biological activity characterized. Has 2% of the antirachitic effect

of vitamin D_3. May be a cancer preventive agent.
Hoppe-Seyler's Z Physiol Chem. 1936, *241*, 116.

728   *Provitamin D_5*
9547701 $C_{29}H_{48}O$

(3S,10R,13S,14R,17R)-17-[(2R,5S)-5-ethyl-6-methyl-
heptan-2-yl]-10,13-dimethyl-
2,3,4,9,11,12,14,15,16,17-decahydro-1H-
cyclopenta[a]phenanthren-3-ol.
LMST03040003; Poriferasta-5,7-dien-3β-ol; 24S-ethyl-
5,7-cholestadien-3-ol; 9547701. *J Amer Chem Soc.*
1982, *104(21)*, 5780-1; USP 7211172.

729   *9,10-Secostigmasta-5Z,7E,10(19)-trien-3β-ol*
71761-06-3 $C_{29}H_{48}O$

(1R,3Z)-3-[(2E)-2-[(1R,3aR,7aS)-1-[(2R,5R)-5-ethyl-6-
methyl-heptan-2-yl]-7a-methyl-2,3,3a,5,6,7-
hexahydro-1H-inden-4-ylidene]ethylidene]-4-
methylidene-cyclohexan-1-ol.
9,10-Secostigmasta-5Z,7E,10(19)-trien-3β-ol; 6440503;
71761-06-3.

730   *6S,19-Ethano-25-hydroxy-6,19-dihydrovitamin D_3*
9547524 $C_{29}H_{48}O_2$

(1S,5S)-5-[(E)-[(1R,3aR,7aS)-1-[(2R)-6-hydroxy-6-
methyl-heptan-2-yl]-7a-methyl-2,3,3a,5,6,7-hexa-
hydro-1H-inden-4-ylidene]methyl]-1,2,3,4,5,6,7,8-
octahydronaphthalen-1-ol.
LMST03020417;   6S,19-ethano-9,10-seco-5(10),7E-
cholestadiene-3S,25-diol;   6S,19-ethano-25-hydroxy-
6.19-dihydrocholecalciferol;   9547524;   Synthesized
from (5E)-25-hydroxyvitamin D_3. Binding affinity for
the vitamin D receptor in HL-60 cells is < 1/10000 that
of 1α,25-dihydroxyvitamin D_3, differentiation of HL-60
cells is < 1/1000. *J Med Chem*. 1985, *28(9)*, 1148-53.

731   *6R,19-Ethano-25-hydroxy-6,19-dihydrovitamin D_3*
9547525 $C_{29}H_{48}O_2$

(1S,5R)-5-[(E)-[(1R,3aR,7aS)-1-[(2R)-6-hydroxy-6-
methyl-heptan-2-yl]-7a-methyl-2,3,3a,5,6,7-hexa-
hydro-1H-inden-4-ylidene]methyl]-1,2,3,4,5,6,7,8-
octahydronaphthalen-1-ol.
LMST03020418;   6R,19-ethano-9,10-seco-5(10),7E-
cholestadiene-3S,25-diol;   6R,19-ethano-25-hydroxy-
6,19-dihydrocholecalciferol;   9547525;   Synthesized
from (5E)-25-hydroxyvitamin D_3. Binding affinity for
the vitamin D receptor in HL-60 cells is < 1/10000 that
of 1α,25-dihydroxyvitamin D_3, differentiation of HL-60
cells is < 1/1000. *J Med Chem*. 1985, *28(9)*, 1148-53.

732   *1α-Hydroxy-26,27-dimethylvitamin D_3*
9547522 $C_{29}H_{48}O_2$

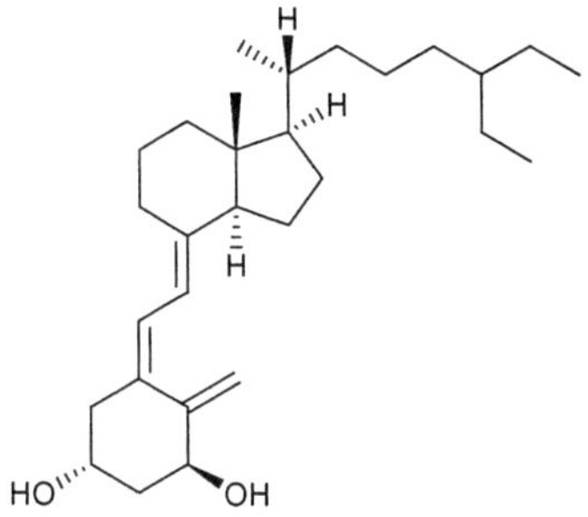

(1R,3S,5Z)-5-[(2E)-2-[(1R,3aR,7aS)-1-[(2R)-6-ethyl-octan-2-yl]-7a-methyl-2,3,3a,5,6,7-hexahydro-1H-inden-4-ylidene]ethylidene]-4-methylidene-cyclo-hexane-1,3-diol.
LMST03020415; 26,27-dimethyl-9,10-seco-5Z,7E,10 (19)-cholestatriene-1S,3R-diol; 1α-hydroxy-26,27-di-methylcholecalciferol; 9547522; Synthesized from 3β-hydroxycholenic acid. Compared to 1α,25-dihydroxy-vitamin D$_3$, binding affinity for serum vitamin D binding protein is < 0.01%, and affinity for the receptor of HL-60 cells is 0.1%; λ$_m$ = 265 nm. **Chem Pharm Bull (Tokyo)**. 1988, **36(7)**, 2303-11; **Steroids**. 1991, **56(3)**, 142-7.

733   25-Hydroxy-26,27-dimethylvitamin D$_3$
116925-40-7   C$_{29}$H$_{48}$O$_2$

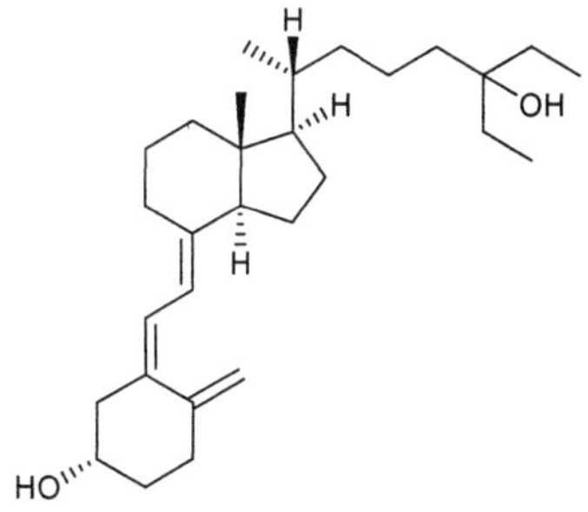

(7R)-7-[(1R,3aR,4E,7aS)-4-[(2Z)-2-[(5S)-5-hydroxy-2-methylidene-cyclohexylidene]ethylidene]-7a-methyl-2,3,3a,5,6,7-hexahydro-1H-inden-1-yl]-3-ethyl-octan-3-ol.
LMST03020416; 26,27-dimethyl-9,10-seco-5Z,7E,10 (19)-cholestatriene-3S,25-diol; 25-hydroxy-26,27-dimethylcholecalciferol; 26,27-Dimethylcalcifediol; 25-Hydroxy-26,27-dimethylvitamin D$_3$; 25-Hydroxy-26, 27-dimethylcholecalciferol; 6450051; 9547523; 116925-40-7; Synthesized from 3β-hydroxycholenic acid or 4-cholenoic acid. Compared to 1α,25-dihydroxyvitamin D$_3$, binding affinity for serum vitamin D binding protein is 66.7%, and affinity for the receptor of HL-60 cells is 0.9%; λ$_m$ = 264 nm (EtOH). **Chem Pharm Bull (Tokyo)**. 1988, **36(7)**, 2303-11; **J Steroid Biochem**. 1988, **31(2)**, 147-60; **Steroids**. 1991, **56(3)**, 142-7.

734   1-Hydroxyvitamin D$_5$
187935-17-7   C$_{29}$H$_{48}$O$_2$

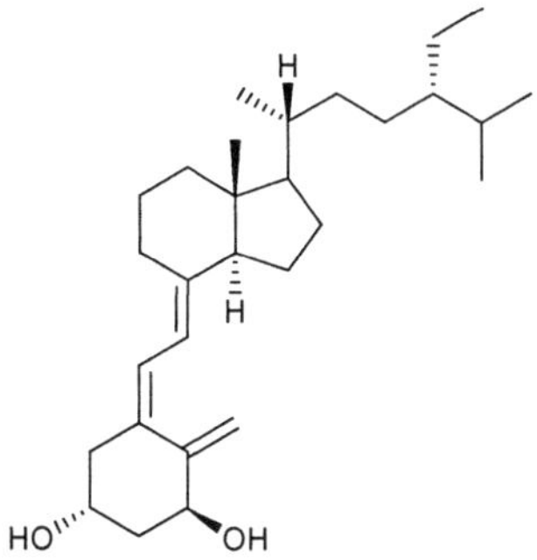

1α,3β-Dihydroxy-9,10-secostigmasta-5Z,7E,10(19)-triene.
1α-Hydroxyvitamin D$_5$; CCRIS 8700; 1α-hydroxy-24-ethylcholecalciferol; 187935-17-7; Synthetic compound. Less calcemic than calcitriol. Reduces the incidence of mammary carcinogenesis **in vivo**. Effective against breast cancer without inducing hypercalcemia. **J Natl Cancer Inst**. 2000, **92(22)**, 1836-40.

735   25-Hydroxy-6,19-dihydro-6,19-ethano-vitamin D$_3$
96616-70-5   C$_{29}$H$_{48}$O$_2$

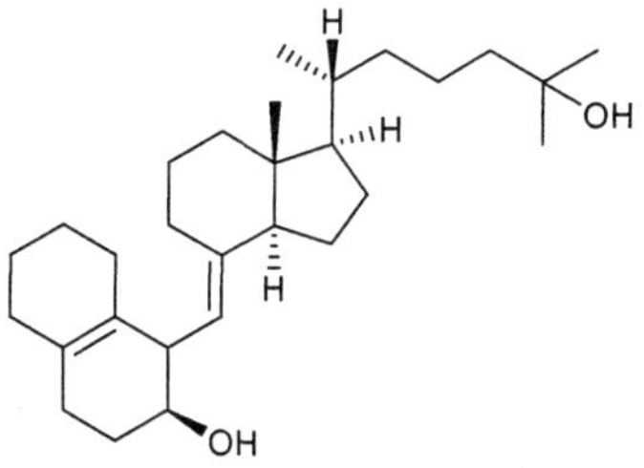

1-[(Z)-[1-(6-hydroxy-6-methyl-heptan-2-yl)-7a-methyl-2,3,3a,4,6,7-hexahydro-1H-inden-5-ylidene]methyl]-1,2,3,4,5,6,7,8-octahydronaphthalen-2-ol.
25Hde-vitamin D$_3$; 25-Hydroxy-6,19-dihydro-6,19-ethanovitamin D$_3$; 1,2,3,4,5,6,7,8-Octahydro-8-((octahydro-1-(5-hydroxy-1,5-dimethylhexyl)-7a-methyl-4H-inden-4-ylidene)methyl)-2-naphthalenol (1R-(1α(R*), 3aβ,4E(2S*,8R*),7aα))-; 2-Naphthalenol, 1,2,3,4,5,6, 7,8-octahydro-8-((octahydro-1-(5-hydroxy-1,5-di-methylhexyl)-7a-methyl-4H-inden-4-ylidene)methyl)-, (1R-(1α(R*),3aβ,4E(2S*,8R*),7aα))-; 96616-70-5; 6438665. **J Med Chem**. 1985, **28(9)**, 1148-53.

736   11α-Ethyl-1α,25-dihydroxyvitamin D$_3$
9547526   C$_{29}$H$_{48}$O$_3$

(1R,3S,5Z)-5-[(2E)-2-[(1R,3aR,6R,7aS)-6-ethyl-1-[(2R)-6-hydroxy-6-methyl-heptan-2-yl]-7a-methyl-2,3,3a,5,6,7-hexahydro-1H-inden-4-ylidene]ethylidene]-4-methylidene-cyclohexane-1,3-diol.
LMST03020419; 11S-ethyl-9,10-seco-5Z,7E,10(19)-cholestatriene-1S,3R,25-triol; 11α-ethyl-1α,25-di-hydroxycholecalciferol; 9547526; Synthesized by Horner coupling of 25-hydroxylated CD-ring ketone. Bioactivity reported. *J Biol Chem.* 1992, *267(5)*, 3044-51.

737   *1α,25-Dihydroxy-26,27-dimethylvitamin D₃*
97473-92-2 $C_{29}H_{48}O_3$

(1R,3S,5Z)-5-[(2E)-2-[(1R,3aR,7aS)-1-[(2R)-6-ethyl-6-hydroxy-octan-2-yl]-7a-methyl-2,3,3a,5,6,7-hexahydro-1H-inden-4-ylidene]ethylidene]-4-methylidene-cyclohexane-1,3-diol.
LMST03020420; 26,27-dimethyl-9,10-seco-5Z,7E,10(19)-cholestatriene-1S,3R,25-triol; 1α,25-dihydroxy-26,27-dimethylcholecalciferol; 26,27-Dimethylcalcitriol; 1,25-Dihydroxy-26,27-dimethylvitamin D₃; 9547527; 97473-92-2; Synthesized from 3β-hydroxycholenic acid or 25-hydroxy-26,27-dimethylvitamin D₃. Compared to 1α, 25-dihydroxyvitamin D₃, affinity for serum vitamin D binding protein is 10%, affinity for receptor of HL-60 cells is 116.7%, ability to induce HL-60 cell differentiation (assayed by NBT reduction) is approximately 400% under serum-supplemented culture conditions, and 100% under serum-free culture conditions. Bone Ca mobilization response and dose response of various 26,27-dialkyl-1,25-(OH)2D3 in bone Ca mobilization is reported. Slighly less active than 1α, 25-dihydroxyvitamin D₃ in increasing serum calcium in vitamin D-deficient rats. Has bioactivity similar to that of 25-hydroxyvitamin D₃. Binds to rat plasma vitamin D binding protein with approximately one-third the affinity of 25-hydroxyvitamin D₃. Induces

osteocalcin promotion, similarly to 1,25-dihydroxy-vitamin D₃. Can effect cellular differentiation but lacks hypercalcemic activity *in vitro;* λ$_m$ = 265 nm (EtOH). *Ann.* 1932, *492, 226; Steroids.* 1991, *56(3)*, 142-7; *Calcif Tissue Int.* 1992, *51(3)*, 218-22; *Chem Pharm Bull (Tokyo).* 1985, *33(2)*, 878-81; 1988. *36(7)*, 2303-11; *J Steroid Biochem.* 1988, *31(2)*, 147-60.

738   *1α,25-Dihydroxy-24a,24b-dihomo-vitamin D₃*
9547528 $C_{29}H_{48}O_3$

(1R,3S,5Z)-5-[(2E)-2-[(1R,3aR,7aS)-1-[(2R)-8-hydroxy-8-methyl-nonan-2-yl]-7a-methyl-2,3,3a,5,6,7-hexahydro-1H-inden-4-ylidene]ethylidene]-4-methylidene-cyclohexane-1,3-diol.
LMST03020421; 24a,24b-dihomo-9,10-seco-5Z,7E,10(19)-cholestatriene-1S,3R,25-triol; 1α,25-dihydroxy-24a,24b-dihomocholecalciferol; 9547528; Prepared from hydroxy-protected (5E)-23,24,25,26,27-penta-norvitamin D₃ 22-bis(methylseleno)acetalhydroxy or from protected (5E)-22-tosyloxy-23,24,25,26,27-penta-norvitamin D₃. Ten times more active than 1α,25-dihydroxyvitamin D₂ in differentiating HL-60 cells, 1000 times less active in mobilizing skeletal calcium; λ$_m$ = 264 nm (iPrOH-hexane). *J Org Chem.* 1988, *53(15)*, 3450-7; *Biochemistry.* 1990, *29(1)*, 190-6.

739   *1α,25-Dihydroxy-24a,24b-dihomo-20-epivitamin D₃*
9547529 $C_{29}H_{48}O_3$

(1R,3S,5Z)-5-[(2E)-2-[(1R,3aR,7aS)-1-[(2S)-8-hydroxy-8-methyl-nonan-2-yl]-7a-methyl-2,3,3a,5,6,7-hexahydro-1H-inden-4-ylidene]ethylidene]-4-methylidene-cyclo-hexane-1,3-diol.
LMST03020422; (20S)-24a,24b-dihomo-9,10-seco-5Z,7E,10(19)-cholestatriene-1S,3R,25-triol; 1α,25-di-

hydroxy-24a,24b-dihomo-20-epicholecalciferol; 9547529; Synthesized from protected (5E)-1α-hydroxy-22-p-toluenesulfonyloxy-23,24,25,26,27-pentanor-vitamin D$_3$. Compared to 1α,25-dihydroxyvitamin D$_3$, inhibition of proliferation of U937 cells is 5000%, induction of differentiation of U937 cells is 10000% and calciuric effects on normal rats is 40%. Gene Regulation, Structure-Function Analysis and Clinical Application. Proceedings of the Eighth Workshop on Vitamin D Paris, France July 5-10. The 20-Epi Modification in The Vitamin D Series : Selective Enhancement of 'Non-Classical' Receptpr-Mediated Effects. (Norman, A. W., Bouillon, R., Thomasset, M., eds), pp163-164, Walter de Gruyter Berlin New York (1991).

740   *24,24-Dihomo-1,25-dihydroxyvitamin D$_3$*

114694-09-6   C$_{29}$H$_{48}$O$_3$

(1R,3S,5E)-5-[(2E)-2-[(1R,3aR,7aS)-1-(8-hydroxy-8-methyl-nonan-2-yl)-7a-methyl-2,3,3a,5,6,7-hexahydro-1H-inden-4-ylidene]ethylidene]-4-methylidene-cyclohexane-1,3-diol.
MC 1147; 24,24-Dihomo-1,25-dihydroxyvitamin D$_3$; 24,24-Dihomo-1,25-dihydroxycholecalciferol; 1,3-Cyclohexanediol, 4-methylene-5-((octahydro-1-(7-hydroxy-1,7-dimethyloctyl)-7a-methyl-4H-inden-4-ylidene)ethylidene)-, (1R-(1α(R*),3aβ,4E(1R*,3S*,5Z), 7aα))-; LMST03020421; 6444150; 114694-09-6; Approximately 100-fold less active than 1α,25-(OH)2D$_3$ as evaluated by serum calcium and osteocalcin concentrations, as well as by duodenal calbindin D28K and bone calcium content. Has 3.5 - 25% of the affinity of 1,25-dihydroxyvitamin D$_3$ for the vitamin D receptor from chick and rat duodenum or from human peripheral blood mononuclear cells or HL-60 cells. *Steroids.* 1997, *62(7)*, 546-53.

741   *1α,25-Dihydroxy-26,27-dimethyl-24a-homo-22-thiavitamin D$_3$*

9547530   C$_{29}$H$_{48}$O$_3$S

(1R,3S,5Z)-5-[(2E)-2-[(1S,3aR,7aR)-1-[(1S)-1-(4-ethyl-4-hydroxy-hexyl)sulfanylethyl]-7a-methyl-2,3,3a,5,6,7-hexahydro-1H-inden-4-ylidene]ethylidene]-4-methylidene-cyclohexane-1,3-diol.
LMST03020423; 26,27-dimethyl-24a-homo-9,10-seco-22-thia-5Z,7E,10(19)-cholestatriene-1S,3R,25-triol; 1α, 25-dihydroxy-24a-homo-26,27-dimethyl-22-thiacholecalciferol; 9547530; Synthesized from dehydroepiandrosterone *via* the 5,7-diene. *Bioorg Med Chem Lett.* 1995, *5(3)*, 279-82.

742   *1α,25-Dihydroxy-26,27-dimethyl-24a-homo-22-thia-20-epivitamin D$_3$*

9547531   C$_{29}$H$_{48}$O$_3$S

(1R,3S,5Z)-5-[(2E)-2-[(1S,3aR,7aR)-1-[(1R)-1-(4-ethyl-4-hydroxy-hexyl)sulfanylethyl]-7a-methyl-2,3,3a,5,6,7-hexahydro-1H-inden-4-ylidene]ethylidene]-4-methylidene-cyclohexane-1,3-diol.
LMST03020424; (20R)-26,27-dimethyl-24a-homo-9, 10-seco-22-thia-5Z,7E,10(19)-cholestatriene-1S,3R,25-triol; 1α,25-dihydroxy-26,27-dimethyl-24a-homo-22-thia-20-epicholecalciferol; 9547531; Synthesized from dehydroepiandrosterone *via* the 5,7-diene. *Bioorg Med Chem Lett.* 1995, *5(3)*, 279-82.

743   *1α,25-Dihydroxy-24a,24b,24c-trihomo-22-thiavitamin D$_3$*

9547532   C$_{29}$H$_{48}$O$_3$S

(1R,3S,5Z)-5-[(2E)-2-[(1S,3aR,7aR)-1-[(1S)-1-(6-
hydroxy-6-methyl-heptyl)sulfanylethyl]-7a-methyl-
2,3,3a,5,6,7-hexahydro-1H-inden-4-ylidene]
ethylidene]-4-methylidene-cyclohexane-1,3-diol.
LMST03020425; 24a,24b,24c-trihomo-9,10-seco-22-
thia-5Z,7E,10(19)-cholestatriene-1S,3R,25-triol; 1α,25-
dihydroxy-24a,24b,24c-trihomo-22-thiacholecalc-
iferol; 9547532; Synthesized from dehydroepiandrosterone *via* the 5,7-diene. *Bioorg Med Chem Lett.*
1995, *5(3)*, 279-82.

744   *1α,25-Dihydroxy-24a,24b,24c-trihomo-
       22-thia-20-epivitamin D₃*

9547533   C₂₉H₄₈O₃S

(1R,3S,5Z)-5-[(2E)-2-[(1S,3aR,7aR)-1-[(1R)-1-(6-
hydroxy-6-methyl-heptyl)sulfanylethyl]-7a-methyl-
2,3,3a,5,6,7-hexahydro-1H-inden-4-ylidene]
ethylidene]-4-methylidene-cyclohexane-1,3-diol.
LMST03020426;   (20R)-24a,24b,24c-trihomo-9,10-
seco-22-thia-5Z,7E,10(19)-cholestatriene-1S,3R,25-
triol; 1α,25-dihydroxy-24a,24b,24c-trihomo-22-thia-
20-epicholecalciferol; 9547533; Synthesized from dehydroepiandrosterone *via* the 5,7-diene. *Bioorg Med
Chem Lett.* 1995, *5(3)*, 279-82.

745   *1α-Hydroxy-2β-(2-hydroxyethoxy)-
       vitamin D₃*

9547534   C₂₉H₄₈O₄

(1R,2R,3R,5Z)-5-[(2E)-2-[(1R,3aR,7aS)-7a-methyl-1-
[(2R)-6-methylheptan-2-yl]-2,3,3a,5,6,7-hexahydro-
1H-inden-4-ylidene]ethylidene]-2-(2-hydroxyethoxy)-
4-methylidene-cyclohexane-1,3-diol.
LMST03020427; 2R-(2-hydroxyethoxy)-9,10-seco-5Z,
7E,10(19)-cholestatriene-1R,3R-diol; 1α-hydroxy-2β-
(2-hydroxyethoxy)cholecalciferol; 9547534; Synthesized photochemically from the 5,7-diene. Effect on
plasma calcium levels in rats fed with a low
calcium/vitamin D deficient diet reported; λₘ = 262.5
nm (EtOH). *Chem Pharm Bull (Tokyo).* 1993, *41(6)*,
1111-13.

746   *1α,25-Dihydroxy-20S-methoxy-24a-
       homovitamin D₃*

9547535   C₂₉H₄₈O₄

(1R,3S,5Z)-5-[(2E)-2-[(1S,3aR,7aR)-1-[(2S)-7-hydroxy-2-
methoxy-7-methyl-octan-2-yl]-7a-methyl-2,3,3a,5,6,7-
hexahydro-1H-inden-4-ylidene]ethylidene]-4-
methylidene-cyclohexane-1,3-diol.
LMST03020428; 1α,25-dihydroxy-20S-methoxy-24a-
homocholecalciferol;   20S-methoxy-24a-homo-9,10-
seco-5Z,7E,10(19)-cholestatriene-1S,3R,25-triol;
9547535; Synthesized from (5E)-1α-hydroxy-20-keto-
22,23,24,25,26,27-hexanorvitamin D₃. Compared to
1α,25-dihydroxyvitamin D₃, inhibition of U937 cell
(human histiocytic lymphoma cell line) proliferation is
10000%, binding to the vitamin D receptor from
rachitic chicken intestine is < 0.5%, calcemic activity
determined by the increase in urinary calcium
excretion in rats is 3%. Vitamin D. A Pluripotent
Steroid Hormone: Structural Studies, Molecular
Endocrinology and Clinical Applications. Proceedings
of the Ninth Workshop on Vitamin D, Orlando, Florida
(USA) May 28-June 2. Synthesis and Biological Activity
of 20-Hydroxylated Vitamin D Analogues. (Norman, A.

W., Bouillon, R., Thomasset, M., eds), pp95-96, Walter de Gruyter Berlin New York (1994).

747 ***1α,25-Dihydroxy-24a,24b,24c-trihomo-22-oxavitamin D₃***
9547536 $C_{29}H_{48}O_4$

(1R,3S,5Z)-5-[(2E)-2-[(1S,3aR,7aR)-1-[(1R)-1-(6-hydroxy-6-methyl-heptoxy)ethyl]-7a-methyl-2,3,3a,5,6,7-hexahydro-1H-inden-4-ylidene]ethylidene]-4-methylidene-cyclohexane-1,3-diol.
LMST03020429; 24a,24b,24c-trihomo-22-oxa-9,10-seco-5Z,7E,10(19)-cholestatriene-1S,3R,25-triol; 1α,25-dihydroxy-24a,24b,24c-trihomo-22-oxacholecalciferol; 9547536; The activity inducing differentiation of human myeloid leukemia cells (HL-60) is reported; λ_m = 262 nm (EtOH). ***Chem Pharm Bull (Tokyo).*** 1992, ***40(6)***, 1494-9.

748 ***1α,25-Dihydroxy-26,27-dimethyl-24a-homo-22-oxavitamin D₃***
9547537 $C_{29}H_{48}O_4$

(1R,3S,5Z)-5-[(2E)-2-[(1S,3aR,7aR)-1-[(1S)-1-(4-ethyl-4-hydroxy-hexoxy)ethyl]-7a-methyl-2,3,3a,5,6,7-hexahydro-1H-inden-4-ylidene]ethylidene]-4-methylidene-cyclohexane-1,3-diol.
LMST03020430; 26,27-dimethyl-24a-homo-22-oxa-9,10-seco-5Z,7E,10(19)-cholestatrien-1S,3R,25-triol; 1α,25-dihydroxy-26,27-dimethyl-24a-homo-22-oxacholecalciferol; 9547537; Compared to 1α,25-dihydroxyvitamin D₃, inhibition of U937 cell (human histiocytic lymphoma cell line) proliferation is 10000%, induction of U937 cell differentiation is 2000%, binding to the vitamin D receptor from rachitic chicken intestine is 32%, calcemic activity determined by the increase in urinary calcium excretion in rats is 30%, and inhibitory

effect on murine thymocyte activation is 900%. ***Biochem Pharmacol.*** 1991, ***42(8)***, 1569-75.

749 ***1α,25-Dihydroxy-26,27-dimethyl-24a-homo-22-oxa-20-epivitamin D₃***
131875-08-6 $C_{29}H_{48}O_4$

(1R,3S,5Z)-5-[(2E)-2-[(1S,3aR,7aR)-1-[(1R)-1-(4-ethyl-4-hydroxy-hexoxy)ethyl]-7a-methyl-2,3,3a,5,6,7-hexahydro-1H-inden-4-ylidene]ethylidene]-4-methylidene-cyclohexane-1,3-diol.
LMST03020432; Lexacalcitol; KH 1060; KH-1060; (20R)-26,27-dimethyl-24a-homo-22-oxa-9,10-seco-5Z,7E,10(19)-cholestatriene-1S,3R,25-triol; 20R-((4-Ethyl-4-hydroxyhexyl)oxy)-9,10-secopregna-5Z,7E, 10 (19)-triene-1α,3β-diol; 5288670; 131875-08-6. ***Endocrinology.*** 1999, ***140(10)***, 4779-88; ***Biochem Pharmacol.*** 1991, ***42(8)***, 1569-75; ***J Biol Chem.*** 1995, ***270(18)***, 10551-8; ***Mol Pharm.*** 2002, ***62(4)***, 788-94. ***Leo AB.***

750 ***1α,25-Dihydroxy-20,26,27-trimethyl-23-oxavitamin D₃***
9547538 $C_{29}H_{48}O_4$

(1R,3S,5Z)-5-[(2E)-2-[(1R,3aR,7aR)-1-[1-(2-ethyl-2-hydroxy-butoxy)-2-methyl-propan-2-yl]-7a-methyl-2,3,3a,5,6,7-hexahydro-1H-inden-4-ylidene]ethylidene]-4-methylidene-cyclohexane-1,3-diol.
LMST03020431; 20,26,27-trimethyl-23-oxa-9,10-seco-5Z,7E,10(19)-cholestatriene-1S,3R,25-triol; 1α,25-dihydroxy-20,26,27-trimethyl-23-oxacholecalciferol;
9547538; Compared to 1α,25-dihydroxyvitamin D₃, affinity for pig intestinal nuclear receptor is 50%, affinity for human vitamin D binding protein is < 0.02%, differentiation of HL-60 cells is 50%, and calciuric effect is 20%. Vitamin D. A Pluripotent Steroid Hormone: Structural Studies, Molecular Endocrinology and Clinical Applications. Proceedings

of the Ninth Workshop on Vitamin D, Orlando, Florida (USA) May 28-June 2. Synthesis and Biological Activity of 23-Oxa Vitamin D Analogues. (Norman, A. W., Bouillon, R., Thomasset, M., eds), pp93-94, Walter de Gruyter Berlin New York (1994).

751   1α,25-Dihydroxy-11α-(2-hydroxyethyl)-vitamin D₃

9547539 $C_{29}H_{48}O_4$

(1R,3S,5Z)-5-[(2E)-2-[(1R,3aR,6R,7aS)-6-(2-hydroxy-ethyl)-1-[(2R)-6-hydroxy-6-methyl-heptan-2-yl]-7a-methyl-2,3,3a,5,6,7-hexahydro-1H-inden-4-ylidene] ethylidene]-4-methylidene-cyclohexane-1,3-diol.
LMST03020433;   11S-(2-hydroxyethyl)-9,10-seco-5Z, 7E,10(19)-cholestatriene-1S,3R,25-triol;   1α,25-di-hydroxy-11α-(2-hydroxyethyl)cholecalciferol;
9547539; Synthesized by Horner coupling of 25-hydroxylated CD-ring ketone. Compared to 1α,25-dihydroxyvitamin D₃, affinity for rat duodenum receptor and human vitamin D binding protein are <0.1% and 115%, differentiation of HL-60 cells is 7%, inhibition of the proliferation of peripheral blood mononuclear cells is <1%, bone resorption is <1% (3 day), <1 (6 day), biological activity in rachitic chick (serum calcium, bone calcium, serum osteocalcin and calbindin D-28K) are all <1%. *J Biol Chem*. 1992, *267(5)*, 3044-51.

752   1α,25-Dihydroxy-11β-(2-hydroxyethyl)-vitamin D₃

9547540 $C_{29}H_{48}O_4$

(1R,3S,5Z)-5-[(2E)-2-[(1R,3aR,6S,7aS)-6-(2-hydroxy-ethyl)-1-[(2R)-6-hydroxy-6-methyl-heptan-2-yl]-7a-methyl-2,3,3a,5,6,7-hexahydro-1H-inden-4-ylidene] ethylidene]-4-methylidene-cyclohexane-1,3-diol.
LMST03020434;   11R-(2-hydroxyethyl)-9,10-seco-5Z,7E,10(19)-cholestatriene-1S,3R,25-triol;   1α,25-di-

hydroxy-11β-(2-hydroxyethyl)cholecalciferol;
9547540; Prepared by Horner coupling of the 25-hydroxylated CD-ring ketone. Compared to 1α,25-dihydroxyvitamin D₃, affinity for rat duodenum receptor and human vitamin D binding protein are 0.2% and 40%, differentiation of HL-60 cells is 5%, inhibition of the proliferation of peripheral blood mononuclear cells is <1%, biological activity in rachitic chick (serum calcium, bone calcium, serum osteocalcin, calbindin D-28K) are all <1%. *J Biol Chem*. 1992, *267(5)*, 3044-51.

753   1α,22R,25-Trihydroxy-24a,24b-dihomo-20-epivitamin D₃

9547541 $C_{29}H_{48}O_4$

(7R,8R)-8-[(1R,3aR,4E,7aS)-4-[(2Z)-2-[(3S,5R)-3,5-dihydroxy-2-methylidene-cyclohexylidene]ethylidene]-7a-methyl-2,3,3a,5,6,7-hexahydro-1H-inden-1-yl]-2-methyl-nonane-2,7-diol.
LMST03020435;   1α,22R,25-trihydroxy-24a,24b-di-homo-20-epicholecalciferol;   (20R)-24a,24b-dihomo-9,10-seco-5Z,7E,10(19)-cholestatriene-1S,3R,22R,25-tetrol; 9547541; Synthesized from protected (5E)-1α-hydroxy-22-oxo-23,24,25,26,27-pentanor-20-epi-vitamin D₃. Compared to 1α,25-dihydroxyvitamin D₃, induction of differentiation of U937 cells is 2000%, inhibition of proliferation is 2100%, VDR (rachitic chicken intestinal receptor) binding affinity is 0.1%, and calciuric effect in normal rats is 10%. Vitamin D. A Pluripotent Steroid Hormone: Structural Studies, Molecular Endocrinology and Clinical Applications. Proceedings of the Ninth Workshop on Vitamin D, Orlando, Florida (USA) May 28-June 2. Chemistry and Biology of Highly Active 22-Oxy Analogs of 20-Epi Calcitriol with very low Binding Affinity to the Vitamin D Receptor. (Norman, A. W., Bouillon, R., Thomasset, M., eds), pp85-86, Walter de Gruyter Berlin New York (1994).

754   1α,20S,25-Trihydroxy-24a,24b-dihomovitamin D₃

9547542 $C_{29}H_{48}O_4$

(2S)-2-[(1S,3aR,4E,7aR)-4-[(2Z)-2-[(3S,5R)-3,5-dihydroxy-2-methylidene-cyclohexylidene]ethylidene]-7a-methyl-2,3,3a,5,6,7-hexahydro-1H-inden-1-yl]-8-methyl-nonane-2,8-diol.
LMST03020436; 1α,20S,25-trihydroxy-24a,24b-di-homocholecalciferol; 24a,24b-dihomo-9,10-seco-5Z, 7E,10(19)-cholestatriene-1S,3R,20S,25-tetrol; 9547542; Prepared from a (5E)-1α-hydroxy-20-keto-22,23,24,25, 26,27-hexanorvitamin D_3 derivative. In inhibition of U937 cell (human histiocytic lymphoma cell line) proliferation is <10% as effective as 1α,25-dihydroxyvitamin D_3. Vitamin D. A Pluripotent Steroid Hormone: Structural Studies, Molecular Endocrinology and Clinical Applications. Proceedings of the Ninth Workshop on Vitamin D, Orlando, Florida (USA) May 28-June 2. Synthesis and Biological Activity of 20-Hydroxylated Vitamin D Analogues. (Norman, A. W., Bouillon, R., Thomasset, M., eds), pp95-96, Walter de Gruyter Berlin New York (1994).

755 *14α,20S,25-Trihydroxy-26,27-dimethyl-vitamin D₃*
9547543 $C_{29}H_{48}O_4$

(2S)-2-[(1S,3aR,4E,7aR)-4-[(2Z)-2-[(3S,5R)-3,5-dihydroxy-2-methylidene-cyclohexylidene]ethylidene]-7a-methyl-2,3,3a,5,6,7-hexahydro-1H-inden-1-yl]-6-ethyl-octane-2,6-diol.
LMST03020437; 1α,20S,25-trihydroxy-26,27-di-methylcholecalciferol; 26,27-dimethyl-9,10-seco-5Z, 7E,10(19)-cholestatriene-1S,3R,20S,25-tetrol; 9547543. Vitamin D. A Pluripotent Steroid Hormone: Structural Studies, Molecular Endocrinology and Clinical Applications. Proceedings of the Ninth Workshop on Vitamin D, Orlando, Florida (USA) May 28-June 2. Synthesis and Biological Activity of 20-Hydroxylated Vitamin D Analogues. (Norman, A. W., Bouillon, R., Thomasset, M., eds), pp95-96, Walter de Gruyter Berlin New York (1994).

756 *1α,22S,25-Trihydroxy-26,27-dimethyl-vitamin D₃*
9547545 $C_{29}H_{48}O_4$

(2S,3S)-2-[(1R,3aR,4E,7aS)-4-[(2Z)-2-[(3S,5R)-3,5-dihydroxy-2-methylidene-cyclohexylidene]ethylidene]-7a-methyl-2,3,3a,5,6,7-hexahydro-1H-inden-1-yl]-6-ethyl-octane-3,6-diol.
LMST03020439; 1α,22S,25-trihydroxy-26,27-di-methylcholecalciferol; 26,27-dimethyl-9,10-seco-5Z, 7E,10(19)-cholestatriene-1S,3R,22S,25-tetrol; 9547545.

757 *1α,22R,25-Trihydroxy-26,27-dimethyl-20-epivitamin D₃*
9547544 $C_{29}H_{48}O_4$

(2R,3R)-2-[(1R,3aR,4E,7aS)-4-[(2Z)-2-[(3S,5R)-3,5-dihydroxy-2-methylidene-cyclohexylidene]ethylidene]-7a-methyl-2,3,3a,5,6,7-hexahydro-1H-inden-1-yl]-6-ethyl-octane-3,6-diol.
LMST03020438; 1α,22R,25-trihydroxy-26,27-di-methyl-20-epicholecalciferol; (20R)-26,27-dimethyl-9, 10-seco-5Z,7E,10(19)-cholestatriene-1S,3R,22R,25-tetrol; 9547544; Synthesized from protected (5E)-1α-hydroxy-22-oxo-23,24,25,26,27-pentanor-20-epi-vitamin D_3. Compared to 1α,25-dihydroxyvitamin D_3, inhibition of U937 cell (human histiocytic lymphoma cell line) proliferation is 10000%, inhibition of proliferation is 24000%, VDR (rachitic chicken intestinal receptor) binding affinity is 6% and calciuric effect in normal rats is 710%. Vitamin D. A Pluripotent Steroid Hormone: Structural Studies, Molecular Endocrinology and Clinical Applications. Proceedings of the Ninth Workshop on Vitamin D, Orlando, Florida (USA) May 28-June 2. Chemistry and Biology of Highly Active 22-Oxy Analogs of 20-Epi Calcitriol with very low Binding Affinity to the Vitamin D Receptor. (Norman, A. W., Bouillon, R., Thomasset, M., eds),

pp85-86, Walter de Gruyter Berlin New York (1994).

758 **1α,20S,25-Trihydroxy-26,27-dimethyl-24a-homovitamin D₃**
9547580 $C_{29}H_{48}O_4$

(1S)-1-[(1S,3aR,4E,7aR)-4-[(2Z)-2-[(3S,5R)-3,5-dihydroxy-2-methylidene-cyclohexylidene]ethylidene]-7a-methyl-2,3,3a,5,6,7-hexahydro-1H-inden-1-yl]-6-ethyl-octane-1,6-diol.
LMST03020475;   1α,20S,25-trihydroxy-26,27-di-methyl-24a-homocholecalciferol; 26,27-dimethyl-24a-homo-9,10-seco-5Z,7E,10(19)-cholestatriene-1S,3R,20S,25-tetrol; 9547580; Synthesized from a (5E)-1α-hydroxy-20-keto-22,23,24,25,26,27-hexanorvitamin D₃ derivative. Compared to 1α,25-dihydroxyvitamin D₃, inhibition of U937 cell (human histiocytic lymphoma cell line) proliferation is 900%, binding to the vitamin D receptor from rachitic chicken intestine is 0.6%, and calcemic activity determined by the increase in urinary calcium excretion in rats is < 0.5%. Vitamin D. A Pluripotent Steroid Hormone: Structural Studies, Molecular Endocrinology and Clinical Applications. Proceedings of the Ninth Workshop on Vitamin D, Orlando, Florida (USA) May 28-June 2. Synthesis and Biological Activity of 20-Hydroxylated Vitamin D Analogues. (Norman, A. W., Bouillon, R., Thomasset, M., eds), pp95-96, Walter de Gruyter Berlin New York (1994).

759 **1α-Hydroxy-18-(5-hydroxy-5-methylhex-yloxy)-23,24,25,26,27-pentanorvitamin D₃**
9547683 $C_{29}H_{48}O_4$

(1R,3S,5Z)-5-[(2E)-2-[(1R,3aR,7aR)-7a-[(5-hydroxy-5-methyl-hexoxy)methyl]-1-propan-2-yl-2,3,3a,5,6,7-hexahydro-1H-inden-4-ylidene]ethylidene]-4-methylidene-cyclohexane-1,3-diol.
LMST03020588; 1α-hydroxy-18-(5-hydroxy-5-methyl-

hexyloxy)-23,24,25,26,27-pentanorcholecalciferol; 18-(5-hydroxy-5-methylhexyloxy)-23,24-dinor-9,10-seco-5Z,7E,10(19)-cholatriene-1S,3R-diol; 9547683.

760 **1α,25-Dihydroxy-2β-(2-hydroxyethoxy)-vitamin D₃**
9547546 $C_{29}H_{48}O_5$

(1R,2R,3R,5Z)-5-[(2E)-2-[(1R,3aR,7aS)-1-[(2R)-6-hydroxy-6-methyl-heptan-2-yl]-7a-methyl-2,3,3a,5,6,7-hexahydro-1H-inden-4-ylidene]ethylidene]-2-(2-hydroxyethoxy)-4-methylidene-cyclohexane-1,3-diol.
LMST03020440; 2R-(2-hydroxyethoxy)-9,10-seco-5Z,7E,10(19)-cholestatriene-1R,3R,25-triol;   1α,25-di-hydroxy-2β-(2-hydroxyethoxy)cholecalciferol; 9547546; Synthesized photochemically from the 5,7-diene. Spinal bone mineral density in pre-osteoporosis model rats and binding affinity to calf thymus vitamin D receptor reported; λ_m = 264 nm (EtOH). *Chem Pharm Bull (Tokyo).* 1997, *45(10)*, 1626-30.

761 **10R-Ethoxy-10,19-dihydrovitamin D₃**
9547549 $C_{29}H_{50}O_2$

(1S,3Z,4R)-3-[(2E)-2-[(1R,3aR,7aS)-7a-methyl-1-[(2R)-6-methylheptan-2-yl]-2,3,3a,5,6,7-hexahydro-1H-inden-4-ylidene]ethylidene]-4-ethoxy-4-methyl-cyclohexan-1-ol.
LMST03020443;   10R-ethoxy-9,10-seco-5Z,7E-chol-estadien-3S-ol; toxisterol₃ B3; 10R-ethoxy-10,19-dihydrocholecalciferol; 9547549; One of the overirradiation products of 7-dehydrocholesterol in ethanol; λ_m = 250 nm (ε = 17000 EtOH). *J R Neth Chem Soc.* 1977, *96*, 104.

762 **(5E)-10R-Ethoxy-10,19-dihydrovitamin D₃**

9547547

$C_{29}H_{50}O_2$

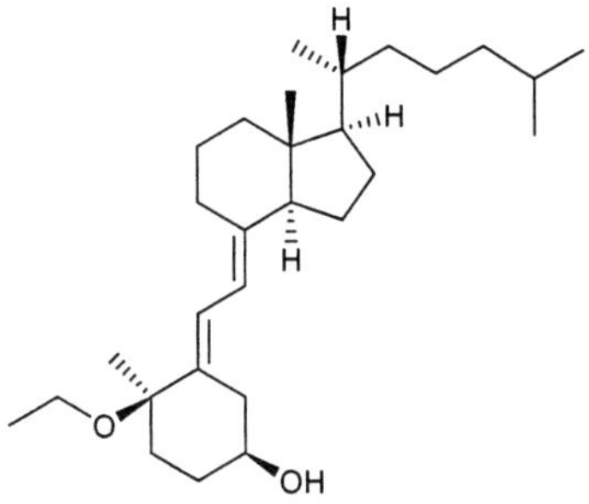

(1S,3E,4R)-3-[(2E)-2-[(1R,3aR,7aS)-7a-methyl-1-[(2R)-6-methylheptan-2-yl]-2,3,3a,5,6,7-hexahydro-1H-inden-4-ylidene]ethylidene]-4-ethoxy-4-methyl-cyclohexan-1-ol.
LMST03020441; 10R-ethoxy-9,10-seco-5E,7E-cholestadien-3S-ol; toxisterol₃ B1; (5E)-10R-ethoxy-10,19-dihydrocholecalciferol; 9547547; One of the overirradiation products of 7-dehydrocholesterol in ethanol; λ_m = 251 nm (ε = 33000 EtOH). *Tet Lett.* 1975, *7*, 427; *J R Neth Chem Soc.* 1977, *96*, 104.

763 **(5E)-10S-Ethoxy-10,19-dihydrovitamin D₃**

9547548

$C_{29}H_{50}O_2$

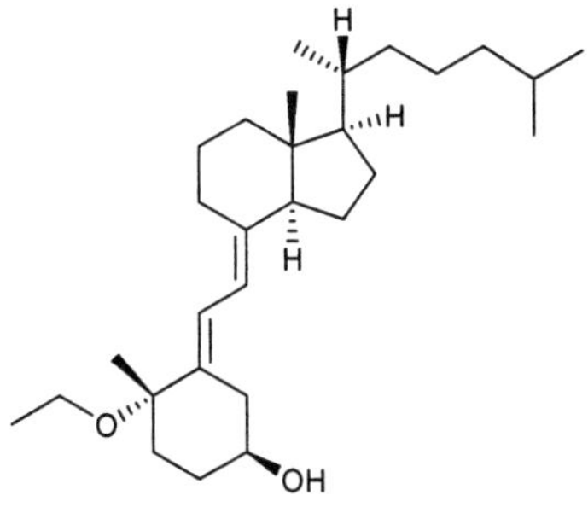

(1S,3E,4S)-3-[(2E)-2-[(1R,3aR,7aS)-7a-methyl-1-[(2R)-6-methylheptan-2-yl]-2,3,3a,5,6,7-hexahydro-1H-inden-4-ylidene]ethylidene]-4-ethoxy-4-methyl-cyclohexan-1-ol.
LMST03020442; 10S-ethoxy-9,10-seco-5E,7E-cholestadien-3S-ol; toxisterol₃ B2; (5E)-10S-ethoxy-10,19-dihydrocholecalciferol; 9547548; One of the overirradiation products of 7-dehydrocholesterol in ethanol; λ_m = 251 nm (ε = 29000 EtOH), 244, 260nm. *J R Neth Chem Soc.* 1977, *96*, 104.

764 **2α-(3-Hydroxypropyl)-1α,25-dihydroxy-19-norvitamin D₃**

$C_{29}H_{50}O_4$

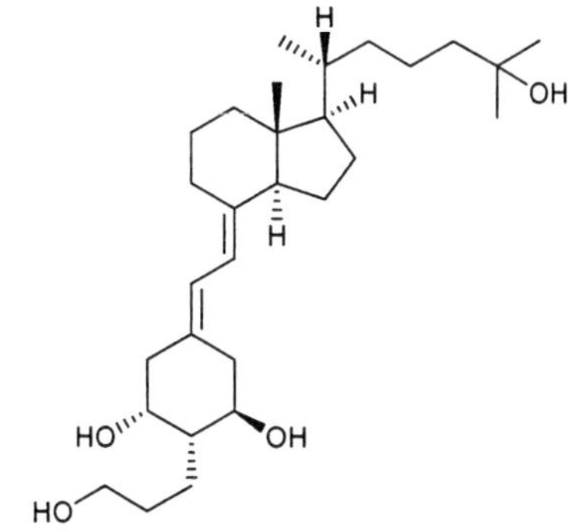

Synthetic analog of 19-norvitamin D₃. About 36 times more potent than calcitriol in binding to the bovine vitamin D receptor. *J Org Chem.* 2003, *68(19)*, 7407-7415.

765 **1α,25-Dihydroxy-2α-(3-hydroxypropoxy)-19-norvitamin D₃**

9547651

$C_{29}H_{50}O_5$

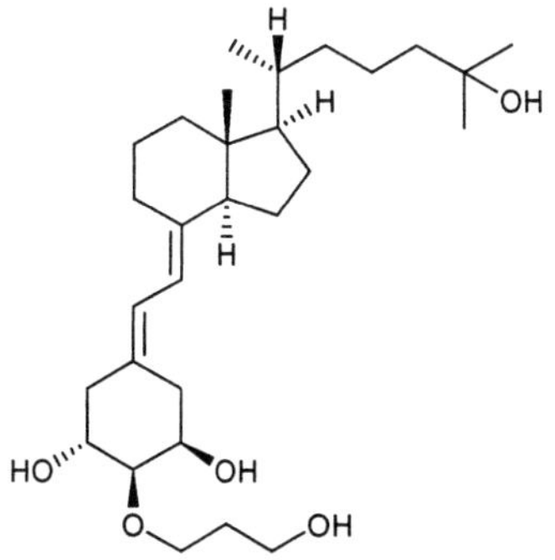

(1R,2S,3R)-5-[(2E)-2-[(1R,3aR,7aS)-1-[(2R)-6-hydroxy-6-methyl-heptan-2-yl]-7a-methyl-2,3,3a,5,6,7-hexahydro-1H-inden-4-ylidene]ethylidene]-2-(3-hydroxypropoxy)cyclohexane-1,3-diol.
LMST03020549; 1α,25-dihydroxy-2α-(3-hydroxypropoxy)-19-norcholecalciferol; 2S-(3-hydroxypropoxy)-19-nor-9,10-seco-5,7E-cholestadiene-1R,3R,25-triol; 9547651; Synthesized by combining A-ring synthon obtained from (-)-quinic acid with 25-hydroxylated Grundmann type ketone. Has low calcemic activity but high cell differentiating activity nearly as potent as that of 1α,25-dihydroxyvitamin D₃; λ_m = 243, 251.5, 261 nm (EtOH). *J Med Chem.* 1994, *37(22)*, 3730-8.

766 **1α,25-Dihydroxy-2β-(3-hydroxypropoxy)-19-norvitamin D₃**

9547651

$C_{29}H_{50}O_5$

(1R,2R,3R)-5-[(2E)-2-[(1R,3aR,7aS)-1-[(2R)-6-hydroxy-6-methyl-heptan-2-yl]-7a-methyl-2,3,3a,5,6,7-hexahydro-1H-inden-4-ylidene]ethylidene]-2-(3-hydroxypropoxy)cyclohexane-1,3-diol.

LMST03020550; 1α,25-dihydroxy-2β-(3-hydroxyprop-oxy)-19-norcholecalciferol; 2R-(3-hydroxypropoxy)-19-nor-9,10-seco-5,7E-cholestadiene-1R,3R,25-triol; 9547651; Synthesized by combining the A-ring synthon obtained from (-)-quinic acid with 25-hydroxylated Grundmann type ketone. Has low calcemic activity *in vivo*; λ$_m$ = 243, 251.5, 261 nm (EtOH). *J Med Chem.* 1994, **37(22)**, 3730-8.

767   **VD 2716**

$C_{30}H_{42}O_3S$

9,10-Seco-(1α,3β-dihydroxy-10-methylene-5,7-dien-17β-(2'-methylenethio-(m-2-hydroxy-i-propylphenyl)-androstane.

GS 1558; VD 2716; Increases osteocalcin mRNA and osteocalcin secretion more efficiently than calcitriol. *Eur J Biochem.* 1999, **261(3)**, 706-713; *Steroids.* 2001 **66(3-5)**, 223-5; *J Cell Biochem.* 2000, **76(4)**, 548-58; *Mol Pharm.* 2002, **62(4)**, 788-94. *Leo AB.*

768   **VD 2656**

$C_{30}H_{42}O_3S$

1α-Hydroxy-25,26,27-trinorvitamin D$_3$; VD 2656; Increases osteocalcin mRNA and osteocalcin secretion more efficiently than calcitriol. *Eur J Biochem.* 1999, **261(3)**, 706-713; *Mol Pharm.* 2002, **62(4)**, 788-94. *Leo AB.*

769   **1α,25-Dihydroxy-26,27-dimethyl-17E, 20,22,22,23,23-hexadehydro-24a-homovitamin D$_3$**

9547550 $C_{30}H_{44}O_3$

(1R,3S,5Z)-5-[(2E)-2-[(1E,3aR,7aS)-1-(7-ethyl-7-hydroxy-non-3-yn-2-ylidene)-7a-methyl-2,3,3a,5,6,7-hexahydroinden-4-ylidene]ethylidene]-4-methylidene-cyclohexane-1,3-diol.

LMST03020444; 1α,25-dihydroxy-26,27-dimethyl-17E,20,22,22,23,23-hexadehydro-24a-homochole-calciferol; 26,27-dimethyl-24a-homo-9.10-seco-5Z,7E,10(19),17E(20)-cholestatetraen-22-yne-1S,3R,25-triol; 9547550; Synthesized from hydroxy-protected (5E)-1α-hydroxy-20-oxo-22,23,24,25,26,27-hexanorvitamin D$_3$. Compared to 1α,25-dihydroxyvitamin D$_3$, inhibition of proliferation of U937 cells is 250%, affinity for chicken intestinal vitamin D receptor is 0.1%. Vitamin D. A Pluripotent Steroid Hormone: Structural Studies, Molecular Endocrinology and Clinical Applications. Proceedings of the Ninth Workshop on Vitamin D, Orlando, Florida (USA) May 28-June 2. Chemistry and Biology of 22,23-Yne Analogs of Calcitriol. (Norman, A. W., Bouillon, R., Thomasset, M., eds), pp73-74, Walter de Gruyter Berlin New York (1994).

770   **1α,25-Dihydroxy-26,27-dimethyl-17Z, 20,22,22,23,23-hexadehydro-24a-homovitamin D$_3$**

9547551 $C_{30}H_{44}O_3$

(1R,3S,5Z)-5-[(2E)-2-[(1Z,3aR,7aS)-1-(7-ethyl-7-hydroxy-non-3-yn-2-ylidene)-7a-methyl-2,3,3a,5,6,7-hexahydroinden-4-ylidene]ethylidene]-4-methylidene-cyclohexane-1,3-diol.

LMST03020445; 1α,25-dihydroxy-26,27-dimethyl-17Z,20,22,22,23,23-hexadehydro-24a-homochole-calciferol; 26,27-dimethyl-24a-homo-9,10-seco-5Z,7E,10(19),17Z(20)-cholestatetraen-22-yne-1S,3R,25-triol; 9547551; Synthesized from hydroxy-protected (5E)-1α-hydroxy-20-oxo-22,23,24,25,26,27-hexanorvitamin D$_3$. Compared to 1α,25-dihydroxyvitamin D$_3$, inhibition of proliferation of U937 cells is 71000%, affinity for chicken intestinal vitamin D receptor is 17% and calciuric activity is 71%. Vitamin D. A Pluripotent Steroid Hormone: Structural Studies, Molecular Endocrinology and Clinical Applications. Proceedings of the Ninth Workshop on Vitamin D, Orlando, Florida (USA) May 28-June 2. Chemistry and Biology of 22,23-Yne Analogs of Calcitriol. (Norman, A. W., Bouillon, R., Thomasset, M., eds), pp73-74, Walter de Gruyter Berlin New York (1994).

771   *1α,25-Dihydroxy-26,27-dimethyl-20,21, 22,22,23,23-hexadehydro-24a-homovitamin D$_3$*

9547552 $C_{30}H_{44}O_3$

(1R,3S,5Z)-5-[(2E)-2-[(1S,3aR,7aS)-1-(7-ethyl-7-hydroxy-non-1-en-3-yn-2-yl)-7a-methyl-2,3,3a,5,6,7-hexahydro-1H-inden-4-ylidene]ethylidene]-4-methylidene-cyclohexane-1,3-diol.

LMST03020446; 26,27-dimethyl-24a-homo-9,10-seco-5Z,7E,10(19),20-cholestatetraen-22-yne-1S,3R,25-triol;

1α,25-dihydroxy-26,27-dimethyl-20,21,22,22,23,23-hexahydro-24a-homocholecalciferol; 9547552; Synthesized from hydroxy-protected (5E)-1α-hydroxy-20-oxo-22,23,24,25,26,27-hexanorvitamin D$_3$. Compared to 1α,25-dihydroxyvitamin D$_3$, inhibition of proliferation of U937 cells is 120%, affinity for chicken intestinal vitamin D receptor is 3%. Vitamin D. A Pluripotent Steroid Hormone: Structural Studies, Molecular Endocrinology and Clinical Applications. Proceedings of the Ninth Workshop on Vitamin D, Orlando, Florida (USA) May 28-June 2. Chemistry and Biology of 22,23-Yne Analogs of Calcitriol. (Norman, A. W., Bouillon, R., Thomasset, M., eds), pp73-74, Walter de Gruyter Berlin New York (1994).

772   *1α,25-Dihydroxy-22E,23,24E,24a,24bE, 24c-hexadehydro-24a,24b,24c-trihomovitamin D$_3$*

9547553 $C_{30}H_{44}O_3$

(1R,3S,5Z)-5-[(2E)-2-[(1R,3aR,7aS)-1-[(2S,3E,5E,7E)-9-hydroxy-9-methyl-deca-3,5,7-trien-2-yl]-7a-methyl-2,3,3a,5,6,7-hexahydro-1H-inden-4-ylidene]ethylidene]-4-methylidene-cyclohexane-1,3-diol.

LMST03020447; 1α,25-dihydroxy-22E,23,24E,24a,24bE,24c-hexadehydro-24a,24b,24c-trihomocholecalciferol; 24a,24b,24c-trihomo-9,10-seco-5Z,7E,10(19),22E,24E,24bE-cholestahexaene-1S,3R-25-triol; 9547553; Synthesized from protected (5E)-1α-hydroxy-22-oxo-23,24,25,26,27-pentanorvitamin D$_3$. Compared to 1α,25-dihydroxyvitamin D$_3$, inhibition of proliferation of U937 cells is < 80% and induction of differentiation of U937 cells is < 5%. Gene Regulation, Structure-Function Analysis and Clinical Application. Proceedings of the Eighth Workshop on Vitamin D Paris, France July 5-10. Synthesis and Biological Activity of 1α-Hydroxylated Vitamin D$_3$ Analogues with Hydroxylated Side Chains, Multi-Homologated in The 24- or 24,26,27-Positions. (Norman, A. W., Bouillon, R., Thomasset, M., eds), pp159-160, Walter de Gruyter Berlin (1991).

773   **VD 2728**

            $C_{30}H_{44}O_3$

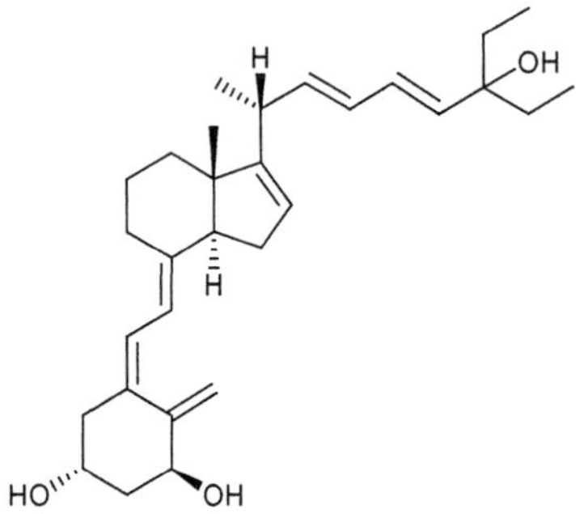

Increases osteocalcin mRNA and osteocalcin secretion more efficiently than calcitriol. *Eur J Biochem*. 1999, *261(3)*, 706-713; *Mol Pharm*. 2002, *62(4)*, 788-94. *Leo AB*.

774   **VD 2736**

$C_{30}H_{44}O_3$

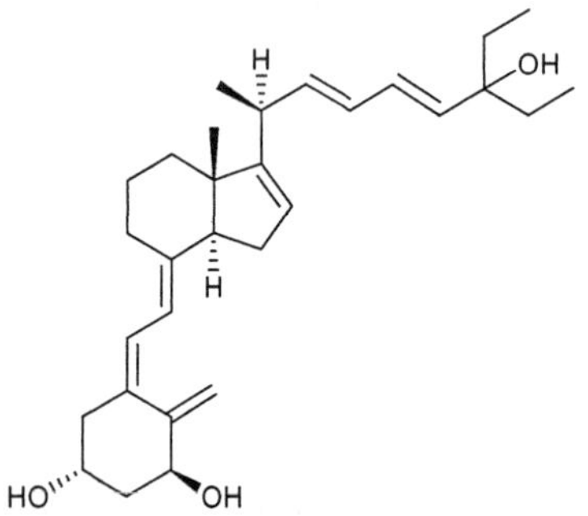

Increases osteocalcin mRNA and osteocalcin secretion more efficiently than calcitriol. *Eur J Biochem*. 1999, *261(3)*, 706-713; *Mol Pharm*. 2002, *62(4)*, 788-94. *Leo AB*.

775   **1α,25-Dihydroxy-11-(3-hydroxy-1-propynyl)-9,11-didehydrovitamin D₃**
9547554
$C_{30}H_{44}O_4$

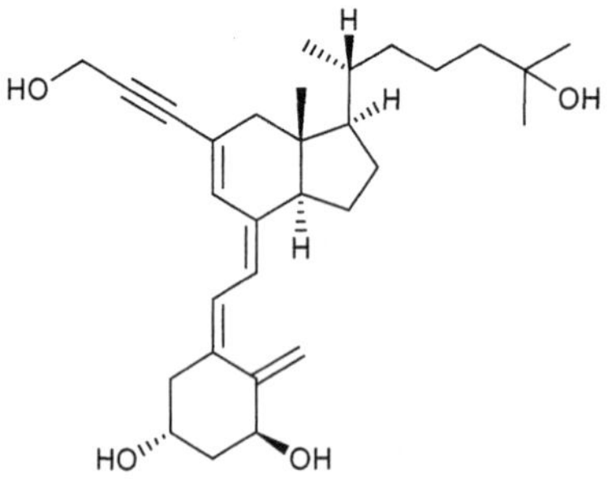

(1R,3S,5Z)-5-[(2Z)-2-[(1R,3aR,7aS)-1-[(2R)-6-hydroxy-6-methyl-heptan-2-yl]-6-(3-hydroxyprop-1-ynyl)-7a-methyl-2,3,3a,7-tetrahydro-1H-inden-4-ylidene]ethylidene]-4-methylidene-cyclohexane-1,3-diol. LMST03020448; 11-(3-hydroxy-1-propynyl)-9,10-seco-5Z,7E,9(11),10(19)-cholestatetraene-1S,3R,25-triol; 1α,25-dihydroxy-11-(3-hydroxy-1-propynyl)-9,11-

didehydrocholecalciferol; 9547554.

776   **1α-Hydroxymethyl-16-ene-24,24-difluoro-26,27-bishomo-25-hydroxyvitamin D₃**
$C_{30}H_{46}F_2O_3$

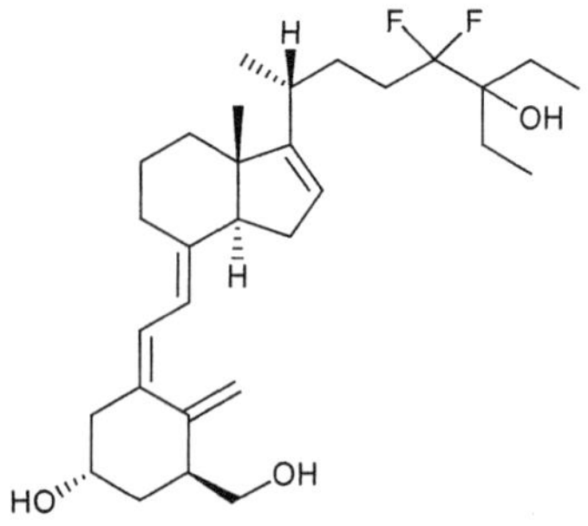

QW1624F2-2; Has antitumor activity against human neuroblastoma. Less calcemic than Seocalcitol (EB-1089). *J Cell Biochem*. 2006, *97(1)*, 198-206.

777   **1α,24-Dihydroxy-22-ene-24-cyclopropylvitamin D₃**
$C_{30}H_{46}O_3$

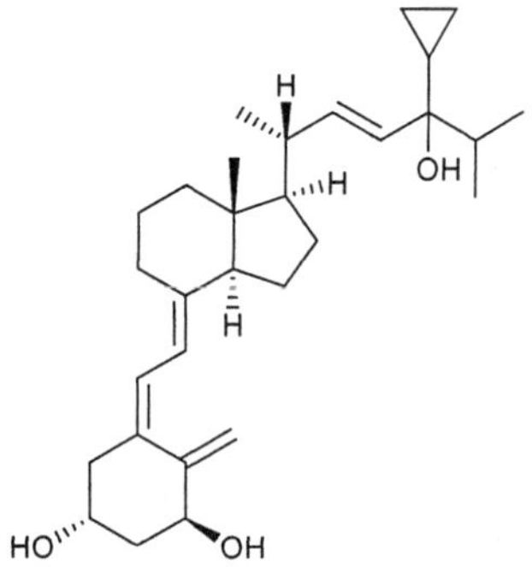

MC 903; Primes leukemia cells for monocytic differentiation. *Endocrinology*. 1999, *140(10)*, 4779-88.

778   **1α,25-Dihydroxy-26,27-dimethyl-22E,23,24E,24a-tetradehydro-24a-homovitamin D₃**
134404-52-7
$C_{30}H_{46}O_3$

(1R,3S,5Z)-5-[(2E)-2-[(1R,3aR,7aS)-1-[(2S,3E,5E)-7-ethyl-7-hydroxy-nona-3,5-dien-2-yl]-7a-methyl-2,3,3a,5,6,7-hexahydro-1H-inden-4-ylidene]ethylidene]-4-methylidene-cyclohexane-1,3-diol. LMST03020449; Seocalcitol; CB-1089; EB-1089; 22-24-Diene-24A,26A,27A,trihomo-1α,25-dihydroxyvitamin D_3; 1α,25-dihydroxy-26,27-dimethyl-22E,23,24E,24a-tetradehydro-24a-homocholecalciferol; 26,27-dimethyl-24a-homo-9,10-seco-5Z,7E,10(19),22E,24E-cholestapentaene-1S,3R,25-triol; EB1; 5288149; 6435810; 134404-52-7; Synthesized from protected (5E)-1α-hydroxy-22-oxo-23,24,25,26,27-pentanor-vitamin D_3. Calcitriol analog. Compared to 1α,25-dihydroxyvitamin D_3, inhibition of proliferation of U937 cells is 6800%, induction of differentiation of U937 cells is 6700%, and calciuric activity is 40%. Inhibition of proliferation of breast cancer cell MCF-7 is 1000% times more potent than 1α,25-dihydroxyvitamin D_3. Inhibits the growth of NMU (N-methyl-nitrosourea) induced rat mammary tumors *in vivo* in a dose-dependent manner. Used systemically to treat osteoporosis. Highly potent inhibitor of cell proliferation. Entered clinical trials for cancer treatment in 1995. Phase I trials established a safe dose level of 12 μg/m^2/day. Currently (2006) in Phase II/III trials for breast cancer, pancreatic cancer and leukemia. Metabolized to the 26-hydroxy EB1089 compound as the major product and 26α-hydroxy compound and two other minor metabolites. *Endocrinology*. 2002, *143(7)*, 2508-14; *Biochem Pharmacol*. 1992, *44(12)*, 2273-80; 1997, *53(6)*, 783-93. *Leo AB*.

779   *1α,25-Dihydroxy-26,27-dimethyl-22,22,23,23-tetradehydro-24a-homovitamin D_3*
9547555 $C_{30}H_{46}O_3$

(1R,3S,5Z)-5-[(2E)-2-[(1R,3aR,7aS)-1-[(2S)-7-ethyl-7-hydroxy-non-3-yn-2-yl]-7a-methyl-2,3,3a,5,6,7-hexahydro-1H-inden-4-ylidene]ethylidene]-4-methylidene-cyclohexane-1,3-diol. LMST03020450; 26,27-dimethyl-24a-homo-9,10-seco-5Z,7E,10(19)-cholestatrien-22-yne-1S,3R,25-triol; 1α,25-dihydroxy-26,27-dimethyl-22,22,23,23-tetradehydro-24a-homocholecalciferol; 9547555; Synthesized from hydroxy-protected (5E)-1α-hydroxy-22,22,23,23-tetradehydro-24,25,26,27-tetranorvitamin D_3. Is 180% as effective as 1α,25-dihydroxyvitamin D_3 in induction of proliferation of U937 cells. Vitamin D. A Pluripotent Steroid Hormone: Structural Studies, Molecular Endocrinology and Clinical Applications. Proceedings of the Ninth Workshop on Vitamin D, Orlando, Florida (USA) May 28-June 2. Chemistry and Biology of 22,23-Yne Analogs of Calcitriol. (Norman, A. W., Bouillon, R., Thomasset, M., eds), pp73-74, Walter de Gruyter Berlin New York (1994).

780   *1α,25-Dihydroxy-26,27-dimethyl-22,22,23,23-tetradehydro-24a-homo-20-epivitamin D_3*
9547556 $C_{30}H_{46}O_3$

(1R,3S,5Z)-5-[(2E)-2-[(1R,3aR,7aS)-1-[(2R)-7-ethyl-7-hydroxy-non-3-yn-2-yl]-7a-methyl-2,3,3a,5,6,7-hexahydro-1H-inden-4-ylidene]ethylidene]-4-methylidene-cyclohexane-1,3-diol. LMST03020451; (20R)-26,27-dimethyl-24a-homo-9,10-seco-5Z,7E,10(19)-cholestatrien-22-yne-1S,3R,25-triol; 1α,25-dihydroxy-26,27-dimethyl-22,22,23,23-tetradehydro-24a-homo-20-epicholecalciferol; 9547556; Synthesized from hydroxy-protected (5E)-1α-hydroxy-22,22,23,23-tetradehydro-24,25,26,27-tetranorvitamin D_3. Is 28000% as effective as 1α,25-dihydroxyvitamin D_3 in induction of proliferation of U937 cells and 44% in affinity for chicken intestinal vitamin D receptor. Vitamin D. A Pluripotent Steroid Hormone: Structural Studies, Molecular Endocrinology and Clinical Applications. Proceedings of the Ninth Workshop on Vitamin D, Orlando, Florida (USA) May 28-June 2. Chemistry and Biology of 22,23-Yne Analogs of Calcitriol. (Norman, A. W., Bouillon, R., Thomasset, M., eds), pp73-74, Walter de Gruyter Berlin New York (1994).

781   *20-Epi-22,24-diene-24a,26a,27a-trihomo-1α,25-dihydroxyvitamin D_3*

$C_{30}H_{46}O_3$

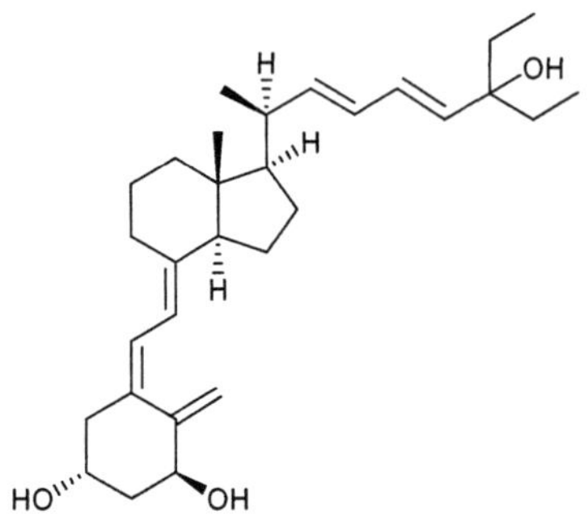

1,3-Cyclohexanediol, 5-((1-(6-ethyl-6-hydroxy-1β-methyl-2,4-octadienyl)octahydro-7a-methyl-4H-inden-4-ylidene)ethylidene)-4-methylene-, (1R-(1α(1R",2E,4E),3aβ,4E(1R",3S*,5Z),7aα))-; EB 1129; 20-epi-22,24-diene-24a,26a,27a-trihomo-1α,25-dihydroxyvitamin D$_3$; Increases osteocalcin mRNA and osteocalcin secretion more efficiently than calcitriol, but less effectively than its 20-epimer, EB 1089. *Eur J Biochem.* 1999, **261**(3), 706-13. *Leo AB*.

782   ***22,24-Diene-24a,26a,27a-trihomo-1α, 25-dihydroxyvitamin D₃***

$C_{30}H_{46}O_3$

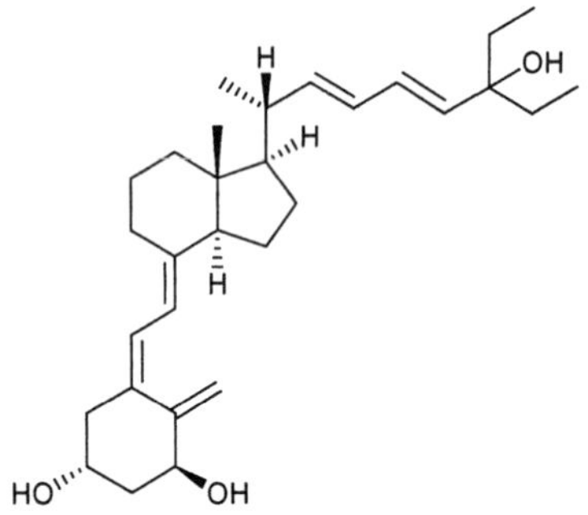

1,3-Cyclohexanediol, 5-((1-(6-ethyl-6-hydroxy-1α-methyl-2,4-octadienyl)octahydro-7a-methyl-4H-inden-4-ylidene)ethylidene)-4-methylene-, (1R-(1α(1R*,2E,4E),3aβ,4E(1R*,3S*,5Z),7aα))-; EB 1089; 22,24-diene-24a,26a,27a-trihomo-1α,25-dihydroxyvitamin D$_3$; (See also 778); Increases osteocalcin mRNA and osteocalcin secretion more efficiently than calcitriol. *Eur J Biochem.* 1999, **261**(3), 706-13. *Leo AB*.

783   ***1α,25-Dihydroxy-22R-methoxy-26,27-di-methyl-23,23,24,24-tetradehydrovitamin D₃***
9547557 $C_{30}H_{46}O_4$

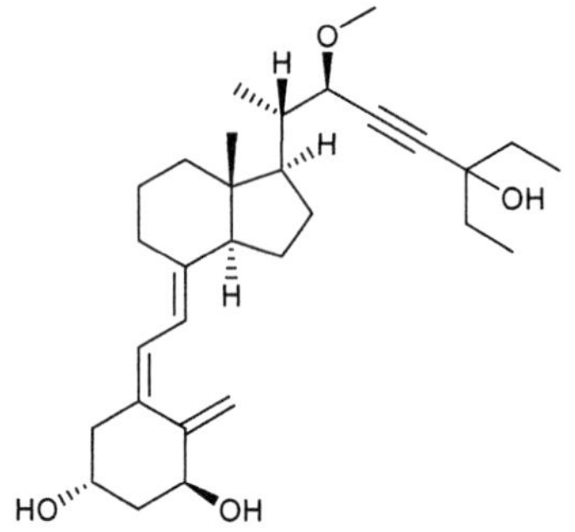

(1R,3S,5Z)-5-[(2E)-2-[(1R,3aR,7aS)-1-[(2S,3R)-6-ethyl-6-hydroxy-3-methoxy-oct-4-yn-2-yl]-7a-methyl-2,3,3a,5,6,7-hexahydro-1H-inden-4-ylidene]ethylidene]-4-methylidene-cyclohexane-1,3-diol. LMST03020452; 1α,25-dihydroxy-22R-methoxy-26, 27-dimethyl-23,23,24,24-tetradehydrocholecalciferol; 22R-methoxy-26,27-dimethyl-9,10-seco-5Z,7E,10(19)-cholestatrien-23-yne-1S,3R,25-triol; 9547557.

784   ***1α,25-Dihydroxy-22S-methoxy-26,27-dimethyl-23,24-tetradehydro-20-epivitamin D₃***
9547558 $C_{30}H_{46}O_4$

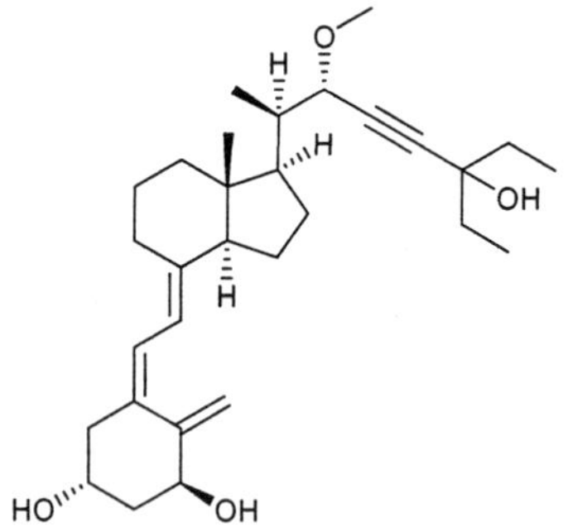

(1R,3S,5Z)-5-[(2E)-2-[(1R,3aR,7aS)-1-[(2R,3S)-6-ethyl-6-hydroxy-3-methoxy-oct-4-yn-2-yl]-7a-methyl-2,3,3a,5,6,7-hexahydro-1H-inden-4-ylidene]ethylidene]-4-methylidene-cyclohexane-1,3-diol. LMST03020453; 1α,25-dihydroxy-22S-methoxy-26,27-dimethyl-23,24-tetradehydro-20-epicholecalciferol; 20R)-26,27-dimethyl-22S-methoxy-9,10-seco-5Z,7E,10(19)-cholestatrien-23-yne-1S,3R,25-triol; 9547558; Synthesized from protected (5E)-1α-hydroxy-22-oxo-23,24,25,26,27-pentanor-20-epivitamin D$_3$. Compared to 1α,25-dihydroxyvitamin D$_3$, induction of differentiation of .U937 cells is 100000%, inhibition of proliferation is 125000%, VDR (rachitic chicken intestinal receptor) binding affinity is 38% and calciuric effect in normal rats is 1200%. Vitamin D. A Pluripotent Steroid Hormone: Structural Studies, Molecular Endocrinology and Clinical Applications. Proceedings of the Ninth Workshop on Vitamin D, Orlando, Florida (USA) May 28-June 2. Chemistry and Biology of Highly Active 22-Oxy Analogs of 20-Epi Calcitriol with very low Binding Affinity to the Vitamin D Receptor. (Norman, A. W., Bouillon, R., Thomasset, M., eds), pp85-86, Walter de Gruyter Berlin New York

(1994).

785   *1α,22S,25-Trihydroxy-26,27-dimethyl-23,23,24,24-tetradehydro-24a-homovitamin D₃*
9547560 $C_{30}H_{46}O_4$

(2S,3S)-2-[(1R,3aR,4E,7aS)-4-[(2Z)-2-[(3S,5R)-3,5-dihydroxy-2-methylidene-cyclohexylidene]ethylidene]-7a-methyl-2,3,3a,5,6,7-hexahydro-1H-inden-1-yl]-7-ethyl-non-4-yne-3,7-diol.
LMST03020455;   1α,22S,25-trihydroxy-26,27-di-methyl-23,23,24,24-tetradehydro-24a-homochole-calciferol; 26,27-dimethyl-24a-homo-9,10-seco-5Z,7E,10(19)-cholestatrien-23-yne-1S,3R,22S,25-tetrol;
9547560.

786   *1α,22R,25-Trihydroxy-26,27-dimethyl-23,24-tetradehydro-24a-homo-20-epivitamin D₃*
9547559 $C_{30}H_{46}O_4$

(2R,3R)-2-[(1R,3aR,4E,7aS)-4-[(2Z)-2-[(3S,5R)-3,5-dihydroxy-2-methylidene-cyclohexylidene]ethylidene]-7a-methyl-2,3,3a,5,6,7-hexahydro-1H-inden-1-yl]-7-ethyl-non-4-yne-3,7-diol.
LMST03020454;   1α,22R,25-trihydroxy-26,27-di-methyl-23,24-tetradehydro-24a-homo-20-epichole-calciferol; (20R)-26,27-dimethyl-24a-homo-9,10-seco-5Z,7E,10(19)-cholestatrien-23-yne-1S,3R,22R,25-tetrol;
9547559; Synthesized from protected (5E)-1α-hydroxy-22-oxo-23,24,25,26,27-pentanor-20-epivitamin D₃.

Compared to 1α,25-dihydroxyvitamin D₃, induction of differentiation of U937 cells is 10%, inhibition of proliferation is 20%, VDR (rachitic chicken intestinal receptor) binding affinity is 0.5%. Vitamin D. A Pluripotent Steroid Hormone: Structural Studies, Molecular Endocrinology and Clinical Applications.

Proceedings of the Ninth Workshop on Vitamin D, Orlando, Florida (USA) May 28-June 2. Chemistry and Biology of Highly Active 22-Oxy Analogs of 20-Epi Calcitriol with very low Binding Affinity to the Vitamin D Receptor. (Norman, A. W., Bouillon, R., Thomasset, M., eds), pp85-86, Walter de Gruyter Berlin New York (1994).

787   *1α,22S,25-Trihydroxy-26,27-dimethyl-23,24-tetradehydro-24a-homo-20-epivitamin D₃*
9547561 $C_{30}H_{46}O_4$

(2R,3S)-2-[(1R,3aR,4E,7aS)-4-[(2Z)-2-[(3S,5R)-3,5-dihydroxy-2-methylidene-cyclohexylidene]ethylidene]-7a-methyl-2,3,3a,5,6,7-hexahydro-1H-inden-1-yl]-7-ethyl-non-4-yne-3,7-diol.
LMST03020456;   1α,22S,25-trihydroxy-26,27-di-methyl-23,24-tetradehydro-24a-homo-20-epichole-calciferol; (20R)-26,27-dimethyl-24a-homo-9,10-seco-5Z,7E,10(19)-cholestatrien-23-yne-1S,3R,22S,25-tetrol;
9547561; Synthesized from protected (5E)-1α-hydroxy-22-oxo-23,24,25,26,27-pentanor-20-epivitamin D₃.

Compared to 1α,25-dihydroxyvitamin D₃, induction of differentiation of U937 cells is 20000%, inhibition of proliferation is 28000%, VDR (rachitic chicken intestinal receptor) binding affinity is 6%, and calciuric effect in normal rats is 190%. Vitamin D. A Pluripotent Steroid Hormone: Structural Studies, Molecular Endocrinology and Clinical Applications. Proceedings of the Ninth Workshop on Vitamin D, Orlando, Florida (USA) May 28-June 2. Chemistry and Biology of Highly Active 22-Oxy Analogs of 20-Epi Calcitriol with very low Binding Affinity to the Vitamin D Receptor. (Norman, A. W., Bouillon, R., Thomasset, M., eds), pp85-86, Walter de Gruyter Berlin New York (1994).

788   *1α-Hydroxy-18-(4-hydroxy-4-ethyl-2-hexynyloxy)-23,24,25,26,27-pentanorvitamin D₃*
9547687 $C_{30}H_{46}O_4$

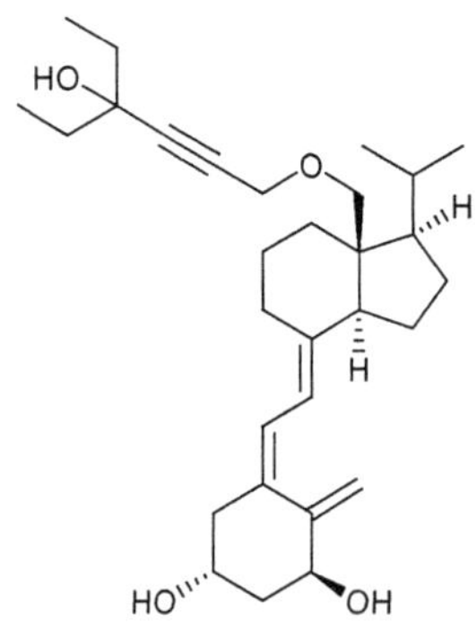

(1R,3S,5Z)-5-[(2E)-2-[(1R,3aR,7aR)-7a-[(4-ethyl-4-hydroxy-hex-2-ynoxy)methyl]-1-propan-2-yl-2,3,3a,5,6,7-hexahydro-1H-inden-4-ylidene]ethylidene]-4-methylidene-cyclohexane-1,3-diol. LMST03020592; 18-(4-hydroxy-4-ethyl-2-hexynyloxy)-23,24-dinor-9,10-seco-5Z,7E,10(19)-cholatriene-1S.3R-diol; 1α-hydroxy-18-(4-hydroxy-4-ethyl-2-hexynyloxy)-23.24,25,26,27-pentanorcholecalciferol; 9547687.

789 **24,24-Difluoro-1α,25-dihydroxy-26,27-dimethyl-24a-homovitamin D$_3$**

9547562

$C_{30}H_{48}F_2O_3$

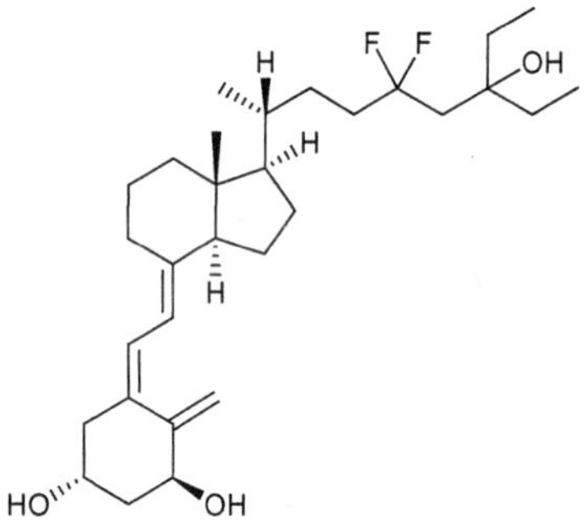

(1R,3S,5Z)-5-[(2E)-2-[(1R,3aR,7aS)-1-[(2R)-7-ethyl-5,5-difluoro-7-hydroxy-nonan-2-yl]-7a-methyl-2,3,3a,5,6,7-hexahydro-1H-inden-4-ylidene]ethylidene]-4-methylidene-cyclohexane-1,3-diol. LMST03020457; 24,24-difluoro-26,27-dimethyl-24a-homo-9,10-seco-5Z,7E,10(19)-cholestatriene-1S,3R,25-triol; 24,24-difluoro-1α,25-dihydroxy-26,27-dimethyl-24a-homocholecalciferol; 9547562.

790 **26,27-Dihomo-1α-hydroxyvitamin D$_2$**

$C_{30}H_{48}O_2$

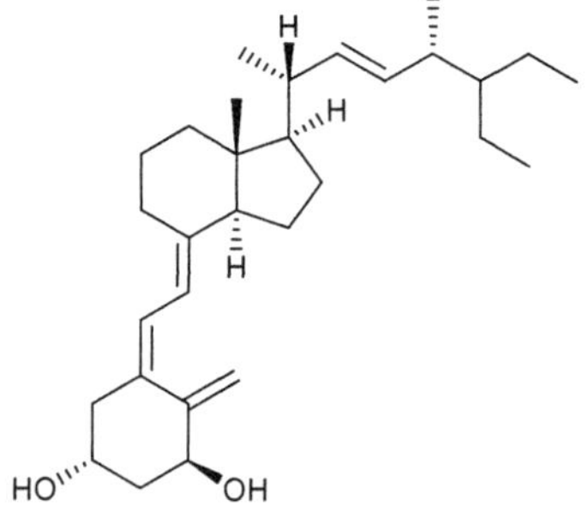

26,27-dihomo-1,25-(OH)$_2$-D$_2$; Synthetic vitamin D$_2$ derivative. Is not oxidised by vitamin D 25-hydroxylase. *J Biol Chem*. 1994, **269(39)**, 24014-19.

791 **26,27-Dihomo-1α-hydroxy-24-epi-vitamin D$_2$**

$C_{30}H_{48}O_2$

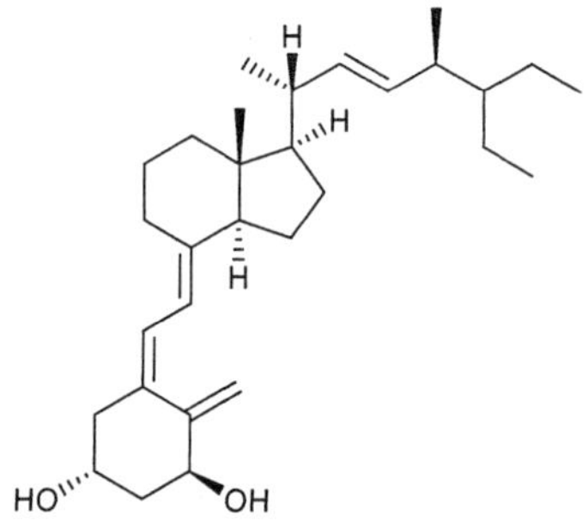

26,27-dihomo-24-epi-1,25-(OH)$_2$-D$_2$; Synthetic vitamin D$_2$ derivative. Is not oxidised by vitamin D 25-hydroxylase. *J Biol Chem*. 1994, **269(39)**, 24014-19.

792 **1α,25-Dihydroxy-22,23-didehydro-24a,24b,24c-trihomovitamin D$_3$**

123963-52-0

$C_{30}H_{48}O_3$

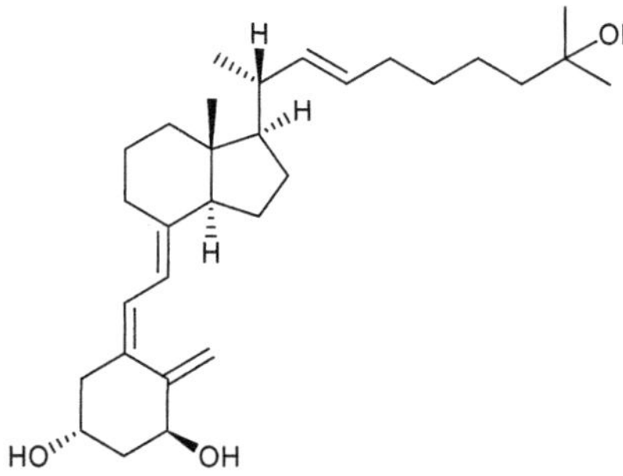

(1R,3S,5Z)-5-[(2E)-2-[(1R,3aR,7aS)-1-[(E,2S)-9-hydroxy-9-methyl-dec-3-en-2-yl]-7a-methyl-2,3,3a,5,6,7-hexahydro-1H-inden-4-ylidene]ethylidene]-4-methylidene-cyclohexane-1,3-diol. LMST03020458; 24a,24b,24c-trihomo-9,10-seco-5Z,7E,10(19),22E-cholestatetraene-1S,3R,25-triol; 1α,25-dihydroxy-22,23-didehydro-24a,24b,24c-trihomocholecalciferol; 1,25-Dihydroxy-24-trihomo-22-ene-vitamin D₃; 1,25-Dihydroxy-24-trihomo-22-ene-cholecalciferol; 1,25-Dte-D₃; 6439173; 9547563; 123963-52-0; Synthesized from 1α-hydroxylated pentanorvitamin D 22-carboxaldehyde and a C(8) side chain fragment. Half as active as 1α,25-dihydroxyvitamin D₃ in differentiating HL-60 cells but it has no activity in mobilizing bone calcium. The ethyl side chains reduce slightly the monocytic differentiation-inducing activity compared to that of 1α,25-dihydroxyvitamin D₃; λ_m = 264 nm (EtOH). ***Biochemistry.*** 1990, ***29(1)***, 190-6.

793   *1α,25-Dihydroxy-26,27-dimethyl-22E,23-didehydro-24a-homovitamin D₃*

9547564 $C_{30}H_{48}O_3$

(1R,3S,5Z)-5-[(2E)-2-[(1R,3aR,7aS)-1-[(E,2S)-7-ethyl-7-hydroxy-non-3-en-2-yl]-7a-methyl-2,3,3a,5,6,7-hexahydro-1H-inden-4-ylidene]ethylidene]-4-methylidene-cyclohexane-1,3-diol.
LMST03020459; 1α,25-dihydroxy-26,27-dimethyl-22E,23-didehydro-24a-homocholecalciferol; 26,27-dimethyl-24a-homo-9,10-seco-5Z,7E,10(19),22E-cholestatetraene-1S,3R,25-triol; 9547564.

794   *1α,25-Dihydroxy-26,27-dimethyl-22E,23-didehydro-24a-homo-20-epivitamin D₃*

9547565 $C_{30}H_{48}O_3$

(1R,3S,5Z)-5-[(2E)-2-[(1R,3aR,7aS)-1-[(E,2R)-7-ethyl-7-hydroxy-non-3-en-2-yl]-7a-methyl-2,3,3a,5,6,7-hexahydro-1H-inden-4-ylidene]ethylidene]-4-methylidene-cyclohexane-1,3-diol.
LMST03020460; 1α,25-dihydroxy-26,27-dimethyl-22E,23-didehydro-24a-homo-20-epicholecalciferol;

(20S)-26,27-dimethyl-24a-homo-9,10-seco-5Z,7E,10(19),22E-cholestatetraene-1S,3R,25-triol; 9547565; Synthesized from protected (5E)-1α-hydroxy-22-oxo-23,24,25,26,27-pentanorvitamin D₃. Compared to 1α,25-dihydroxyvitamin D₃, inhibition of proliferation of U937 cells is 15000%, induction of differentiation of U937 cells is 100000%, and calciuric effects on normal rats is 120%. Gene Regulation, Structure-Function Analysis and Clinical Application. Proceedings of the Eighth Workshop on Vitamin D Paris, France July 5-10. The 20-Epi Modification in The Vitamin D Series: Selective Enhancement of 'Non-Classical' Receptpr-Mediated Effects. (Norman, A. W., Bouillon, R., Thomasset, M., eds), pp163-164, Walter de Gruyter Berlin New York (1991).

795   *1α,25-Dihydroxy-26,27-dimethyl-22Z,23-didehydro-24a-homo-20-epivitamin D₃*

9547566 $C_{30}H_{48}O_3$

(1R,3S,5Z)-5-[(2E)-2-[(1R,3aR.7aS)-1-[(Z,2R)-7-ethyl-7-hydroxy-non-3-en-2-yl]-7a-methyl-2,3,3a,5,6,7-hexahydro-1H-inden-4-ylidene]ethylidene]-4-methylidene-cyclohexane-1,3-diol.
LMST03020461; 1α,25-dihydroxy-26,27-dimethyl-22Z,23-didehydro-24a-homo-20-epicholecalciferol; (20S)-26,27-dimethyl-24a-homo-9,10-seco-5Z,7E,10(19),22Z-cholestatetraene-1S,3R,25-triol; 9547566; Synthesized from protected (5E)-1α-hydroxy-22-oxo-23,24,25,26,27-pentanorvitamin D₃. Compared to 1α,25-dihydroxyvitamin D₃, inhibition of proliferation of U937 cells is 340000%, induction of differentiation of U937 cells is 1000000%, and calciuric effects on normal rats is 330%. Gene Regulation, Structure-Function Analysis and Clinical Application. Proceedings of the Eighth Workshop on Vitamin D Paris. France July 5-10. The 20-Epi Modification in The Vitamin D Series : Selective Enhancement of 'Non-Classical' Receptpr-Mediated Effects. (Norman, A. W., Bouillon, R., Thomasset, M., eds), pp163-164, Walter de Gruyter Berlin New York (1991).

796   *25-Hydroxy-1α-hydroxymethyl-26,27-dimethyl-16,17-didehydrovitamin D₃*

9547675 $C_{30}H_{48}O_3$

(7R)-7-[(3aR,4E,7aS)-4-[(2Z)-2-[(3S,5S)-5-hydroxy-3-(hydroxymethyl)-2-methylidene-cyclohexylidene]ethylidene]-7a-methyl-3a,5,6,7-tetrahydro-3H-inden-1-yl]-3-ethyl-octan-3-ol.
LMST03020580; 1S-hydroxymethyl-26,27-dimethyl-9,10-seco-5Z,7E,10(19),16-cholestatetraene-3S,25-diol; 25-hydroxy-1α-hydroxymethyl-26,27-dimethyl-16,17-didehydrocholecalciferol; 9547675.

797   *25-Hydroxy-1β-hydroxymethyl-26,27-di-methyl-16,17-didehydro-3-epivitamin D₃*
9547676 $C_{30}H_{48}O_3$

(7R)-7-[(3aR,4E,7aS)-4-[(2Z)-2-[(3R,5R)-5-hydroxy-3-(hydroxymethyl)-2-methylidene-cyclohexylidene]ethylidene]-7a-methyl-3a,5,6,7-tetrahydro-3H-inden-1-yl]-3-ethyl-octan-3-ol.
LMST03020581; 1R-hydroxymethyl-26,27-dimethyl-9,10-seco-5Z,7E,10(19),16-cholestatetraene-3R,25-diol; 25-hydroxy-1β-hydroxymethyl-26,27-dimethyl-16,17-didehydro-3-epicholecalciferol; 9547676.

798   *25-Hydroxy-1α-hydroxymethyl-26,27-dimethyl-16,17-didehydro-3-epivitamin D₃*
                                           $C_{30}H_{48}O_3$

(1S,3S,5Z)-5-[(2E)-2-[(3aR,7aS)-1-[(5S)-9-hydroxy-9-methyl-decan-5-yl]-7a-methyl-3a,5,6,7-tetrahydro-3H-inden-4-ylidene]ethylidene]-4-methylidene-cyclohexane-1,3-diol.
LMST03020580; Ro 26-9228; JK-1626-2; Synthetic low calcemic vitamin D₃ analog which prevents metastatic bone lesions by inhibiting the mitogenic response of osteoblasts to growth factors produced by MDA-PCa 2b cells. Causes inhibition of prostate cancer-mediated osteoblastic bone lesions. *J Steroid Biochem Mol Biol.* 2005, **97(1-2)**, 203-11.

799   *20S-Butyl-1α,25-dihydroxy-16,17-didehydro-21-norvitamin D₃*
9547679 $C_{30}H_{48}O_3$

(1R,3S,5Z)-5-[(2E)-2-[(3aR,7aS)-1-[(5S)-9-hydroxy-9-methyl-decan-5-yl]-7a-methyl-3a,5,6,7-tetrahydro-3H-inden-4-ylidene]ethylidene]-4-methylidene-cyclohexane-1,3-diol.
LMST03020584; 20S-butyl-1α,25-dihydroxy-16,17-didehydro-21-norcholecalciferol; 20S-butyl-21-nor-9,10-seco-5Z,7E,10(19),16-cholestatetraene-1S,3R,25-triol; 9547679.

800   *26,27-Diethyl-1α,25-dihydroxy-20,21-didehydro-23-oxavitamin D₃*
9547567 $C_{30}H_{48}O_4$

(1R,3S,5Z)-5-[(2E)-2-[(1S,3aR,7aS)-1-[3-(2-hydroxy-2-propyl-pentoxy)prop-1-en-2-yl]-7a-methyl-2,3,3a,5,6,7-hexahydro-1H-inden-4-ylidene]ethylidene]-4-methylidene-cyclohexane-1,3-diol.
LMST03020462; 26,27-diethyl-23-oxa-9,10-seco-5Z,7E,10(19),20-cholestatetraene-1S,3R,25-triol; 26,27-diethyl-1α,25-dihydroxy-20,21-didehydro-23-oxacholecalciferol; 9547567; Compared to 1α,25-dihydroxy-vitamin D₃, affinity for pig intestinal nuclear receptor is 25%, affinity for human vitamin D binding protein is

<0.02%, differentiation of HL-60 cells is 25%, and calciuric effect is 1%. Vitamin D. A Pluripotent Steroid Hormone: Structural Studies, Molecular Endocrinology and Clinical Applications. Proceedings of the Ninth Workshop on Vitamin D, Orlando, Florida (USA) May 28-June 2. Synthesis and Biological Activity of 23-Oxa Vitamin D Analogues. (Norman, A. W., Bouillon, R., Thomasset, M., eds), pp93-94, Walter de Gruyter Berlin New York (1994).

801   1α,25-Dihydroxy-24a,24b,24c-trihomo-vitamin D₃

9547568   C₃₀H₅₀O₃

(1R,3S,5Z)-5-[(2E)-2-[(1R,3aR,7aS)-1-[(2R)-9-hydroxy-9-methyl-decan-2-yl]-7a-methyl-2,3,3a,5,6,7-hexahydro-1H-inden-4-ylidene]ethylidene]-4-methylidene-cyclohexane-1,3-diol.
LMST03020463; 24a,24b,24c-trihomo-9,10-seco-5Z,7E,10(19)-cholestatrien-1S,3R,25-triol; 1α,25-dihydroxy-24a,24b,24c-trihomocholecalciferol;
9547568; Synthesized from hydroxy-protected (5E)-23,24,25,26,27-pentanorvitamin D₃ 22-bis(methylseleno)acetal. Compared to 1α,25-dihydroxyvitamin D₃, inhibition of proliferation of U937 cells is 20%, induction of differentiation of U937 cells is 40%, and calciuric activity is 2%. Gene Regulation, Structure-Function Analysis and Clinical Application. Proceedings of the Eighth Workshop on Vitamin D Paris, France July 5-10. Synthesis and Biological Activity of 1α-Hydroxylated Vitamin D₃ Analogues with Hydroxylated Side Chains, Multi-Homologated in The 24- or 24,26,27-Positions. (Norman, A. W., Bouillon, R., Thomasset, M., eds), pp159-160, Walter de Gruyter Berlin (1991).

802   1α,25-Dihydroxy-26,27-dimethyl-24a-homovitamin D₃

9547569   C₃₀H₅₀O₃

(1R,3S,5Z)-5-[(2E)-2-[(1R,3aR,7aS)-1-[(2R)-7-ethyl-7-hydroxy-nonan-2-yl]-7a-methyl-2,3,3a,5,6,7-hexahydro-1H-inden-4-ylidene]ethylidene]-4-methylidene-cyclohexane-1,3-diol.
LMST03020464; 26,27-dimethyl-24a-homo-9,10-seco-5Z,7E,10(19)-cholestatrien-1S,3R,25-triol; 1α,25-dihydroxy-26,27-dimethyl-24a-homocholecalciferol; 9547569; Synthesized from hydroxy-protected (5E)-23,24,25,26,27-pentanorvitamin D₃ 22-bis(methylseleno)acetal. Compared to 1α,25-dihydroxyvitamin D₃, inhibition of U937 cell (human histiocytic lymphoma cell line) proliferation is 800%, induction of U937 cell differentiation is 500%, binding to the vitamin D receptor from rachitic chicken intestine is 18%, calcemic activity determined by the increase in urinary calcium excretion in rats is 20%, inhibitory effect on murine thymocyte activation is 1900%. *Biochem Pharmacol.* 1991, *42(8)*, 1569-1575., Gene Regulation, Structure-Function Analysis and Clinical Application. Proceedings of the Eighth Workshop on Vitamin D Paris, France July 5-10. Synthesis and Biological Activity of 1α-Hydroxylated Vitamin D₃ Analogues with Hydroxylated Side Chains, Multi-Homologated in The 24- or 24,26,27-Positions. (Norman, A. W., Bouillon, R., Thomasset, M., eds), pp159-160, Walter de Gruyter Berlin (1991).

803   1α,25-Dihydroxy-26,27-dimethyl-24a-homo-20-epivitamin D₃

9547570   C₃₀H₅₀O₃

(1R,3S,5Z)-5-[(2E)-2-[(1R,3aR,7aS)-1-[(2S)-7-ethyl-7-hydroxy-nonan-2-yl]-7a-methyl-2,3,3a,5,6,7-hexahydro-1H-inden-4-ylidene]ethylidene]-4-methylidene-cyclohexane-1,3-diol.
LMST03020465; (20S)-26,27-dimethyl-24a-homo-9,10-seco-5Z,7E,10(19)-cholestatriene-1S,3R,25-triol;

1α,25-dihydroxy-26,27-dimethyl-24a-homo-20-epi-cholecalciferol; 9547570; Synthesized from hydroxy-protected (5E)-23,24,25,26,27-pentanorvitamin D_3 22-bis(methylseleno)acetal. Compared to 1α,25-dihydroxyvitamin D_3, inhibition of U937 cell (human histicytic lymphoma cell line) proliferation is 800%, induction of U937 cell differentiation is 500%, binding to the vitamin D receptor from rachitic chicken intestine is 18%, calcemic activity determined by the increase in urinary calcium excretion in rats is 20%, and inhibitory effect on murine thymocyte activation is 1900%. *Biochem Pharmacol.* 1991, *42(8)*, 1569-1575. Gene Regulation, Structure-Function Analysis and Clinical Application. Proceedings of the Eighth Workshop on Vitamin D Paris, France July 5-10. Synthesis and Biological Activity of 1α-Hydroxylated Vitamin D_3 Analogues with Hydroxylated Side Chains, Multi-Homologated in The 24- or 24,26,27-Positions. (Norman, A. W., Bouillon, R., Thomasset, M., eds), pp159-160, Walter de Gruyter Berlin (1991).

804   *1α,25-Dihydroxy-26,27-dimethyl-22,23-didehydro-24a,24b-dihomo-19-norvitamin D_3*

9547647 $C_{30}H_{50}O_3$

(1R,3R)-5-[(2E)-2-[(1R,3aR,7aS)-1-[(E,2S)-8-ethyl-8-hydroxy-dec-3-en-2-yl]-7a-methyl-2,3,3a,5,6,7-hexahydro-1H-inden-4-ylidene]ethylidene]-cyclohexane-1,3-diol.
LMST03020544; 26,27-dimethyl-24a,24b-dihomo-19-nor-9,10-seco-5,7E,22E-cholestatriene-1R,3R,25-triol; 1α,25-dihydroxy-26,27-dimethyl-22,23-didehydro-24a,24b-dihomo-19-norcholecalciferol; 9547647; Synthetic vitamin D derivative; λ_m = 243, 251.5, 261 nm (EtOH). *Tet Lett.* 1992, *33*, 2937-40.

805   *24a-Homo-26,27-dimethyl-1α,25-di-hydroxyvitamin D_3*

128312-71-0 $C_{30}H_{50}O_3$

(1R,3S,5E)-5-[(2E)-2-[(1R,3aR,7aS)-1-(7-ethyl-7-hydroxy-nonan-2-yl)-7a-methyl-2,3,3a,5,6,7-hexa-hydro-1H-inden-4-ylidene]ethylidene]-4-methylidene-cyclohexane-1,3-diol.
CB 966; CB-966; LMST03020464; 1,3-Cyclo-hexanediol, 5-((1-(6-ethyl-6-hydroxy-1-methyloctyl)-octahydro-7a-methyl-4H-inden-4-ylidene)ethylidene)-4-methylene-, (1R-(1α(R*), 3aβ,4E(1R*,3S*,5Z),7aα))-; 6439266; 128312-71-0; Inhibits acute myelogenous leukemia progenitor proliferation by suppressing interleukin-1β production. *J Clin Invest.* 1997, *100(7)*, 1716-24; *J Steroid Biochem Mol Biol.* 1995, *55(3-4)*, *337-46; Cancer Res.* 1994, *54(21),* 5711-7.

806   *24a-Homo-26,27-dimethyl-1α,25-dihydroxy-20-epivitamin D_3*

134523-85-6 $C_{30}H_{50}O_3$

(1R,3S,5E)-5-[(2E)-2-[(3aR,7aS)-1-(7-ethyl-7-hydroxy-nonan-2-yl)-7a-methyl-2,3,3a,5,6,7-hexahydro-1H-inden-4-ylidene]ethylidene]-4-methylidene-cyclohexane-1,3-diol.
(1R-(1α(S*),3aβ,4E(1R*,3S*,5Z),7aα))-5-((1-(6-Ethyl-6-hydroxy-1-methyloctyl)octahydro-7a-methyl-4H-inden-4-ylidene)ethylidene)-4-methylene-1,3-cyclohexane-diol; 5-((1-(6-Ethyl-6-hydroxy-1-methyloctyl)octahydro-7a-methyl-4H-inden-4-ylidene)ethylidene)-4-methylene-1,3-cyclohexanediol (1R-(1α(S*),3aβ,4E(1R*,3S*,5Z),7aα))-; 20-epi-24a-Homo-26,27-dimethyl-1α,25-dihydroxycholecalc-iferol; LMST03020465; MC-1301; 1,3-Cyclo-hexanediol, 5-((1-(6-ethyl-6-hydroxy-1-methyloctyl)-octahydro-7a-methyl-4H-inden-4-ylidene)ethylidene)-4-methylene-, (1R-(1α(S*),3aβ, 4E(1R*,3S*,5Z),7aα))-; 6439378; 134523-85-6; More potent than 1,25(OH)2D₃ in inhibiting clonal keratinocyte growth. *Br J Pharmacol.* 1997, *120(6),*

1119-27; *Biochem Pharmacol.* 1995, **49(5)**, 621-4.

807   *26,27-Diethyl-1α,25-dihydroxy-22-thiavitamin D₃*

9547571 $C_{30}H_{50}O_3S$

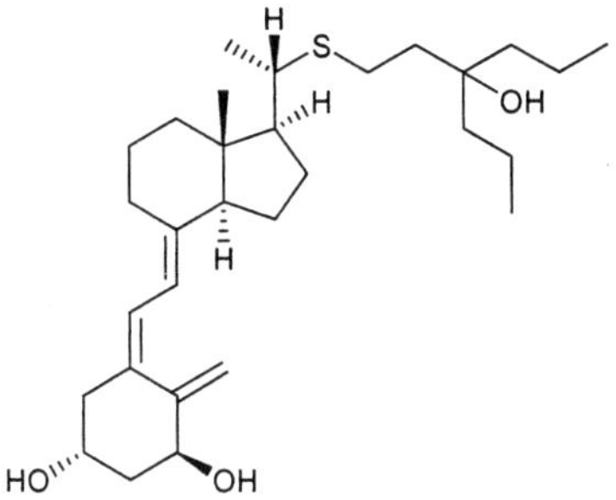

(1R,3S,5Z)-5-[(2E)-2-[(1S,3aR,7aR)-1-[(1S)-1-(3-hydroxy-3-propyl-hexyl)sulfanylethyl]-7a-methyl-2,3,3a,5,6,7-hexahydro-1H-inden-4-ylidene]ethylidene]-4-methylidene-cyclohexane-1,3-diol. LMST03020466; 26,27-diethyl-9,10-seco-22-thia-5Z,7E,10(19)-cholestatriene-1S,3R,25-triol; 26,27-diethyl-1α,25-dihydroxy-22-thiacholecalciferol; 9547571; Synthesized from dehydroepiandrosterone. *Bioorg Med Chem Lett.* 1995, **5(3)**, 279-82.

808   *26,27-Diethyl-1α,25-dihydroxy-22-thia-20-epivitamin D₃*

9547572 $C_{30}H_{50}O_3S$

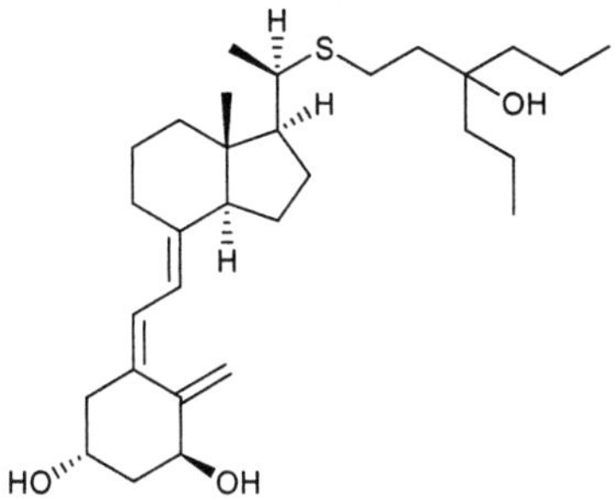

(1R,3S,5Z)-5-[(2E)-2-[(1S,3aR,7aR)-1-[(1R)-1-(3-hydroxy-3-propyl-hexyl)sulfanylethyl]-7a-methyl-2,3,3a,5,6,7-hexahydro-1H-inden-4-ylidene]ethylidene]-4-methylidene-cyclohexane-1,3-diol. LMST03020467; (20R)-26,27-diethyl-9,10-seco-22-thia-5Z,7E,10(19)-cholestatriene-1S,3R,25-triol; 26,27-diethyl-1α,25-dihydroxy-22-thia-20-epicholecalciferol; 9547572; Synthesized from dehydroepiandrosterone. *Bioorg Med Chem Lett.* 1995, **5(3)**, 279-82.

809   *1α-Hydroxy-2β-(3-hydroxypropoxy)-vitamin D₃*

9547573 $C_{30}H_{50}O_4$

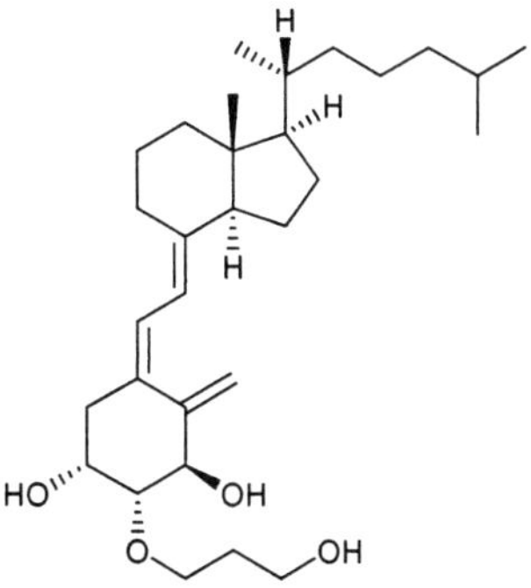

(1R,2R,3R,5Z)-5-[(2E)-2-[(1R,3aR,7aS)-7a-methyl-1-[(2R)-6-methylheptan-2-yl]-2,3,3a,5,6,7-hexahydro-1H-inden-4-ylidene]ethylidene]-2-(3-hydroxypropoxy)-4-methylidene-cyclohexane-1,3-diol. LMST03020468;   2R-(3-hydroxypropoxy)-9,10-seco-5Z,7E,10(19)-cholestatriene-1R,3R-diol;   1α-hydroxy-2β-(3-hydroxypropoxy)cholecalciferol; 9547573; Synthesized photochemically from the 5,7-diene. Plasma calcium levels in rats fed with a low calcium/vitamin D deficient diet reported; λ_m = 263 nm (EtOH). *Chem Pharm Bull (Tokyo).* 1993, **41(6)**, 1111-13.

810   *1β-Hydroxy-2β-(3-hydroxypropoxy)-vitamin D₃*

9547574 $C_{30}H_{50}O_4$

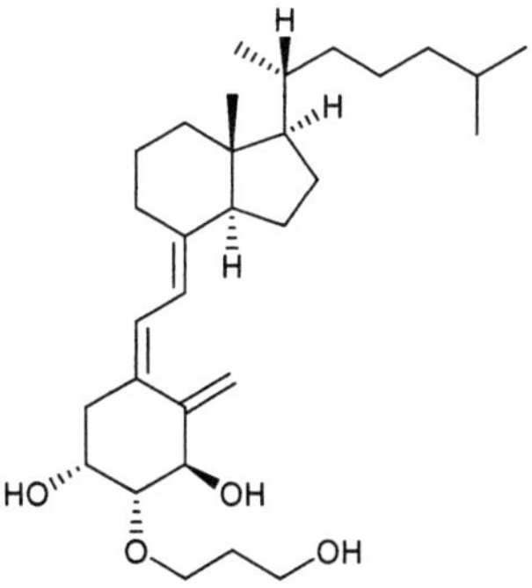

(1R,2R,3S,5Z)-5-[(2E)-2-[(1R,3aR,7aS)-7a-methyl-1-[(2R)-6-methylheptan-2-yl]-2,3,3a,5,6,7-hexahydro-1H-inden-4-ylidene]ethylidene]-2-(3-hydroxypropoxy)-4-methylidene-cyclohexane-1,3-diol. LMST03020469;   2R-(3-hydroxypropyl)-9,10-seco-5Z,7E,10(19)-cholestatriene-1S,3R,25-triol;   1α,25-dihydroxy-2β-(3-hydroxypropyl)cholecalciferol;   ED-71; 9547574; Synthesized photochemically from the 5,7-diene. Spinal bone mineral density in pre-osteoporosis model rats and binding affinity to calf thymus vitamin D receptor reported. Increases plasma calcium levels in rats on a low Ca/D deficient diet more significantly than 1α,25-dihydroxyvitamin D₃; λ_m = 264 nm (EtOH). *Chem Pharm Bull (Tokyo).* 1997, **45(10)**, 1626-30.

811   *26,27-Diethyl-1α,25-dihydroxy-22-oxa-vitamin D₃*

9547575　　　　　　　　　　　　　$C_{30}H_{50}O_4$

(1R,3S,5Z)-5-[(2E)-2-[(1S,3aR,7aR)-1-[(1S)-1-(3-hydroxy-3-propyl-hexoxy)ethyl]-7a-methyl-2,3,3a,5,6,7-hexahydro-1H-inden-4-ylidene]ethylidene]-4-methylidene-cyclohexane-1,3-diol.
LMST03020470;　26,27-diethyl-22-oxa-9,10-seco-5Z,7E,10(19)-cholestatriene-1S,3R,25-triol;　26,27-diethyl-1α,25-dihydroxy-22-oxavitamin D_3;　26,27-diethyl-1α,25-dihydroxy-22-oxacholecalciferol;　9547575;
Equipotent with 1α,25-dihydroxyvitamn D_3 in inducing differentiation of human myeloid leukemia cells (HL-60) into macrophases *in vitro* estimated by superoxide anion generation; λ_m = 263 nm (EtOH). **Chem Pharm Bull (Tokyo)**. 1992, **40(6)**, 1494-9.

812　**26,27-Diethyl-1α,25-dihydroxy-23-oxavitamin D_3**

9547576　　　　　　　　　　　　　$C_{30}H_{50}O_4$

(1R,3S,5Z)-5-[(2E)-2-[(1R,3aR,7aS)-1-[(2S)-1-(2-hydroxy-2-propyl-pentoxy)propan-2-yl]-7a-methyl-2,3,3a,5,6,7-hexahydro-1H-inden-4-ylidene]ethylidene]-4-methylidene-cyclohexane-1,3-diol.
LMST03020471;　26,27-diethyl-23-oxa-9,10-seco-5Z,7E,10(19)-cholestatriene-1S,3R,25-triol;　26,27-diethyl-1α,25-dihydroxy-23-oxacholecalciferol;　9547576;
Synthesized from protected (5E)-1α,22-dihydroxy-23,24,25,26,27-pentanorvitamin D_3. Compared to 1α,25-dihydroxyvitamin D_3, affinity for pig intestinal nuclear receptor is 13%, affinity for human vitamin D binding protein is <0.02%, differentiation of HL-60 cells is 25%, and calciuric effect is 1%. **Tet Lett**. 1991, **32(38)** 5073-6; Schering AG, Int. Pat. Appl. No. WO 94/00428.

813　**1α,25-Dihydroxy-20S-methoxy-26,27-dimethylvitamin D_3**

9547577　　　　　　　　　　　　　$C_{30}H_{50}O_4$

(1R,3S,5Z)-5-[(2E)-2-[(1S,3aR,7aR)-1-[(2S)-6-ethyl-6-hydroxy-2-methoxy-octan-2-yl]-7a-methyl-2,3,3a,5,6,7-hexahydro-1H-inden-4-ylidene]ethylidene]-4-methylidene-cyclohexane-1,3-diol.
LMST03020472;　1α,25-dihydroxy-20S-methoxy-26,27-dimethylcholecalciferol;　20S-methoxy-26,27-dimethyl-9,10-seco-5Z,7E,10(19)-cholestatriene-1S,3R,25-triol;　9547577;　Synthesized from a (5E)-1α-hydroxy-20-keto-22,23,24,25,26,27-hexanorvitamin D_3 derivative. Compared to 1α,25-dihydroxyvitamin D_3, inhibition of U937 cell (human histiocytic lymphoma cell line) proliferation is 800000%, binding to the vitamin D receptor from rachitic chicken intestine is 50%, calcemic activity determined by the increase in urinary calcium excretion in rats is 32%. Vitamin D. A Pluripotent Steroid Hormone: Structural Studies, Molecular Endocrinology and Clinical Applications. Proceedings of the Ninth Workshop on Vitamin D, Orlando, Florida (USA) May 28-June 2. Synthesis and Biological Activity of 20-Hydroxylated Vitamin D Analogues. (Norman, A. W., Bouillon, R., Thomasset, M., eds), pp95-96, Walter de Gruyter Berlin New York (1994).

814　**1α,25-Dihydroxy-26,27-dimethyl-24a,24b-dihomo-22-oxa-20-epivitamin D_3**

9547578　　　　　　　　　　　　　$C_{30}H_{50}O_4$

(1R,3S,5Z)-5-[(2E)-2-[(1S,3aR,7aR)-1-[(1R)-1-(5-ethyl-5-hydroxy-heptoxy)ethyl]-7a-methyl-2,3,3a,5,6,7-hexahydro-1H-inden-4-ylidene]ethylidene]-4-methylidene-cyclohexane-1,3-diol.
LMST03020473;　(20R)-26,27-dimethyl-24a,24b-di-homo-22-oxa-9,10-seco-5Z,7E,10(19)-cholestatrien-1S,3R,25-triol;　1α,25-dihydroxy-26,27-dimethyl-24a,24b-dihomo-22-oxa-20-epicholecalciferol;　9547578;
Compared to 1α,25-dihydroxyvitamin D_3, inhibition of U937 cell (human histiocytic lymphoma cell line)

proliferation is 28000%, induction of U937 cell differentiation is 40000%, binding to the vitamin D receptor from rachitic chicken intestine is 61%, calcemic activity determined by the increase in urinary calcium excretion in rats is 80%. ***Biochem Pharmacol.*** 1991, ***42(8)***, 1569-75.

815   **1α,22R,25-Trihydroxy-26,27-dimethyl-24a-homo-20-epivitamin D₃**

9547579 $C_{30}H_{50}O_4$

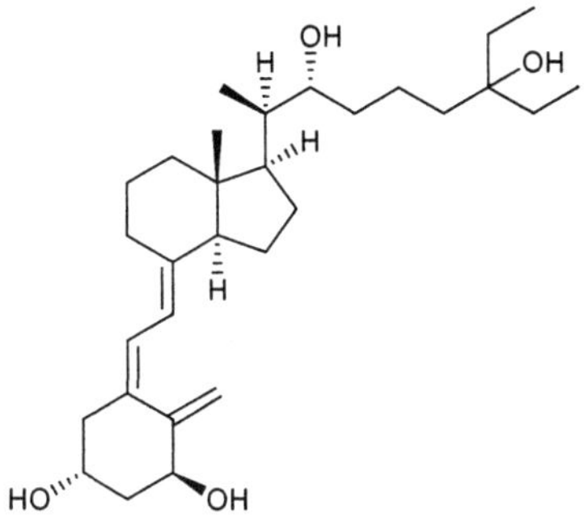

(2R,3R)-2-[(1R,3aR,4E,7aS)-4-[(2Z)-2-[(3S,5R)-3,5-dihydroxy-2-methylidene-cyclohexylidene]ethylidene]-7a-methyl-2,3,3a,5,6,7-hexahydro-1H-inden-1-yl]-7-ethyl-nonane-3,7-diol.
LMST03020474; 1α,22R,25-trihydroxy-26,27-dimethyl-24a-homo-20-epicholecalciferol; (20R)-26,27-dimethyl-24a-homo-9,10-seco-5Z,7E,10(19)-cholestatriene-1S,3R,22R,25-tetrol; 9547579; Synthesized from protected (5E)-1α-hydroxy-22-oxo-23,24,25,26,27-pentanor-20-epivitamin D₃. Compared to 1α,25-dihydroxyvitamin D₃, induction of differentiation of U937 cells is 2000%, inhibition of proliferation is 2800%, VDR (rachitic chicken intestinal receptor) binding affinity is 2%, calciuric effect in normal rats is 80%. Vitamin D. A Pluripotent Steroid Hormone: Structural Studies, Molecular Endocrinology and Clinical Applications. Proceedings of the Ninth Workshop on Vitamin D, Orlando, Florida (USA) May 28-June 2. Chemistry and Biology of Highly Active 22-Oxy Analogs of 20-Epi Calcitriol with very low Binding Affinity to the Vitamin D Receptor. (Norman, A. W., Bouillon, R., Thomasset, M., eds), pp85-86, Walter de Gruyter Berlin New York (1994).

816   **1α,22S,25-Trihydroxy-26,27-dimethyl-24a-homovitamin D₃**

9547581

                                        $C_{30}H_{50}O_4$

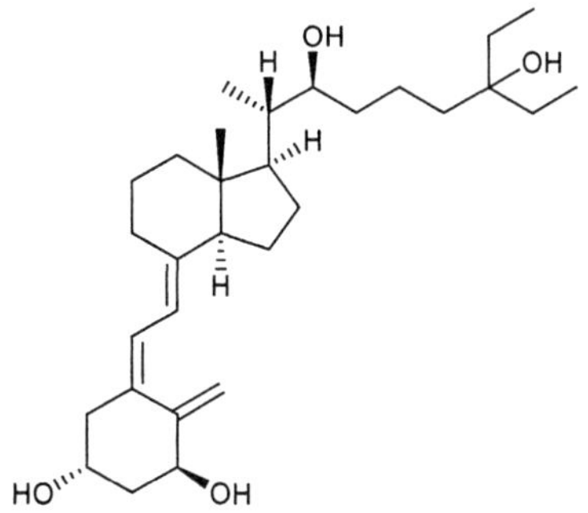

(2S,3S)-2-[(1R,3aR,4E,7aS)-4-[(2Z)-2-[(3S,5R)-3,5-dihydroxy-2-methylidene-cyclohexylidene]ethylidene]-7a-methyl-2,3,3a,5,6,7-hexahydro-1H-inden-1-yl]-7-ethyl-nonane-3,7-diol.
LMST03020476; 1α,22S,25-trihydroxy-26,27-dimethyl-24a-homocholecalciferol; 26,27-dimethyl-24a-homo-9,10-seco-5Z,7E,10(19)-cholestatriene-1S,3R,22S,25-tetrol; 9547581.

817   **1α-Hydroxy-18-(4-hydroxy-4-ethylhexyl-oxy)-23,24,25,26,27-pentanorvitamin D₃**

9547682 $C_{30}H_{50}O_4$

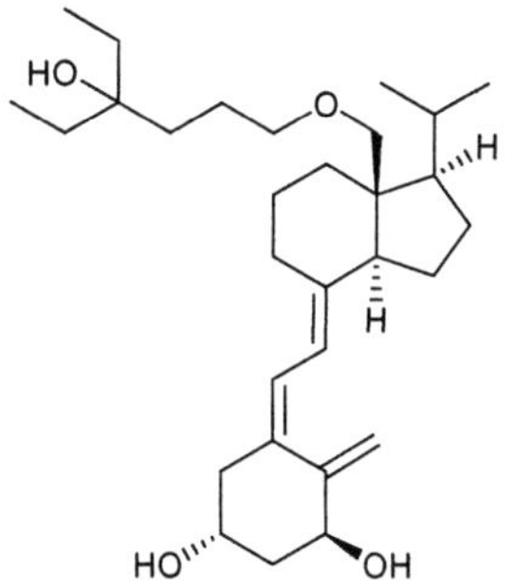

(1R,3S,5Z)-5-[(2E)-2-[(1R,3aR,7aR)-7a-[(4-ethyl-4-hydroxy-hexoxy)methyl]-1-propan-2-yl-2,3,3a,5,6,7-hexahydro-1H-inden-4-ylidene]ethylidene]-4-methylidene-cyclohexane-1,3-diol.
LMST03020587; 18-(4-hydroxy-4-ethylhexyloxy)-23,24-dinor-9,10-seco-5Z,7E,10(19)-cholatriene-1S,3R-diol; 1α-hydroxy-18-(4-hydroxy-4-ethylhexyloxy)-23,24,25,26,27-pentanorcholecalciferol; 9547682.

818   **1α,25-Dihydroxy-2α-(3-hydroxypropyl)vitamin D₃**

                                        $C_{30}H_{50}O_4$

9,10-Seco-5Z,7E,10(19)-20R-cholestatriene-2S-(2-hydroxypropyl)-1S,3R,25-triol; Synthesized from D-xylose. Has 300% of the affinity of 1,25-dihydroxyvitamin D_3 for the vitamin D receptor; White solid; $[\alpha]_D^{20}$- +161.29° (CHCl$_3$, c = 0.00186); λ_m = 268, 227 nm (EtOH). *J Org Chem*. 2001, **66(26)**, 8760-71.

819   *1α,25-Dihydroxy-2β-(3-hydroxypropoxy)vitamin D_3*

6918141 $C_{30}H_{50}O_5$

(1R,2R,3R,5Z)-5-[(2E)-2-[(1R,3aR,7aS)-1-[(2R)-6-hydroxy-6-methyl-heptan-2-yl]-7a-methyl-2,3,3a,5,6,7-hexahydro-1H-inden-4-ylidene]ethylidene]-2-(3-hydroxypropoxy)-4-methylidene-cyclohexane-1,3-diol. LMST03020477; ED-71; 2R-(3-hydroxypropoxy)-9,10-seco-5Z,7E,10(19)-cholestatriene-1R,3R,25-triol; 1α,25-dihydroxy-2β-(3-hydroxypropoxy)cholecalciferol; 6918141; Synthesized photochemically from the 5,7-diene. Compared to 1α,25-dihydroxyvitamin D_3, binding potency to vitamin D binding protein of ED-71 is 270%, binding potency to vitamin D receptor is 12%. Improves bone mineral density and mechanical bone strength in the pre-osteoporosis model rats made by ovariectomy more effectively than 1α,25-dihydroxyvitamin D_3; λ_m = 263 nm (EtOH). *Chem Pharm Bull (Tokyo)*.1997, **45(10)**, 1626-30; *Bioorg Med Chem Lett*. 1993, **3(9)**, 1815-9.

820   *1β,25-Dihydroxy-2β-(3-hydroxyprop-oxy)vitamin D_3*

9547582 $C_{30}H_{50}O_5$

(1R,2R,3S,5Z)-5-[(2E)-2-[(1R,3aR,7aS)-1-[(2R)-6-hydroxy-6-methyl-heptan-2-yl]-7a-methyl-2,3,3a,5,6,7-hexahydro-1H-inden-4-ylidene]ethylidene]-2-(3-hydroxypropoxy)-4-methylidene-cyclohexane-1,3-diol. LMST03020478; 2R-(3-hydroxypropoxy)-9,10-seco-5Z, 7E,10(19)-cholestatriene-1S,3R,25-triol; 1β,25-di-hydroxy-2β-(3-hydroxypropoxy)cholecalciferol; 9547582; Prepared from pro-ED-71. Binding affinity for vitamin D receptor and vitamin D binding protein is 0.3% and 670% that of 1α,25-dihydroxyvitamin D_3; λ_m = 264 nm (EtOH). *Bioorg Med Chem Lett*. 1994, **4(12)**, 1523-6.

821   *24α,25-Dihydroxy-1α-hydroxymethyl-26,27-dimethyl-24a-homo-22-oxavitamin D_3*

9547665 $C_{30}H_{50}O_5$

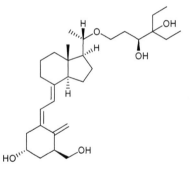

(3S)-1-[(1S)-1-[(1S,3aR,4E,7aR)-4-[(2Z)-2-[(3S,5S)-5-hydroxy-3-(hydroxymethyl)-2-methylidene-cyclohexylidene]ethylidene]-7a-methyl-2,3,3a,5,6,7-hexahydro-1H-inden-1-yl]ethoxy]-4-ethyl-hexane-3,4-diol. LMST03020570; 1S-hydroxymethyl-26,27-dimethyl-24aS-homo-22-oxa-9,10-seco-5Z,7E,10(19)-cholesta-triene-3S,24a,25-triol; 24α,25-dihydroxy-1α-hydroxy-methyl-26,27-dimethyl-24a-homo-22-oxacholecalcif-erol; 9547665.

822   *24α,25-Dihydroxy-1β-hydroxymethyl-26,27-dimethyl-24a-homo-22-oxa-3-epivitamin D_3*

9547666 $C_{30}H_{50}O_5$

(3S)-1-[(1S)-1-[(1S,3aR,4E,7aR)-4-[(2Z)-2-[(3R,5R)-5-hydroxy-3-(hydroxymethyl)-2-methylidene-cyclo-hexylidene]ethylidene]-7a-methyl-2,3,3a,5,6,7-hexa-hydro-1H-inden-1-yl]ethoxy]-4-ethyl-hexane-3,4-diol. LMST03020571; 1R-hydroxymethyl-26,27-dimethyl-24a-homo-22-oxa-9,10-seco-5Z,7E,10(19)-cholesta-triene-3R,24aS,25-triol; 24α,25-dihydroxy-1β-hydroxy-methyl-26,27-dimethyl-24a-homo-22-oxa-3-epicholecalciferol; 9547666.

823   1α,25-dihydroxy-2α-(3-hydroxyprop-oxy)vitamin D₃

104121-92-8   C₃₀H₅₀O₅

(1S,2S,3S,5Z)-5-[(2E)-2-[(1R,3aR,7aS)-1-[(2R)-6-hydroxy-6-methyl-heptan-2-yl]-7a-methyl-2,3,3a,5,6,7-hexahydro-1H-inden-4-ylidene]ethylidene]-2-(3-hydroxypropoxy)-4-methylidene-cyclohexane-1,3-diol. ED-71; 2-HOPr-2OHVitD3; 2-(3-Hydroxypropoxy)-1,25-dihydroxyvitamin D₃; 2-(3-Hydroxypropoxy)calcitriol; 2-(3-Hydroxypropoxy)-1, 25-dihydroxyvitamin D₃; 9,10-Secocholesta-5Z,7E,10(19)-triene-1α,3β,25-triol, 2β-(3-hydroxypropoxy)-; 6438982; 104121-92-8; Synthetic derivative of vitamin D₃, calcitriol analog. Has 500 times the calcium mobilizing activity of 1,25-dihydroxyvitamin D₃. Increases bone mass in osteoporotic patients under vitamin D supplementation. Used to treat osteoporosis. *Bioorg Med Chem Lett.* 2000, *10(10)*, 1129-32.

824   3'-O-Aminopropyl-25-hydroxyvitamin D₃
C₃₀H₅₁NO₂

Used as an affinity matrix to purify human vitamin D-binding protein. *Prot Exp Purif.* 1985, *6(2)*, 185-8; *Biochemistry.* 1991, *30(19)*, 4809-13.

825   1α,25-Dihydroxyvitamin   D₃   25-trimethylsilyl ether

5364601   C₃₀H₅₂O₃Si

(5E)-5-[(2E)-2-[7a-methyl-1-(6-methyl-6-trimethylsilyl-oxy-heptan-2-yl)-2,3,3a,5,6,7-hexahydro-1H-inden-4-ylidene]ethylidene]-4-methylidene-cyclohexane-1,3-diol. 25-[(Trimethylsilyl)oxy]-9,10-secocholesta-5Z,7E,10-triene-1,3-diol; 9,10-Secocholesta-5Z,7E,10(19)-triene-1α,3β,25-triol, 2β-(3-hydroxypropoxy)-; 5364601; 6438982.

826   1α-Hydroxy-22-[3-(1-hydroxy-1-methyl-ethyl)phenyl]-23,24,25,26,27-pentanor-5Z-vitamin D₃

9547583   C₃₁H₄₄O₃

(1R,3S,5Z)-5-[(2E)-2-[(1R,3aR,7aS)-1-[(2R)-1-[3-(2-hydroxypropan-2-yl)phenyl]propan-2-yl]-7a-methyl-

2,3,3a,5,6,7-hexahydro-1H-inden-4-ylidene]
ethylidene]-4-methylidene-cyclohexane-1,3-diol.
LMST03020479; 22-[3-(1-hydroxy-1-methylethyl)-
phenyl]-23,24-dinor-9,10-seco-5Z,7E,10(19)-chola-
triene-1S,3R-diol; 1α-hydroxy-22-[3-(1-hydroxy-1-
methylethyl)phenyl]-23,24,25,26,27-pentanorchole-
calciferol; 9547583; Compared to 1α,25-dihydroxy-
vitamin D$_3$, intestinal calcium absorption is 13%, bone
calcium mobilization is 7.7%, affinities for chick
intestinal receptor, HL-60 cell receptor and serum
vitamin D binding protein are 62%, 32% and 25%,
and inhibition of 1α-hydroxylase activity is 99%,
differentiation of HL-60 cells is 80%. Analogs of 1,25-
(OH)2 D3. In Vitamin D Gene Regulation Structure-
Function Analysis and Clinical Application (Norman,
A.W., Bouillon R. and Thomasset, M., eds), pp 165-
166, Walter de Gruyter, Berlin/New York (1991).

827   *1α-Hydroxy-22-[3-(1-hydroxy-1-methyl-ethyl)phenyl]-23,24,25,26,27-pentanor-5E-vitamin D$_3$*

133910-11-9   C$_{31}$H$_{44}$O$_3$

(1R,3S,5E)-5-[(2E)-2-[(1R,3aR,7aS)-1-[1-[3-(2-
hydroxypropan-2-yl)phenyl]propan-2-yl]-7a-methyl-
2,3,3a,5,6,7-hexahydro-1H-inden-4-ylidene]
ethylidene]-4-methylidene-cyclohexane-1,3-diol.
LMST03020479; 133910-11-9; Affinity for the vitamin
D receptor is 62% that of calcitriol, but has slightly
better antiproliferative activity. *Mol Endocrinol*. 2000,
14(11), 1788-96.

828   *GS 1500*

C$_{31}$H$_{44}$O$_3$S

9,10-Seco-(1α,3β-dihydroxy-10-methylene-5,7-dien-
17α-(2'-methylenethio-(m-2-hydroxy-i-propylphenyl)
androstane.

GS 1500; Increases osteocalcin mRNA levels more
than calcitriol. Causes fast increase in intracellular
calcium levels in skeletal muscle. Binds to the vitamin
D receptor more effectively than calcitriol. *Steroids*.
1992, *57(7)*, 344-7; *Br J Pharmacol*. 1991, *126(8)*,
1815-23; *Eur J Biochem*. 1991, *261(3)*, 706-13.

829   *GS 1558*

C$_{31}$H$_{44}$O$_3$S

9,10-Seco-(1α,3β-dihydroxy-10-methylene-5,7-dien-
17β-(2'-methylenethio-(m-2-hydroxy-i-propylphenyl)-
androstane.
GS 1558; Increases osteocalcin mRNA levels only half
as much as calcitriol. Binds to the vitamin D receptor
less effectively than calcitriol. *Eur J Biochem*. 1999,
261(3), 706-13.

830   *EB 1213*

167357-73-5   C$_{31}$H$_{44}$O$_4$

(5E)-5-[(2E)-2-[1-[1-[3-(2-hydroxypropan-2-yl)phen-
oxy]propan-2-yl]-7a-methyl-2,3,3a,5,6,7-hexahydro-
1H-inden-4-ylidene]ethylidene]-4-methylidene-
cyclohexane-1,3-diol.
EB 1213; EB-1213; 4-Methylene-5-((octahydro-1-(2-(3-
(1-hydroxy-1-methylethyl)phenoxy)-1-methylethyl)-7a-
methyl-4H-inden-4-ylidene)ethylidene)-1,3-cyclo-
hexanediol; 6439508; 167357-73-5; Inhibits cell
growth of HaCaT keratinocytes after 48 hrs. of
incubation and triggers the hydrolysis of sphingomyelin. Has been tested as a non-hypercalcaemic
agent in the treatment of uraemic secondary
hyperparathyroidism but with inconclusive results.
Affinity for the intracellular vitamin D receptor
comparable to that of calcitriol. Induces sphingomyelin
hydrolysis and apoptosis in human keratinocyte cells
and is promising as a non-hypercalcemic agent in the

treatment of uremic secondary hyperparathyroidism. *Cell Mol Biol (Noisy-le-grand)*. 2000, *46(1)*, 111-19; *Nephrol Dial Transplant*. 1996, *11(9)*, 1781-6; *J Investig Dermatol Symp Proc*. 1996, *1(1)*, 44-8.

831 *1α,25-Dihydroxy-26,27-dimethyl-17Z, 20,22,22,23,23-hexadehydro-24a,24b-dihomovitamin D₃*

9547584 $C_{31}H_{46}O_3$

(1R,3S,5Z)-5-[(2E)-2-[(1E,3aR,7aS)-1-(8-ethyl-8-hydroxy-dec-3-yn-2-ylidene)-7a-methyl-2,3,3a,5,6,7-hexahydroinden-4-ylidene]ethylidene]-4-methylidene-cyclohexane-1,3-diol.
LMST03020480; 1α,25-dihydroxy-26,27-dimethyl-17Z,20,22,22,23,23-hexadehydro-24a,24b-dihomocholecalciferol; 26,27-dimethyl-24a,24b-dihomo-9,10-seco-5Z,7E,10(19),17Z(20)-cholestatetraen-22-yne-1S,3R,25-triol; 9547584; Synthesized from hydroxyprotected (5E)-1α-hydroxy-20-oxo-22,23,24,25,26,27-hexanorvitamin D₃. Compared to 1α,25-dihydroxyvitamin D₃, inhibition of proliferation of U937 cells is 29000%, affinity for chicken intestinal vitamin D receptor is 2%, and calciuric activity is 25%. Vitamin D. A Pluripotent Steroid Hormone: Structural Studies, Molecular Endocrinology and Clinical Applications. Proceedings of the Ninth Workshop on Vitamin D, Orlando, Florida (USA) May 28-June 2. Chemistry and Biology of 22,23-Yne Analogs of Calcitriol. (Norman, A. W., Bouillon, R., Thomasset, M., eds), pp73-74, Walter de Gruyter Berlin New York (1994).

832 *1α,25-Dihydroxy-(26,26)-(27,27)-diethanovitamin D₃*

9547585 $C_{31}H_{48}O_3$

(1R,3S,5Z)-5-[(2E)-2-[(1R,3aR,7aS)-1-[(2R)-6,6-dicyclopropyl-6-hydroxy-hexan-2-yl]-7a-methyl-

2,3,3a,5,6,7-hexahydro-1H-inden-4-ylidene] ethylidene]-4-methylidene-cyclohexane-1,3-diol.
LMST03020481; (26,26)-(27,27)-diethano-9,10-seco-5Z,7E,10(19)-cholestatriene-1S,3R,25-triol; 1α,25-dihydroxy-(26,26)-(27,27)-diethanocholecalciferol; 9547585. *J Org Chem*. 1993, *58(1)*, 118-23.

833 *1α,25-Dihydroxy-26,27-dimethyl-22E, 23,24E,24a-tetradehydro-24a,24b-dihomovitamin D₃*

9547586 $C_{31}H_{48}O_3$

(1R,3S,5Z)-5-[(2E)-2-[(1R,3aR,7aS)-1-[(2S,3E,5E)-8-ethyl-8-hydroxy-deca-3,5-dien-2-yl]-7a-methyl-2,3,3a,5,6,7-hexahydro-1H-inden-4-ylidene] ethylidene]-4-methylidene-cyclohexane-1,3-diol.
LMST03020482; 1α,25-dihydroxy-26,27-dimethyl-22E,23,24E,24a-tetradehydro-24a,24b-dihomocholecalciferol; 26,27-dimethyl-24a,24b-dihomo-9,10-seco-5Z,7E,10(19),22E,24E-cholestapentaene-1S,3R,25-triol; 9547586; Synthesized from protected (5E)-1α-hydroxy-22-oxo-23,24,25,26,27-pentanorvitamin D₃. Compared to 1α,25-dihydroxyvitamin D₃, inhibition · of proliferation of U937 cells is 700%, induction of differentiation of U937 cells is 200%, and calciuric activity is 20%. Gene Regulation, Structure-Function Analysis and Clinical Application. Proceedings of the Eighth Workshop on Vitamin D Paris, France July 5-10. Synthesis and Biological Activity of 1α-Hydroxylated Vitamin D₃ Analogues with Hydroxylated Side Chains, Multi-Homologated in The 24- or 24,26,27-Positions. (Norman, A. W., Bouillon, R., Thomasset, M., eds), pp159-160, Walter de Gruyter Berlin (1991).

834 *1α,25-Dihydroxy-26,27-dimethyl-22,22, 23,23-tetradehydro-24a,24b-dihomovitamin D₃*

9547587 $C_{31}H_{48}O_3$

(1R,3S,5Z)-5-[(2E)-2-[(1R,3aR,7aS)-1-[(2S)-8-ethyl-8-hydroxy-dec-3-yn-2-yl]-7a-methyl-2,3,3a,5,6,7-hexahydro-1H-inden-4-ylidene]ethylidene]-4-methylidene-cyclohexane-1,3-diol.
LMST03020483; 26,27-dimethyl-24a,24b-dihomo-9,10-seco-5Z,7E,10(19)-cholestatrien-22-yne-1S,3R,25-triol; 1α,25-dihydroxy-26,27-dimethyl-22,22,23,23-tetradehydro-24a,24b-dihomocholecalciferol; 9547587; Synthesized from hydroxy-protected (5E)-1α-hydroxy-22,22,23,23-tetradehydro-24,25,26,27-tetra-norvitamin D$_3$. Compared to 1α,25-dihydroxyvitamin D$_3$, inhibition of proliferation of U937 cells is 10%, affinity for chicken intestinal vitamin D receptor is 2%. Vitamin D. A Pluripotent Steroid Hormone: Structural Studies, Molecular Endocrinology and Clinical Applications. Proceedings of the Ninth Workshop on Vitamin D, Orlando, Florida (USA) May 28-June 2. Chemistry and Biology of 22,23-Yne Analogs of Calcitriol. (Norman, A. W., Bouillon, R., Thomasset, M., eds), pp73-74, Walter de Gruyter Berlin New York (1994).

835   **1α,25-Dihydroxy-26,27-dimethyl-22,22,23,23-tetradehydro-24a,24b-dihomo-20-epivitamin D$_3$**

9547588   C$_{31}$H$_{48}$O$_3$

(1R,3S,5Z)-5-[(2E)-2-[(1R,3aR,7aS)-1-[(2R)-8-ethyl-8-hydroxy-dec-3-yn-2-yl]-7a-methyl-2,3,3a,5,6,7-hexahydro-1H-inden-4-ylidene]ethylidene]-4-methylidene-cyclohexane-1,3-diol.
LMST03020484; (20R)-26,27-dimethyl-24a,24b-di-homo-9,10-seco-5Z,7E,10(19)-cholestatrien-22-yne-1S,3R,25-triol; 1α,25-dihydroxy-26,27-dimethyl-22,22,23,23-tetradehydro-24a,24b-dihomo-20-epivitamin D$_3$; 1α,25-dihydroxy-26,27-dimethyl-22,22,23,23-tetradehydro-24a,24b-dihomo-20-epicholecalciferol; 9547588; Synthesized from hydroxy-protected (5E)-1α-hydroxy-22,22,23,23-tetradehydro-24,25,26,27-tetra-norvitamin D$_3$. Compared to 1α,25-dihydroxyvitamin D$_3$, inhibition of proliferation of U937 cells is 49000%, affinity for chicken intestinal vitamin D receptor is 7%, calciuric activity is 63%. Vitamin D. A Pluripotent Steroid Hormone: Structural Studies, Molecular Endocrinology and Clinical Applications. Proceedings of the Ninth Workshop on Vitamin D, Orlando, Florida (USA) May 28-June 2. Chemistry and Biology of 22,23-Yne Analogs of Calcitriol. (Norman, A. W., Bouillon, R., Thomasset, M., eds), pp73-74, Walter de Gruyter Berlin New York (1994).

836   **25-Hydroxy-1α-hydroxymethyl-26,27-di-methyl-24a-homo-22,23,24,24a-tetradehydro-vitamin D$_3$**

9547669   C$_{31}$H$_{48}$O$_3$

(4E,6E,8S)-8-[(1R,3aR,4E,7aS)-4-[(2Z)-2-[(3S,5S)-5-hydroxy-3-(hydroxymethyl)-2-methylidene-cyclohexylidene]ethylidene]-7a-methyl-2,3,3a,5,6,7-hexahydro-1H-inden-1-yl]-3-ethyl-nona-4,6-dien-3-ol.
LMST03020574; 1S-hydroxymethyl-26,27-dimethyl-24a-homo-9,10-seco-5Z,7E,10(19),22E,24E-cholesta-pentaene-3S,25-diol; 25-hydroxy-1α-hydroxymethyl-26,27-dimethyl-24a-homo-22,23,24,24a-tetradehydro-cholecalciferol; 9547669.

837   **25-Hydroxy-1β-hydroxymethyl-26,27-di-methyl-24a-homo-22,23,24,24a-tetradehydro-3-epivitamin D$_3$**

9547670   C$_{31}$H$_{48}$O$_3$

(4E,6E,8S)-8-[(1R,3aR,4E,7aS)-4-[(2Z)-2-[(3R,5R)-5-hydroxy-3-(hydroxymethyl)-2-methylidene-cyclohexylidene]ethylidene]-7a-methyl-2,3,3a,5,6,7-hexahydro-1H-inden-1-yl]-3-ethyl-nona-4,6-dien-3-ol.
LMST03020575; 1R-hydroxymethyl-26,27-dimethyl-24a-homo-9,10-seco-5Z,7E,10(19),22E,24E-cholesta-pentaene-3R,25-diol; 25-hydroxy-1β-hydroxymethyl-26,27-dimethyl-24a-homo-22,23,24,24a-tetradehydro-3-epicholecalciferol; 9547670. Vitamin D. A Pluripotent Steroid Hormone: Structural Studies, Molecular Endocrinology and Clinical Applications. Proceedings of the Ninth Workshop on Vitamin D, Orlando, Florida (USA) May 28-June 2. Chemistry and Biology of Highly Active 22-Oxy Analogs of 20-Epi Calcitriol with very low Binding Affinity to the Vitamin D Receptor. (Norman, A. W., Bouillon, R., Thomasset, M., eds), pp85-86, Walter de Gruyter Berlin New York

(1994).

838 1α,25-Dihydroxy-22R-methoxy-26,27-di-methyl-23,23,24,24-tetradehydro-24a-homo-vitamin D₃

9547589 $C_{31}H_{48}O_4$

(1R,3S,5Z)-5-[(2E)-2-[(1R,3aR,7aS)-1-[(2S,3R)-7-ethyl-7-hydroxy-3-methoxy-non-4-yn-2-yl]-7a-methyl-2,3,3a,5,6,7-hexahydro-1H-inden-4-ylidene]ethylidene]-4-methylidene-cyclohexane-1,3-diol.
LMST03020485; 1α,25-dihydroxy-22R-methoxy-26,27-dimethyl-23,23,24,24-tetradehydro-24a-homo-cholecalciferol; 22R-methoxy-26,27-dimethyl-24a-homo-9,10-seco-5Z,7E,10(19)-cholestatrien-23-yne-1S,3R,25-triol; 9547589.

839 1α,25-Dihydroxy-22S-methoxy-26,27-dimethyl-23,24-tetradehydro-24a-homo-20-epivitamin D₃

9547590 $C_{31}H_{48}O_4$

(1R,3S,5Z)-5-[(2E)-2-[(1R,3aR,7aS)-1-[(2R,3S)-7-ethyl-7-hydroxy-3-methoxy-non-4-yn-2-yl]-7a-methyl-2,3,3a,5,6,7-hexahydro-1H-inden-4-ylidene]ethylidene]-4-methylidene-cyclohexane-1,3-diol.
LMST03020486; 1α,25-dihydroxy-22S-methoxy-26,27-dimethyl-23,24-tetradehydro-24a-homo-20-epicholecalciferol; (20R)-26,27-dimethyl-22S-methoxy-24a-homo-9,10-seco-5Z,7E,10(19)-cholestatrien-23-yne-1S,3R,25-triol; 9547590; Synthesized from protected (5E)-1α-hydroxy-22-oxo-23,24,25,26,27-pentanor-20-epivitamin D₃. Compared to 1α,25-dihydroxyvitamin D₃, induction of differentiation of U937 cells is 20000%, inhibition of proliferation is

62000%, VDR (rachitic chicken intestinal receptor) binding affinity is 14%, and calciuric effect in normal rats is 110%. Vitamin D A Pluripotent Steroid Hormone : Structural Studies, Molecular Endocrinology and Clinical Applications. Proceedings of the Ninth Workshop on Vitamin D, Orlando, Florida (USA) May 28-June 2. Chemistry and Biology of Highly Active 22-Oxy Analogs of 20-Epi Calcitriol with very low Binding Affinity to the Vitaamin D Receptor. (Norman, A. W.; Bouillon, R.; Thomasset, M.; eds), pp85-86, Walter de Gruyter Berlin New York (1994).

840 1α,25-Dihydroxy-22S-ethoxy-26,27-di-methyl-23,24-tetradehydro-20-epivitamin D₃

9547591 $C_{31}H_{48}O_4$

(1R,3S,5Z)-5-[(2E)-2-[(1R,3aR,7aS)-1-[(2R,3S)-3-ethoxy-6-ethyl-6-hydroxy-oct-4-yn-2-yl]-7a-methyl-2,3,3a,5,6,7-hexahydro-1H-inden-4-ylidene]ethylidene]-4-methylidene-cyclohexane-1,3-diol.
LMST03020487; 1α,25-dihydroxy-22S-ethoxy-26,27-dimethyl-23,24-tetradehydro-20-epicholecalciferol; (20R)-26,27-dimethyl-22S-ethoxy-9,10-seco-5Z,7E,10(19)-cholestatrien-23-yne-1S,3R,25-triol; 9547591; Synthesized from protected (5E)-1α-hydroxy-22-oxo-23,24,25,26,27-pentanor-20-epivitamin D₃. Compared to 1α,25-dihydroxyvitamin D₃, induction of differentiation of U937 cells is 100000%, inhibition of proliferation is 79000%, VDR (rachitic chicken intestinal receptor) binding affinity is 19%, and calciuric effect in normal rats is 100%. Vitamin D. A Pluripotent Steroid Hormone: Structural Studies, Molecular Endocrinology and Clinical Applications. Proceedings of the Ninth Workshop on Vitamin D, Orlando, Florida (USA) May 28-June 2. Chemistry and Biology of Highly Active 22-Oxy Analogs of 20-Epi Calcitriol with very low Binding Affinity to the Vitamin D Receptor. (Norman, A. W., Bouillon, R., Thomasset, M., eds), pp85-86, Walter de Gruyter Berlin New York (1994).

841 1α,22R,25-Trihydroxy-26,27-dimethyl-23,23,24,24-tetradehydro-24a,24b-dihomo-vitamin D₃

9547592

$C_{31}H_{48}O_4$

(2S,3R)-2-[(1R,3aR,4E,7aS)-4-[(2Z)-2-[(3S,5R)-3,5-dihydroxy-2-methylidene-cyclohexylidene]ethylidene]-7a-methyl-2,3,3a,5,6,7-hexahydro-1H-inden-1-yl]-8-ethyl-dec-4-yne-3,8-diol.
LMST03020488;   1α,22R,25-trihydroxy-26,27-dimethyl-23,23,24,24-tetradehydro-24a,24b-dihomo-cholecalciferol; 26,27-dimethyl-24a,24b-dihomo-9,10-seco-5Z,7E.10(19)-cholestatrien-23-yne-1S,3R,22R,25-tetrol; 9547592.

842   **1α,22S,25-Trihydroxy-26,27-dimethyl-23,23,24,24-tetradehydro-24a,24b-dihomo-vitamin D₃**

9547593 $C_{31}H_{48}O_4$

(2S,3S)-2-[(1R,3aR,4E,7aS)-4-[(2Z)-2-[(3S,5R)-3,5-dihydroxy-2-methylidene-cyclohexylidene]ethylidene]-7a-methyl-2,3,3a,5,6,7-hexahydro-1H-inden-1-yl]-8-ethyl-dec-4-yne-3,8-diol.
LMST03020489;   1α,22S,25-trihydroxy-26,27-dimethyl-23,23,24,24-tetradehydro-24a,24b-dihomo-vitamin D₃; 1α,22S,25-trihydroxy-26,27-dimethyl-23,23,24,24-tetradehydro-24a,24b-dihomocholecalciferol; 26,27-dimethyl-24a,24b-dihomo-9,10-seco-5Z,7E,10(19)-cholestatrien-23-yne-1S,3R,22S,25-tetrol; 9547593.

843   **1α,22S,25-Trihydroxy-26,27-dimethyl-23,24-tetradehydro-24a,24b-dihomo-20-epi-vitamin D₃**

9547594 $C_{31}H_{48}O_4$

(2R,3S)-2-[(1R,3aR,4E,7aS)-4-[(2Z)-2-[(3S,5R)-3,5-dihydroxy-2-methylidene-cyclohexylidene]ethylidene]-7a-methyl-2,3,3a,5,6,7-hexahydro-1H-inden-1-yl]-8-ethyl-dec-4-yne-3,8-diol.
LMST03020490;   1α,22S,25-trihydroxy-26,27-dimethyl-23,24-tetradehydro-24a,24b-dihomo-20-epi-cholecalciferol;   (20R)-26,27-dimethyl-24a,24b-di-homo-9,10-seco-5Z,7E,10(19)-cholestatrien-23-yne-1S,3R,22S,25-tetrol; 9547594; Synthesized from protected (5E)-1α-hydroxy-22-oxo-23,24,25.26,27-pentanor-20-epivitamin D₃. Compared to 1α,25-dihydroxyvitamin D₃, induction of differentiation of U937 cells is 10000%, inhibition of proliferation is 1200%, VDR (rachitic chicken intestinal receptor) binding affinity is 0.5%. Vitamin D. A Pluripotent Steroid Hormone: Structural Studies, Molecular Endocrinology and Clinical Applications. Proceedings of the Ninth Workshop on Vitamin D, Orlando, Florida (USA) May 28-June 2. Chemistry and Biology of Highly Active 22-Oxy Calcitriol Analogs of 20-Epi Calcitriol with very low Binding Affinity to the Vitamin D Receptor. (Norman, A. W., Bouillon, R., Thomasset, M., eds), pp85-86, Walter de Gruyter Berlin New York (1994).

844   **1α,22R,25-Trihydroxy-26,27-dimethyl-23,24-tetradehydro-24a,24b-dihomo-20-epi-vitamin D₃**

9547595 $C_{31}H_{48}O_4$

(2R,3R)-2-[(1R,3aR,4E,7aS)-4-[(2Z)-2-[(3S,5R)-3,5-dihydroxy-2-methylidene-cyclohexylidene]ethylidene]-7a-methyl-2,3,3a,5,6,7-hexahydro-1H-inden-1-yl]-8-ethyl-dec-4-yne-3,8-diol.
LMST03020491;   1α,22R,25-trihydroxy-26,27-dimethyl-23,24-tetradehydro-24a,24b-dihomo-20-epi-cholecalciferol;   (20R)-26,27-dimethyl-24a,24b-di-homo-9,10-seco-5Z,7E,10(19)-cholestatrien-23-yne-1S,3R,22R,25-tetrol;   5947595;   Synthesized from

protected (5E)-1α-hydroxy-22-oxo-23,24,25,26,27-pentanor-20-epivıtamın D_3. Compared to 1α,25-dihydroxyvitamin D_3, induction of differentiation of U937 cells is 500%, inhibition of proliferation is <10%, VDR (rachitic chicken intestinal receptor) binding affinity is <0.5%. Vitamin D. A Pluripotent Steroid Hormone: Structural Studıes, Molecular Endocrinology and Clinical Applications. Proceedings of the Ninth Workshop on Vıtamın D, Orlando, Florida (USA) May 28-June 2. Chemistry and Biology of Hıghly Active 22-Oxy Analogs of 20-Epı Calcitriol with very low Binding Affinity to the Vitamin D Receptor. (Norman, A. W., Bouillon, R., Thomasset, M., eds), pp85-86, Walter de Gruyter Berlin New York (1994).

845　1-Hydroxyvitamin D₃ diacetate
41461-12-5　　　　　　　　　　$C_{31}H_{48}O_4$

[(1S,3Z,5R)-3-[(2E)-2-[(3aR,7aS)-7a-methyl-1-[(2R)-6-methylheptan-2-yl]-2,3,3a,5,6,7-hexahydro-1H-inden-4-ylidene]ethylidene]-5-acetyloxy-4-methylidene-cyclohexyl] acetate.
1HV D(3) Diacetate; 1-Hydroxyvitamin D_3 diacetate; 1α-Hydroxyvitamin D_3 diacetate; 9,10-Secocholesta-5Z,7E,10(19)-triene-1α,3β-diol, diacetate, (1α,3β,5Z, 7E)-; 6441633; 41461-12-5; 132031-91-5; Formed by thermal isomerization at 60° of 1α-hydroxyprevitamin D_3 (1α-OH-previtamin D_3) diacetate in various solvents. *J Nutr Sci Vitaminol (Tokyo)*. 1990, *36(4)*, 299-309.

846　1-Hydroxyprevitamin D₃ diacetate
54712-17-3　　　　　　　　　　$C_{31}H_{48}O_4$

[(1S,5R)-3-[(Z)-2-[(3aR,7aS)-7a-methyl-1-[(2R)-6-methylheptan-2-yl]-1,2,3,3a,6,7-hexahydroinden-4-yl]ethenyl]-5-acetyloxy-4-methyl-1-cyclohex-3-enyl] acetate.
1-Hydroxyprevitamin D_3 diacetate; 1-HPV D(3) diacetate; 1α-Hydroxyprevitamin D_3 diacetate; 9,10-

Secocholesta-5(10),6Z,8-triene-1α,3β-diol, diacetate; 6441689; 54712-17-3; Heated at 60°, forms 24-Nor-25-hydroxyvitamin D_3. *J Nutr Sci Vitaminol (Tokyo)*. 1990, *36(4)*, 299-309.

847　1α,25-Dihydroxy-22-oxavitamin D₃ 3-hemiglutarate
9547596　　　　　　　　　　$C_{31}H_{48}O_7$

5-[(1R,3Z,5S)-3-[(2E)-2-[(1S,3aR,7aR)-1-[(1S)-1-(3-hydroxy-3-methyl-butoxy)ethyl]-7a-methyl-2,3,3a,5,6,7-hexahydro-1H-inden-4-ylidene]ethylidene]-5-hydroxy-4-methylidene-cyclohexyl]oxy-5-oxo-pentanoic acid.
LMST03020492; 22-oxa-9,10-seco-5Z,7E,10(19)-cholestatrıene-1S,3R,25-triol 3-hemiglutarate; 1α,25-dihydroxy-22-oxacholecalciferol 3-hemiglutarate; 5947596; Synthesized from (5Z,7E)-(1S,3R)-1,25-dihydroxy-22-oxa-9,10-secocholesta-5,7,10(19)-trıen-3-yl (2,2'2'-trichloroethylglutarate) by deprotection. Used in production of anti-22-oxacalcitriol antisera. *Chem Pharm Bull (Tokyo)*. 1992, *40(6)*, 1520-2.

848　22-Oxacalcitriol-3-hemiglutarate
143773-34-6　　　　　　　　　　$C_{31}H_{48}O_7$

5-[(1S,3E,5S)-3-[(2E)-2-[(3aS,7aS)-7a-methyl-2,3,3a,5,6,7-hexahydro-1H-inden-4-ylidene]ethylidene]-5-hydroxy-4-methylidene-cyclohexyl]oxy-5-oxo-pentanoic acid.
LMST03020492; Oct-3-hemiglutarate; Oct-3-HG; 22-Oxacalcitriol-3-hemiglutarate; 1S,25-dihydroxy-22-oxa-9,10-secocholesta-5Z,7E,10(19)-trien-3-yl

hemiglutarate; 6444275; 143773-34-6; 103309-75-7;
A haptenic derivative of 22-oxacalcitriol. Has been
used in a radioimmunoassay designed for the
pharmacokinetic study of 22-oxacalcitriol. **Chem
Pharm Bull (Tokyo)**. 1992, **40(6)**, 1520-2.

849   **Vitamin D$_3$ butyrate**
31316-20-8 $C_{31}H_{50}O_2$

[(1R,3Z)-3-[(2E)-2-[(3aR,7aS)-7a-methyl-1-[(2R)-6-
methylheptan-2-yl]-2,3,3a,5,6,7-hexahydro-1H-inden-
4-ylidene]ethylidene]-4-methylidene-cyclohexyl]
butanoate.
Vitamin D$_3$ butyrate; 9,10-Secocholesta-5Z,7E,10(19)-
trien-3β-yl butyrate; 9,10-Secocholesta-5Z,7E,10(19)-
trien-3β-ol, butanoate; EINECS 250-567-5; 6441537;
31316-20-8. **Int J Vitam Nutr Res.** 1981, **51(4)**, 353-8.

850   **(5E,10E)-19-(3-Carboxylpropyl)vitamin
       D$_3$**
9547598 $C_{31}H_{50}O_3$

(5E)-5-[(2E,4S)-2-[(2E)-2-[(1R,3aR,7aS)-7a-methyl-1-
[(2R)-6-methylheptan-2-yl]-2,3,3a,5,6,7-hexahydro-
1H-inden-4-ylidene]ethylidene]-4-hydroxy-
cyclohexylidene]pentanoic acid.
LMST03020494; (5E,10E)-19-(3-carboxylpropyl)chole-
calciferol; 19-(3-carboxylpropyl)-9,10-seco-5E,7E,10E
(19)-cholestatrien-3S-ol; 9547598; Synthesized from
vitamin D$_3$ **via** electrophilic substitution of its sulfur
dioxide adduct; λ_m = 264 nm (95% EtOH). **J Org
Chem**. 1983, **48(20)**, 3483-8.

851   **19-(3-Carboxylpropyl)-10E-vitamin D$_3$**
9547597 $C_{31}H_{50}O_3$

(5E)-5-[(2Z,4S)-2-[(2E)-2-[(1R,3aR,7aS)-7a-methyl-1-
[(2R)-6-methylheptan-2-yl]-2,3,3a,5,6,7-hexahydro-
1H-inden-4-ylidene]ethylidene]-4-hydroxy-cyclohex-
ylidene]pentanoic acid.
LMST03020493; (10E)-19-(3-carboxylpropyl)chole-
calciferol; 19-(3-carboxylpropyl)-9,10-seco-5Z,7E,10E
(19)-cholestatrien-3S-ol; 9547597; Synthesized from
vitamin D$_3$ by electrophilic substitution of its sulfur
dioxide adduct. **J Org Chem**. 1983, **48(20)**, 3483-8.

852   **22R-Ethyl-1α,25-dihydroxy-23,24-
       dehydro-24a,24b-dihomo-20-epivitamin D$_3$**
                                     $C_{31}H_{50}O_3$

(5Z,7E,23E)-(1S,3R,20S,22S)-22β-Ethyl-24a,24b-
dihomo-9,10-seco-5,7,10(19),23-cholestatetraene-
1,3,25-triol; Synthetic derivative of vitamin D$_3$. Is about
100 times more efficient than calcitriol in cell
differentiation. Affinity for the vitamin D receptor is
only 14% that of calcitriol; λ_m = 263 nm (95% EtOH).
J Med Chem. 2002, **45(9)**, 1825-34.

853   **22S-Ethyl-1α,25-dihydroxy-23,24-
       dehydro-24a,24b-dihomo-20-epivitamin D$_3$**
                                     $C_{31}H_{50}O_3$

(5Z,7E,23E)-(1S,3R,20S,22S)-22α-Ethyl-24a,24b-dihomo-9,10-seco-5,7,10(19),23-cholestatetraene-1,3,25-triol; Synthetic derivative of vitamin D_3. Is about as efficient as calcitriol in cell differentiation; λ_m = 263 nm (95% EtOH). *J Med Chem.* 2002, **45(9)**, 1825-34.

854   **26,27-Diethyl-1α,25-dihydroxy-20,21-methano-23-oxavitamin D_3**

9547599 $C_{31}H_{50}O_4$

(1R,3S,5Z)-5-[(2E)-2-[(1S,3aR,7aS)-1-[1-[(2-hydroxy-2-propyl-pentoxy)methyl]cyclopropyl]-7a-methyl-2,3,3a,5,6,7-hexahydro-1H-inden-4-ylidene]ethylidene]-4-methylidene-cyclohexane-1,3-diol. LMST03020495; 26,27-diethyl-20,21-methano-23-oxa-9,10-seco-5Z,7E,10(19)-cholestatriene-1S,3R,25-triol; 26,27-diethyl-1α,25-dihydroxy-20,21-methano-23-oxacholecalciferol; 9547599; Compared to 1α,25-dihydroxyvitamin D_3, affinity for pig intestinal nuclear receptor is 50%, affinity for human vitamin D binding protein is <0.02%, differentiation of HL-60 cells is 12.5%, and calciuric effect is 0.3%. Vitamin D. A Pluripotent Steroid Hormone: Structural Studies, Molecular Endocrinology and Clinical Applications. Proceedings of the Ninth Workshop on Vitamin D, Orlando, Florida (USA) May 28-June 2. Synthesis and Biological Activity of 23-Oxa Vitamin D Analogues. (Norman, A. W., Bouillon, R., Thomasset, M., eds), pp93-94, Walter de Gruyter Berlin New York (1994).

855   **1α-(2-Fluoroethyl)-25-hydroxy-26,27-dimethyl-24a-homo-22-oxavitamin D_3**

9547667 $C_{31}H_{51}FO_3$

6-[(1S)-1-[(1S,3aR,4E,7aR)-4-[(2Z)-2-[(3R,5S)-3-(2-fluoroethyl)-5-hydroxy-2-methylidene-cyclohexylidene]ethylidene]-7a-methyl-2,3,3a,5,6,7-hexahydro-1H-inden-1-yl]ethoxy]-3-ethyl-hexan-3-ol. LMST03020572;   1R-(2-fluoroethyl)-26,27-dimethyl-24a-homo-22-oxa-9,10-seco-5Z,7E,10(19)-cholesta-triene-3S,25-diol; 1α-(2-fluoroethyl)-25-hydroxy-26,27-dimethyl-24a-homo-22-oxacholecalciferol; 9547667.

856   **1β-(2-Fluoroethyl)-25-hydroxy-26,27-dimethyl-24a-homo-22-oxa-3-epivitamin D_3**

9547668 $C_{31}H_{51}FO_3$

6-[(1S)-1-[(1S,3aR,4E,7aR)-4-[(2Z)-2-[(3S,5S)-3-(2-fluoroethyl)-5-hydroxy-2-methylidene-cyclohexylidene]ethylidene]-7a-methyl-2,3,3a,5,6,7-hexahydro-1H-inden-1-yl]ethoxy]-3-ethyl-hexan-3-ol. LMST03020573;   1S-(2-fluoroethyl)-26,27-dimethyl-24a-homo-22-oxa-9,10-seco-5Z,7E,10(19)-cholesta-triene-3R,25-diol;   1β-(2-fluoroethyl)-25-hydroxy-26,27-dimethyl-24a-homo-22-oxa-3-epicholecalciferol; 9547668.

857   **1β-Butyl-1α,25-dihydroxyvitamin D_3**

9547600 $C_{31}H_{52}O_3$

(1S,3R,5Z)-5-[(2E)-2-[(1R,3aR,7aS)-1-[(2R)-6-hydroxy-6-methyl-heptan-2-yl]-7a-methyl-2,3,3a,5,6,7-hexahydro-1H-inden-4-ylidene]ethylidene]-1-butyl-6-methylidene-cyclohexane-1,3-diol.
LMST03020496;   1S-butyl-9,10-seco-5Z,7E,10(19)-cholestatriene-1S,3R,25-triol;   1β-butyl-1α,25-dihydroxycholecalciferol; 9547600; Synthesized from 1α,25-dihydroxyvitamin D$_3$ by oxidation of the 1-hydroxyl group, treatment with BuLi and thermal isomerization (both C(1) epimers are formed). Binding affinity for calf thymus vitamin D receptor is 1/1000 that of 1α,25-dihydroxyvitamin D$_3$; λ$_m$ = 262, 252 nm (95% EtOH). *J Org Chem*. 1993, *58*(7), 1895-9.

858   *1α-Butyl-1β,25-dihydroxyvitamin D$_3$*
9547601   C$_{31}$H$_{52}$O$_3$

(1R,3R,5Z)-5-[(2E)-2-[(1R,3aR,7aS)-1-[(2R)-6-hydroxy-6-methyl-heptan-2-yl]-7a-methyl-2,3,3a,5,6,7-hexa-hydro-1H-inden-4-ylidene]ethylidene]-1-butyl-6-methylidene-cyclohexane-1,3-diol.
LMST03020497;   1R-butyl-9,10-seco-5Z,7E,10(19)-cholestatriene-1R,3R,25-triol;   1α-butyl-1β,25-dihydroxycholecalciferol; 9547601; Synthesized from 1α,25-dihydroxyvitamin D$_3$ by oxidation of the 1-hydroxyl group, treatment with BuLi and thermal isomerization (both C(1) epimers are formed). Binding affinity for calf thymus vitamin D·receptor is 1/1000 that of 1α,25-dihydroxyvitamin D$_3$; λ$_m$ = 263 nm (95% EtOH). *J Org Chem*. 1993, *58(7)*, 1895-9.

859   *1α,25-Dihydroxy-26,27-diethylvitamin D$_3$*
106372-51-4   C$_{31}$H$_{52}$O$_3$

(1R,3S,5Z)-5-[(2E)-2-[(1R,3aR,7aS)-1-[(2R)-6-hydroxy-6-propyl-nonan-2-yl]-7a-methyl-2,3,3a,5,6,7-hexahydro-1H-inden-4-ylidene]ethylidene]-4-methylidene-cyclohexane-1,3-diol.
LMST03020498;   26,27-diethyl-9,10-seco-5Z,7E,10 (19)-cholestatriene-1S,3R,25-triol; 26,27-diethyl-1α,25-dihydroxycholecalciferol; 1,25-Ddpv-D$_3$; 1,25-Di-hydroxy-26.27-dipropylcholecalciferol; 1,25-Di-hydroxy-26,27-diethylcholecalciferol; 26,27-diethyl-1,25-dihydroxyvitamin D$_3$; 26,27-DDHVD3; 26,27-diethyl-1α,25-dihydroxycholecalciferol; 6439071; 9547602; 106372-51-4; Synthesized from cholenic acid. The propyl side chains greatly reduce the monocytic differentiation-inducing activity compared to that of 1α,25-dihydroxyvitamin D$_3$. Has reduced binding affinity to vitamin D receptor protein. Modulates monocytic differentiation of promyelocytic leukemia (HL-60) cells. Compared to 1α,25-dihydroxyvitamin D$_3$, inhibition of proliferation of HL-60 cells is >1000%, binding affinity for serum vitamin D binding protein and for the receptor of HL-60 cells is <1% and 21%, ability to induce HL-60 cell differentiation (assayed by NBT reduction) is approximately 100% under serum-supplemented culture conditions, but is <100% under serum-free culture conditions; λ$_m$ = 265 nm. *Chem Pharm Bull (Tokyo)*. 1987, *35(10)*, 4362-5; *Steroids*. 1991, *56(3)*, 142-7.

860   *1α,25-Dihydroxy-26,27-dimethyl-24a, 24b-dihomovitamin D$_3$*
9547603   C$_{31}$H$_{52}$O$_3$

(1R,3S,5Z)-5-[(2E)-2-[(1R,3aR,7aS)-1-[(2R)-8-ethyl-8-hydroxy-decan-2-yl]-7a-methyl-2,3,3a,5,6,7-hexahydro-1H-inden-4-ylidene]ethylidene]-4-methylidene-cyclohexane-1,3-diol.
LMST03020499;   26,27-dimethyl-24a,24b-dihomo-9,10-seco-5Z,7E,10(19)-cholestatrien-1S,3R,25-triol;

1α,25-dihydroxy-26,27-dimethyl-24a,24b-dihomo-cholecalciferol; 9547603; Prepared from hydroxy-protected (5E)-23,24,25,26,27-pentanorvitamin D_3 22-bis(methylseleno)acetal. Compared to 1α,25-di-hydroxyvitamin D_3, inhibition of proliferation of U937 cells is 40%, induction of differentiation of U937 cells is 10%, and calciuric activity is 2%. Gene Regulation, Structure-Function Analysis and Clinical Application. Proceedings of the Eighth Workshop on Vitamin D Paris, France July 5-10. Synthesis and Biological Activity of 1α-Hydroxylated Vitamin D_3 Analogues with Hydroxylated Side Chains, Multi-Homologated in The 24- or 24,26,27-Positions. (Norman, A. W., Bouillon, R., Thomasset, M., eds), pp159-160, Walter de Gruyter Berlin (1991).

861 *1α,25-Dihydroxy-2β-butoxyvitamin D_3*

9547613 $C_{31}H_{52}O_4$

(1R,2R,3R,5Z)-5-[(2E)-2-[(1R,3aR,7aS)-1-[(2R)-6-hydroxy-6-methyl-heptan-2-yl]-7a-methyl-2,3,3a,5,6,7-hexahydro-1H-inden-4-ylidene]ethylidene]-2-butoxy-4-methylidene-cyclohexane-1,3-diol. LMST03020509; 2R-butoxy-9,10-seco-5Z,7E,10(19)-cholestatriene-1R,3R,25-triol; 1α,25-dihydroxy-2β-butoxyvitamin D_3; 1α,25-dihydroxy-2β-butoxycholecalciferol; 9547613; Synthesized photochemically from the 5,7-diene. Spinal bone mineral density in pre-osteoporosis model rats and binding affinity to calf thymus vitamin D receptor reported; λ_m = 264 nm (EtOH). *Chem Pharm Bull (Tokyo)*. 1997, *45(10)*, 1626-30.

862 *1α,25-Dihydroxy-2β-(4-hydroxybutyl)-vitamin D_3*

9547605 $C_{31}H_{52}O_4$

(1R,2R,3S,5Z)-5-[(2E)-2-[(1R,3aR,7aS)-1-[(2R)-6-hydroxy-6-methyl-heptan-2-yl]-7a-methyl-2,3,3a,5,6,7-hexahydro-1H-inden-4-ylidene]ethylidene]-2-(4-hydroxybutyl)-4-methylidene-cyclohexane-1,3-diol. LMST03020501; 2R-(4-hydroxybutyl)-9,10-seco-5Z,7E,10(19)-cholestatriene-1S,3R,25-triol; 1α,25-dihydroxy-2β-(4-hydroxybutyl)cholecalciferol; 9547605; Synthesized photochemically from the 5,7-diene. Spinal bone mineral density in pre-osteoporosis model rats and binding affinity to calf thymus vitamin D receptor reported; λ_m = 264 nm (EtOH). *Chem Pharm Bull (Tokyo)*. 1997, *45(10)*, 1626-30.

863 *1α,25-Dihydroxy-20S-methoxy-26,27-di-methyl-24a-homovitamin D_3*

9547606 $C_{31}H_{52}O_4$

(1R,3S,5Z)-5-[(2E)-2-[(1S,3aR,7aR)-1-[(2S)-7-ethyl-7-hydroxy-2-methoxy-nonan-2-yl]-7a-methyl-2,3,3a,5,6,7-hexahydro-1H-inden-4-ylidene]ethylidene]-4-methylidene-cyclohexane-1,3-diol. LMST03020502; 1α,25-dihydroxy-20S-methoxy-26,27-dimethyl-24a-homocholecalciferol; 20S-methoxy-26,27-dimethyl-24a-homo-9,10-seco-5Z,7E,10(19)-cholestatriene-1S,3R,25-triol; 9547606; Synthesized from a (5E)-1α-hydroxy-20-keto-22,23,24,25,26,27-hexanorvitamin D_3 derivative. Compared to 1α,25-dihydroxyvitamin D_3, inhibition of U937 cell (human histiocytic lymphoma cell line) proliferation is 30000%, binding to the vitamin D receptor from rachitic chicken intestine is 1%, calcemic activity determined by the increase in urinary calcium excretion in rats is 6%. Vitamin D. A Pluripotent Steroid Hormone: Structural Studies, Molecular Endocrinology and Clinical Applications. Proceedings of the Ninth Workshop on Vitamin D, Orlando, Florida

(USA) May 28-June 2. Synthesis and Biological Activity of 20-Hydroxylated Vitamin D Analogues. (Norman, A. W., Bouillon, R., Thomasset, M., eds), pp95-96, Walter de Gruyter Berlin New York (1994).

864   *1α,25-Dihydroxy-20S-ethoxy-26,27-di-methylvitamin D₃*

9547609   C₃₁H₅₂O₄

(1R,3S,5Z)-5-[(2E)-2-[(1S,3aR,7aR)-1-[(2S)-2-ethoxy-6-ethyl-6-hydroxy-octan-2-yl]-7a-methyl-2,3,3a,5,6,7-hexahydro-1H-inden-4-ylidene]ethylidene]-4-methylidene-cyclohexane-1,3-diol.
LMST03020505; 1α,25-dihydroxy-20-ethoxy-26,27-di-methylcholecalciferol; 20S-ethoxy-26,27-dimethyl-9,10-seco-5Z,7E,10(19)-cholestatriene-1S,3R,25-triol; 9547609; Synthesized from a (5E)-1α-hydroxy-20-keto-22,23,24,25,26,27-hexanorvitamin D₃ derivative.

Compared to 1α,25-dihydroxyvitamin D₃, inhibition of U937 cell (human histiocytic lymphoma cell line) proliferation is 500000%, binding to the vitamin D receptor from rachitic chicken intestine is 7%, and calcemic activity determined by the increase in urinary calcium excretion in rats is 14%. Vitamin D. A Pluripotent Steroid Hormone: Structural Studies, Molecular Endocrinology and Clinical Applications. Proceedings of the Ninth Workshop on Vitamin D, Orlando, Florida (USA) May 28-June 2. Synthesis and Biological Activity of 20-Hydroxylated Vitamin D Analogues. (Norman, A. W., Bouillon, R., Thomasset, M., eds), pp95-96, Walter de Gruyter Berlin New York (1994).

865   *1α,25-Dihydroxy-26,27-diethyl-24a-homo-22-oxa-20-epivitamin D₃*

9547607   C₃₁H₅₂O₄

(1R,3S,5Z)-5-[(2E)-2-[(1S,3aR,7aR)-1-[(1R)-1-(4-hydroxy-4-propyl-heptoxy)ethyl]-7a-methyl-2,3,3a,5,6,7-hexahydro-1H-inden-4-ylidene]ethylidene]-4-methylidene-cyclohexane-1,3-diol.
LMST03020503; (20R)-26,27-diethyl-24a-homo-22-oxa-9,10-seco-5Z,7E,10(19)-cholestatriene-1S,3R,25-triol; 26,27-diethyl-1α,25-dihydroxy-24a-homo-22-oxa-20-epicholecalciferol; 9547607.

866   *1α-Hydroxy-2β-(4-hydroxybutoxy)-vitamin D₃*

9547604   C₃₁H₅₂O₄

(1R,2R,3R,5Z)-5-[(2E)-2-[(1R,3aR,7aS)-7a-methyl-1-[(2R)-6-methylheptan-2-yl]-2,3,3a,5,6,7-hexahydro-1H-inden-4-ylidene]ethylidene]-2-(4-hydroxybutoxy)-4-methylidene-cyclohexane-1,3-diol.
LMST03020500; 2R-(4-hydroxybutoxy)-9,10-seco-5Z,7E,10(19)-cholestatriene-1R,3R-diol; 1α-hydroxy-2β-(4-hydroxybutoxy)cholecalciferol; 9547604; Synthesized photochemically from the 5,7-diene; λₘ = 263.5 nm (EtOH); *Chem Pharm Bull (Tokyo)*. 1993, *41(6)*, 1111-13.

867   *1α,22S,25-Trihydroxy-26,27-dimethyl-24a,24b-dihomovitamin D₃*

9547608   C₃₁H₅₂O₄

(2S,3S)-2-[(1R,3aR,4E,7aS)-4-[(2Z)-2-[(3S,5R)-3,5-dihydroxy-2-methylidene-cyclohexylidene]ethylidene]-7a-methyl-2,3,3a,5,6,7-hexahydro-1H-inden-1-yl]-8-ethyl-decane-3,8-diol.
LMST03020504; 1α,22S,25-trihydroxy-26,27-dimethyl-24a,24b-dihomocholecalciferol; 26,27-dimethyl-24a,24b-dihomo-9,10-seco-5Z,7E,10(19)-

cholestatriene-1S,3R,22S,25-tetrol; 9547608.

868 1α,22R,25-Trihydroxy-26,27-dimethyl-24a,24b-dihomo-20-epivitamin D₃

9547611 $C_{31}H_{52}O_4$

(2R,3R)-2-[(1R,3aR,4E,7aS)-4-[(2Z)-2-[(3S,5R)-3,5-dihydroxy-2-methylidene-cyclohexylidene]ethylidene]-7a-methyl-2,3,3a,5,6,7-hexahydro-1H-inden-1-yl]-8-ethyl-decane-3,8-diol.
LMST03020507; 1α,22R,25-trihydroxy-26,27-dimethyl-24a,24b-dihomo-20-epicholecalciferol; (20R)-26,27-dimethyl-24a,24b-dihomo-9,10-seco-5Z,7E,10(19)-cholestatriene-1S,3R,22R,25-tetrol; 9547611; Synthesized from a (5E)-1α-hydroxy-20-keto-22,23,24,25,26,27-hexanorvitamin D₃ derivative. Compared to 1α,25-dihydroxyvitamin D₃, inhibition of differentiation of U937 cell (human histiocytic lymphoma cell line) is 100%, inhibition of proliferation is 210%, binding to the vitamin D receptor from rachitic chicken intestine is <0.005%. Vitamin D. A Pluripotent Steroid Hormone: Structural Studies, Molecular Endocrinology and Clinical Applications. Proceedings of the Ninth Workshop on Vitamin D, Orlando, Florida (USA) May 28-June 2. Chemistry and Biology of Highly Active 22-Oxy Analogs of 20-Epi Calcitriol with very low Binding Affinity to the Vitamin D Receptor. (Norman, A. W., Bouillon, R., Thomasset, M., eds), pp85-86, Walter de Gruyter Berlin New York (1994).

869 1α,20S,25-Trihydroxy-26,27-diethyl-vitamin D₃

9547610 $C_{31}H_{52}O_4$

(2S)-2-[(1S,3aR,4E,7aR)-4-[(2Z)-2-[(3S,5R)-3,5-di-hydroxy-2-methylidene-cyclohexylidene]ethylidene]-7a-methyl-2,3,3a,5,6,7-hexahydro-1H-inden-1-yl]-6-propyl-nonane-2,6-diol.
LMST03020506; 1α,20S,25-trihydroxy-26,27-

diethylcholecalciferol; 26,27-diethyl-9,10-seco-5Z,7E,10(19)-cholestatriene-1S,3R,20S,25-tetrol; 9547610; Synthesized from a (5E)-1α-hydroxy-20-keto-22,23,24,25,26,27-hexanorvitamin D₃ derivative. Compared to 1α,25-dihydroxyvitamin D₃, inhibition of U937 cell (human histiocytic lymphoma cell line) proliferation is < 10%, binding to the vitamin D receptor from rachitic chicken intestine is 0.5%, calcemic activity was not determined. Vitamin D. A Pluripotent Steroid Hormone: Structural Studies, Molecular Endocrinology and Clinical Applications. Proceedings of the Ninth Workshop on Vitamin D, Orlando, Florida (USA) May 28-June 2. Synthesis and Biological Activity of 20-Hydroxylated Vitamin D Analogues. (Norman, A. W., Bouillon, R., Thomasset, M., eds), pp95-96, Walter de Gruyter Berlin New York (1994).

870 1α,25-Dihydroxy-2β-(4-hydroxybutoxy)-vitamin D₃

9547612 $C_{31}H_{52}O_5$

(1R,2R,3R,5Z)-5-[(2E)-2-[(1R,3aR,7aS)-1-[(2R)-6-hydroxy-6-methyl-heptan-2-yl]-7a-methyl-2,3,3a,5,6,7-hexahydro-1H-inden-4-ylidene]ethylidene]-2-(4-hydroxybutoxy)-4-methylidene-cyclohexane-1,3-diol.
LMST03020508; 2R-(4-hydroxybutoxy)-9,10-seco-5Z,7E,10(19)-cholestatriene-1R,3R,25-triol; 1α,25-dihydroxy-2β-(4-hydroxybutoxy)cholecalciferol; 9547612; Synthesized photochemically from the 5,7-diene. Spinal bone mineral density in pre-osteoporosis model rats and binding affinity to calf thymus vitamin D receptor reported; λ_m = 264 nm (EtOH). **Chem Pharm Bull (Tokyo)**. 1997, **45(10)**, 1626-30.

871 1α,25-Dihydroxy-2-[3'-(methoxymethoxy)propylidene]-19-norvitamin D₃

$C_{31}H_{52}O_5$

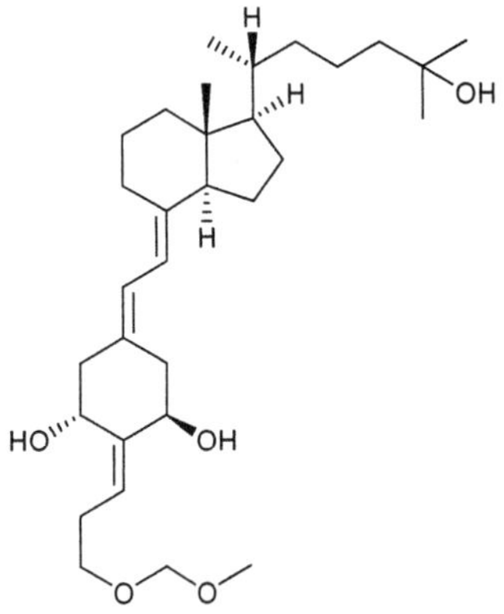

Synthetic analog of calcitriol. Has higher calcemic activity than calcitriol. *J Med Chem.* 2006, *49(10)*, 2909-20.

872 *21-(3-Methyl-3-hydroxybutyl)-19-nor-vitamin D$_3$*

$C_{31}H_{54}O_4$

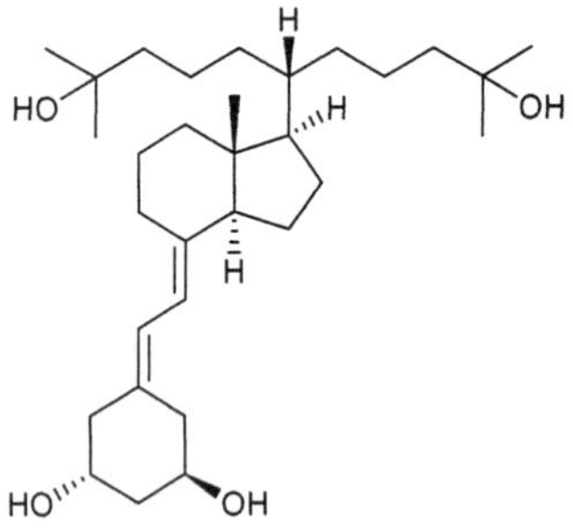

A vitamin D agonist. *Mol Pharmacol.* 2003, *63(6)*, 1230-7.

873 *(20R)-24-Hydroxy-19-norgeminivitamin D$_3$*

$C_{31}H_{54}O_5$

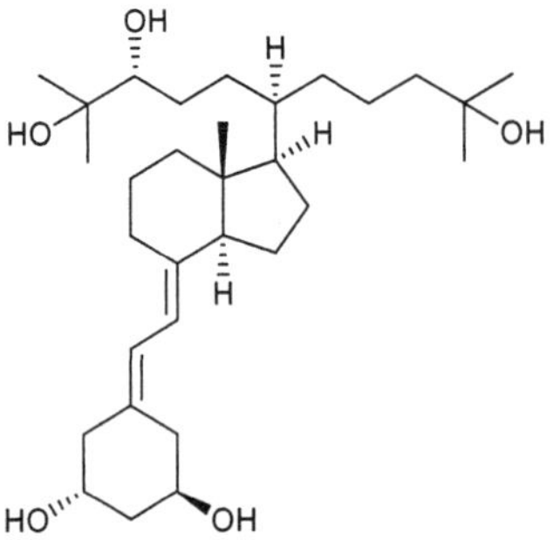

(3R,6R)-6-{(1R,3aS,7aR)-4-[2-((R)-3-(R)-Hydroxy-5-hydroxy-cyclohexylidene)-(E)-ethylidene]-7a-methyloctahydroinden-1-yl}-2,10-dimethylundecane-2,3,10-triol.
24-hydroxy-19-norgeminicalcitriol; Synthetic vitamin

D$_3$ analog. *J Med Chem.* 2004, *47(26)*, 6476-84. **Hoffmann-LaRoche Inc., Hoffmann-LaRoche Ltd.**

874 *(20S)-24-Hydroxy-19-norgeminivitamin D$_3$*

$C_{31}H_{54}O_5$

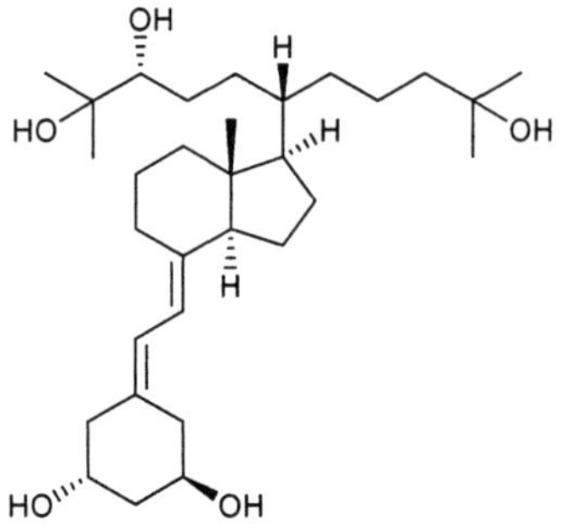

(3R,6S)-6-{(1R,3aS,7aR)-4(E)-[2-((R)-3-(S)-Hydroxy-5-hydroxy-2-methylene-cyclohexylidene)-E)-ethylidene]-7a-methyloctahydroinden-1-yl}-2,10-dimethylundec-ane-2,3,10-triol.
24-hydroxy-19-nor-20-epi-geminicalcitriol; Synthetic vitamin D$_3$ analog. *J Med Chem.* 2004, *47(26)*, 6476-84.

875 *1α-Hydroxy-23-[3-(1-hydroxy-1-methyl-ethyl)phenyl]-22,22,23,23-tetradehydro-24,25,26,27-tetranorvitamin D$_3$*
9547614

$C_{32}H_{42}O_3$

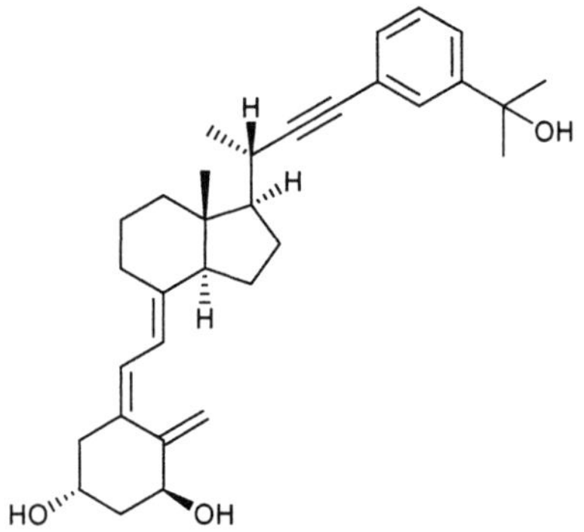

(1R,3S,5Z)-5-[(2E)-2-[(1R,3aR,7aS)-1-[(2S)-4-[3-(2-hydroxypropan-2-yl)phenyl]but-3-yn-2-yl]-7a-methyl-2,3,3a,5,6,7-hexahydro-1H-inden-4-ylidene]ethylidene]-4-methylidene-cyclohexane-1,3-diol.
LMST03020510; 23-[3-(1-hydroxy-1-methylethyl)-phenyl]-24-nor-9,10-seco-5Z,7E,10(19)-cholatrien-22-yne-1S,3R-diol; 1α-hydroxy-23-[3-(1-hydroxy-1-methylethyl)phenyl]-22,22,23,23-tetradehydro-24,25,26,27-tetranorcholecalciferol; 9547614.

876 *1,25-Dihydroxy-20S-21-(3-hydroxy-3-methylbutyl)-23-yne-26,27-hexafluorovitamin D$_3$*

$C_{32}H_{44}F_6O_4$

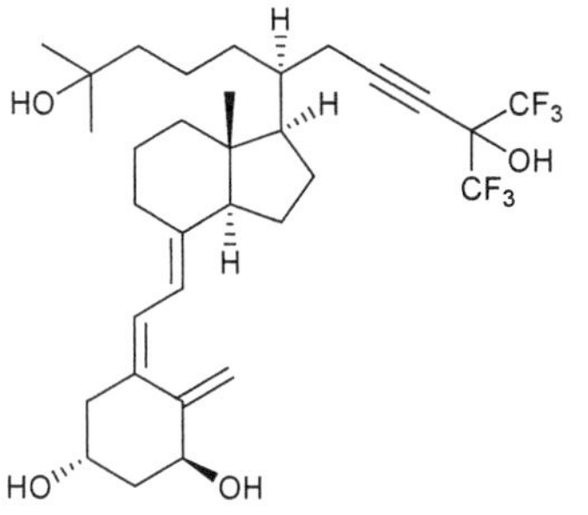

Gemini-23-yne-26,27-hexafluoro-D$_3$; Ro 43 83586; USP 6030962; Highly potent in inhibiting clonal growth of HL-60, MCF-7 and LNCaP tumor cells. *J Steroid Biochem Mol Biol*. 2006, *100(4-5)*, 107-116; *Mol Pharm*. 2003, *63(6)*, 1230-7. **Hoffmann-LaRoche Inc.**

877   1,25-Dihydroxy-21-(3-hydroxy-3-methyl-butyl)-23-yne-26,27-hexafluoro-vitamin D$_3$

C$_{32}$H$_{44}$F$_6$O$_4$

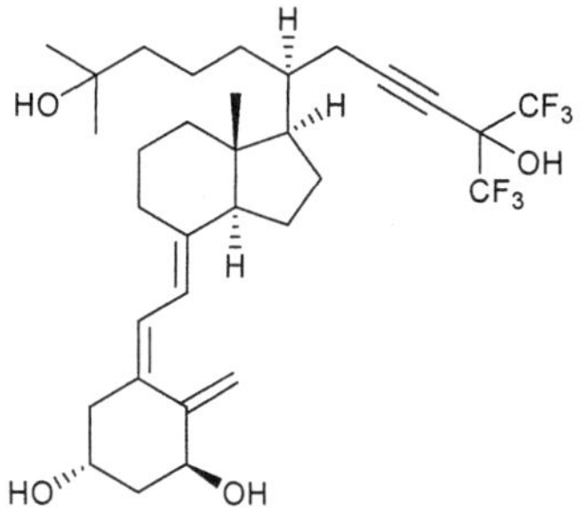

21-(3-hydroxy-3-methylbutyl)-23-yne-26,27-hexafluoro-1,25-dihydroxyvitamin D$_3$; Ro 43-83582; 21-(3-OH-methylbutyl)-23-yne-26,26-F6-1α,25(OH)-2D$_3$; Synthetic analog of vitamin D$_3$. Binds to the vitamin D receptor in the same expandable site used by Vitamin D$_3$. May prevent progression of early stage breast cancer. *J Biol Chem*. 2006, *281(15)*, 10516-26; *Mol Pharmacol*. 2003, *63(6)*, 1230-7.

878   1α,25-Dihydroxy-20S-phenyl-16,17-di-dehydro-21-norvitamin D$_3$

9547677

C$_{32}$H$_{44}$O$_3$

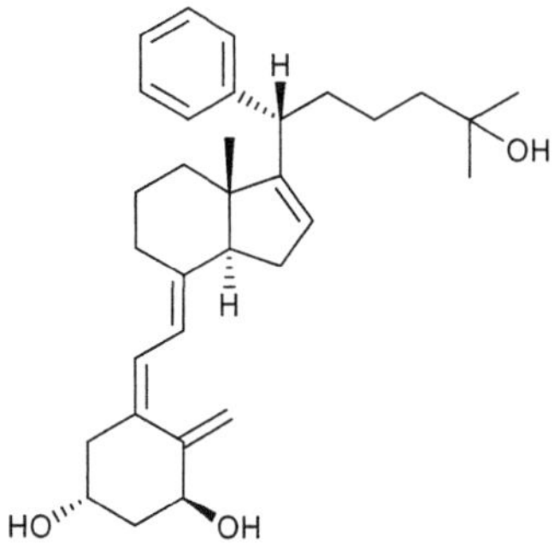

(1R,3S,5Z)-5-[(2E)-2-[(3aR,7aS)-1-[(1R)-5-hydroxy-5-methyl-1-phenyl-hexyl]-7a-methyl-3a,5,6,7-tetrahydro-3H-inden-4-ylidene]ethylidene]-4-methylidene-cyclo-hexane-1,3-diol.
LMST03020582; 1α,25-dihydroxy-20S-phenyl-16,17-didehydro-21-norcholecalciferol; 20S-phenyl-21-nor-9,10-seco-5Z,7E,10(19),16-cholestatetraene-1S,3R,25-triol; 9547677.

879   1α,25-Dihydroxy-20R-phenyl-16,17-di-dehydro-21-norvitamin D$_3$

9547680

C$_{32}$H$_{44}$O$_3$

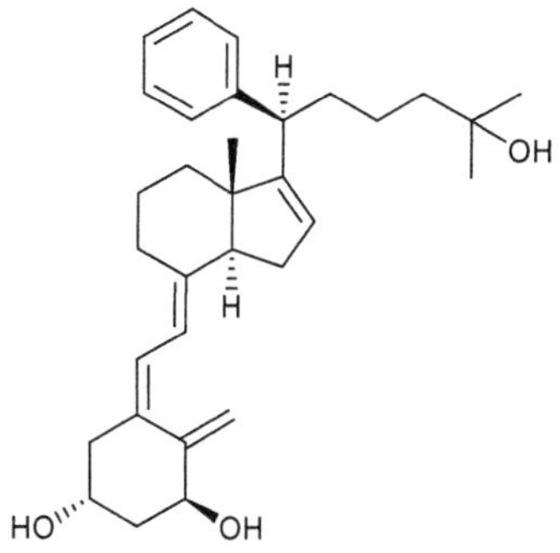

(1R,3S,5Z)-5-[(2E)-2-[(3aR,7aS)-1-[(1S)-5-hydroxy-5-methyl-1-phenyl-hexyl]-7a-methyl-3a,5,6,7-tetrahydro-3H-inden-4-ylidene]ethylidene]-4-methylidene-cyclo-hexane-1,3-diol.
LMST03020585; 1α,25-dihydroxy-20R-phenyl-16,17-didehydro-21-norcholecalciferol; 20R-phenyl-21-nor-9,10-seco-5Z,7E,10(19),16-cholestatetraene-1S,3R,25-triol; 9547680.

880   1α-Hydroxy-18-[m-(1-hydroxy-1-methyl-ethyl)-benzyloxy]-23,24,25,26,27-pentanor-vitamin D$_3$

9547684

C$_{32}$H$_{46}$O$_4$

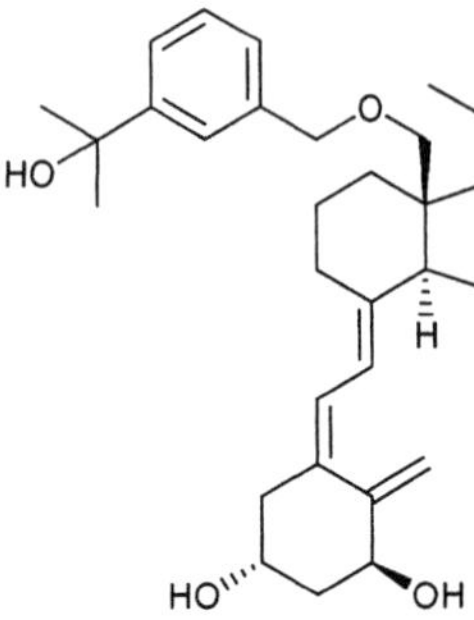

(1R,3S,5Z)-5-[(2E)-2-[(1R,3aR,7aR)-7a-[[3-(2-hydroxypropan-2-yl)phenyl]methoxymethyl]-1-propan-2-yl-2,3,3a,5,6,7-hexahydro-1H-inden-4-ylidene] ethylidene]-4-methylidene-cyclohexane-1,3-diol. LMST03020589; 18-[m-(1-hydroxy-1-methylethyl)-benzyloxy]-23,24-dınor-9,10-seco-5Z,7E,10(19)-cholatriene-1S,3R-diol; 1α-hydroxy-18-[m-(1-hydroxy-1-methylethyl)-benzyloxy]-23,24,25,26,27-pentanorcholecalciferol; 9547684.

881   *Atocalcitol*

302904-82-1 $C_{32}H_{46}O_4$

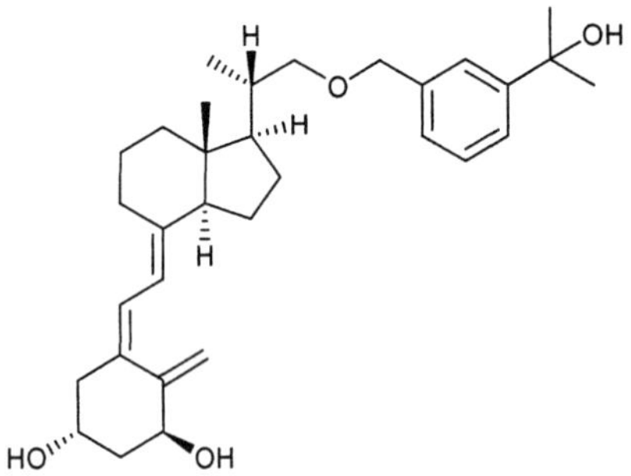

(1R,3S,5Z)-5-[(2E)-2-[(1R,3aR,7aS)-1-[(2R)-1-[[3-(2-hydroxypropan-2-yl)phenyl]methoxy]propan-2-yl]-7a-methyl-2,3,3a,5,6,7-hexahydro-1H-inden-4-ylidene] ethylidene]-4-methylidene-cyclohexane-1,3-diol. Atocalcitol; Atocalcitolum; Atokalcitol; (1S,3R,5Z,7E,20R)-20-(3-(2-Hydroxypropan-2-yl)benzyloxymethyl)-9,10-seco-pregna-5,7,10(19)-triene-1α,3β-diol; 6445225; 302904-82-1.

882   *11-(3-Acetoxy-1-propynyl)-1α,25-di-hydroxy-9,11-didehydrovitamin D_3*

9547615 $C_{32}H_{46}O_5$

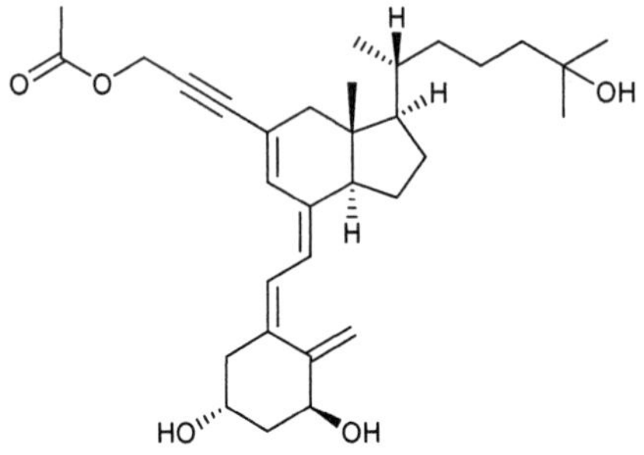

3-[(3R,3aS,7Z,7aR)-7-[(2Z)-2-[(3S,5R)-3,5-dihydroxy-2-methylidene-cyclohexylidene]ethylidene]-3-[(2R)-6-hydroxy-6-methyl-heptan-2-yl]-3a-methyl-2,3,4,7a-tetrahydro-1H-inden-5-yl]prop-2-ynyl acetate. LMST03020511; 11-(3-acetoxy-1-propynyl)-9,10-seco-5Z,7E,9(11),10(19)-cholestatetraene-1S,3R,25-triol; 11-(3-acetoxy-1-propynyl)-1α,25-dihydroxy-9,11-didehydrocholecalciferol; 9547615.

883   *ZK 168281*

$C_{32}H_{46}O_5$

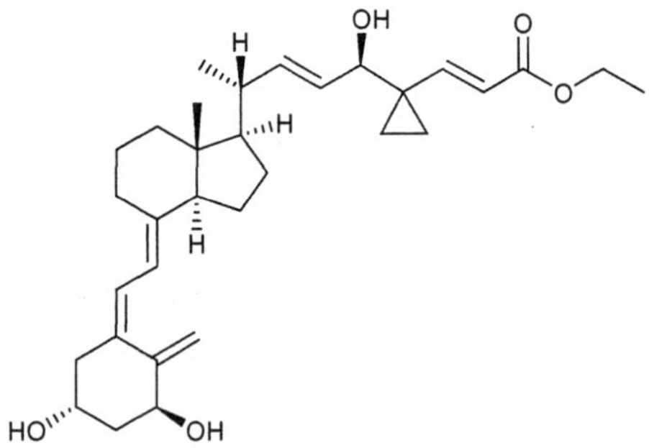

ZK 168281; Full antagonist of 1,25-dihydroxyvitamin D_3. *Chem Biol.* 2000, *7(11)*, 885-94; *J Med Chem.* 2004, *47(8)*, 1956-61; *Mol Pharm.* 2000, *58(5)*, 1067-74. *Schering AG.*

884   *17α-24-cyclopropyl-24-carbobutoxy-9, 10-seco-chola-5Z,7E,10,22E-tetraene-1S,3-diol-23-one*

$C_{32}H_{46}O_5$

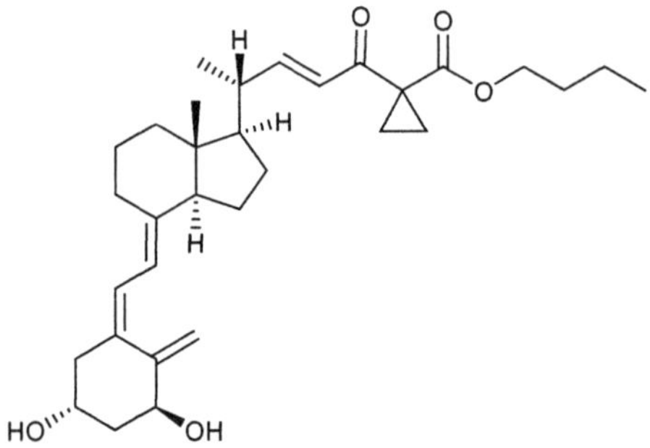

17α-24-cyclopropyl-24-carbobutoxy-9,10-seco-chola-5Z,7E,10,22E-tetraene-1S,3-dıol-23-one; Shows little or no antagonism towards 1α,25-dihydroxyvitamin D_3; *Mol Pharmacol.* 2000, *58(5)*, 1067-74. *Schering AG.*

885   *1α,25-Dihydroxy-26,27-dimethyl-22E, 23,24E,24a,24bE,24c-hexadehydro-24a,24b, 24c-trihomovitamin D_3*

9547616 $C_{32}H_{48}O_3$

(1R,3S,5Z)-5-[(2E)-2-[(1R,3aR,7aS)-1-[(2R,3E,5E,7E)-9-ethyl-9-hydroxy-undeca-3,5,7-trien-2-yl]-7a-methyl-2,3,3a,5,6,7-hexahydro-1H-inden-4-ylidene]ethylidene]-4-methylidene-cyclohexane-1,3-diol. LMST03020512; 1α,25-dihydroxy-26,27-dimethyl-22E,23,24E,24a,24bE,24c-hexadehydro-24a,24b,24c-trihomocholecalciferol; 26,27-dimethyl-24a,24b,24c-trihomo-9,10-seco-5Z,7E,10(19),22E,24E,24bE-cholestahexaene-1S,3R,25-triol; 9547616; Synthesized from protected (5E)-1α-hydroxy-22-oxo-23,24,25,26,27-pentanorvitamin D$_3$. Compared to 1α,25-dihydroxyvitamin D$_3$, inhibition of proliferation of U937 cells is < 80%, and induction of differentiation of U937 cells is < 5%. Gene Regulation, Structure-Function Analysis and Clinical Application. Proceedings of the Eighth Workshop on Vitamin D Paris, France July 5-10. Synthesis and Biological Activity of 1α-Hydroxylated Vitamin D$_3$ Analogues with Hydroxylated Side Chains, Multi-Homologated in The 24- or 24,26,27-Positions. (Norman, A. W., Bouillon, R., Thomasset, M., eds), pp159-160, Walter de Gruyter Berlin (1991).

886   ***ZK 159222***

$C_{32}H_{48}O_5$

Butyl-(5Z,7E,22E)-(1S,3R,24R)-1,3,24-trihydroxy-26,27-cyclo-9,10-secocholesta-5,7,10(19),22-tetraene-25-carboxylate.
ZK 159222; 17α-24-cyclopropyl-24-carbobutoxy-9,10-seco-chola-5Z,7E,10,22E-tetraene-1S,3,24-triol; Partial antagonist of 1,25-dihydroxyvitamin D$_3$. ***J Biol Chem***. 2000, ***275(22)***, 16506-12; ***J Med Chem***. 2004, ***47(8)***, 1956-61; ***Mol Pharm***. 2000, ***58(5)***, 1067-74. ***Schering AG***.

887   **1α,25-Dihydroxy-26,27-diethyl-22E,23,24E,24a-tetradehydro-24a-homovitamin D$_3$**
9547617 $C_{32}H_{50}O_3$

(1R,3S,5Z)-5-[(2E)-2-[(1R,3aR,7aS)-1-[(2S,3E,5E)-7-hydroxy-7-propyl-deca-3,5-dien-2-yl]-7a-methyl-2,3,3a,5,6,7-hexahydro-1H-inden-4-ylidene]ethylidene]-4-methylidene-cyclohexane-1,3-diol. LMST03020513; 1α,25-dihydroxy-26,27-diethyl-22E,23,24E,24a-tetradehydro-24a-homocholecalciferol; 26,27-diethyl-24a-homo-9,10-seco-5Z,7E,10(19),22E,24E-cholestapentaene-1S,3R,25-triol; 9547617; Synthesized from protected (5E)-1α-hydroxy-22-oxo-23,24,25,26,27-pentanorvitamin D$_3$. Compared to 1α,25-dihydroxyvitamin D$_3$, inhibition of proliferation of U937 cells is < 70%, induction of differentiation of U937 cells is < 10%. Gene Regulation, Structure-Function Analysis and Clinical Application. Proceedings of the Eighth Workshop on Vitamin D Paris, France July 5-10. Synthesis and Biological Activity of 1α-Hydroxylated Vitamin D$_3$ Analogues with Hydroxylated Side Chains, Multi-Homologated in The 24- or 24,26,27-Positions. (Norman, A. W., Bouillon, R., Thomasset, M., eds), pp159-160, Walter de Gruyter Berlin (1991).

888   **1α,25-Dihydroxy-26,27-dimethyl-22,22,23,23-tetradehydro-24a,24b,24c-trihomo-20-epivitamin D$_3$**
9547618 $C_{32}H_{50}O_3$

(1R,3S,5Z)-5-[(2E)-2-[(1R,3aR,7aS)-1-[(2R)-9-ethyl-9-hydroxy-undec-3-yn-2-yl]-7a-methyl-2,3,3a,5,6,7-

hexahydro-1H-inden-4-ylidene]ethylidene]-4-methylidene-cyclohexane-1,3-diol.
LMST03020514; (20R)-26,27-dimethyl-24a,24b,24c-trihomo-9,10-seco-5Z,7E,10(19)-cholestatrien-22-yne-1S,3R,25-triol; 1α,25-dihydroxy-26,27-dimethyl-22,22,23,23-tetrahydro-24a,24b,24c-trihomo-20-epicholecalciferol; 9547618; Synthesized from hydroxy-protected (5E)-1α-hydroxy-22,22,23,23-tetradehydro-24,25,26,27-tetranor-20-epivitamin D₃. Compared to 1α,25-dihydroxyvitamin D₃, inhibition of proliferation of U937 cells is 2500% and affinity for chicken intestinal vitamin D receptor is 76%. Vitamin D. A Pluripotent Steroid Hormone: Structural Studies, Molecular Endocrinology and Clinical Applications. Proceedings of the Ninth Workshop on Vitamin D, Orlando, Florida (USA) May 28-June 2. Chemistry and Biology of 22,23-Yne Analogs of Calcitriol. (Norman, A. W., Bouillon, R., Thomasset, M., eds), pp73-74, Walter de Gruyter Berlin New York (1994).

889   *1α,25-Dihydroxy-22-ethoxy-23-yne-24, 26,27-trihomo-20-epivitamin D₃*

$C_{32}H_{50}O_4$

1S,3R-Dihydroxy-20R-(1-ethoxy-5-ethyl-5-hydroxy-2-heptyn-1-yl)-9,10-secopregna-5Z,7E,10(19)triene.
CB 1093; LMST03020515; 20-epi-22-ethoxy-23-yne-24,26,27-trihomo-1,25-dihydroxyvitamin D₃; 22-ethoxy-23-yne-24,26, 27-trihomo-1α,25-dihydroxyvitamin D₃; Prevents depletion of nerve growth factor in diabetic rats. Causes fast increase in intracellular calcium levels in skeletal muscle. *Diabetologia*. 1999, *42(11)*, 1308-13; *Br J Pharmacol*. 1999, *126(8)*, 1815-23.

890   *1α,25-Dihydroxy-22S-ethoxy-26,27-di-methyl-23,24-tetrahydro-24a-homo-20-epi-vitamin D₃*
9547619

$C_{32}H_{50}O_4$

(1R,3S,5Z)-5-[(2E)-2-[(1R,3aR,7aS)-1-[(2R,3S)-3-ethoxy-7-ethyl-7-hydroxy-non-4-yn-2-yl]-7a-methyl-2,3,3a,5,6,7-hexahydro-1H-inden-4-ylidene]ethylidene]-4-methylidene-cyclohexane-1,3-diol.
LMST03020515; 1α,25-dihydroxy-22-ethoxy-26,27-dimethyl-23,24-tetradehydro-24a-homo-20-epicholecalciferol; (20R)-26,27-dimethyl-22S-ethoxy-24a-homo-9,10-seco-5Z,7E,10(19)-cholestatrien-23-yne-1S,3R,25-triol; 9547619; Synthesized from protected (5E)-1α-hydroxy-22-oxo-23,24,25,26,27-pentanor-20-epivitamin D₃. Compared to 1α,25-dihydroxyvitamin D₃, induction of differentiation of U937 cells is > 50000%, inhibition of proliferation is 84000%, VDR (rachitic chicken intestinal receptor) binding affinity is 0.2%, and calciuric effect in normal rats is 20%. Vitamin D. A Pluripotent Steroid Hormone: Structural Studies, Molecular Endocrinology and Clinical Applications. Proceedings of the Ninth Workshop on Vitamin D, Orlando, Florida (USA) May 28-June 2. Chemistry and Biology of Highly Active 22-Oxy Analogs of 20-Epi Calcitriol with very low Binding Affinity to the Vitamin D Receptor. (Norman, A. W., Bouillon, R., Thomasset, M., eds), pp85-86, Walter de Gruyter Berlin New York (1994).

891   *1α,22R,25-Trihydroxy-26,27-dimethyl-23,23,24,24-tetradehydro-24a,24b,24c-tri-homovitamin D₃*
9547621

$C_{32}H_{50}O_4$

(2S,3R)-2-[(1R,3aR,4E,7aS)-4-[(2Z)-2-[(3S,5R)-3,5-dihydroxy-2-methylidene-cyclohexylidene]ethylidene]-7a-methyl-2,3,3a,5,6,7-hexahydro-1H-inden-1-yl]-9-ethyl-undec-4-yne-3,9-diol.
LMST03020517; 1α,22R,25-trihydroxy-26,27-dimethyl-23,23,24,24-tetradehydro-24a,24b,24c-tri-homocholecalciferol; 26,27-dimethyl-24a,24b,24c-trihomo-9,10-seco-5Z,7E,10(19)-cholestatrien-23-yne-

1S,3R,22R,25-tetrol; 9547621.

892 *1α,22R,25-Trihydroxy-26,27-dimethyl-23,24-tetradehydro-24a,24b,24c-trihomo-20-epivitamin D₃*

9547620 $C_{32}H_{50}O_4$

(2R,3R)-2-[(1R,3aR,4E,7aS)-4-[(2Z)-2-[(3S,5R)-3,5-dihydroxy-2-methylidene-cyclohexylidene]ethylidene]-7a-methyl-2,3,3a,5,6,7-hexahydro-1H-inden-1-yl]-9-ethyl-undec-4-yne-3,9-diol.
LMST03020516; 1α,22R,25-trihydroxy-26,27-di-methyl-23,24-tetradehydro-24a,24b,24c-trihomo-20-epicholecalciferol; (20R)-26,27-dimethyl-24a,24b,24c-trihomo-9,10-seco-5Z,7E,10(19)-cholestatrien-23-yne-1S,3R,22R,25-tetrol; 9547620; Synthesized from protected (5E)-1α-hydroxy-22-oxo-23,24,25,26,27-pentanor-20-epivitamin D₃. Compared to 1α,25-di-hydroxyvitamin D₃, induction of differentiation of U937 cells is <1%; inhibition of proliferation is 7%, VDR (rachitic chicken intestinal receptor) binding affinity is <0.3%. Vitamin D. A Pluripotent Steroid Hormone: Structural Studies, Molecular Endocrinology and Clinical Applications. Proceedings of the Ninth Workshop on Vitamin D, Orlando, Florida (USA) May 28-June 2. Chemistry and Biology of Highly Active 22-Oxy Analogs of 20-Epi Calcitriol with very low Binding Affinity to the Vitamin D Receptor. (Norman, A. W., Bouillon, R., Thomasset, M., eds), pp85-86, Walter de Gruyter Berlin New York (1994).

893 *1α,22S,25-Trihydroxy-26,27-dimethyl-23,23,24,24-tetradehydro-24a,24b,24c-tri-homovitamin D₃*

9547623 $C_{32}H_{50}O_4$

(2S,3S)-2-[(1R,3aR,4E,7aS)-4-[(2Z)-2-[(3S,5R)-3,5-dihydroxy-2-methylidene-cyclohexylidene]ethylidene]-7a-methyl-2,3,3a,5,6,7-hexahydro-1H-inden-1-yl]-9-ethyl-undec-4-yne-3,9-diol.
LMST03020519; 1α,22S,25-trihydroxy-26,27-di-methyl-23,23,24,24-tetradehydro-24a,24b,24c-trihomocholecalciferol; 26,27-dimethyl-24a,24b,24c-trihomo-9,10-seco-5Z,7E,10(19)-cholestatrien-23-yne-1S,3R,22S,25-tetrol; 9547623.

894 *1α,22S,25-Trihydroxy-26,27-dimethyl-23,24-tetradehydro-24a,24b,24c-trihomo-20-epivitamin D₃*

9547622 $C_{32}H_{50}O_4$

(2R,3S)-2-[(1R,3aR,4E,7aS)-4-[(2Z)-2-[(3S,5R)-3,5-dihydroxy-2-methylidene-cyclohexylidene]ethylidene]-7a-methyl-2,3,3a,5,6,7-hexahydro-1H-inden-1-yl]-9-ethyl-undec-4-yne-3,9-diol.
LMST03020518; 1α,22S,25-trihydroxy-26,27-di-methyl-23,24-tetradehydro-24a,24b,24c-trihomo-20-epicholecalciferol; (20R)-26,27-dimethyl-24a,24b,24c-trihomo-9,10-seco-5Z,7E,10(19)-cholestatrien-23-yne-1S,3R,22S,25-tetrol; 9547622; Synthesized from protected (5E)-1α-hydroxy-22-oxo-23,24,25,26,27-pentanor-20-epivitamin D₃. Compared to 1α,25-dihydroxyvitamin D₃, induction of differentiation of U937 cells is 100%; inhibition of proliferation is 20%, VDR (rachitic chicken intestinal receptor) binding affinity is < 0.3%. Vitamin D. A Pluripotent Steroid Hormone: Structural Studies, Molecular Endocrinology and Clinical Applications. Proceedings of the Ninth Workshop on Vitamin D, Orlando, Florida (USA) May 28-June 2. Chemistry and Biology of Highly Active 22-Oxy Analogs of 20-Epi Calcitriol with very low Binding Affinity to the Vitamin D Receptor. (Norman, A. W., Bouillon, R., Thomasset, M., eds), pp85-86, Walter de Gruyter Berlin New York (1994).

895 *17α-24,24-Dimethyl-24-carbobutoxy-9,10-secochola-5Z,7E,10,22E-tetraene-1S,3,23-triol*

$C_{32}H_{50}O_5$

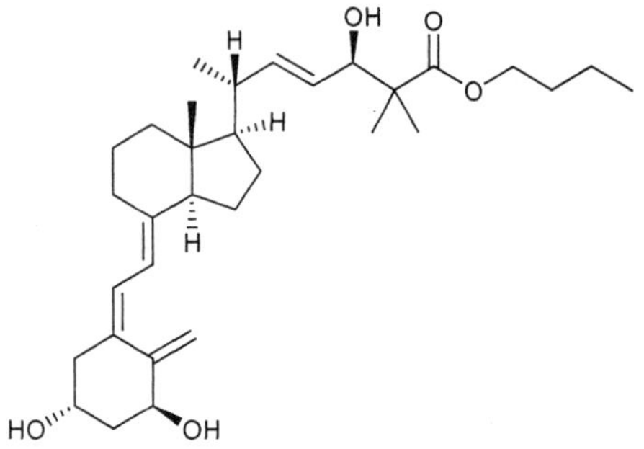

17α-24,24-dimethyl-24-carbobutoxy-9,10-secochola-5Z,7E,10,22E-tetraene-1S,3,23-triol; Shows little or no antagonism towards 1α,25-dihydroxyvitamin D_3. *Mol Pharmacol.* 2000, *58(5)*, 1067-74.

896   *16-Glutaryloxy-1α,25-dihydroxyvitamin D_3*

$C_{32}H_{50}O_7$

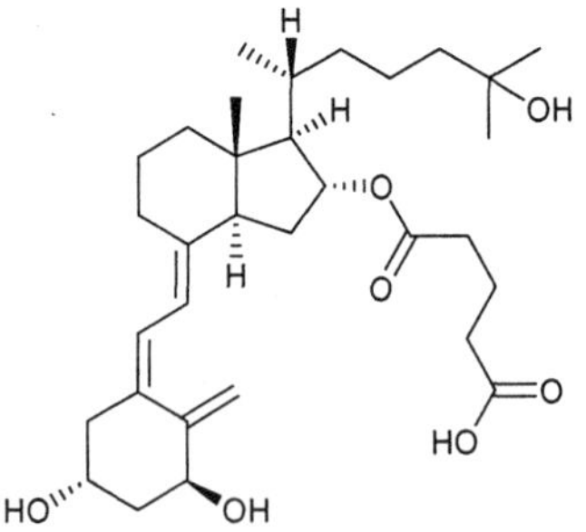

(5Z,7E)-(1R,3S,20R)-9,10-Secocholesta-5,7,10(19)-triene-1,3,25-triol-16α-yl hydrogen glutarate; Synthesized as a hapten for 1α,25-dihydroxyvitamin D_3. *J Org Chem.* 2003, *68(4)*, 1367-75. *Leo AB.*

897   *16-Glutaryloxy-1α,25-dihydroxy-20-epi-vitamin D_3*

$C_{32}H_{50}O_7$

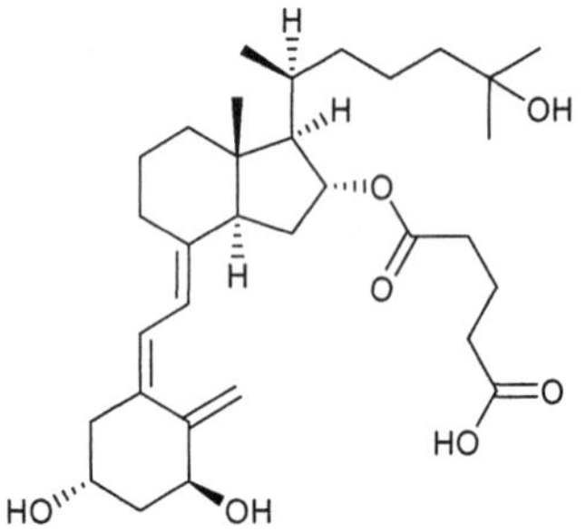

(5Z,7E)-(1R,3S,20S)-9,10-Secocholesta-5,7,10(19)-triene-1,3,25-triol-16α-yl hydrogen glutarate; Synthesized as a hapten for 1α,25-dihydroxy-10-epivitamin D_3. *J Org Chem.* 2003, *68(4)*, 1367-75. *Leo AB.*

898   *1α,25-Dihydroxy-2β-pentylvitamin D_3*
9547624 $C_{32}H_{54}O_3$

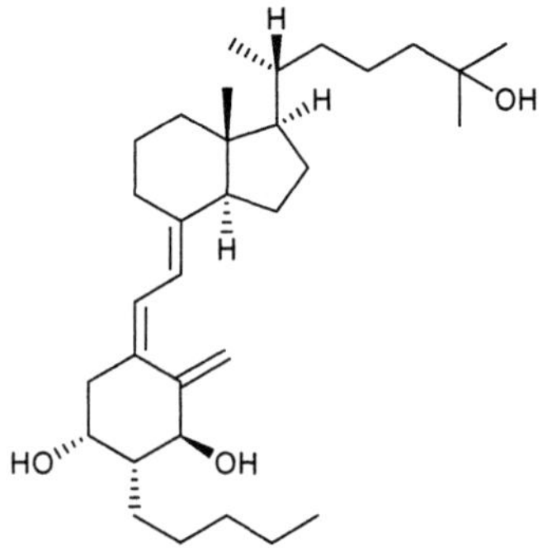

(1R,2R,3S,5Z)-5-[(2E)-2-[(1R,3aR,7aS)-1-[(2R)-6-hydroxy-6-methyl-heptan-2-yl]-7a-methyl-2,3,3a,5,6,7-hexahydro-1H-inden-4-ylidene]ethylidene]-4-methylidene-2-pentyl-cyclohexane-1,3-diol.
LMST03020520; 2R-pentyl-9,10-seco-5Z,7E,10(19)-cholestatriene-1S,3R,25-triol; 1α,25-dihydroxy-2β-pentylcholecalciferol; 9547624; Synthesized photochemically from the 5,7-diene. Spinal bone mineral density in pre-osteoporosis model rats and binding affinity to calf thymus vitamin D receptor reported; λ_m = 264 nm (EtOH). *Chem Pharm Bull (Tokyo).*1997, *45(10)*, 1626-30.

899   *1α,25-Dihydroxy-26,27-diethyl-24a-homovitamin D_3*
9547625 $C_{32}H_{54}O_3$

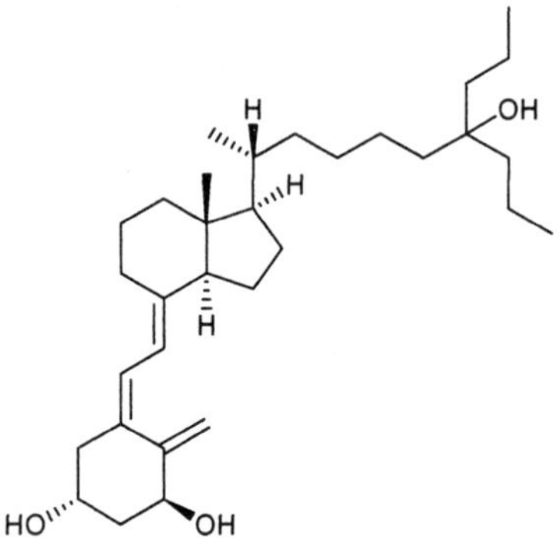

(1R,3S,5Z)-5-[(2E)-2-[(1R,3aR,7aS)-1-[(2R)-7-hydroxy-7-propyl-decan-2-yl]-7a-methyl-2,3,3a,5,6,7-hexahydro-1H-inden-4-ylidene]ethylidene]-4-methylidene-cyclohexane-1,3-diol.
LMST03020521; 26,27-diethyl-24a-homo-9,10-seco-5Z,7E,10(19)-cholestatriene-1S,3R,25-triol; 1α,25-dihydroxy-26,27-diethyl-24a-homocholecalciferol; 9547625; Synthesized from hydroxy-protected (5E)-22-tosyloxy-23,24,25,26,27-pentanorvitamin D_3. Compared to 1α,25-dihydroxyvitamin D_3, inhibition of proliferation of U937 cells is 90% and induction of differentiation of U937 cells is 50%. Gene Regulation, Structure-Function Analysis and Clinical Application. Proceedings of the Eighth Workshop on Vitamin D

Paris, France July 5-10. Synthesis and Biological Activity of 1α-Hydroxylated Vitamin D$_3$ Analogues with Hydroxylated Side Chains, Multi-Homologated in The 24- or 24,26,27-Positions. (Norman, A. W., Bouillon, R., Thomasset, M., eds), pp159-160, Walter de Gruyter Berlin (1991).

900 *26,27-Diethyl-1α,25-dihydroxy-24a,24b-dihomo-23-oxa-20-epivitamin D$_3$*

9547630   C$_{32}$H$_{54}$O$_4$

(1R,3S,5Z)-5-[(2E)-2-[(1R,3aR,7aS)-1-[(2R)-1-(4-hydroxy-4-propyl-heptoxy)propan-2-yl]-7a-methyl-2,3,3a,5,6,7-hexahydro-1H-inden-4-ylidene]ethylidene]-4-methylidene-cyclohexane-1,3-diol. LMST03020526; (20R)-26,27-diethyl-24a,24b-dihomo-23-oxa-9,10-seco-5Z,7E,10(19)-cholestatriene-1S,3R,25-triol; 26,27-diethyl-1α,25-dihydroxy-24a,24b-dihomo-23-oxa-20-epicholecalciferol; 9547630.

901 *1α,25-Dihydroxy-2β-(5-hydroxypentyl)-vitamin D$_3$*

9547627   C$_{32}$H$_{54}$O$_4$

(1R,2R,3R,5Z)-5-[(2E)-2-[(1R,3aR,7aS)-1-[(2R)-6-hydroxy-6-methyl-heptan-2-yl]-7a-methyl-2,3,3a,5,6,7-hexahydro-1H-inden-4-ylidene]ethylidene]-2-(5-hydroxypentyl)-4-methylidene-cyclohexane-1,3-diol. LMST03020523; 2R-(5-hydroxypentyl)-9,10-seco-5Z,7E,10(19)-cholestatriene-1S,3R,25-triol; 1α,25-dihydroxy-2β-(5-hydroxypentyl)cholecalciferol; 9547627; Synthesized photochemically from the 5,7-diene. Spinal bone mineral density in pre-osteoporosis model rats and binding affinity to calf thymus vitamin D receptor reported; λ$_m$ = 264 nm (EtOH). *Chem Pharm Bull (Tokyo)*. 1997, *45(10)*, 1626-30.

902 *1α,25-Dihydroxy-21-(3-hydroxy-3-methylbutyl)vitamin D$_3$*

C$_{32}$H$_{54}$O$_4$

21-(3-Hydroxy-3-methylbutyl)-9,10-secocholesta-5Z,7E,10(19)-triene-1α,3β,25,triol. Gemini; Ro 27-2310; Has increased antitumor activity compared to calcitriol. Binds to the vitamin D receptor but with less affinity than 20-epi-1,la,25-dihydroxyvitamin D$_3$. *J Med Chem*. 2000, *43(14)*, 2719-30; *J Steroid Biochem Mol Biol*. 2006, *100(4-5)*, 107-16. **Hoffmann-LaRoche Inc.**

903 *1α,25-Dihydroxy-20S-methoxy-26,27-diethylvitamin D$_3$*

9547628   C$_{32}$H$_{54}$O$_4$

(1R,3S,5Z)-5-[(2E)-2-[(1S,3aR,7aR)-1-[(2S)-6-hydroxy-2-methoxy-6-propyl-nonan-2-yl]-7a-methyl-2,3,3a,5,6,7-hexahydro-1H-inden-4-ylidene]ethylidene]-4-methylidene-cyclohexane-1,3-diol. LMST03020524; 1α,25-dihydroxy-20S-methoxy-26,27-diethylcholecalciferol; 20S-methoxy-26,27-diethyl-9,10-seco-5Z,7E,10(19)-cholestatriene-1S,3R,25-triol; 9547628; Synthesized from a (5E)-1α-hydroxy-20-keto-22,23,24,25,26,27-hexanorvitamin D$_3$ derivative.

Compared to 1α,25-dihydroxyvitamin D$_3$, inhibition of U937 cell (human histiocytic lymphoma cell line) proliferation is 10%, binding to the vitamin D receptor from rachitic chicken intestine is 0.9%, calcemic activity was not determined. Vitamin D. A Pluripotent Steroid Hormone: Structural Studies, Molecular Endocrinology and Clinical Applications. Proceedings of the Ninth Workshop on Vitamin D, Orlando, Florida (USA) May 28-June 2. Synthesis and Biological Activity of 20-Hydroxylated Vitamin D Analogues. (Norman, A. W., Bouillon, R., Thomasset, M., eds), pp95-96, Walter de Gruyter Berlin New York (1994).

904 **1α,25-Dihydroxy-20S-ethoxy-26,27-di-methyl-24a-homovitamin D₃**
9547629 $C_{32}H_{54}O_4$

(1R,3S,5Z)-5-[(2E)-2-[(1S,3aR,7aR)-1-[(2S)-2-ethoxy-7-ethyl-7-hydroxy-nonan-2-yl]-7a-methyl-2,3,3a,5,6,7-hexahydro-1H-inden-4-ylidene]ethylidene]-4-methylidene-cyclohexane-1,3-diol.
LMST03020525; 1α,25-dihydroxy-20-ethoxy-26,27-dimethyl-24a-homocholecalciferol; 20S-ethoxy-26,27-dimethyl-24a-homo-9,10-seco-5Z,7E,10(19)-cholestatriene-1S,3R,25-triol; 9547629; Synthesized from a (5E)-1α-hydroxy-20-keto-22,23,24,25,26,27-hexanor-vitamin D₃ derivative. Compared to 1α,25-dihydroxyvitamin D₃, inhibition of U937 cell (human histiocytic lymphoma cell line) proliferation is 30000%, binding to the vitamin D receptor from rachitic chicken intestine is 1%, calcemic activity was not determined. Vitamin D. A Pluripotent Steroid Hormone: Structural Studies, Molecular Endocrinology and Clinical Applications. Proceedings of the Ninth Workshop on Vitamin D, Orlando, Florida (USA) May 28-June 2. Synthesis and Biological Activity of 20-Hydroxylated Vitamin D Analogues. (Norman, A. W., Bouillon, R., Thomasset, M., eds), pp95-96, Walter de Gruyter Berlin New York (1994).

905 **1α-Hydroxy-2β-(5-hydroxypentoxy)-vitamin D₃**
9547626 $C_{32}H_{54}O_4$

(1R,2R,3S,5Z)-5-[(2E)-2-[(1R,3aR,7aS)-7a-methyl-1-[(2R)-6-methylheptan-2-yl]-2,3,3a,5,6,7-hexahydro-1H-inden-4-ylidene]ethylidene]-2-(5-hydroxypentoxy)-4-methylidene-cyclohexane-1,3-diol.
LMST03020522; 2R-(5-hydroxypentoxy)-9,10-seco-5Z,

7E,10(19)-cholestatriene-1R,3R-diol; 1α-hydroxy-2β-(5-hydroxypentoxy)cholecalciferol; 9547626; Synthesized photochemically from the 5,7-diene. Plasma calcium levels reported in rats fed with a low calcium/vitamin D deficient diet after administration; λ_m = 263.5 nm (EtOH). *Chem Pharm Bull (Tokyo)*. 1993, *41(6)*, 1111-13.

906 **1α,25-Dihydroxy-2β-(5-hydroxypent-oxy)vitamin D₃**
9547631 $C_{32}H_{54}O_5$

(1R,2R,3R,5Z)-5-[(2E)-2-[(1R,3aR,7aS)-1-[(2R)-6-hydroxy-6-methyl-heptan-2-yl]-7a-methyl-2,3,3a,5,6,7-hexahydro-1H-inden-4-ylidene]ethylidene]-2-(5-hydroxypentoxy)-4-methylidene-cyclohexane-1,3-diol.
LMST03020527; 2R-(5-hydroxypentoxy)-9,10-seco-5Z,7E,10(19)-cholestatriene-1R,3R,25-triol; 1α,25-dihydroxy-2β-(5-hydroxypentoxy)cholecalciferol; 9547631; Synthesized photochemically from the 5,7-diene. Spinal bone mineral density in pre-osteoporosis model rats and binding affinity to calf thymus vitamin D receptor reported; λ_m = 264 nm (EtOH). *Chem Pharm Bull (Tokyo)*. 1997, *45(10)*, 1626-30.

907 **(20R)-24-Hydroxygeminivitamin D₃**
$C_{32}H_{54}O_5$

(3R,6R)-6-{(1R,3aS,7aR)-4-[2-((R)-3-(S)-Hydroxy-5-hydroxy-2-methylene-cyclohexylidene)-(E)-ethylidene]-7amethyloctahydroinden-1-yl}-2,10-dimethylundecane-2,3,10-triol.

24-hydroxygeminicalcitriol; Synthetic vitamin D₃ analog. *J Med Chem*. 2004, *47(26)*, 6476-84.

908 **(20S)-24-Hydroxygeminivitamin D$_3$**

$C_{32}H_{54}O_5$

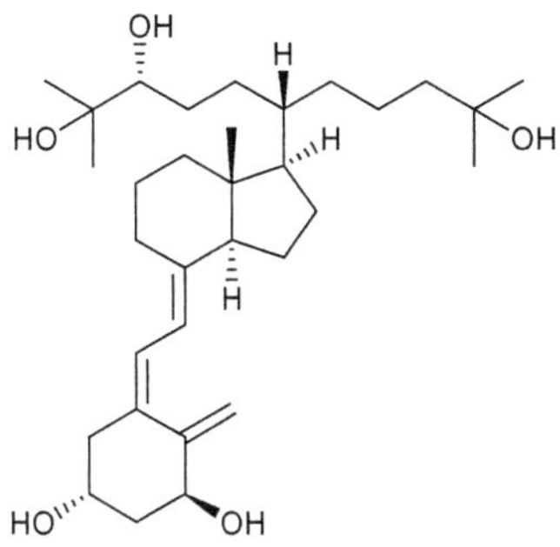

(3R,6S)-6-{(1R,3aS,7aR)-4(E)-[2-((R)-3-(R)-Hydroxy-5-hydroxy-cyclohexylidene)-(E)-ethylidene]-7a-methyl-octahydroinden-1-yl}-2,10-dimethylundecane-2,3,10-triol.
24-hydroxygemini-20-epicalcitriol; Synthetic vitamin D$_3$ analog. *J Med Chem*. 2004, *47(26)*, 6476-84.

909 **1α,25-Dihydroxy-11α-phenylvitamin D$_3$**
9547632

$C_{33}H_{48}O_3$

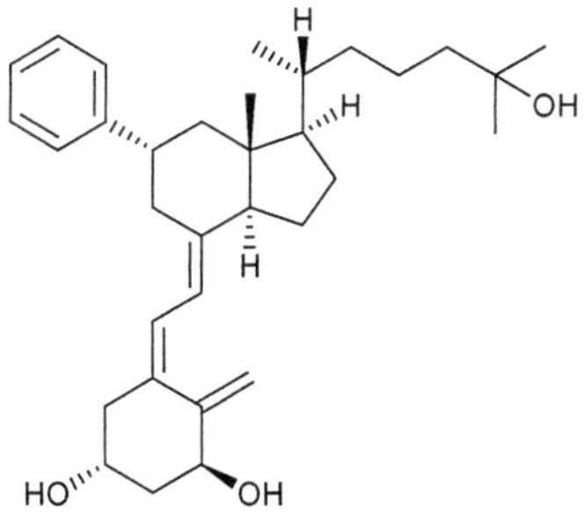

(1R,3S,5Z)-5-[(2E)-2-[(1R,3aR,6R,7aS)-1-[(2R)-6-hydroxy-6-methyl-heptan-2-yl]-7a-methyl-6-phenyl-2,3,3a,5,6,7-hexahydro-1H-inden-4-ylidene]ethylidene]-4-methylidene-cyclohexane-1,3-diol.
LMST03020528; 11S-phenyl-9,10-seco-5Z,7E,10(19)-cholestatriene-1S,3R,25-triol; 1α,25-dihydroxy-11α-phenylcholecalciferol; 9547632; Prepared by convergent synthesis by Horner coupling of 25-hydroxylated CD-ring ketone. Biological activity reported. *J Biol Chem*. 1992, *267(5)*, 3044-51.

910 **1α,25-Dihydroxy-11β-phenylvitamin D$_3$**
9547633

$C_{33}H_{48}O_3$

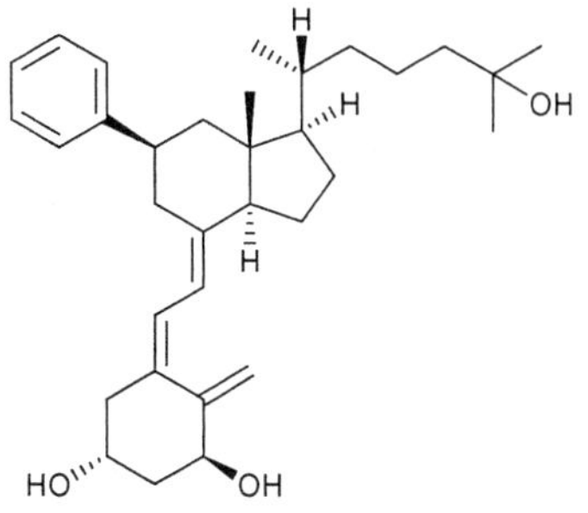

(1R,3S,5Z)-5-[(2E)-2-[(1R,3aR,6S,7aS)-1-[(2R)-6-hydroxy-6-methyl-heptan-2-yl]-7a-methyl-6-phenyl-2,3,3a,5,6,7-hexahydro-1H-inden-4-ylidene]ethylidene]-4-methylidene-cyclohexane-1,3-diol.
LMST03020529; 11R-phenyl-9,10-seco-5Z,7E,10(19)-cholestatriene-1S,3R,25-triol; 1α,25-dihydroxy-11β-phenylvitamin D$_3$; 1α,25-dihydroxy-11β-phenylcholecalciferol; 9547633; Prepared by convergent synthesis by Horner coupling of 25-hydroxylated CD-ring ketone. Biological activity reported. *J Biol Chem*. 1992, *267(5)*, 3044-51.

911 **23S,25-Dihydroxy-24-oxovitamin D$_3$ 23-(β-glucuronide)**
9547634

$C_{33}H_{50}O_{10}$

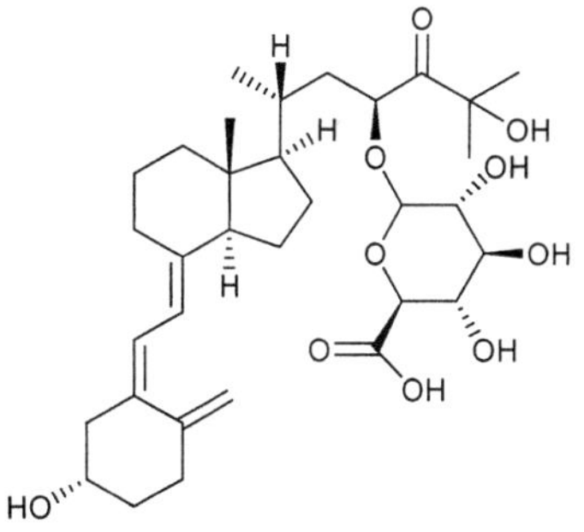

(2S,3S,4S,5R)-6-[(4S,6R)-6-[(1R,3aR,4E,7aS)-4-[(2Z)-2-[(5S)-5-hydroxy-2-methylene-cyclohexylidene]ethyl-idene]-7a-methyl-2,3,3a,5,6,7-hexahydro-1H-inden-1-yl]-2-hydroxy-2-methyl-3-oxo-heptan-4-yl]oxy-3,4,5-trihydroxy-oxane-2-carboxylic acid.
LMST03020530; 23S,25-dihydroxy-24-oxocholecalciferol 23-(β-glucuronide); 24-oxo-9,10-seco-5Z,7E,10(19)-cholestatriene-3S,23S,25-triol 23-(β-glucuronide); 9547634; Isolated and identified from bile of dogs given 24R,25-dihydroxyvitamin D$_3$; λ$_m$ = 265 nm. *Biochem Biophys Acta*. 1997, *1346(2)*, 147-57.

912 **1α,25-Dihydroxy-2α-(Benzyloxy)-19-norvitamin D$_3$**
9547652

$C_{33}H_{50}O_4$

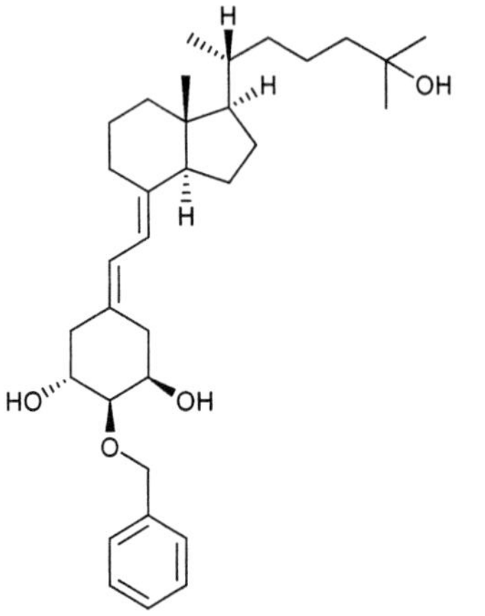

(1R,3R)-5-[(2E)-2-[(1R,3aR,7aS)-1-[(2R)-6-hydroxy-6-methyl-heptan-2-yl]-7a-methyl-2,3,3a,5,6,7-hexa-hydro-1H-inden-4-ylidene]ethylidene]-2-phenylmethoxy-cyclohexane-1,3-diol.
LMST03020551; 2S-(benzyloxy)-19-nor-9,10-seco-5Z,7E-cholestadiene-1R,3R,25-triol; 2α-(benzyloxy)-1α,25-dihydroxy-19-norvitamin D_3; 2α-(benzyloxy)-1α,25-dihydroxy-19-norcholecalciferol; 19-nor-1α,25-(OH)$_2$α-(benzyloxy)D_3; 9547652; Vitamin D_3 analogue synthesized by combining A-ring synthon obtained from (-)-quinic acid with 25-hydroxylated Grundmann type ketone. Compared to calcitriol, promotes intestinal calcium transport but not bone calcium mobilization. Intestinal calcium transport activity is equivalent to that of 1α,25-dihydroxyvitamin D_3 but mobilizes calcium from bone poorly. Cell (HL-60) differentiating activity is about 10% of that of 1α,25-dihydroxyvitamin D_3; λ_m = 243, 251.5, 261 nm (EtOH). *J Med Chem*. 1994, **37(22)**, 3730-8.

913   17α-24-Cyclobutyl-24-carbobutoxy-9, 10-secochola-5Z,7E,10,22E-tetraene-1S,3,23-triol

$C_{33}H_{50}O_5$

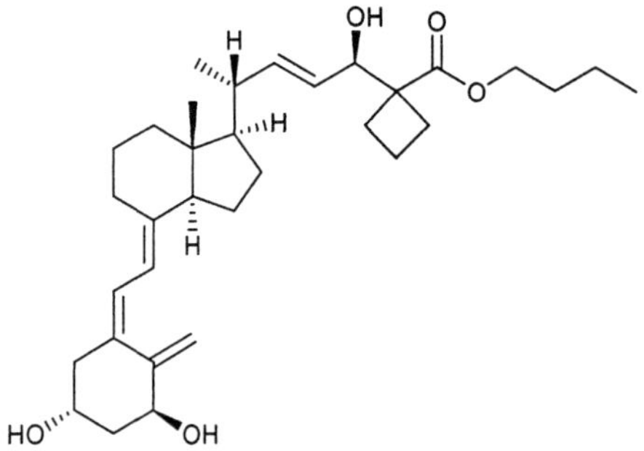

17α-24-cyclobutyl-24-carbobutoxy-9,10-secochola-5Z,7E,10,22E-tetraene-1S,3,23-triol; Shows little or no antagonism towards 1α,25-dihydroxyvitamin D_3. *Mol Pharmacol*. 2000, **58(5)**, 1067-74.

914   11α-Hemiglutaryloxy-1,25-dihydroxy-vitamin D_3

$C_{32}H_{32}O_6$

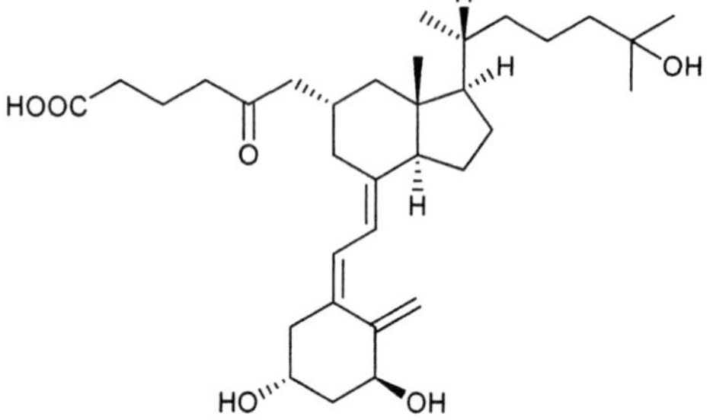

Used in immunoaffinity chromatography method for determination of 1α,25-dihydroxyvitamin D_3 in plasma. *Anal Biochem*. 1997, **244(2)**, 374-83.

915   Vitamin D_3 glucosiduronate

$C_{33}H_{52}O_7$

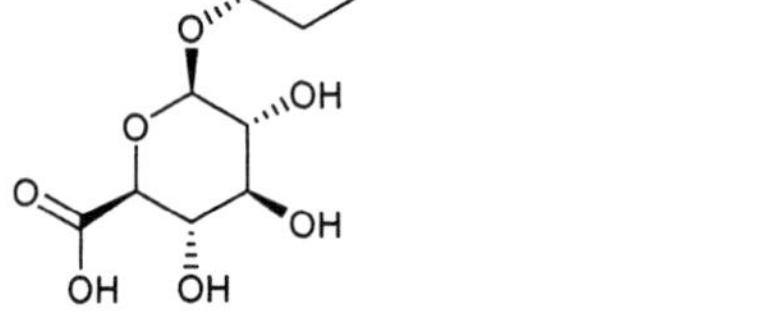

β-D-Glucopyranosiduronic acid, (3β,5Z,7E)-9,10-secocholesta-5,7,10(19)-trien-3-yl.
Vitamin D_3 3-glucuronide; vitamin D_3 3β-glucosiduronate; Isolated from rat bile. Synthesized, is degraded to vitamin D_3 *in vivo*. *Biol Pharm Bull (Tokyo)*. 1996, **19(4)**, 491-4; *J Clin Invest*. 1980, 66(6), 1274-80; 1984, 74, 303.

916   Vitamin D_3 3-β-D-glucopyranoside
24003-73-4 $C_{33}H_{52}O_7$

(2S,3S,4S,5R,6R)-6-[(3E)-3-[(2E)-2-[(1R,7aS)-7a-methyl-
1-[(2R)-6-methylheptan-2-yl]-2,3,3a,5,6,7-hexahydro-
1H-inden-4-ylidene]ethylidene]-4-methylidene-cyclo-
hexyl]oxy-3,4,5-trihydroxy-oxane-2-carboxylic acid.

VD_3-Glucopyranoside; VD_3G; Vitamin D_3 3β-D-
glucopyranoside; β-D-Glucopyranosiduronic acid,
(3β)-9,10-secocholesta-5,7,10(19)-trien-3-yl; 24003-
73-4; 24003-85-8; 6449995; Activity is similar to that
of vitamin D_3. *Int J Vitam Nutr Res.* 1984, *54(1)*, 25-
34.

917   *1-Hydroxyvitamin D_3 3-D-glucopyrano-side*

91037-29-5 $C_{33}H_{52}O_8$

(2S,3S,4S,5R,6S)-6-[(3E)-3-[(2E)-2-[(1R,7aS)-7a-methyl-
1-[(2R)-6-methylheptan-2-yl]-2,3,3a,5,6,7-hexahydro-
1H-inden-4-ylidene]ethylidene]-5-hydroxy-4-
methylidene-cyclohexyl]oxy-3,4,5-trihydroxy-oxane-2-
carboxylic acid.

HD_3 Glucopyranoside; HD_3G; 1-Hydroxyvitamin D_3 3-
D-glucopyranoside; 1α-Hydroxyvitamin D_3 3β-D-
glucopyranoside; 6439890; 91037-29-5; β-D-
Glucopyranosiduronic acid,-1α-hydroxy-9,10-seco-

cholesta-5Z,7E,10(19)-trien-3β-yl, endo-(+-)-; 91037-
29-5; Has about 10% activity of free 1α-
hydroxyvitamin D_3. *Int J Vitam Nutr Res.* 1984, *54(1)*,
25-34.

918   *1,25-Dihydroxyvitamin D_3 monoglucur-onide*

82694-73-3 $C_{33}H_{54}O_{10}$

(1R,3S.5E)-5-[(2E)-2-[(1R,3aR,7aS)-1-[(2R)-6-hydroxy-6-
methyl-heptan-2-yl]-7a-methyl-2,3,3a,5,6,7-
hexahydro-1H-inden-4-ylidene]ethylidene]-4-
methylidene-cyclohexane-1,3-diol; (2S,3S,4S,5R,6R)-
3,4,5,6-tetrahydroxyoxane-2-carboxylic acid.

1,25-Dvdmg; 1,25-Dihydroxyvitamin D_3 monogluc-
uronide; β-D-Glucopyranosiduronic acid, (1α,3β)-
dihydroxy-9,10-secocholesta-5Z,7E,10(19)-trienyl;
6439627; 82694-73-3; Isolated from rat bile. *J Biol
Chem.* 1982, *257(13)*, 7491-4.

919   *1,25-Dihydroxyvitamin D_3 3-glycoside*

$C_{33}H_{54}O_8$

β-D-Glucoside, (3β, 5Z,7E)-9,10-secocholesta-
5,7,10(19)-trien-3-yl.
1,25-Dihydroxycholecalciferol 3-glycoside; Synthesized. Had no effect on serum calcium, bone weight or
calcium binding protein. *Int J Vitam Nutr Res.* 1985,
55(3), 263-7.

920   *1α,25-Dihydroxy-26,27-dipropylvitamin*
         D₃
9547635 $C_{33}H_{56}O_3$

(1R,3S,5Z)-5-[(2E)-2-[(1R,3aR,7aS)-1-[(2R)-6-butyl-6-
hydroxy-decan-2-yl]-7a-methyl-2,3,3a,5,6,7-
hexahydro-1H-inden-4-ylidene]ethylidene]-4-
methylidene-cyclohexane-1,3-diol.
LMST03020531;   26,27-dipropyl-9,10-seco-5Z,7E,
10(19)-cholestatriene-1S,3R,25-triol; 1α,25-dihydroxy-
26,27-dipropylcholecalciferol; 9547635; Compared to
1α,25-dihydroxyvitamin D₃, binding affinity for serum
vitamin D binding protein and for the receptor of HL-
60 cells is <1% and 3.4%, fails to induce HL-60 cell
differentiation under either serum-supplemented or
serum-free conditions at a concentration of 48 nM.
Steroids. 1991, *56(3)*, 142-7.

921   *1α,25-Dihydroxy-26,26,26,27,27,27-*
         hexamethylvitamin D₃
9547636 $C_{33}H_{56}O_3$

(1R,3S,5Z)-5-[(2E)-2-[(1R,3aR,7aS)-1-[(2R)-6-hydroxy-
7,7-dimethyl-6-tert-butyl-octan-2-yl]-7a-methyl-
2,3,3a,5,6,7-hexahydro-1H-inden-4-ylidene]
ethylidene]-4-methylidene-cyclohexane-1,3-diol.
LMST03020532; 26,26,26,27,27,27-hexamethyl-9,10-
seco-5Z,7E,10(19)-cholestatriene-1S,3R,25-triol;
1α,25-dihydroxy-26,26,26,27,27,27-hexamethyl
cholecalciferol; 9547636. *J Org Chem*. 1993, *58(1)*,
118-23.

922   *26,27-Diethyl-1α,25-dihydroxy-24a,24b-*
         dihomovitamin D₃
9547637 $C_{33}H_{56}O_3$

(1R,3S,5Z)-5-[(2E)-2-[(1R,3aR,7aS)-1-[(2R)-8-hydroxy-
8-propyl-undecan-2-yl]-7a-methyl-2,3,3a,5,6,7-hexa-
hydro-1H-inden-4-ylidene]ethylidene]-4-methylidene-
cyclohexane-1,3-diol.
LMST03020533; 26,27-diethyl-24a,24b-dihomo-9,10-
seco-5Z,7E,10(19)-cholestatriene-1S,3R,25-triol; 26,27-
diethyl-1α,25-dihydroxy-24a,24b-dihomochole-
calciferol; 9547637; Synthesized from hydroxy-
protected (5E)-23,24,25,26,27-pentanorvitamin D₃ 22-
bis(methylseleno)acetal. Compared to 1α,25-di-
hydroxyvitamin D₃, inhibition of proliferation of U937
cells is < 10%, induction of differentiation of U937
cells is 1%. Gene Regulation, Structure-Function
Analysis and Clinical Application. Proceedings of the
Eighth Workshop on Vitamin D Paris, France July 5-10.
Synthesis and Biological Activity of 1α-Hydroxylated
Vitamin D₃ Analogues with Hydroxylated Side Chains,
Multi-Homologated in The 24- or 24,26,27-Positions.
(Norman, A. W., Bouillon, R., Thomasset, M., eds),
pp159-160, Walter de Gruyter Berlin (1991).

923   *1α,25-Dihydroxy-2β-(6-hydroxyhexyl)-*
         vitamin D₃
9547638 $C_{33}H_{56}O_4$

(1R,2R,3S,5Z)-5-[(2E)-2-[(1R,3aR,7aS)-1-[(2R)-6-
hydroxy-6-methyl-heptan-2-yl]-7a-methyl-2,3,3a,5,6,7-
hexahydro-1H-inden-4-ylidene]ethylidene]-2-(6-
hydroxyhexyl)-4-methylidene-cyclohexane-1,3-diol.
LMST03020534;   2R-(6-hydroxyhexyl)-9,10-seco-
5Z,7E,10(19)-cholestatriene-1S,3R,25-triol;   1α,25-
dihydroxy-2β-(6-hydroxyhexyl)cholecalciferol;
9547638; Synthesized photochemically from the 5,7-
diene. Spinal bone mineral density in pre-osteoporosis
model rats and binding affinity to calf thymus vitamin
D receptor reported; λ_m = 262 nm (EtOH). *Chem*

Pharm Bull (Tokyo). 1997, **45(10)**, 1626-30.

924   **25-(5-Butyloxazol-2-yl)-26,27-cyclo-9,
10-secocholesta-5Z-7E-10(19),22E-tetraene-1,
3,24-triol**

$C_{34}H_{47}NO_4$

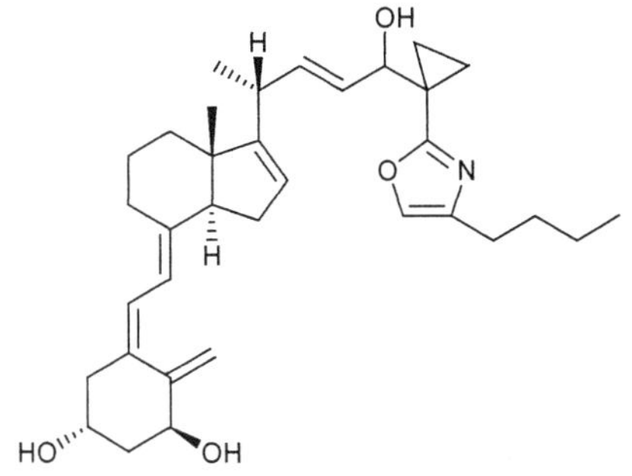

ZK 191874; 25-(5-butyloxazol-2-yl)-26,27-cyclo-9,10-secocholesta-5Z-7E-10(19),22E-tetraene-1,3,24-triol; A vitamin D antagonist. Has a therapeutic advantage over 1,25-dihydroxyvitamin D_3 by inducing immuno-suppressive effects at concentrations that do not cause hypercalcemia. *J Invest Dermatol*. 2002, **119(6)**, 1434-42.

925   **1α,25-Dihydroxy-11-(4-hydroxymethyl-
phenyl)-9,11-didehydrovitamin D_3**
9547639 $C_{34}H_{48}O_4$

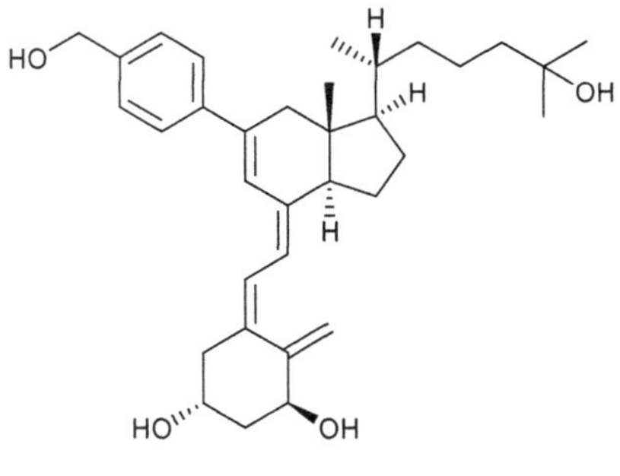

(1R,3S,5Z)-5-[(2Z)-2-[(1R,3aR,7aS)-1-[(2R)-6-hydroxy-6-methyl-heptan-2-yl]-6-[4-(hydroxymethyl)phenyl]-7a-methyl-2,3,3a,7-tetrahydro-1H-inden-4-ylidene]ethylidene]-4-methylidene-cyclohexane-1,3-diol. LMST03020535; 11-(4-hydroxymethylphenyl)-9,10-seco-5Z,7E,9(11),10(19)-cholestatetraene-1S,3R,25-triol; 1α,25-dihydroxy-11-(4-hydroxymethylphenyl)-9,11-didehydrocholecalciferol; 9547639.

926   **1α-Hydroxy-18-[m-(1-hydroxy-1-ethyl-
propyl)-benzyloxy]-23,24,25,26,27-pentanor-
vitamin D_3**
9547685
$C_{34}H_{50}O_4$

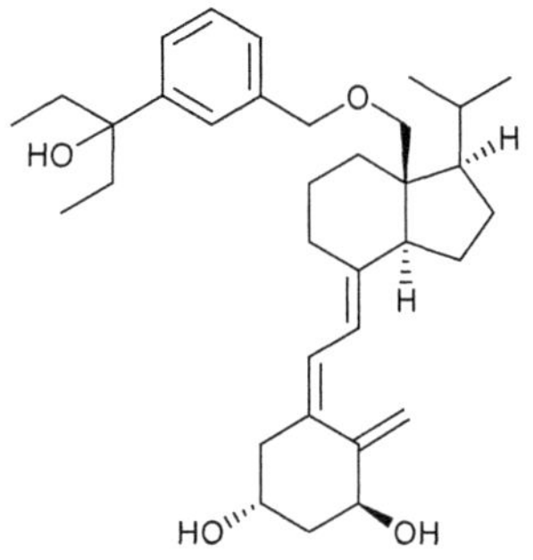

(1R,3S,5Z)-5-[(2E)-2-[(1R,3aR,7aR)-7a-[[3-(3-hydroxypentan-3-yl)phenyl]methoxymethyl]-1-propan-2-yl-2,3,3a,5,6,7-hexahydro-1H-inden-4-ylidene]ethylidene]-4-methylidene-cyclohexane-1.3-diol. LMST03020590;   18-[m-(1-hydroxy-1-ethylpropyl)-benzyloxy]-23,24-dinor-9,10-seco-5Z,7E,10(19)-cholatriene-1S,3R-diol;   1α-hydroxy-18-[m-(1-hydroxy-1-ethylpropyl)-benzyloxy]-23,24,25,26,27-pentanor-cholecalciferol; 9547685.

927   **17α-24-Cyclopentyl-24-carbobutoxy-
9,10-seco-chola-5Z,7E,10,22E-tetraene-
1S,3,23-triol**

$C_{34}H_{52}O_5$

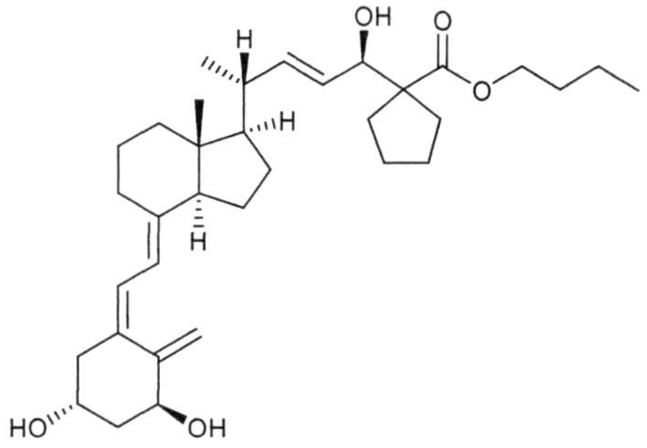

17α-24-cyclopentyl-24-carbobutoxy-9,10-seco-chola-5Z,7E,10,22E-tetraene-1S,3,23-triol; Shows little or no antagonism towards 1α,25-dihydroxyvitamin D_3. *Mol Pharmacol*. 2000, **58(5)**, 1067-74.

928   **Vitamin D_2 glucosiduronate**
57803-96-0 $C_{34}H_{52}O_7$

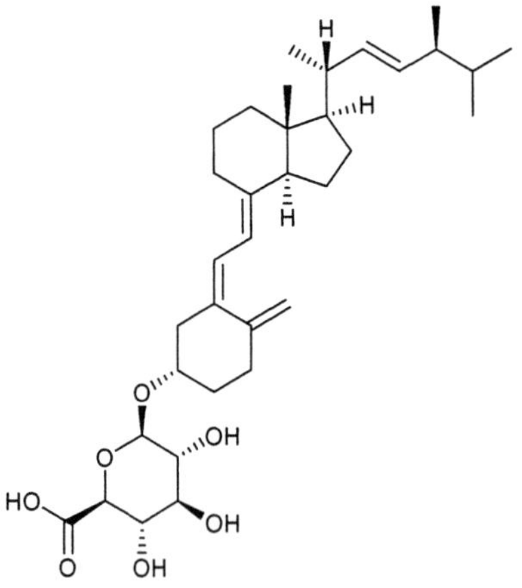

β-D-Glucopyranosiduronic acid, 9,10-secoergosta-5Z,7E,10(19),22-tetraen-3β-yl.

Vitamin D_2 3-glucuronide; Vitamin D_2 glucosiduronate; Vitamin D_2 3β-glucosiduronic; β-D-Glucopyranosiduronic acid, 9,10-secoergosta-5Z,7e,10(19), 22e-tetraen-3β-yl; 57803-96-0; Characterized by HPLC-MS and synthesized. Promotes calcium mobilization, but less effectively than vitamin D_2. *J Chromatogr B Biomed Sci Appl.* 1997, *690(1-2)*, 348-54; *J Steroid Biochem*. 1982, *17(5)*, 495-502.

929   *25-Hydroxyvitamin D_2 3-glucuronide*
76020-77-4 $C_{34}H_{52}O_8$

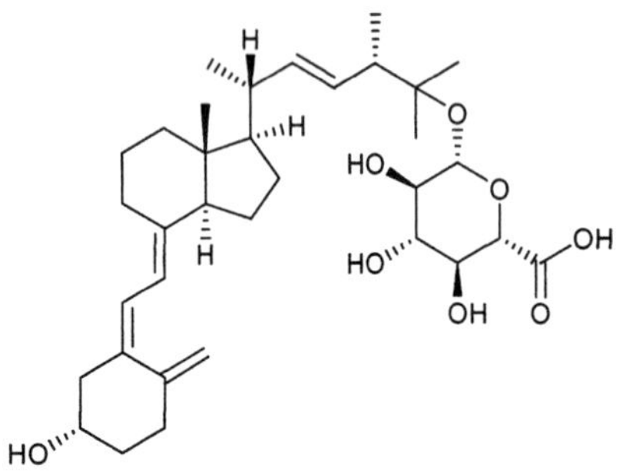

(E,3S,6S)-6-[(1R,3aR,4E,7aS)-4-[(2Z)-2-[(5S)-5-hydroxy-2-methylidene-cyclohexylidene]ethylidene]-7a-methyl-2,3,3a,5,6,7-hexahydro-1H-inden-1-yl]-2,3-dimethyl-hept-4-en-2-ol; (2S,3S,4S,5R,6R)-3,4,5,6-tetrahydroxy-oxane-2-carboxylic acid.

25-Hydroxyvitamin D_2 3-glucuronide; 25-Hydroxy-ergocalciferol 3-glucuronide; 25-Hydroxyvitamin D 3-glucuronide; 9,10-Secoergosta-5Z,7E,10(19),22E-tetra-ene-3β.25-diol 3-glucuronide; 25-OHD2 3-(b-gluc-uronide); 25-Hvd2-3-gur; LMST03010060; 9547260; 76020-77-4 ; A major biliary metabolite of vitamin D_2 isolated from chicks and later characterized by EI and chemical ionization MS. *Chromatogr B Biomed Sci Appl.* 1997, *690(1-2)*, 348-54; *Biochemistry*. 1981, *20(1)*, 222-6.

930   *25-Hydroxyvitamin D_2 25-(β-glucuron-ide)*

9547260 $C_{34}H_{52}O_8$

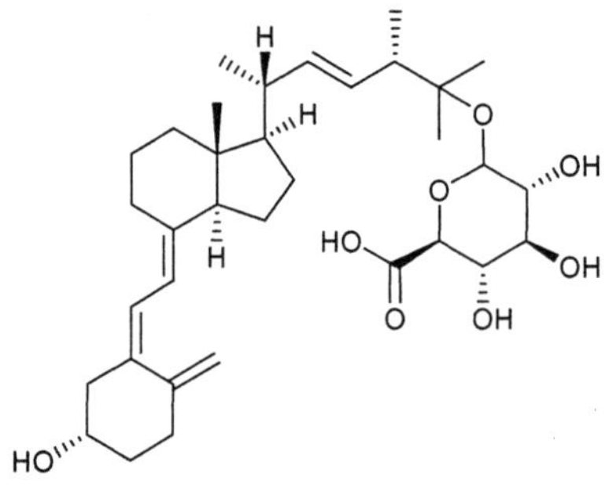

(2S,3S,4S,5R)-6-[(E,3S,6S)-6-[(1R,3aR,4E,7aS)-4-[(2Z)-2-[(5S)-5-hydroxy-2-methylidene-cyclohexylidene]ethylidene]-7a-methyl-2,3,3a,5,6.7-hexahydro-1H-inden-1-yl]-2,3-dimethyl-hept-4-en-2-yl]oxy-3,4,5-trihydroxy-oxane-2-carboxylic acid.
LMST03010060; 9,10-seco-5Z,7E,10(19),22E-ergosta-tetraene-3S,25-diol 25-(β-glucuronide); 25 hydroxy-ergocalciferol 25-(β-glucuronide); 9547260; *Biochemistry*. 1981, *20(1)*, 222-6.

931   *Vitamin D_3 6R,19-(4-phenyl-1,2,4-triazo-line-3,5-dione) adduct*
9547640 $C_{35}H_{49}N_3O_3$

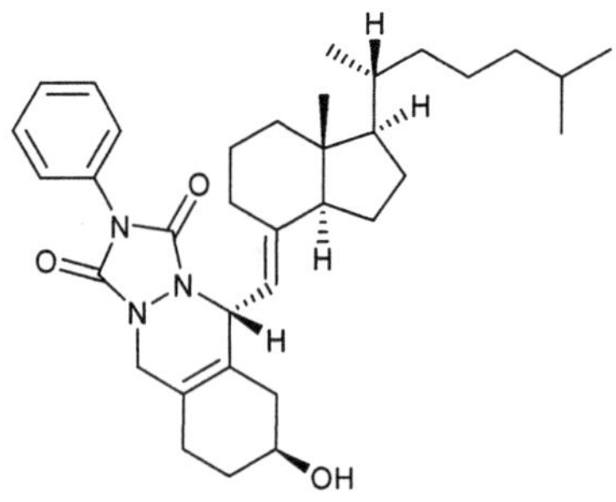

(7E)-(3S,6R)-9,10-seco-5,7,10(19)-cholestatrien-3-ol 6,19-(4-phenyl-1,2,4-triazoline-3,5-dione) adduct.
LMST03020536; cholecalciferol 6R,19-(4-phenyl-1,2,4-triazoline-3,5-dione) adduct; 9,10-seco-5,7E,10(19)-cholestatrien-3S-ol 6R,19-(4-phenyl-1,2,4-triazoline-3,5-dione) adduct; 9547640; Formed as a minor product by reaction of 4-phenyl-1,2,4-triazoline-3,5-dione with vitamin D_3; $[\alpha]_D^{23}$ = -173° (c = 0.81 CHCl$_3$). *J Org Chem.* 1976, *41(12)*, 2098-102.

932   *Vitamin D_3 6S,19-(4-phenyl-1,2,4-triazo-line-3,5-dione) adduct*
9547641 $C_{35}H_{49}N_3O_3$

(7E)-(3S,6S)-9,10-seco-5,7,10(19)-cholestatrien-3-ol
6,19-(4-phenyl-1,2,4-triazoline-3,5-dione) adduct.
LMST03020537; cholecalciferol 6S,19-(4-phenyl-
1,2,4-triazoline-3,5-dione) adduct; (7E)-(3S,6S)-9,10-
seco-5,7E,10(19)-cholestatrien-3S-ol 6S,19-(4-phenyl-
1,2,4-triazoline-3,5-dione) adduct; 9547641; Formed
as a major product by reaction of 4-phenyl-1,2,4-
triazoline-3,5-dione with vitamin D_3; $[\alpha]_D^{23}$ = +191° (c =
1.105 CHCl$_3$). *J Org Chem.* 1976, *41(12)*, 2098-102.

933   25-Hydroxyvitamin D_3 3-(N-(4-azido-2-nitrophenyl)glycinate)

101396-04-7 $C_{35}H_{49}N_5O_5$

[(1S,3Z)-3-[(2E)-2-[(1R,7aS)-1-[(2R)-6-hydroxy-6-
methyl-heptan-2-yl]-7a-methyl-2,3,3a,5,6,7-hexa-
hydro-1H-inden-4-ylidene]ethylidene]-4-methylidene-
cyclohexyl] 2-[(4-azido-2-nitro-phenyl)amino]acetate.
Calcifediol-3-anpg; 25-Hydroxyvitamin D_3 3-(N-(4-
azido-2-nitrophenyl)glycinate); 101396-04-7; Glycine,
N-(4-azido-2-nitrophenyl)-, 25-hydroxy-9,10-secochol-
esta-5Z,7E,10(19)-trien-3β-yl ester; 6439938; 101396-
04-7; Used for photoaffinity labeling of the rat plasma
vitamin D binding protein. About 10 times less active
than 25-OH-D_3 in terms of binding but is covalently
linked to the rat vitamin DBP. *Biochemistry.* 1986,
25(17), 4729-33; 1991, *30(19)*, 4809-13.

934   Calcitriol-ang

98728-28-0 $C_{35}H_{49}N_5O_6$

[(3Z,5S)-3-[(2Z)-2-[(1R,3aR,7aS)-1-[(2R)-6-hydroxy-6-
methyl-heptan-2-yl]-7a-methyl-2,3,3a,5,6,7-hexa-
hydro-1H-inden-4-ylidene]ethylidene]-5-hydroxy-4-
methylidene-cyclohexyl] 2-[(4-azido-2-nitro-phenyl)-
amino]acetate.
1,25-Dihydroxyvitamin D_3-3-(N-(4-azido-2-nitro-
phenyl)glycinate); Calcitriol-ang; 1,25-ang; N-(4-
Azido-2-nitrophenyl)glycine-1α,25-dihydroxy-9,10-
secocholesta-5Z,7E,10(19)-trien-3β-yl ester; Glycine,
N-(4-azido-2-nitrophenyl)-1α,25-dihydroxy-9,10-
secocholesta-5Z,7E,10(19)-trien-3β-yl ester; 6438811;
98728-28-0; Used as a radiolabeled photoaffinity
analog of 1,25-dihydroxyvitamin D_3. *Biochem Biophys
Res Commun.* 1985, *132(1)*, 198-203; *Steroids.* 1988,
51(5-6), 623-30; 1993, *58(10)*, 462-5.

935   11α-(4-Dimethylaminophenyl)-1α,25-dihydroxyvitamin D_3

9547642 $C_{35}H_{53}NO_3$

(1R,3S,5Z)-5-[(2E)-2-[(1R,3aR,6R,7aS)-6-(4-dimethyl-
aminophenyl)-1-[(2R)-6-hydroxy-6-methyl-heptan-2-
yl]-7a-methyl-2,3,3a,5,6,7-hexahydro-1H-inden-4-
ylidene]ethylidene]-4-methylidene-cyclohexane-1,3-
diol.
LMST03020538; 11S-(4-dimethylaminophenyl)-9,10-
seco-5Z,7E,10(19)-cholestatriene-1S,3R,25-triol; 11α-

(4-dimethylaminophenyl)-1α,25-di-hydroxychole-calciferol; 9547642.

936   **11-(4-Acetoxymethylphenyl)-1α,25-dihydroxy-9,11-didehydrovitamin D₃**

9547643 $C_{36}H_{50}O_5$

[4-[(3R,3aS,7Z,7aR)-7-[(2Z)-2-[(3S,5R)-3,5-dihydroxy-2-methylidene-cyclohexylidene]ethylidene]-3-[(2R)-6-hydroxy-6-methyl-heptan-2-yl]-3a-methyl-2,3,4,7a-tetrahydro-1H-inden-5-yl]phenyl]methyl acetate. LMST03020539; 11-(4-acetoxymethylphenyl)-9,10-seco-5Z,7E,9(11),10(19)-cholestatetraene-1S,3R,25-triol; 11-(4-acetoxymethylphenyl)-1α,25-dihydroxy-9,11-didehydrocholecalciferol; 9547643.

937   **25-Hydroxyvitamin D₃ 3β-3'-[N-(4-azido-2-nitrophenyl)amino]propyl ether**

133191-08-9 $C_{36}H_{53}N_5O_4$

vitamin D₃ 25-ANE; Synthesized from 25-hydroxyvitamin D₃ and used as a second generation photoaffinity analog of vitamin D₃; 6439348; 133191-08-9 *Biochemistry*. 1991, *30(19)*, 4809-13.

938   **1α,25-Dihydroxy-25,25-diphenyl-26,27-dinorvitamin D₃**

9547646 $C_{37}H_{48}O_3$

(1R,3S,5Z)-5-[(2E)-2-[(1R,3aR,7aS)-1-[(2R)-6-hydroxy-6,6-diphenyl-hexan-2-yl]-7a-methyl-2,3,3a,5,6,7-hexa-hydro-1H-inden-4-ylidene]ethylidene]-4-methylidene-cyclohexane-1,3-diol. LMST03020543; 25,25-diphenyl-26,27-dinor-9,10-seco-5Z,7E,10(19)-cholestatriene-1S,3R,25-triol; 1α,25-dihydroxy-25,25-diphenyl-26,27-dinorvitamin D₃; 1α,25-dihydroxy-25,25-diphenyl-26,27-dinor-cholecalciferol; 9547646. *J Org Chem*. 1993, *58(1)*, 118-23.

939   **1-Hydroxyvitamin D₃ cellobioside**

89457-70-5 $C_{39}H_{64}O_{12}$

(±)-1α-hydroxy-9,10-secocholesta-5Z,7E,10(19)-trien-3β-yl 4-O-β-D-glucopyranosyl-β-D-glucopyranoside. 1-Hydroxyvitamin D₃-cellobioside; 1α-Hydroxyvitamin D₃-cellobioside; HD₃-Cellobioside; HD₃C; 89457-70-5; Has no vitaminic activity in chickens. *Int J Vitam Nutr Res*. 1984, *54(1)*, 25-34.

940   **Vitamin D₂-phenobarbital complex**

120063-23-2 $C_{40}H_{56}N_2O_4$

(1R,3Z)-3-[(2E)-2-[(1R,3aR,7aS)-1-[(E,2S,5R)-5,6-dimethylhept-3-en-2-yl]-7a-methyl-2,3,3a,5,6,7-hexahydro-1H-inden-4-ylidene]ethylidene]-4-methylidene-cyclohexan-1-ol; 5-ethyl-5-phenyl-1,3-diazinane-2,4,6-trione. 6443422; 120063-23-2. Phenobarbital, used chronically, inhibits production of and release from the liver of 1α,25 dihydroxyvitamin D$_3$. *Res Exp Med (Berl)*. 1972, **158(3)**, 194-204.

941   ***Vitamin D$_3$ 6R,19-[4-{2-(6,7-dimethoxy-4-methyl-3-oxo-3,4-dihydroquinoxalinyl)ethyl]-1,2,4-triazoline-3,5-dione] adduct***

9547644 $C_{42}H_{59}N_5O_6$

9,10-seco-5,7E,10(19)-cholestatrien-3S-ol 6R,19-[4-{2-(6,7-dimethoxy-4-methyl-3-oxo-3,4-dihydroquinoxal-inyl)ethyl}-1,2,4-triazoline-3,5-dione] adduct. LMST03020541; cholecalciferol 6R,19-[4-{2-(6,7-dimethoxy-4-methyl-3-oxo-3,4-dihydroquinoxalin-yl)ethyl}-1,2,4-triazoline-3,5-dione] adduct; 9,10-seco-5,7E,10(19)-cholestatrien-3S-ol 6R,19-[4-{2-(6,7-dimethoxy-4-methyl-3-oxo-3,4-dihydroquinoxalin-yl)ethyl}-1,2,4-triazoline-3,5-dione] adduct; 9547644; Fluorescently tagged vitamin D derivative. Synthesized from vitamin D$_3$ as the minor product by reaction with 4-{2-(6,7-dimethoxy-4-methyl-3-oxo-3,4-dihydroquin-oxalinyl)ethyl}-1,2,4-triazoline-3,5-dione (DMEQ-TAD) . in CH$_2$Cl$_2$ at room temperature. *Bioorg Med Chem Lett.* 1993, **3(9)**, 1809-14.

942   ***Vitamin D$_3$ 6S,19-[4-{2-(6,7-dimethoxy-4-methyl-3-oxo-3,4-dihydroquinoxalinyl)ethyl]-1,2,4-triazoline-3,5-dione] adduct***

9547645 $C_{42}H_{59}N_5O_6$

9,10-seco-5,7E,10(19)-cholestatrien-3S-ol 6S,19-[4-{2-(6,7-dimethoxy-4-methyl-3-oxo-3,4-dihydroquinoxal-inyl)ethyl}-1,2,4-triazoline-3,5-dione] adduct. LMST03020542; cholecalciferol 6S,19-[4-{2-(6,7-dimethoxy-4-methyl-3-oxo-3,4-dihydroquinoxal-inyl)ethyl}-1,2,4-triazoline-3,5-dione] adduct; 9,10-seco-5,7E,10(19)-cholestatrien-3S-ol 6S,19-[4-{2-(6,7-dimethoxy-4-methyl-3-oxo-3,4-dihydroquinoxalinyl)-ethyl}-1,2,4-triazoline-3,5-dione] adduct; 9547645; Fluorescently tagged vitamin D derivative. Synthesized from vitamin D$_3$ as the major product by reaction with 4-{2-(6,7-dimethoxy-4-methyl-3-oxo-3,4-dihydroquinoxalinyl)ethyl}-1,2,4-triazoline-3,5-dione (DMEQ-TAD) in CH$_2$Cl$_2$ at room temperature. *Bioorg Med Chem Lett.* 1993, **3**, 1809.

943   ***Vitamin D$_3$ palmitate***

13403-10-6 $C_{43}H_{74}O_2$

[(1R,3Z)-3-[(2E)-2-[(1R,3aR,7aS)-7a-methyl-1-[(2R)-6-methylheptan-2-yl]-2,3,3a,5,6,7-hexahydro-1H-inden-4-ylidene]ethylidene]-4-methylidene-cyclohexyl] hexadecanoate.

Vitamin D$_3$ palmitate; Cholecalciferol palmitate; 9,10-Secocholesta-5Z,7E,10(19)-trien-3β-ol, hexadecanoate; 6439029; 13403-10-6; Tested as a dietary source of vitamin D$_3$ but is not suitable as a less calcinogenic form of the vitamin. *J Clin Endocrinol Metab*. 1996, **81(4)**, 1385-8; *Int J Vitam Nutr Res*. 1981, **51(4)**, 359-64.

944   ***Chloroquine & 25-OH Vitamin D***

6474052 $C_{45}H_{70}ClN_3O_2$

6-[(4E)-4-[(2Z)-2-[(5S)-5-hydroxy-2-methylidene-cyclohexylidene]ethylidene]-7a-methyl-2,3,3a,5,6.7-hexahydro-1H-inden-1-yl]-2-methyl-heptan-2-ol.

AIDS009831; Chloroquine & 25-OH Vitamin D_3; AIDS-009831; 6474052; Chloroquine reported to enhance the anti-TB effectiveness of vitamin D_3. *J Clin Endocrinol Metab.* 1999, *84(2)*, 799-801.

945   *Frubiase*

56391-76-5 $C_{52}H_{87}Ca_2O_{31}P$

Dicalcium; (1R,3Z)-3-[(2E)-2-[(1R,3aR,7aS)-1-[(E,2S,5R)-5,6-dimethylhept-3-en-2-yl]-7a-methyl-2,3,3a,5,6,7-hexahydro-1H-inden-4-ylidene]ethylidene]-4-methylidene-cyclohexan-1-ol; (2R)-2-[(1S)-1,2-di-hydroxyethyl]-4,5-dihydroxy-furan-3-one; 2-hydroxy-propanoate; (2R,3S,4R,5R)-2,3,4,5,6-pentahydroxy-hexanoate; phosphoric acid.

Frubiase; D-Gluconic acid, calcium salt (2:1), mixt. with L-ascorbic acid, 2-hydroxypropanoic acid calcium salt (2:1), phosphoric acid and 9,10-secoergosta-5Z,7E,10(19),22E-tetraen-3β-ol; 6443806; 56391-76-5; Has been tested for antiallergic properties. Marketed as a calcium supplement. *Fortschr Med.* 1983, *101(42)*, 1939-43.

946   *Cholest-5-en-3β-ol, compound with 9, 10-secocholesta-5Z,7E,10(19)-trien-3β-ol*

2138-18-3 $C_{54}H_{90}O_2$

(1R,3Z)-3-[(2E)-2-[(1R,3aR,7aS)-7a-methyl-1-[(2R)-6-methylheptan-2-yl]-2,3,3a,5,6,7-hexahydro-1H-inden-4-ylidene]ethylidene]-4-methylidene-cyclohexan-1-ol; (3S,8S,9S,10R,13R,14S,17R)-10,13-dimethyl-17-[(2R)-6-methylheptan-2-yl]-2,3,4,7,8,9,11,12,14,15,16,17-dodecahydro-1H-cyclopenta[a]phenanthren-3-ol.

Cholest-5-en-3β-ol, compound with 9,10-secocholesta-5Z,7E,10(19)-trien-3β-ol; 6451357; 2138-18-3. Cholesterol has little effect on vitamin D_3 metabolism in neonates. *J Pediatr Gastroenterol Nutr.* 2002, *35(2)*, 180-4.

947   *Calcitriol dimer*

$C_{60}H_{98}O_6$

Synthesized and investigated as a possible chemical inducer of vitamin D receptor dimerization. *Org Lett.* 1999, *1(7)*,1005-7.

SECTION II

Indexes

CAS Registry Number Index

CAS RN	Record No.	CAS RN	Record No.	CAS RN	Record No.
50-14-6	531	59783-84-5	419	96999-67-6	462
57-87-4	521	60965-80-2	384	96999-68-7	257
67-96-9	604	62077-06-9	482	97473-92-2	737
67-97-0	351	62743-72-0	63	97903-37-2	23
115-61-7	523	63283-36-3	490	98040-59-6	212
434-16-2	350	63819-58-9	308	98353-78-7	477
474-69-1	522	63819-59-0	221	98728-28-0	934
474-73-7	483	63819-60-3	220	98830-20-7	613
511-28-4	600	63819-61-4	107	99447-30-0	476
1173-13-3	341	64164-40-5	314	100496-04-6	580
1178-00-3	500	65445-14-9	360	100643-18-2	471
1406-16-2	351	65878-49-1	491	101396-04-7	933
1715-86-2	223	69556-15-6	355, 376	101558-90-1	201
1784-46-9	588	69879-46-5	309	103305-10-8	589
2138-18-3	946	70574-97-9	222	103305-11-9	590
3308-52-9	588	71183-99-8	564	103309-75-7	848
3965-99-9	351	71204-89-2	15	103638-37-5	307
6609-90-1	588	71302-34-6	181	103656-40-2	645
8024-19-9	351	71699-09-7	327	103732-08-7	567
8050-67-7	351	71761-06-3	729	103909-75-7	93
10529-43-8	467	71848-98-1	132	104121-92-8	823
10529-44-9	466	72606-49-2	213	104211-64-5	709
11048-08-1	534	74041-09-1	318	104211-73-6	719
13403-10-6	943	74886-61-6	260	104870-37-3	583
19356-17-3	370	75303-43-4	144	105687-81-8	648
21343-34-0	543	75946-87-1	131	106315-28-0	36
21343-40-8	543	76020-77-4	929	106372-51-4	859
22481-38-5	682	76026-39-6	726	106647-61-4	705
24003-73-4	916	76338-50-6	283	106647-71-6	706
24003-85-8	916	76355-23-2	457	107793-48-6	505
25312-65-6	29	77372-59-9	449	108345-00-2	331
25631-39-4	484	77733-16-5	410	108387-51-5	562
27460-27-1	483	78609-64-0	202	109947-25-3	264
29261-12-9	414	80463-19-0	235	110536-31-7	305
29864-49-1	543	80463-20-3	453	110927-46-3	378
31316-20-8	849	81203-50-1	191	111687-67-3	75
32222-06-3	386	81446-12-0	406	112828-00-9	168
32511-63-0	387	81515-15-3	455, 461	112965-21-6	169
36149-00-5	373	82095-23-6	284	113490-37-2	118
36415-31-3	65	82095-24-7	294	113490-39-4	68
40013-87-4	422	82694-73-3	918	114489-80-4	150
41461-12-5	845	83136-06-5	285	114694-09-6	740
41461-13-6	379	83353-84-8	259	115540-42-6	520
42763-68-8	529	83805-11-2	147	115586-24-8	518
43217-89-6	383	84070-69-9	189	115586-24-8	518
50351-34-3	543	84164-55-6	289	116925-40-7	733
50392-32-0	492	85925-89-9	224	118694-43-2	152
50649-94-7	445, 460	85925-90-2	73, 74	119839-97-3	508
51504-03-1	614	86120-56-1	3	120063-23-2	940
53776-52-6	534	86677-62-5	325	120336-94-9	236
53839-02-4	299	86701-33-9	442	123000-44-2	568
54473-74-4	606	86852-07-5	74	123836-13-5	706
54573-75-0	538	87147-48-6	295	123963-52-0	792
54712-17-3	846	87407-70-3	108	123992-85-8	584
55248-15-2	553	87480-00-0	88	123992-86-9	585
55700-58-8	421	87678-01-1	296	124043-51-2	570
55721-11-4	411	87680-15-7	290	124409-58-1	258
56142-94-0	448	88200-28-6	17	128312-71-0	805
56391-76-5	945	89321-96-0	375	130447-37-9	119
56720-87-7	368	89457-70-5	939	131875-08-6	749
57102-09-7	403	91037-29-5	917	131918-61-1	418
57333-96-7	400	91625-75-1	306	132014-43-8	670
57651-82-8	351	93129-94-3	385, 400	132031-91-5	845
57803-96-0	928	95270-41-0	256	132788-52-4	1
58050-56-9	542	95480-84-5	64	133910-11-9	827
58239-34-2	377	95826-03-2	206	134404-52-7	778
58542-37-3	336	95841-71-7	372	134508-36-4	714
58702-12-8	7	96616-70-5	735	134523-85-6	806

CAS RN	Record No.	CAS RN	Record No.	CAS RN	Record No.
135776-86-2	328	143625-40-5	679	154356-84-0	715
135821-90-8	616	143773-34-6	848	167357-73-5	830
140387-52-6	94	143982-69-8	223	187935-17-7	734
141300-55-2	86	144300-56-1	282, 287	199798-84-0	688
142508-67-6	620	144699-06-9	80	214678-00-9	334
142508-68-7	619	145459-22-9	625	214678-06-5	333
143625-39-2	499	150337-94-3	704	302904-82-1	881

NLM PubChem CID Index

CID No.	Record No.	CID No.	Record No.	CID No.	Record No.
1590	417	5283689	240	5283759	452
2522	168	5283690	241	5283760	454
2523	513	5283691	264	5283761	456
2524	392	5283692	242	5283762	458
2735	352	5283693	243	5283763	472
3249	531	5283694	244	5283764	479
3323	146	5283695	245	5283765	509
5751	531	5283696	246	5283766	535
65316	29	5283697	247	5283767	536
77996	418	5283698	248	5283768	547
121948	483	5283699	253	5283769	237
123705	414	5283700	254	5283770	548
124941	63	5283702	274	5283771	596
125859	372	5283703	275	5283772	597
131554	1	5283704	276	5283773	598
132570	499	5283705	278	5283774	599
132571	679	5283706	279	5283775	612
166583	380	5283707	280	5283776	626
191124	492	5283708	281	5283777	627
191986	132	5283709	337	5283778	628
193829	383	5283710	338	5283779	630
439423	350	5283711	339	5283780	631
440558	223	5283712	340	5283781	632
443981	385	5283713	342	5283782	633
444679	521	5283714	343	5283783	634 650
446421	481	5283715	345	5283784	635
486588	417	5283716	347 348	5283785	636
3080581	500	5283717	356	5283786	637
3080672	466	5283718	357	5283787	638
3947358	178	5283719	358	5283788	639
3965999	352	5283720	359	5283790	641
3966620 (ChemDB)	146	5283721	360	5283791	642
4479094	538	5283722	361	5283792	643
4525692	446 459	5283723	362	5283793	647
4525693	237	5283724	363	5283794	648 649
4636600	418	5283725	364	5283795	443
5280447	370	5283726	365	5283796	676
5280453	386	5283727	366	5283797	629
5280793	531	5283728	367	5283798	498
5280795	351	5283729	368	5283799	497
5281010	604	5283730	369	5283803	30
5281058	341	5283731	370	5284358	531
5281104	418	5283732	371	5288149	778
5281107	538 540	5283733	399	5288670	749
5282134	169	5283734	400	5288783	167
5282177	385	5283735	401	5289504	52
5282181	355	5283737	390	5289547	16
5282190	147	5283738	404	5289548	486 487
5282368	490	5283739	391	5289549	416
5283594	648	5283740	393	5293702	274
5283628	609	5283741	388	5293708	281
5283669	608	5283742	389	5293741	388
5283672	225	5283743	405	5293766	535
5283674	173	5283744	407	5311071	607
5283675	219	5283745	408	5314030	565
5283676	226	5283746	409	5314031	344
5283677	227	5283747	410	5315257	531
5283678	228	5283748	411	5353325	370
5283679	229 237	5283749	412	5353610	531
5283680	163	5283750	414	5356615	531
5283681	230	5283751	437	5363181	422
5283682	225	5283752	438	5364601	825
5283683	231	5283753	439	5364803	414
5283684	232	5283754	440	5366081	651
5283685	233	5283755	441	5370881	480
5283686	234	5283756	447	5371993	394
5283687	238	5283757	451	5372246	537
5283688	239	5283758	450	5372266	420

CID No.	Record No.	CID No.	Record No.	CID No.	Record No.
5460703	600 602	6439053	648	6443613	309
5478815	93	6439055	236	6443806	945
5702050	531	6439066	36	6443813	542
5702762	531	6439071	859	6443822	221
5710148	543	6439141	568	6443824	314
5823716	347	6439159	706	6443843	327
5823764	479	6439162	584	6443860	318
5823789	640	6439163	570	6443868	260
5947595	844	6439165	331	6443891	305
5947596	847	6439173	792	6444030	131
6368826	373	6439174	562	6444044	457
6398761	93	6439178	258	6444050	449
6432478	531	6439195	462	6444057	202
6433735	374	6439242	75	6444144	118
6434253	422	6439245	150	6444145	68
6434322	355	6439248	520	6444150	740
6435783	168	6439266	805	6444257	848
6435810	778	6439295	119	6445225	881
6436131	169	6439321	264	6446280	448
6436868	523	6439324	670	6446381	377
6436872	522	6439377	714	6449826	378
6437079	387	6439378	806	6449838	518
6437387	384	6439508	830	6449866	616
6437855	553	6439568	235	6449938	585
6438325	538	6439569	453	6449995	916
6438336	445 460	6439590	406	6450023	379
6438368	191	6439591	455 461	6450051	733
6438384	152	6439608	284	6450185	600
6438386	181	6439609	294	6450185	600
6438393	564	6439627	918	6450246	682
6438507	403	6439640	285	6451357	946
6438511	7	6439646	259	6452575	534
6438585	491	6439679	421	6452902	299
6438598	256	6439689	289	6453648	336
6438606	64	6439724	224	6454339	482
6438648	206	6439790	325	6474052	944
6438665	735	6439793	442	6476068	214
6438681	257	6439797	74	6504811	401
6438701	213	6439809	295	6505210	286
6438738	94	6439812	108	6506392	380
6438742	86	6439813	88	6506519	586
6438752	620	6439818	296	6536972	533
6438753	619	6439865	375	6540551	423
6438767	282	6439890	917	6540731	531
6438777	625	6439938	933	6560150	352
6438783	477	6439944	705	6604177	531
6438785	23	6439950	406	6604608	532
6438798	212	6440503	729	6604662	353
6438808	283	6440542	726	6708477	605
6438811	934	6440809	467	6708595	339
6438813	613	6441337	211	6708745	531
6438836	476	6441383	490	6713937	524
6438871	410	6441537	849	6713938	340
6438872	580	6441581	715	6850801	539
6438879	471	6441633	845	6861541	531
6438910	201	6441689	846	6918141	819
6438960	589	6441761	588	7067801	531
6438961	590	6441773	355 376	9547223	332
6438974	307	6441889	419	9547224	501
6438975	645	6442094	583	9547225	502
6438976	567	6443307	484	9547226	507
6438982	823	6443422	940	9547227	516
6438985	709	6443532	529	9547228	517
6438986	719	6443542	614	9547229	530
6439023	73	6443589	308	9547230	525
6439029	943	6443590	220	9547231	526
6439039	508	6443591	107	9547232	527

CID No.	Record No.
9547233	528
9547234	545
9547235	544
9547236	574
9547237	578
9547238	579
9547239	575
9547241	576
9547242	577
9547243	554
9547244	555
9547245	556
9547246	557
9547247	558
9547248	559
9547249	560
9547250	561
9547251	566
9547252	569
9547253	581
9547254	582
9547255	592
9547256	589
9547257	610
9547258	710
9547259	719
9547260	929 930
9547261	563
9547262	3
9547263	4
9547264	5
9547265	6
9547266	8
9547267	9
9547268	10
9547269	11
9547270	12
9547271	13
9547272	14
9547273	15
9547274	18
9547275	19
9547276	20
9547277	21
9547278	22
9547279	17
9547280	24
9547281	26
9547282	27
9547283	28
9547284	51
9547285	32
9547286	33
9547287	34
9547288	35
9547289	37
9547290	38
9547291	39
9547292	46
9547293	49
9547294	50
9547295	54
9547296	55
9547297	58
9547298	59
9547299	60
9547300	69

CID No.	Record No.
9547301	70
9547302	71
9547303	83
9547304	84
9547305	85
9547306	80
9547307	86
9547308	90
9547309	91
9547310	89
9547311	95
9547312	96
9547313	97
9547314	98
9547315	99
9547316	100
9547317	103
9547318	104
9547319	105
9547320	106
9547321	107
9547322	109
9547323	110
9547324	111
9547325	112
9547326	120
9547327	121
9547328	134
9547329	135
9547330	136
9547331	137
9547332	138
9547333	139
9547334	140
9547335	141
9547336	142
9547337	143
9547338	148
9547339	145
9547340	149
9547341	151
9547342	153
9547343	154
9547344	155
9547345	156
9547346	159
9547347	160
9547348	161
9547349	164
9547350	165
9547351	166
9547352	172
9547353	170
9547354	174
9547355	175
9547356	176
9547357	177
9547358	178
9547359	179
9547360	180
9547361	183
9547362	184
9547363	185
9547364	186
9547365	187
9547366	188
9547367	189

CID No.	Record No.
9547368	190
9547369	192
9547370	193
9547371	402
9547372	197
9547373	198
9547374	199
9547375	200
9547376	194
9547377	195
9547378	196
9547379	204
9547380	205
9547381	203
9547382	207
9547383	208
9547384	209
9547385	210
9547386	211
9547387	215
9547388	216
9547389	217
9547389	217
9547390	218
9547391	267
9547392	268
9547393	269
9547394	249
9547395	250
9547396	251
9547397	252
9547399	270
9547400	271
9547401	272
9547402	273
9547403	277
9547404	291
9547405	292
9547406	293
9547407	297
9547408	306
9547409	301
9547410	302
9547411	303
9547412	304
9547413	305
9547414	310
9547415	311
9547416	312
9547417	313
9547418	314
9547419	315
9547420	316
9547421	317
9547422	320
9547423	321
9547424	322
9547425	323
9547426	324
9547427	325
9547428	346
9547429	349
9547430	395
9547431	396
9547432	424
9547433	425
9547434	397
9547435	398

CID No.	Record No.	CID No.	Record No.	CID No.	Record No.
9547436	82	9547503	695	9547570	803
9547437	426	9547504	691	9547571	807
9547438	427	9547505	693	9547572	808
9547439	413	9547506	694	9547573	809
9547440	429	9547507	696	9547574	810
9547441	430	9547508	697	9547575	811
9547442	464	9547509	698	9547576	812
9547443	465	9547510	699	9547577	813
9547444	431	9547511	706	9547578	814
9547445	432	9547512	712	9547579	815
9547446	433	9547513	713	9547580	758
9547447	434	9547514	714	9547581	816
9547448	435	9547515	715	9547582	820
9547449	436	9547516	716	9547583	826
9547450	444	9547517	720	9547584	831
9547451	469	9547518	721	9547585	832
9547452	470	9547519	722	9547586	833
9547453	474	9547520	723 724	9547587	834
9547454	475	9547521	725	9547588	835
9547455	473	9547522	732	9547589	838
9547456	478	9547523	733	9547590	839
9547457	485	9547524	730	9547591	840
9547458	503	9547525	731	9547592	841
9547459	504	9547526	736	9547593	842
9547460	506	9547527	737	9547594	843
9547461	510	9547528	738	9547595	844
9547462	514	9547529	739	9547596	847
9547463	519	9547530	741	9547597	851
9547464	546	9547531	742	9547598	850
9547465	549	9547532	743	9547599	854
9547466	550	9547533	744	9547600	857
9547467	662	9547534	745	9547601	858
9547468	593	9547535	746	9547602	859
9547469	594	9547536	747	9547603	860
9547470	595	9547537	748	9547604	866
9547471	617	9547538	750	9547605	862
9547472	656	9547539	751	9547606	863
9547473	657	9547540	752	9547607	865
9547474	618	9547541	753	9547608	867
9547475	621	9547542	754	9547609	864
9547476	622	9547543	755	9547610	869
9547477	616	9547544	757	9547611	868
9547478	623	9547545	756	9547612	870
9547479	624	9547546	760	9547613	861
9547480	644	9547547	762	9547614	875
9547481	645	9547548	763	9547615	882
9547482	646	9547549	761	9547616	885
9547483	658	9547550	769	9547617	887
9547484	659	9547551	770	9547618	888
9547485	660	9547552	771	9547619	890
9547486	661	9547553	772	9547620	892
9547487	663	9547554	775	9547621	891
9547488	664	9547555	779	9547622	894
9547489	665	9547556	780	9547623	893
9547490	667	9547557	783	9547624	898
9547491	666	9547558	784	9547625	899
9547492	668	9547559	786	9547626	905
9547493	669	9547560	785	9547627	901
9547494	671	9547561	787	9547628	903
9547495	672	9547562	789	9547629	904
9547496	673	9547563	792	9547630	900
9547497	674	9547564	793	9547631	906
9547498	675	9547565	794	9547632	909
9547499	684	9547566	795	9547633	910
9547500	685	9547567	800	9547634	911
9547501	686	9547568	801	9547635	920
9547502	687	9547569	802	9547636	921

CID No.	Record No.	CID No.	Record No	CID No.	Record No
9547638	923	9547663	44	9547687	788
9547639	925	9547664	45	9547688	700
9547640	931	9547665	821	9547689	701
9547641	932	9547666	822	9547690	717
9547642	935	9547667	855	9547691	718
9547643	936	9547668	856	9547692	157
9547644	941	9547669	836	9547693	158
9547645	942	9547670	837	9547694	298
9547646	938	9547671	511	9547695	326
9547647	804	9547672	512	9547696	678
9547648	48	9547673	551	9547697	2
9547649	47	9547674	552	9547698	162
9547650	128 129	9547675	796	9547700	727
9547651	765 766	9547676	797	9547701	728
9547652	912	9547677	878	9547702	707
9547653	415	9547678	692	9547703	708
9547655	488 489 493 494	9547679	799	9547704	603
9547656	495 496	9547680	879	9548497	674
9547657	114	9547681	677	9548797	29
9547658	115	9547682	817	9574456	478
9547659	116	9547683	759	9574699	601
9547660	117	9547685	926	11953890	604
9547661	42	9547686	515	11970180	222
9547662	43	9547684	880		

Name and Synonym Index

Name, Synonym	Record No.	Name, Synonym	Record No.
LMST03020152	163	LMST03020221	340
LMST03020153	230	LMST03020222	341
LMST03020154	225	LMST03020223	342
LMST03020155	231	LMST03020224	343
LMST03020156	232	LMST03020225	344
LMST03020157	233	LMST03020226	345
LMST03020158	234	LMST03020227	346
LMST03020159	238	LMST03020228	347
LMST03020160	267	LMST03020229	348
LMST03020161	268	LMST03020230	349
LMST03020162	269	LMST03020231	355
LMST03020163	239	LMST03020232	356
LMST03020164	240	LMST03020233	357
LMST03020165	241	LMST03020234	358
LMST03020166	264	LMST03020235	359
LMST03020167	242	LMST03020236	360
LMST03020168	243	LMST03020237	361
LMST03020169	244, 256	LMST03020238	362
LMST03020170	245	LMST03020239	363
LMST03020171	246, 265	LMST03020240	364
LMST03020172	247	LMST03020241	365
LMST03020173	248	LMST03020242	366
LMST03020174	249	LMST03020243	367
LMST03020175	250	LMST03020244	368
LMST03020176	251	LMST03020245	369, 377
LMST03020177	252	LMST03020246	370, 374
LMST03020178	253	LMST03020247	371
LMST03020179	254	LMST03020248	395
LMST03020180	255	LMST03020249	396
LMST03020181	270	LMST03020250	424
LMST03020182	271	LMST03020251	425
LMST03020183	272	LMST03020252	397
LMST03020184	273	LMST03020253	398
LMST03020185	274	LMST03020254	399
LMST03020186	275	LMST03020255	82
LMST03020187	276, 289	LMST03020256	384, 400
LMST03020188	277, 285	LMST03020257	401
LMST03020189	278	LMST03020258	386, 387
LMST03020190	279	LMST03020260	390
LMST03020191	282	LMST03020261	404
LMST03020192	280, 281	LMST03020262	391
LMST03020193	290, 291	LMST03020263	426
LMST03020194	292, 295	LMST03020264	427
LMST03020195	293	LMST03020265	392, 393
LMST03020196	297	LMST03020266	388
LMST03020197	306	LMST03020267	389
LMST03020198	301	LMST03020268	405
LMST03020199	302	LMST03020269	406, 407
LMST03020200	303	LMST03020270	408
LMST03020201	304	LMST03020271	409
LMST03020202	305	LMST03020272	410
LMST03020203	310	LMST03020273	411
LMST03020204	311	LMST03020274	412, 417
LMST03020205	312	LMST03020275	413
LMST03020206	313	LMST03020276	414
LMST03020207	314	LMST03020277	429
LMST03020208	315	LMST03020278	430
LMST03020209	316	LMST03020279	464
LMST03020210	317	LMST03020280	465
LMST03020211	320	LMST03020281	431
LMST03020212	321, 329	LMST03020282	432
LMST03020213	322, 330	LMST03020283	433
LMST03020214	323	LMST03020284	434
LMST03020215	324	LMST03020285	435
LMST03020216	325	LMST03020286	436
LMST03020217	337	LMST03020287	437
LMST03020219	338	LMST03020288	438
LMST03020220	339	LMST03020289	439

Name, Synonym	Record No.	Name, Synonym	Record No.
LMST03020429	747	LMST03020497	858
LMST03020430	748	LMST03020498	859
LMST03020431	750	LMST03020499	860
LMST03020432	749	LMST03020500	866
LMST03020433	751	LMST03020501	862
LMST03020434	752	LMST03020502	863
LMST03020435	753	LMST03020503	865
LMST03020436	754	LMST03020504	867
LMST03020437	755	LMST03020505	864
LMST03020438	757	LMST03020506	869
LMST03020439	756	LMST03020507	868
LMST03020440	760	LMST03020508	870
LMST03020441	762	LMST03020509	861
LMST03020442	763	LMST03020510	875
LMST03020443	761	LMST03020511	882
LMST03020444	769	LMST03020512	885
LMST03020445	770	LMST03020513	887
LMST03020446	771	LMST03020514	888
LMST03020447	772	LMST03020515	889, 890
LMST03020448	775	LMST03020516	892
LMST03020449	778	LMST03020517	891
LMST03020450	779	LMST03020518	894
LMST03020451	780	LMST03020519	893
LMST03020452	783	LMST03020520	898
LMST03020453	784	LMST03020521	899
LMST03020454	786	LMST03020522	905
LMST03020455	785	LMST03020523	901
LMST03020456	787	LMST03020524	903
LMST03020457	789	LMST03020525	904
LMST03020458	792	LMST03020526	900
LMST03020459	793	LMST03020527	906
LMST03020460	794	LMST03020528	909
LMST03020461	795	LMST03020529	910
LMST03020462	800	LMST03020530	911
LMST03020463	801	LMST03020531	920
LMST03020464	802, 805	LMST03020532	921
LMST03020465	803, 806	LMST03020533	922
LMST03020466	807	LMST03020534	923
LMST03020467	808	LMST03020535	925
LMST03020468	809	LMST03020536	931
LMST03020469	810	LMST03020537	932
LMST03020470	811	LMST03020538	935
LMST03020471	812	LMST03020539	936
LMST03020472	813	LMST03020540	629
LMST03020473	814	LMST03020541	941
LMST03020474	815	LMST03020542	942
LMST03020475	758	LMST03020543	938
LMST03020476	816	LMST03020544	804
LMST03020477	819	LMST03020545	48
LMST03020478	820	LMST03020546	47
LMST03020479	826, 827	LMST03020547	127, 128
LMST03020480	831	LMST03020548	129
LMST03020481	832	LMST03020549	765
LMST03020482	833	LMST03020550	766
LMST03020483	834	LMST03020551	912
LMST03020484	835	LMST03020552	415
LMST03020485	838	LMST03020553	486
LMST03020486	839	LMST03020554	487
LMST03020487	840	LMST03020555	493
LMST03020488	841	LMST03020556	494
LMST03020489	842	LMST03020557	416
LMST03020490	843	LMST03020558	488
LMST03020491	844	LMST03020559	489
LMST03020492	847, 848	LMST03020560	495
LMST03020493	851	LMST03020561	496
LMST03020494	850	LMST03020562	114
LMST03020495	854	LMST03020563	115
LMST03020496	857	LMST03020564	116